Lecture Notes in Geoinformation and Cartography

The Lecture Notes in Geoinformation and Cartography series provides a contemporary view of current research and development in Geoinformation and Cartography, including GIS and Geographic Information Science. Publications with associated electronic media examine areas of development and current technology. Editors from multiple continents, in association with national and international organizations and societies bring together the most comprehensive forum for Geoinformation and Cartography.

The scope of Lecture Notes in Geoinformation and Cartography spans the range of interdisciplinary topics in a variety of research and application fields. The type of material published traditionally includes:

- proceedings that are peer-reviewed and published in association with a conference;
- post-proceedings consisting of thoroughly revised final papers; and
- research monographs that may be based on individual research projects.

The Lecture Notes in Geoinformation and Cartography series also includes various other publications, including:

- tutorials or collections of lectures for advanced courses;
- contemporary surveys that offer an objective summary of a current topic of interest; and
- emerging areas of research directed at a broad community of practitioners.

More information about this series at http://www.springer.com/series/7418

María Teresa Camacho Olmedo
Martin Paegelow
Jean-François Mas
Francisco Escobar
Editors

Geomatic Approaches for Modeling Land Change Scenarios

Springer

Editors
María Teresa Camacho Olmedo
Departamento de Análisis Geográfico Regional y Geografía Física
Universidad de Granada
Granada
Spain

Martin Paegelow
Département de Géographie, Aménagement, Environnement
Université Toulouse Jean Jaurès
Toulouse, Garonne (Haute)
France

Jean-François Mas
Centro de Investigaciones en Geografía Ambiental
Universidad Nacional Autónoma de México (UNAM)
Morelia, Michoacán
Mexico

Francisco Escobar
Department of Geology, Geography and Environmental Sciences
University of Alcalá
Alcalá de Henares, Madrid
Spain

ISSN 1863-2246 ISSN 1863-2351 (electronic)
Lecture Notes in Geoinformation and Cartography
ISBN 978-3-319-60800-6 ISBN 978-3-319-60801-3 (eBook)
https://doi.org/10.1007/978-3-319-60801-3

Library of Congress Control Number: 2017943122

Printed on acid-free paper

This Springer imprint is published by Springer Nature
The registered company is Springer International Publishing AG
The registered company address is: Gewerbestrasse 11, 6330 Cham, Switzerland

Foreword

Alexander von Humboldt, the great scientist and proto-geographer of the early nineteenth century, wrote about the *physiognomy of nature*—the face that nature presents to us in any particular region. For Humboldt, physiognomy was the essential appearance of the landscape. It was important to him because he was deeply attracted to the beauty of the landscape in all its variety, but even more so because the landscape is the visible manifestation of all the complex interactions of the natural and human processes that interested him as a scientist (von Humboldt 1975). As science became increasingly specialized and reductionist during the nineteenth and twentieth centuries, it avoided systems that were highly integrated and inherently complex. Thus, landscapes were left to artists and poets, and to descriptive geographers. And it must be admitted that even von Humboldt did not quite know how to make the physiognomy of nature a science. He imagined a discipline that would combine the efforts of natural scientists and visual artists. This was a vision echoed in the mid-twentieth century by the mathematician and physicist Lewis Fry Richardson, who imagined an orchestra of mathematicians being conducted by a maestro of integration in order to make numerical predictions of the behaviour of complex systems like the weather. But now we have computers, and with computers, we can begin to understand complex systems not just descriptively or intuitively, but formally. We can do more than describe landscapes: we can, increasingly, generate them computationally, as predictive LUCC models do. But to develop good models of landscape—that is, models of complex spatial systems—still requires the vision and intuition of an artist or a maestro.

The modern scientific treatment of complex spatial systems has several roots. On the human side, the Chicago school of sociology in the middle third of the twentieth century developed an approach known as human ecology, which was imported into geography and combined with spatial economic models, ultimately to give rise to the field of regional science. Parallel to these developments, transportation engineers developed mathematical and computational models of travel behaviour and traffic flows. On the natural science side, ecologists made intensive studies of local ecosystems to produce data on species composition and species interactions, with space usually treated only implicitly. At the same time, these systems were modelled mathematically, usually in a highly simplified way, using techniques such as Lotka–Volterra equations. Until recently, these models were almost never explicitly spatial. Other fields such as hydrology, climatology, soil science, agronomy, and forestry also developed mathematical and computational models of the phenomena they dealt with.

All of these fields were dealing with phenomena that are inherently spatial, yet until recently the spatial aspect was either ignored or treated in a highly simplified manner—for example by reducing space to a single dimension, or by representing it as a small set of regions, as is done in much of urban and economic geography. There were two reasons for this. The first was that, for mathematical models, the introduction of space makes most sets of equations impossible to solve analytically; consequently, however plausible a model might seem as a representation of a system, it would be difficult or impossible to know what it was saying about that system. The second reason that space tended to be neglected was that there was very little spatial data available. If detailed data was required, laborious survey or field work was required to get it.

Computers solved the first problem. Computational models—simulations—obviate the need to solve the equations mathematically. Once the equations are embedded in an iterative loop, we can see how the variables of interest change their values from one time step to the next, so we can follow the evolution of the system. Frequently, we are not even interested in the mathematical solution, as the solution state may lie far in the future, or the system may be continually perturbed, or may even transform itself, before the solution state is reached. The change from mathematical to computational modelling not only avoids the technical problem of how to solve the equations, but also allows us to treat the system more realistically, and more in line with our practical reasons for wanting to understand its behaviour: we want to know what changes to expect in the coming years so that we can develop relevant plans, policies, and strategies.

Satellites solved the second problem. Once high-resolution remote sensing data became available, the problem was no longer how to get spatial data, but rather what to do with it all. Of course this data covers only a very limited set of phenomena, with land use/land cover being the most important one from our point of view, but it is nevertheless extremely useful for geographers, planners, agronomists, foresters, and others. This is especially so because scientists are great opportunists. If new types of data become available, we will find a use for it. So now the computational models can tell us not only what is happening every year, but also what is happening every year everywhere in the area we are modelling.

Computers and satellites, and more recently all sorts of geo-referenced data, have made possible the kind of work presented in this volume. But in doing so they have led us into a realm where the phenomena to be studied do not quite conform to the assumptions that are the basis of traditional scientific methodology. As a consequence, we are left to find our way in a new scientific country. At first, this land seems familiar, but then we realize that it is largely unknown. We are in the land of the poet Antonio Machado, the land where,

Traveller, there is no path to follow.

The path is made by walking.

Nevertheless, a rough map of the country is emerging. It is a land of three types of systems. The first are those that are simply *complex self-organizing systems*. These are physical and chemical systems that are driven far from their (thermodynamic) equilibrium state by a constant inflow of energy. The atmosphere provides a good example, with its highly organized but complex structure characterized by such features as cyclones and jet streams. These types of systems are the best understood, due to the work of Prigogine and his group (e.g. Prigogine and Stengers 1984), as well as scientists associated with the Santa Fe Institute such as Stuart Kauffman (e.g Kauffman 1993) and Christopher Langton (e.g. Langton 1992). Next are the living systems—what the mathematical biologist Rosen has called *anticipatory systems* (Rosen 1999). Whereas in non-living systems entities are simply themselves (atoms, molecules, rocks) and interact according to the laws of physics and chemistry, living systems all include models of themselves and their environment, and act, in part, on the basis of those models. The models can be anything from a DNA molecule to a LUCC model. These are information-based systems, although they are necessarily also complex self-organizing systems—the information is operationalised by means of the complex dynamics. Finally, the third class consists of *complicated systems*. These are systems that are integrated aggregates of systems, and have only recently been explicitly recognized as a discrete class of systems with their own issues. Most of the systems dealt with in this book, like most geographical systems, are complicated systems, composed of a variety of natural and human systems; i.e. they are functional complexes of self-organizing and anticipatory systems.

It is relatively straightforward to build and run models of these systems and that tends to obscure the fact that in terms of scientific methodology we are in unknown territory. For the most part, therefore, for lack of better alternatives, we continue to use methods that are not entirely adequate or appropriate. At the same time, we develop and experiment with new approaches, looking for better results. We are making our path by walking. While the research presented in this book is interesting and important on its own terms, placing it in this wider scientific and methodological context shows that it also raises deeper questions that are otherwise only implicit.

Complexity

The kinds of systems being modelled in this book are all complex, far-from-thermodynamic-equilibrium systems. Geosystems, both natural (e.g. atmospheric, hydrological, ecological) and human, exist by virtue of the continual inputs of energy that keep them far from their equilibrium state. (For a more complete discussion of the complex systems approach, see White et al. 2015.) These systems pose difficulties for orthodox science. As Prigogine showed (Prigogine and Stengers 1984), the dynamics of far-from-equilibrium systems are characterized by bifurcations—that is, the systems have multiple possible futures. In other words, even when the underlying process is deterministic (which often it is not) and characterized by stationarity, the outcome of the process is not completely determined: there may be a number of quite different possible outcomes. If this is the nature of the system to be modelled, then any good model of it must also be able to generate bifurcations. In other words, for a given set of initial conditions and parameter values, the model must be able to generate multiple outcomes.

This clearly raises a number of methodological problems when it comes to calibration and validation, since these essential modelling steps rely on comparing a model outcome—a prediction—with an actual data set. The problem is that while the model will be able to predict multiple outcomes for a particular time, we can only have one reality for that time, because the other possible ones did not happen. If we calibrate to get the best match to the observed data set, then as by Brown et al. (2005) showed, we will almost certainly eliminate the ability of the model to generate the bifurcations that represent possible alternatives to the observed data. In other words, the apparently optimal calibration will mean that the model fundamentally misrepresents the nature of the process. Because of the bifurcation phenomenon and the open futures nature of reality, the risk of over-calibration is inherent in modelling complex systems. It can never be eliminated, but it can be reduced by using more than one measure during calibration and validation—for example kappa as well as several landscape metrics. It is especially useful to use measures that are known to have high stationarity, such as some fractal dimensions; if an over-calibrated model is run for a long period, it is likely that it will lose its fractal nature.

Another trade-off in calibration that has no clear solution is that between expert judgment (e.g. of what are reasonable parameter values or realistic land-use patterns) and quantitative, automatic approaches. Partly this is a practical matter: using statistical techniques to extract parameter values from data, or automatic approaches (e.g. genetic algorithms) to optimize values is usually faster and easier, a point emphasized by Clarke in Chap. 8. But it also has the advantage of being objective, so if the same techniques are used in multiple applications of a model, the results are comparable. The OSDD technique proposed by Páez and Escobar in Chap. 18 should prove valuable in this respect. Multiple comparable validated applications constitute a sort of meta-validation which strengthens confidence that the model is capturing a general process, and thus can be relied on when used to perform what-if experiments for planning and policy purposes. Multiple applications also help overcome the problem of multiple possible outcomes of which only one can be observed: if the multiple applications are analogous, in effect several of the possible outcomes may be observed. These methodological problems in calibration and validation are examples of the problems which arise when dealing with complex systems. We are at the beginning of the evolution of new scientific methodologies.

In general, unlike the situation in traditional science, in the science of complex systems there are no certainties. The implication for model design is clear: a good model should be both dynamic and spatial, so that it can generate bifurcations. Since bifurcations appear in time, they can only occur in dynamic models. It is possible for them to occur in a-spatial models, but they appear much more naturally and frequently in spatial models. Since all the models presented in this volume are spatial—indeed most of them are cellular automata based, and thus inherently both dynamic and spatial—they almost certainly have the potential for generating bifurcations, even if this is not discussed explicitly. In principle, there must be some element of stochasticity in order for the bifurcations to become manifest (Prigogine speaks of the emergence of order through fluctuations). This can be introduced explicitly by means of a random term, as in several of the models used in this book (e.g. Metronamica, APoLUS, and SLEUTH) or implicitly by means of running the model multiple times with slight variations in the parameter values or initial conditions (e.g. the initial land-use map); this can be done with any of the models.

The fact that a single model predicts multiple, different, futures also somewhat alters the distinction frequently made between predictions and scenarios. For example, in Chap. 17, Maestripieri, Paegelow, and Selleron characterize prediction as "belonging to the world of rationality and accuracy", while scenarios "transcribe the uncertain nature of the studied process". But in the case of bifurcating systems, it is the "world of rationality and accuracy", i.e. the predictive model, that generates the uncertain nature of the process. The various predicted futures may be treated as scenarios, and the model may be used to explore what parameter values might be changed to make a particular possible future more likely, with the aim of developing policies that would have that effect. But it is actually the models with the experimentally altered parameters that are the scenarios, rather than the initial set of predicted possible futures.

Anticipatory Systems

Many of the processes that are important in LUCC modelling are the result of information-driven anticipatory systems—companies seeking a profit, environmental agencies trying to protect a natural area, or planners locating sites for industrial zones. In many LUCC models—for example most Markov-based transition models—these systems are not treated explicitly, and if the underlying processes are stationary during the relevant applications, then this is a legitimate simplification of the model. But frequently, an adequate representation of the situation requires a model that treats the anticipatory system explicitly. This may be something as simple as parameters representing human or institutional behaviour—an example would be the CA transition rules with their parameters representing the neighbourhood influence—or as elaborate as an explicit model of agent behaviour. An excellent example of the latter is the actor interaction process, based on either contextual interaction theory or participatory action research, included in the CA transition mechanism of the APoLUS model, described by Hewitt in Part V. Another example is the use of cognitive mapping by local stakeholders to derive human and social factors underlying land-use change, together with the use of an econometric model to determine land-use demands in the model of LUCC in a region of Jalisco, Mexico, developed by Kolb, Gerritsen, Garduño, Lazos Chavero, Quijas, Balvanera, Álvarez, and Solís (Chap. 12).

Complicated Systems

One of the findings of complexity theory is that highly complex outcomes, for example fractal patterns, can be generated by simple processes. Thus, almost all work in this area focuses on simple models that can produce elaborate outcomes. Modellers in applied fields, however, especially those dealing with geographical and ecological systems, have always known that the relevant processes are not simple. They are complicated. In particular, these systems are always composed of a number of closely connected but often quite different subsystems, and many of these subsystems (which are often, individually, just complex systems) cannot be fully understood in isolation. Thus, modellers working in areas like LUCC have long felt the need to build integrated models, whether models that integrate a variety of phenomena in a single framework, or models that consist of a set of submodels linked more or less closely, either sequentially or dynamically. The Jalisco model of Kolb et al. (Chap. 12) is again a good example, as is the LUCC model of the Cerrado biome in Brazil by Carvalho Lima, Carvalho Ribeiro, and Soares-Filho (Chap. 19) which links a CA land change model with an econometric model. Linking models of various phenomena to produce a single integrated model of a complicated phenomenon like a city or a rural region is an excellent way to achieve effective multi-disciplinarity, because the individual models can be developed by domain experts in the various relevant fields. For that reason, integrated modelling is an approach that is likely to become increasingly important.

The field of complex systems theory has revealed a number of fundamental methodological issues. At present, there is no understanding of how to handle these issues, but it is becoming increasingly apparent that dealing with them will require both new methods (e.g. new statistical techniques) and new standards and criteria as to what constitutes good science. So at present, those of us working in the area of complex, anticipatory, and complicated systems are left to deal with these problems on our own, devising ad hoc solutions as we can. Some of these will work and others will turn out not to, and in this way, the methodology will evolve and become more appropriate, powerful, and useful. The work in this book is an example of this progress. Every project mentions problems that arise and the attempts, more or less satisfactory, to deal with them. Frequently, these problems are manifestations of the basic nature of complex, complicated systems, and the solutions and work-arounds that are adopted are small evolutionary steps towards a more appropriate methodology for these systems.

While the heart of *Geomatic Approaches for Modeling Land Change Scenarios* is a series of papers concerned with LUCC modelling applications, the book is much more than the usual collection of papers only loosely connected by a common theme. The first four chapters (Part I) collectively give an overview of general issues in LUCC modelling—specifically, approaches in calibration, simulation, validation, and the use of scenarios. These are valuable in orienting the reader to the general themes that arise in all the papers of the next section, especially in that they provide an explicit comparison of the various simulation models that are used in the application chapters of that section. The chapter on scenarios by Escobar, van Delden, and Hewitt is useful as it not only provides an historical overview of the subject, but also clarifies a number of issues, both definitional and methodological, that pervade the use of scenarios in LUCC. Since scenario use is a major theme of the book, this discussion is particularly appropriate. Parts II and III are focused on applications. It begins with seven chapters that address modelling techniques—for example the use of multiple training dates for Markov-based models discussed by Paegelow (Chap. 7), the genetic algorithm-based automatic calibration technique proposed by Clarke (Chap. 8), or the examination of the effect of scale in land-use models by Díaz-Pacheco, van Delden, and Hewitt (Chap. 9). Following these are seven chapters focused on case studies, although most of these also include one or more modelling innovations, which adds to their interest. Finally, Part IV provides concise descriptions of essential techniques underlying LUCC modeling, such as cellular automata, multi-layer perceptrons, Markov chains, multi-criteria evaluation, and receiver operating characteristic analysis and following those, each one of the simulation models that appear in the book are presented shortly in Part V. The book thus provides foundations as well as examples and innovations. This is a real strength.

The variety of approaches and applications encompassed in this book might suggest that it lacks focus, but that is not the case. The historian Sweeny (2015) describes the process of uncovering historical truth as akin to painting a cubist portrait by creating a composition from the use of multiple, partly contradictory, and always partial sightings. The final portrait reveals the multiplicity of partial truths to be a complex but coherent whole. The same is true when picturing the future of geographical systems. *Geomatic Approaches for Modeling Land Change Scenarios* gives us a fine cubist portrait of this field.

Roger White
Department of Geography
Memorial University of Newfoundland
St. John's
Canada
e-mail: roger@mun.ca

References

Brown D, Page S, Riolo R, Zellner M, Rand W (2005) Path dependence and the validation of agent-based spatial models of land use. Int J Geogr Inf Sci 19:153–174

Kauffman S (1993) The origins of order. Oxford University Press

Langton C (1992) Life at the edge of chaos. In: Langton C, Taylor C, Farmer F, Rasmussen S (eds) Artificial life II. Addison-Wesley, Redwood City

Prigogine I, Stengers I (1984) Order out of chaos: man's new dialogue with nature. Bantam Books, New York

Rosen R (1999) Essays on life itself. Columbia University Press, New York

Sweeny R (2015) Why did we choose to industrialize? McGill-Queens University Press, Montreal & Kingston

Von Humboldt A (1975) Views of nature. Arno Press, New York (Reprint and translation of the German original first published in 1850)

White R, Engelen G, Uljee I (2015) Modeling cities and regions as complex systems: from theory to planning applications. MIT Press, Cambridge

Acknowledgements

This work was supported by the BIA2013-43462-P project funded by the Spanish Ministry of Economy and Competitiveness and by the FEDER European Regional Fund.

The editors would like to show their appreciation to all of the contributing authors and reviewers. The strength of this book lies in the international cooperation, which has created a rich diversity of modelling approaches.

Special thanks to Roger White who kindly accepted to write the foreword.

We are also grateful to Alexis Vizcaino, Springer editor, and W. Cartwright, G. Gartner, L. Meng, and M.P. Peterson, Springer series editors, for their interest in our proposition.

Our special thanks to Nigel Walkington for his in-depth English revision and to Enrique Fernández Seguí for his thorough work formatting the text.

Finally, the editors would like to acknowledge the support provided by their respective research centres: Departamento de Análisis Geográfico Regional y Geografía Física, Universidad de Granada, Spain; Départment de Géographie, Aménagement, Environnement, Université Toulouse Jean Jaurès, France; Centro de Investigaciones en Geografía Ambiental, CIGA, Universidad Nacional Autónoma de México (UNAM), Mexico; and Departamento de Geología, Geografía y Medio Ambiente, Universidad de Alcalá, Spain.

Contents

Editors and Contributors

About the Editors

María Teresa Camacho Olmedo, Ph.D. is a tenured Professor at the Department of Geographical Regional Analysis and Physical Geography, at the University of Granada, Spain, where she is currently Head of Department. Dr. Camacho Olmedo leads the project entitled *Simulaciones geomáticas para modelizar dinámicas ambientales II. Horizonte 2020*/Geomatic simulations for modeling environmental dynamics II. Horizon 2020, funded by the Spanish Ministry of the Economy and Competitiveness and the FEDER European Regional Development Fund (ref. BIA2013-43462-P http://geofireg.ugr.es/sigeomod/), after successfully completing various other competitive projects since 2003. Her current research includes land-use and land-cover modelling, simulation models and scenarios, and environmental dynamics. She has published numerous scientific papers and books. For more detailed information please visit her Website at http://geofireg.ugr.es/pages/profesorado/Camacho_olmedo.

Martin Paegelow, Ph.D. is a Professor at the Department of Geography, Land Planning and Environment at the University of Toulouse Jean Jaurès, France, and member of GEODE (Geography of Environment) UMR 5602 CNRS laboratory. In 1991, he obtained a Ph.D. in geography and was accredited to supervise research in 2004. He is a specialist in environmental geography and geomatics. His principal research areas are environmental management and geomatic solutions with a special focus on geomatic land change modelling and its validation. He is also strongly committed to the teaching side of his job and since 2004 has been the co-director of the Master's Degree in Geomatics in Toulouse. For more detailed information please visit his website at http://w3.geode.univ-tlse2.fr/permanents/paegelow.php.

Jean-Francois Mas, Ph.D. is a tenured Professor in the Center of Research in Environmental Geography (Centro de Investigaciones en Geografía Ambiental, CIGA) at the National University of Mexico (Universidad Nacional Autónoma de México, UNAM). He specializes in the fields of remote sensing, geographical information science, and spatial modelling. His research interests include land-use/land-cover change monitoring and modelling, accuracy assessment of spatial data, forest inventory, and vegetation cartography. He has published more than 70 peer-reviewed scientific publications, advised 7 Ph.D. dissertations, supervised 12 master's degree students, and participated in 34 research projects. For more detailed information please visit his website at http://www.ciga.unam.mx/index.php/mas.

Francisco Escobar, Ph.D. is a tenured Professor at the Department of Geology, Geography and Environmental Sciences, pioneers in Geospatial Technologies, at the University of Alcalá, Spain. He has previously worked as a researcher at the European Commission, RMIT University, and the University of Melbourne, and as invited professor at the University of Strasbourg and the Luxembourg Institute of Socio-Economic Research. Dr. Escobar leads research on cellular automata-based land-use modelling and on spatial analysis and geo-visualization applied to public health. His current research includes analysis and evaluation of modelling outcomes. He is developing geospatial techniques for the characterization of neighbourhoods in relationship to cardiovascular health, as well as for the analysis of the spatial distribution of rare and work-related diseases. He has published numerous scientific papers, books, and atlases as well as Web-based cartographic tools in these areas. For more detailed information please visit his website at www.geogra.uah.es/patxi.

Contributors

B. Abuelaish Department of GIS, Environment Quality Authority, Gaza, Palestine; Departamento de Análisis Geográfico Regional y Geografía Física, Universidad de Granada, Granada, Spain

N. Álvarez Departamento de Ecología y Recursos Naturales, Centro Universitario de la Costa Sur Universidad de Guadalajara (UdG), Autlán, Mexico

J.P. Antoni Laboratoire the MA UMR 6049 CNRS, University of Burgundy, Franche-Comté, Dijon, France

P. Balvanera Instituto de Investigaciones en Ecosistemas y Sustentabilidad (IIEES), Universidad Nacional Autónoma de México (UNAM), Mexico City, Mexico

V. Basolo Department of Planning, Policy and Design, University of California, Irvine, CA, USA

M.T. Camacho Olmedo Departamento de Análisis Geográfico Regional y Geografía Física, Universidad de Granada, Granada, Spain

T. Carvalho Lima Centro de Sensoriamento Remoto, Universidade Federal de Minas Gerais (UFMG), Belo Horizonte, Brazil

S. Carvalho Ribeiro Centro de Sensoriamento Remoto, Universidade Federal de Minas Gerais (UFMG), Belo Horizonte, Brazil

K.C. Clarke Department of Geography, University of California Santa Barbara, Santa Barbara, USA

J. Díaz-Pacheco Universidad Complutense de Madrid, Madrid, Spain; Cátedra RRD. Ciudades Resilientes, Universidad de La Laguna, San Cristóbal de La Laguna, Spain

J.R. Eastman Clark Labs, Clark University, Worcester, USA

F. Escobar Department of Geology, Geography and Environmental Sciences, University of Alcalá, Alcalá de Henares, Spain

M. Gallardo Department of Geography, University of Murcia, Murcia, Spain

D. García-Álvarez Departamento de Análisis Geográfico Regional y Geografía Física, Universidad de Granada, Granada, Spain

G. Garduño Comisión Nacional para el Conocimiento y Uso de la Biodiversidad (CONABIO), Tlalpan, Mexico

P.R.W. Gerritsen Departamento de Ecología y Recursos Naturales, Centro Universitario de la Costa Sur Universidad de Guadalajara (UdG), Autlán, Mexico

V. Hernández Jiménez Observatorio para una Cultura del Territorio, Madrid, Spain

R. Hewitt Observatorio para una Cultura del Territorio, Madrid, Spain; James Hutton Institute, Dundee, Scotland, UK

R.J. Hewitt Observatorio para una Cultura del Territorio, Madrid, Spain; Information and Computational Sciences Group, James Hutton Institute United Kingdom, Aberdeen, UK

J.R. Hipp Department of Criminology, Law and Society, University of California, Irvine, CA, USA

V. Judge Laboratoire the MA UMR 6049 CNRS, University of Burgundy, Franche-Comté, Dijon, France; LISER - Luxembourg Institute of Socio-Economic Research, Esch-sur-Alzette, Luxembourg

J.H. Kim Department of Planning, Policy and Design, University of California, Irvine, CA, USA

O. Klein LISER Luxembourg Institute of Socio-Economic Research, Esch-sur-Alzette, Luxembourg

M. Kolb Instituto de Geografía, Universidad Nacional Autónoma de México (UNAM), Mexico City, Mexico

E. Lazos Chavero Instituto de Investigaciones Sociales, Universidad Nacional Autónoma de México (UNAM), Mexico City, Mexico

N. Maestripieri GEODE, University of Toulouse Jean Jaurès, Toulouse, France

J. Martínez-Vega Institute of Economics, Geography and Demography, Spanish National Research Council, Madrid, Spain

J.F. Mas Centro de Investigaciones en Geografía Ambiental, Universidad Nacional Autónoma de México (UNAM), Morelia, Michoacán, Mexico

M. Paegelow GEODE, University of Toulouse Jean Jaurès, Toulouse, France; Département de Géographie, Aménagement et Environnement, Université de Toulouse Jean Jaurès, Toulouse, France

D. Páez Department of Civil and Environmental Engineering, University of Los Andes, Bogotá, Colombia

S. Quijas Centro Universitario de la Costa, Universidad de Guadalajara (UdG), Mexico City, Mexico

H. Rodrigues Centro de Sensoriamento Remoto, Universidade Federal de Minas Gerais, Belo Horizonte (UFMG), Belo Horizonte, Brazil

L. Román Bermejo Observatorio para una Cultura del Territorio, Madrid, Spain

G. Selleron GEODE, University of Toulouse Jean Jaurès, Toulouse, France

B. Soares-Filho Centro de Sensoriamento Remoto, Universidade Federal de Minas Gerais (UFMG), Belo Horizonte, Brazil

J. Solís Instituto de Investigaciones Sociales, Universidad Nacional Autónoma de México (UNAM), Mexico City, Mexico

H. Taud Centro de Innovación y Desarrollo Tecnológico en Cómputo, Instituto Politécnico Nacional, Mexico City, Mexico

J. Toledano Clark Labs, Clark University, Worcester, USA

H. van Delden Research Institute for Knowledge Systems, Maastricht, The Netherlands; The University of Adelaide, Adelaide, Australia

J. van Vliet Environmental Geography Group, Faculty of Earth and Life Sciences, VU University Amsterdam, Amsterdam, The Netherlands

R. Vanhout Research Institute for Knowledge Systems, Maastricht, The Netherlands

P.H. Verburg Environmental Geography Group, Faculty of Earth and Life Sciences, VU University Amsterdam, Amsterdam, The Netherlands

G. Vuidel Laboratoire the MA UMR 6049 CNRS, University of Burgundy, Franche-Comté, Dijon, France

List of Reviewers

All the chapters in this book have been peer-reviewed. We would like to thank all reviewers, external and internal, who contributed to improve the quality of this book with their comments, suggestions, and criticism.

Rahim Aguejdad, University of Strasbourg, France
Jose Barredo, Joint Research Centre, Italy
Pablo Barreira, University of Alcalá, Spain
Calum Brown, University of Edinburgh, UK
María Teresa Camacho Olmedo, University of Granada, Spain
Keith C. Clarke, University of California Santa Barbara, USA
Claude Collet, University of Fribourg, Switzerland
Jaime Díaz-Pacheco, University La Laguna, Spain
Francisco Escobar, University of Alcalá, Spain
Marta Gallardo, University of Concepcion, Chile
David García-Álvarez, University of Granada, Spain
Montserrat Gómez, University of Alcalá, Spain
Dagmar Haase, University of Berlin, Germany
Richard J. Hewitt, James Hutton Institute, UK
John R. Hipp, University of California Irvine, USA
Jae Hong Kim, University of California Irvine, USA
Olivier Klein, LISER Luxembourg Institute of Socio-Economic Research, Luxembourg
Javier Martínez-Vega, CSIC, Spain
Jean François Mas, National Autonomous University of Mexico, Mexico
Nicolás Mastripieri, University of Toulouse Jean Jaurès, France
Mike McCall, National Autonomous University of Mexico, Mexico
Martin Paegelow, University of Toulouse Jean Jaurès, France
Daniel Paez, University of Los Andes, Colombia
Azucena Pérez Vega, University of Guanajato, México

Sunhui Sim, University of North Alabama, USA
Anne Ruas, IFSTTAR, France
David Sheeren, INP–ENSAT, France
Roger White, Memorial University of Newfoundland, Canada

Chapter 1
Geomatic Approaches for Modeling Land Change Scenarios. An Introduction

M.T. Camacho Olmedo, M. Paegelow, J.F. Mas and F. Escobar

Abstract Land change models can help scientists and users to understand change processes and design policies to reduce the negative impact of human activities on the earth system at scales ranging from global to local. With the development of increasingly large computing capacities, multiple computer-based models have been created, with the result that the specific domain covered by the umbrella term "modeling" has become rather vague. Even within the context of the spatiotemporal modeling of land use and cover changes (LUCC), the term "modeling" can have many different meanings. There is also an increasing interest in the literature in comparing the different land change models. One of the aims of this book is to contribute to these processes. We focus on geomatic modeling approaches applied in this context to land change, a term that has been used synonymously for a number of years with LUCC and seems to be overtaking it as the generally used term for this phenomenon. The objective of this book is also clear to see from the methods we have chosen and the subjects we address. This book deals first and foremost with spatially explicit data that can be mapped. However, its additional focus on land change and land change scenarios in the wider field of environmental and social dynamics give it a certain consistency with a view to practical applications.

M.T. Camacho Olmedo (✉)
Departamento de Análisis Geográfico Regional y Geografía Física,
Universidad de Granada, Granada, Spain
e-mail: camacho@ugr.es

M. Paegelow
GEODE UMR 5602 CNRS, Université de Toulouse Jean Jaurès, Toulouse, France
e-mail: paegelow@univ-tlse2.fr

J.F. Mas
Centro de Investigaciones en Geografía Ambiental, Universidad Nacional Autónoma de México (UNAM), Morelia, Michoacán, Mexico
e-mail: jfmas@ciga.unam.mx

F. Escobar
Department of Geology, Geography and Environmental Sciences, University of Alcalá, Alcalá de Henares, Spain
e-mail: francisco.escobar@uah.es

M.T. Camacho Olmedo et al. (eds.), *Geomatic Approaches for Modeling Land Change Scenarios*, Lecture Notes in Geoinformation and Cartography,
https://doi.org/10.1007/978-3-319-60801-3_1

Keywords Land change · Land use and cover changes · Modeling · Geomatics · Scenarios

1 General Context

Planet Earth is being altered at an increasingly rapid pace by human activities such as fossil fuel combustion and CO_2 emissions, the increase in agricultural land at the expense of forests, the alteration of the water cycle and weather patterns, and the release of large amounts of nitrogen, phosphorus, plastic and black carbon into the biosphere. The effects of these human activities are being exacerbated by long term global geological processes. To emphasize the enormous impact of human activities on the earth and the atmosphere, Crutzen and Stoermer (2000) proposed the term "anthropocene" for the current geological epoch. Seen by many as a state-change in the earth system, the anthropocene concept bridges the gap between social and natural sciences in open systems of knowledge production. It covers the cumulative history of local and regional social changes and its connections to global processes. These changes have been linked to the development of extractive resource chains, resource use systems, urbanization and infrastructure, and technological diffusion, all of which show path dependency and emergent patterns visible in the landscape (Brondizio et al. 2016). New types of connectivity are still emerging, while the existing networks that underlie the changes continue to transform the landscapes (Lambin and Meyfroidt 2011; Brondizio et al. 2009). Meyfroidt et al. (2013) point out how local processes of land change can affect distant areas and how the influence of demand and policy often be felt in different regions. These authors use the term 'telecoupling' to describe this kind of relationship and propose combining place-based research and global modeling to evaluate the links between flows of products and services and land change. In this context, land change models can help scientists and users to understand change processes and design policies to reduce the negative impact of human activities on the earth system at scales ranging from global to local.

With the development of increasingly large computing capacities, multiple computer-based models have been created, with the result that the specific domain covered by the umbrella term "modeling" has become rather vague. Even within the context of the spatiotemporal modeling of land use and cover changes (LUCC), the term "modeling" can have many different meanings. There is a degree of confusion, for example, between modeling and simulation. Strictly speaking, modeling relates to understanding and expressing system behavior; in other words, modeling implies a simplified description of reality. By contrast, simulation refers to applying a model to a particular case study over a period of time. Many authors use modeling as a synonym for simulation. However, even if we agree that simulation is modeling a future state of a system, this definition is still extraordinarily ambiguous. A simulation may predict or prospectively model what will probably happen, but it may also be used to develop a scenario to support decision-making processes based on specific hypotheses or simple extrapolations.

Several books on land change modeling have been published in the last few years. These include among others, and in chronological order, the review by Verburg et al. (2006) of land-use and land-cover change modeling concepts and approaches, published in Lambin and Geist (2006). Koomen and Stillwell (2007) advanced the field of land use change modeling with contributions about explanatory, optimization, operational and new models. Paegelow and Camacho Olmedo (2008) introduced modeling approaches based on environmental dynamics, followed by a set of case studies. Murayama and Thapa (2011) and Murayama (2012), centering more on spatial analysis than modeling, analyzed the different GIS-based applications in land change models. Deng (2011) explored different models used in research into land system change. Heppenstall et al. (2012) and Arsanjani (2012) investigated agent and multi-based modeling. Lambin (2013) offered a general overview about modeling land use change in relation to environmental modeling in Wainwright and Mulligan (2013). The US National Research Council (2014), with a large number of contributors, published an outstanding report about the current state of land change modeling and the different approaches to it. The report also presented proposals regarding possible improvements. In their book on the monitoring and modeling of global changes, Li and Yang (2015) included among others a theoretical chapter about challenges in land change modeling. Cities and urban regions were at the heart of the theory and modeling approaches presented by White et al. (2015).

2 Our Perspective

The wide range of land change models can be classified according to their theoretical basis and their purpose. As with GIS, which can be divided into raster and vector approaches, land change models can be classified from a conceptual point of view into those based on spatial patterns (pattern-based models, PBM) and those based on the decisions and interactions of economic agents (agent-based models, ABM). The classification of available modeling software is complicated by the fact that most programs are based on hybrid approaches and offer a wide range of options. Models can also be classified on the basis of the techniques they incorporate, such as Markov chains, suitability maps, pattern based indices, neighborhood relationships like Cellular Automata (CA), etc. Other authors group models together on the basis of their purpose, dividing them into models that produce predictions, projections and normative or exploratory scenarios. Other authors distinguish models by focusing on the way they relate simulation to present and past land use/cover (LUC), splitting them into two main categories: path-dependent scenarios (also called trend or business-as-usual (BAU) scenarios) and contrasting scenarios. Several reviews about land change modeling approaches and their classifications can be found in the literature (Baker 1989; Lambin 1997; Irwin and Geoghegan 2001; Agarwal et al. 2002; Parker et al. 2003; Brown et al. 2004;

Verburg et al. 2006; Koomen and Stillwell 2007; Murayama and Thapa 2011; Kelly et al. 2013; Li and Yang 2015).

The result is that it is impossible to propose one single method for classifying models that meets all the needs and takes into account all the different situations. For practical purposes, we therefore decided to choose an existing classification, which is widely accepted by the international researchers in the literature, i.e. the classification proposed by the US National Research Council (2014). The NRC groups the models into five main categories, plus a hybrid category, which includes models that combine different approaches into a single modeling framework:

1. Inductive pattern-based models (PBM). Land change is modeled empirically using statistical and machine learning methods and observations of past land change to calibrate functions which describe the relationship between these changes and the set of drivers.
2. Cellular approach which integrates maps for suitability for land use/cover with neighborhood effects and information about the expected quantity of change.
3. Sector-based economic models based on the supply and demand for land according to economic sectors and activities and trade between regions.
4. Spatially disaggregate economic approach. Models designed to identify the causal economic relationships impacting the spatial equilibrium within land systems.
5. Agent-based models (ABM) which simulate the decisions regarding land change taken by actors that interact with each other and with their environment to make changes in the land system.

The land change models presented in this book are orientated more towards PBM than ABM and according to the NRC classification can be divided into two groups of PBM and constraint CA based models.

There is an increasing interest in the literature in comparing the different land change models (Pontius et al. 2008; Kelly et al. 2013; Mas et al. 2014; van Vliet et al. 2016; Prestele et al. 2016). One of the aims of this book is to contribute to this process. As the title of this book suggests, we focus on geomatic modeling approaches applied in this context to land change, a term that has been used synonymously for a number of years with LUCC and seems to be overtaking it as the generally used term for this phenomenon. The objective of this book is also clear to see from the methods we have chosen and the subjects we address. This book deals first and foremost with spatially explicit data that can be mapped. However its additional focus on land change and land change scenarios in the wider field of environmental and social dynamics give it a certain consistency with a view to practical application.

3 Structure of the Book

This book is divided into five parts: Part I *Concepts and tools*, Part II *Methodological developments,* Part III *Case studies,* Part IV *Technical notes* and Part V *Modeling software.*

Part I is composed of four chapters each of which focuses on a specific stage in modeling: calibration, simulation, validation and scenarios, authored by the book's editors and collaborators. The aim is to present the basic ideas and to offer a general overview of the concepts, methods and techniques used in land change modeling. Although there are no clearly defined boundaries between calibration and simulation, and despite the fact that calibration and validation share several tools, the authors try to set out a clear position regarding the concepts and stages in land change modeling. The chapters on calibration, simulation and validation are therefore all presented in the same way. A theoretical and methodological review analyzes the approaches used to set the parameters in the different steps in the calibration and simulation processes; while the chapter on validation describes the set of concepts and tools used in this process. The first three chapters also contain coordinated flowcharts and diagrams to explain the procedures in graphic form. The different approaches for estimating parameters are complemented by a table with a comparative analysis of the selected LUCC models. These models are practically applied in the chapters in Parts II and III of this book and described and explained in Parts IV and V. The chapter about scenarios aims to provide an insight into the intricate world of scenario planning and serve as a guide to the scenario modeling process. The chapter seeks to clarify the definition of inter-related and often mixed-up terms such as scenario, prediction, forecast and storyline. Given the extraordinary amount of scenario planning techniques and models found in the literature, the chapter content has been structured in such a way as to alleviate what some authors have described as methodological chaos.

Parts II and III are a collection of fourteen chapters written by researchers from Brazil, Colombia, France, Mexico, Spain, the United Kingdom and the United States. A first group of seven chapters (Part II) focuses on recently proposed methodological developments that have enhanced modeling procedures and results. The first chapter assesses the impact of using one or two time points in the calibration process and the second chapter discusses the impact of using multiple training dates in Markov chain models. The next chapter discusses the possibility of including genetic algorithms in the calibration phase as a means of improving it. The group of methodological developments also includes two chapters about the influence of spatial scale in land change modeling, focusing in one chapter on different experiments in data and parameters, and in the second on the comparison between different land use and cover maps. The last two chapters present and discuss the benefits encountered in participatory-based modeling in two very different areas, one in Spain and the other in Mexico.

A second group of chapters (Part III) has been labeled as case studies although they also offer interesting and innovative methodological proposals. The first chapter tackles the extremely complex case of land dynamics in the Gaza Strip. This is followed by a chapter focusing on land-use modeling in a cross-border region at the heart of Europe (formed by the city of Strasbourg in France and the city of Kehl in Germany). Issues and pressures suffered by natural protected areas within or near large metropolitan areas are analyzed in a chapter about the Madrid region and its recently declared Sierra de Guadarrama National Park. From this point, we cross the Atlantic to present three case studies in North and South America. The first deals with the challenges faced when modeling an area under redevelopment (in this case a former air base in California) as opposed to the more common modeling studies of urban expansion processes. This is followed by research from Chile which highlights the importance of its timber sector and explores various issues related to its future and the difficulties in modeling it. This part ends with a chapter dedicated to future intra-urban transport alternatives for the city of Bogota, Colombia. As is the case in most big cities in the developing world, Bogota faces extraordinary challenges when it comes to balancing growth and sustainability, in which public transport can tip the scales in favor of a desirable future or on the contrary towards an unsustainable horizon.

All these chapters deal with spatiotemporal data and use some of the best-known software packages for LUCC modeling. Each contribution follows a similar structure, even if the modeled object, methods, kind of model and purpose vary a great deal.

The book finished with two parts about *Technical Notes* and *Modeling Software*. Both parts have been written by a large number of scientists each contributing with their respective expertise in each of the technical notes and software presented. The Technical Notes section (Part IV) aims to describe in a simple and intuitive way some of the most frequently used methods in the calibration, simulation and validation stages in selected LUCC models. In each technical note, a short description of the method is followed by technical details with highlighted illustrations and, in some cases, with a complementary example or application.

The chapters on Modeling Software in Part V offer a compilation of short presentations of the packages presented in this book, authored by their designers. These include some of the best-known LUCC models: CA_MARKOV, Land Change Modeler (LCM) (both in TerrSet), Dinamica EGO and CLUMondo, as pattern-based models (PBM); and Metronamica, APoLUS, SLEUTH and LucSim as constraint cellular automata-based models (CCAM). The short presentations all follow the same structure. After the introduction, there is a description of the methods implemented in the model, followed by some examples of applications, a final consideration and a technical summary.

Acknowledgments The editors would like to acknowledge support provided by the BIA2013-43462-P project funded by the Spanish Ministry of Economy and Competitiveness and by the FEDER European Regional Fund.

References

Agarwal C, Green GM, Grove JM et al (2002) A review and assessment of land-use change models: dynamics of space, time, and human choice. United States Department of Agriculture

Arsanjani JJ (2012) Dynamic land-use/cover change simulation: geosimulation and multi agent-based modelling. Springer

Baker WL (1989) A review of models of landscape change. Landsc Ecol 2(2):111–133

Brondizio ES, Ostrom O, Young (2009), Connectivity and the governance of multilevel socio-ecological systems: the role of social capital. Annu Rev Environ Resour 34:253–278

Brondizio ES, O'Brien K, Bai X, Biermann et al (2016) Re-conceptualizing the Anthropocene: a call for collaboration. Global Environ Change: Part A: Hum Policy Dimens 39:318–327

Brown D, Walker R, Manson S, Seto K (2004) Modeling land use and land cover change. In: Gutman G, Janetos AC, Justice CO et al (2004) Land change science. Monitoring and understanding trajectories of change on the Earth's surface. Kluwer Academic Publishers, Dordrecht. Springer

Crutzen PJ, Stoermer EF (2000) The anthropocene. Global Change Newslett 41:17–18

Deng X (2011) Modeling the dynamics and consequences of land system change. Springer

Heppenstall AJ, Crooks AT, See LM, Batty M (2012) Agent-based models of geographical systems. Springer

Irwin EG, Geoghegan J (2001) Theory, data, methods: developing spatially explicit economic models of land-use change. Agr Ecosyst Environ 85:7–23

Kelly RA, Jakeman AJ, Barreteau O et al (2013) Selecting among five common modelling approaches for integrated environmental assessment and management. Environ Model Softw 47:159–181

Koomen E, Stillwell J (eds) (2007) Modelling land-use change. Progress and applications. Springer, Berlin

Lambin EF (1997) Modelling and monitoring land-cover change processes in tropical regions. Prog Phys Geogr 21(3):375–393

Lambin EF (2013) Model Land-Use Change. In: Wainwright J, Mulligan M (eds) Environmental modelling: finding simplicity in complexity, 2nd edn. Wiley, pp 245–254

Lambin EF, Geist H (eds) (2006) Land-use and land-cover change. Local processes and global impacts. global change—The IGBP Series (closed). Springer

Lambin EF, Meyfroidt P (2011) Global land use change, economic globalization, and the looming land scarcity. Proc Natl Sci USA (PNAS) 108(9):3465–3472

Li J, Yang X (2015) Monitoring and modeling of global changes: a geomatics perspective. Springer

Mas JF, Kolb M, Paegelow M, Camacho Olmedo MT, Houet T (2014) Inductive pattern-based land use/ cover change models: a comparison of four software packages. Environ Modell Softw 51:94–111 (Publisher By: Elsevier)

Meyfroidt P, Lambin EF, Erb KH, Hertel TW (2013) Globalization of land use: distant drivers of land change and geographic displacement of land use. Curr Opin Environ Sustain 5:438–444

Murayama Y (2012) Progress in geospatial analysis. Springer, New York

Murayama Y, Thapa RB (2011) Spatial analysis and modeling in geographical transformation process: GIS-based Applications. Springer, New York

National Research Council (2014) Advancing land change modeling: opportunities and research requirements. The National Academies Press, Washington DC

Paegelow M, Camacho Olmedo MT (eds) (2008) Modelling environmental dynamics. Advances in geomatic simulations. Series Environmental science. Springer, Berlin

Parker DC, Manson SM, Janssen MA, Hoffmann MJ, Deadman P (2003) Multi-agent system models for the simulation of land-use and land-cover change: a review. Ann Assoc Am Geograph 93(2)

Pontius RG Jr et al (2008) Comparing input, output, and validation maps for several models of land change. Ann Region Sci 42(1):11–47

Prestele R, Alexander P, Rounsevell MDA, Arneth A, Calvin K, Doelman J, Eitelberg DA, Engström K, Fujimori S, Hasegawa T, Havlik P, Humpenöder F, Jain AK, Krisztin T, Kyle P, Meiyappan P, Popp A, Sands RD, Schaldach R, Schüngel J, Stehfest E, Tabeau A, Van Meijl H, van Vliet J, Verburg PH (2016) Hotspots of uncertainty in land-use and land-cover change projections: a global-scale model comparison. Glob Change Biol 22:3967–3983

van Vliet J, Bregt AK, Brown DG, van Delden H, Heckbert S, Verburg PH (2016) A review of current calibration and validation practices in land-change modeling. Environ Model Softw 82:174–182

Verburg PH et al (2006) Modeling land-use and land-cover change. In Lambin EF, Geist H (2006) Land-use and land-cover change: local processes and global impacts. Springer, pp 117–135

Wainwright J, Mulligan M (2013) Environmental modelling: finding simplicity in complexity, 2nd edn. Wiley

White R, Engelen G, Uljee I (2015) Modeling cities and regions as complex systems: from theory to planning applications. MIT Press

Part I
Concepts and Tools

Chapter 2
LUCC Modeling Approaches to Calibration

J.F. Mas, M. Paegelow and M.T. Camacho Olmedo

Abstract In land change modeling, calibration enables the modeler to establish the parameters for the model in order to produce expected outcomes, similar to those observed for the study area over a period in the past or consistent with a given scenario. Depending on the modeling approach, the parameters are set using maps which describe past change or information obtained from experts or stakeholders. These parameters will control the behavior of the model during the simulation with regard to aspects such as the quantity and the spatiotemporal patterns of modeled change. This chapter focuses on different aspects of calibration, such as the selection and transformation of input variables and the different approaches for estimating the parameters of the most common pattern-based models (PBM) and constraint cellular automata-based models (CCAM).

Keywords Calibration · Land use and cover changes · Land change models · Drivers of change · Markov · Change potential maps

J.F. Mas (✉)
Centro de Investigaciones en Geografía Ambiental, Universidad Nacional Autónoma de México (UNAM), Morelia, Michoacán, Mexico
e-mail: jfmas@ciga.unam.mx

M. Paegelow
GEODE UMR 5602 CNRS, Université de Toulouse Jean Jaurès, Toulouse, France
e-mail: paegelow@univ-tlse2.fr

M.T. Camacho Olmedo
Departamento de Análisis Geográfico Regional y Geografía Física, Universidad de Granada, Granada, Spain
e-mail: camacho@ugr.es

M.T. Camacho Olmedo et al. (eds.), *Geomatic Approaches for Modeling Land Change Scenarios*, Lecture Notes in Geoinformation and Cartography,
https://doi.org/10.1007/978-3-319-60801-3_2

1 Introduction

Calibration is the process whereby the modeler sets the parameters of the model so as to enable it to reproduce outcomes similar to those observed for the study area. The information used for calibration should be at or before the date at which the predictive extrapolation begins (Pontius and Malanson 2005). Calibration is different from verification (also called "internal validation") which refers to the process of certifying the correct internal operation of a model, including debugging and at times sensitivity analysis.

The source of the information used to calibrate the model will depend on the modeling approach. In data-driven models, the modeler carries out an analysis of the data, which typically describes land change over a previous period, in order to obtain the expected pattern of change for the simulation period from this analysis. In knowledge-based models, the information about change patterns is obtained from experts or directly from the agents of change (Fig. 1).

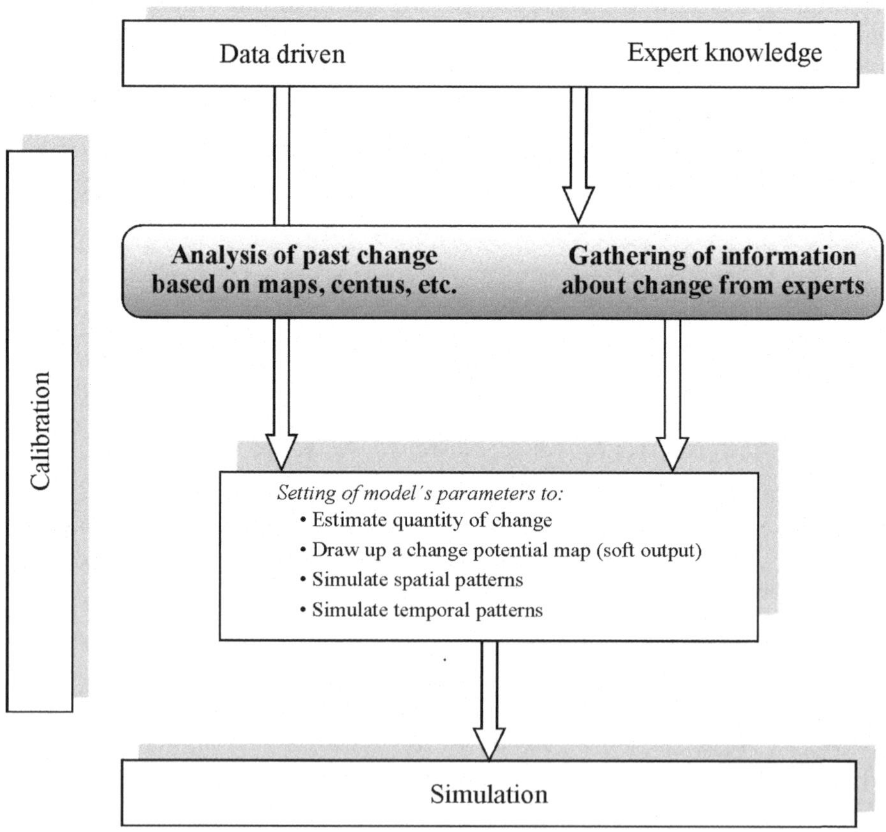

Fig. 1 Flowchart of the general procedure in the calibration stage in LUCC modeling approaches

One of the most important tasks in calibration is the selection and transformation of variables which can explain future changes and the fine tuning of the parameters that control the transition rules. In this chapter, we will review the approaches used to set the parameters used to determine the quantity of modeled change, the relationship between change and its drivers, and the spatial and temporal patterns of change. We will then highlight the related topics and the sources of uncertainty which can affect the process of calibration and the calibration assessment. Authors also describe calibration steps according to LUCC models.[1] Some of them, such as CA_MARKOV, Land Change Modeler (LCM) (both in TerrSet), Dinamica EGO and CLUMondo, are pattern-based models (PBM). By contrast, Metronamica, APoLUS, SLEUTH and LucSim are considered constraint cellular automata-based models (CCAM).

2 Selection of Variables

2.1 LUC Transitions

First, the modeler should choose the land-use change he/she wants to model, the level of detail required in the characterization of the change, the main changes occurring in the study area and the characteristics of the input data. For example, when multi-date maps or classified remotely sensed images are used, the number of map categories will affect the number of mapped transitions. A broad transition (e.g. deforestation) is likely to be mapped more accurately than more detailed transitions such as "pine forest to crop cash agriculture" or "dry forest to temporal agriculture" because most of the confusion between mapped categories occurs between similar land use/cover (forest categories, agriculture categories). In fact, it might be easier to model these two types of deforestation process in a separate way because they respond to different agents, motivations and conditions. The choice of the modeled transitions can also be guided by information from interviews (Voinov et al. 2016). As pointed out by Hewitt (2015), modeled LUC transitions need to be carefully chosen, and a reclassification of LUC categories available in existing cartography should eventually be carried out. Models based on a land systems approach allow us to simulate both LUC conversions and changes in land use intensity (van Asselen and Verburg 2013).

Finally, it is often difficult to model just the few specific transitions that interest the modeler if other transitions also play an important role in the land change dynamic of the entire system. For instance, if a modeler is interested in

[1]See the short presentations in Part V of this book about (in alphabetical order) APoLUS, CA_MARKOV, CLUMondo, Dinamica EGO, Land Change Modeler (LCM), LucSim, Metronamica and SLEUTH. The authors are also grateful to all contributors who helped us understand the different software packages.

deforestation caused by agricultural expansion in a region where important areas of cropland are also being lost to urban expansion, he should also consider the latter transition because it is likely to increase the pressure on forest as a source of land to replace the lost agricultural areas.

Analysis of changes based on LUCC-budget (Pontius et al. 2004, see Technical Notes in Part IV of this book) or intensity analysis (Aldwaik and Pontius 2012) can give useful insights to help the modeler choose the transitions to consider in the model.

2.2 Explanatory Variables of Change

The modeler should select the drivers, or explanatory variables that play a role in the land changes. Even in automated approaches, the selection of initial input variables is based on expert knowledge although data availability is often a major limitation. These variables are diverse and describe different aspects of the study area and its context such as accessibility (distance from human settlements, roads, markets…), suitability of the terrain for diverse human activities (slope, elevation, rainfall, soils…), human activities (agriculture, sawmills and human pressure indices such as population density, marginalization), public policies (protected areas, subsidies for cattle ranching or agriculture). It is worth noting that pattern-based models can produce quite accurate prospective maps using only variables, such as slope and distances that do not explain the causes of the change and focus only on its location. By contrast, process-based approach models will concentrate on variables closer to the causes of the change because they seek to simulate the process of change.

Variables can be divided into static and dynamic variables. Static variables do not change over the course of a simulation. Dynamic variables, which value change during the simulation, include distance to roads that will be built according to a schedule or whose construction is simulated in the model. Such models, called ¨road constructor¨ in some software packages are calibrated by identifying zones of attraction, such as valuable timber areas, and zones of resistance to the path of roads such as flooded or rugged terrains. Other dynamic variables are distances to specific LUC areas, to settlement projects or to conservation units and are usually calibrated using the first date of the calibration period, based on the assumption that the changes observed during the calibration period are explained by the landscape configuration at the beginning of the period.

During the last decade, the amount of available information increased dramatically. Many government agencies have made their information available online, often in a digital GIS compatible format. Remote sensing data is also increasingly, often freely, available. The quality of the imagery has also improved greatly: high spatial resolution images are now common and recently launched satellite constellations enable space agencies to produce images with both high spatial and temporal resolution. Another challenging new source of information is volunteered geographic information which produces a large amount of firsthand information (Goodchild 2007; Jokar Arsanjani et al. 2013).

When selecting the variables to be integrated into the model, different strategies are often carried out in which the drivers are analyzed using statistical indices, expert knowledge, reviews of the literature and workshops with stakeholders. Step-by-step regression models help select the variables with the highest explanatory power. Many other indices are used to evaluate the strength of the relationship between two variables such as for example the average of the absolute value of the weights of evidence (Mendoza Ponce et al. 2017) or the importance of weight (Sangermano et al. 2012). In some models based on the assumption of the independence of the explanatory variables, indices such the Cramer index, chi square, correlation index, Kappa index and joint information uncertainty are used to detect correlated explanatory variables (Mas et al. 2014). Based on these analyses, one or various variables among the correlated variables are discarded from further analysis to reduce correlation. For example, Almeida et al. (2005) used the criterion proposed by Bonham-Carter (1994) and considered two variables as correlated when they had Cramer's Coefficient and Joint Information Uncertainty values of over 0.5.

2.3 Variable Transformation

Variables often have to be adapted into a suitable format for the analysis procedure. For instance, some statistical methods, such as weights of evidence (see Technical Notes in Part IV of this book), require categorical input variables. Thus, continuous variables such as distances should be transformed into bins. By contrast, when using methods such as logistic regression or multilayer perceptron (see Technical Notes in Part IV of this book), modelers try to avoid categorical variables because each category is managed as a dummy binary variable, increasing the dimensionality of the model. Categorical variables can be transformed into continuous ones using the Evidence Likelihood transformation based on the relative frequency of cells belonging to the different categories within areas of change. In logistic regression, the transformation of explanatory variables through algebraic operations such as exponential, quadratic, logarithmic or power, can be done to achieve linear relationships with the logit of the dependent variable. The creation of suitability maps using fuzzy transformation and weighting can also be considered as variables transformation.

3 Parameters to Calibrate

3.1 Quantity of Changes

The main objectives of a land change model generally include the prediction of the quantity of change that may occur in the future.

In past trend-based models, the rate of change is obtained from the analysis of change which occurred during a previous period, the "calibration period". As pointed out by Chen and Pontius (2011), the selection of the calibration period often depends on data availability and can have an important influence on the predictive performance of the model. Broadly speaking, in short period calibration there is a risk of extrapolating change quantity in exceptional moments, while if trends are analyzed over longer intervals, the extreme tendencies tend to be averaged out. For example, Fig. 2 illustrates annual deforested area in the Brazilian Amazon between 1989 and 2015 and average rate computed for periods of three and five years. The rates calculated over longer periods do not present the large fluctuations observed in yearly data. However, there is no fixed rule as to the appropriate calibration period when the rate of change seems erratic. Temporal resolution includes the number of available dates and time intervals. As the most commonly used approaches include only two training dates, the choice of training dates is crucial. The dataset showing the annual deforested area in the Brazilian Amazon (Fig. 2) offers the possibility of computing many rates of change using two training dates. Model output will vary greatly depending on the pair of training dates selected, due to the large fluctuations in the rate of deforestation over time. Paegelow et al. (2014) highlighted the impact of different training dates on the accuracy of a model based on a dataset like this. In this book, Paegelow examines the potential errors resulting from only considering two past dates in Markov projections.

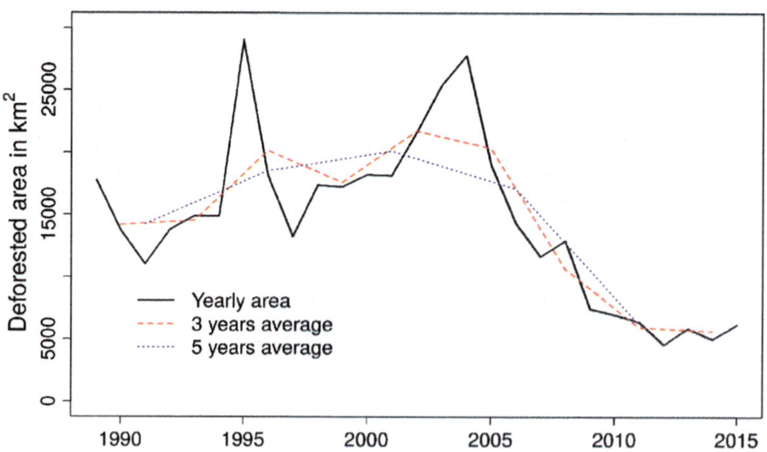

Fig. 2 Deforested area in the Brazilian Amazon (1989–2015). *Source* INPE, Brazil

Many LUCC models are based on Markov chains. As detailed in the technical note in Part IV of this book, a transition matrix for the calibration period can be obtained by overlaying two LUC maps and transforming them into an annual Markov chain probability matrix. This matrix indicates the probability of transition from one category to another over one year and allows us to project the estimated areas of each LUCC transition. There are several methods for obtaining the annual matrix. The method based on the eigenvector and eigenvalues of the original matrix (see equation in Takada et al. 2010 or Mas et al. 2014) can prove problematic: (1) if they produce one or several matrices with complex or negative numbers and (2) when there are two matrices in the results (even if they do not contain complex or negative numbers). These matrices cannot be interpreted as probabilities (Takada et al. 2010).

To overcome the limitations of Markov projection due to the assumption of stationarity of the transition probabilities over the calibration and simulation periods, Collins et al. (1974) calculated dynamic transition probabilities by using different transition matrices at certain time intervals or computing dynamic transition probabilities by postulating rules of behavior for LUC categories.

Markov chains are used in population projection: Population is divided by age and the transition matrix indicates death and birth rates for each group (Shryock and Siegel 1976). It seems logical that the number of births and deaths will depend on the size of the population in each age group and their corresponding birth and death rates. A large population will have more births than a small population with the same birth rate. However, the application of this logic to LUCC rates is far from straightforward. Suppose that there is a large forest region with an annual deforestation rate of 5%. A Markov projection will project a decrease in the total deforested area each year, because the 5% rate will be applied to a diminishing forest area. Nevertheless, the area deforested annually will probably depend on many others factors (e.g. market or population-related) and not on the area of remaining forest, as least until remaining forest areas are very small and confined to inaccessible areas.

Moreover, the Markov assumption that a constant proportion of a given category will present a certain transition at each time step will result in extrapolations reaching a state of equilibrium in terms of the area of each category (Petit et al. 2001), an equilibrium that is rarely observed in true situations. Runfola and Pontius (2013) proposed the Flow matrix, which expresses the sizes of the transitions among categories between two dates as an alternative to the Markov matrix.

If the past-trend-based projection seems to be a risky option due to the large fluctuations in change over time, the modeler can try to model the quantity of change. This can be done by external models using exogenous variables. For example, Barni et al. (2015) calculated the rate of deforestation in a non-spatial numerical model which takes into account planned road building and a migration factor that simulated increased deforestation by expected migrants to the region after road building. This model was calibrated using observed past trends. Vieilledent et al. (2013) also modeled deforestation including the effect of population density.

Models such as CLUMondo represent land-use change in a different way from area-based demand. Input for this model comes in the form of exogenous demands for goods and services, which can be supplied by different land systems characterized by their land cover and their land management intensity. This means that increasing demand for crop products can lead to a combination of expansion in agricultural area and intensification of existing cropland.

Table 1 shows the approaches used by eight popular software packages to estimate the quantity of change. Markov chain is the most common approach, particularly in pattern-based models.

3.2 Function Describing the Relationship Between Change and Drivers

In many pattern-based models, the allocation of change is generally based on maps of change potential which indicate, for each transition, the propensity of change (see Chap. 3 about simulation). This map is usually based on a data-driven analysis of past patterns of change with respect to the explanatory variables. In this way, the map of change potential reflects the changes in the distribution of land-use that occurred during the calibration period. There are many methods used to establish the relationship between the change observed during the calibration period and the variables. The most commonly used are brute force, logistic regression, weights of evidence, decision tree, multilayer perceptron and genetic algorithm (for some of these methods see Technical Notes in Part IV of this book). Some authors combine various methods such as weights of evidence and genetic algorithms (Soares et al. 2013). These methods can be distinguished by their ability to fit non-linear relationships. High flexibility is not always an advantage due to overfitting. When the model is overfitted to specific conditions of the calibration period it is unable to predict the next period (simulation step) correctly. These methods are mainly data-driven. However, the map of change potential can also be partially or totally based on expert knowledge as in Overmars et al. (2007) who drew up their map on the basis of expert advice from agronomists. Some of the methods, such as the weights of evidence in Dinamica EGO, enable users to adjust the importance attributed to expert knowledge from a totally statistical, data driven approach (no modification of the computed values of the weights) to an exclusively expert knowledge approach (complete edition by the expert). A hybrid approach, combining data-driven and expert knowledge, can be obtained with a partial modification of the weights (Mas et al. 2014).

One alternative to the change potential map is a suitability map that expresses the appropriateness of a location for each type of LUC. This map is frequently created using a multi-criteria analysis (see Technical Note in Part IV of this book). The chapter on simulation (see Chap. 3) provides a complete discussion of both the change potential and suitability approaches.

Table 1 Approaches used to estimate the quantity of change in eight software packages

	Pattern-based models (PBM)				Constraint CA-based models (CCAM)		
	CA_Markov TerrSet	LCM TerrSet	Dinamica EGO	CLUMondo	Metronamica and APoLUS	SLEUTH	LucSim
LUC/ continuous variable	LUC	LUC	LUC	LUC	LUC	LUC and Urban Growth	LUC/ Continuous variables
Time points	1 or 2	2	2	1	2	Min 4, no maximum	1 or 2
Estimation of change quantity	Markov	Markov, external	Markov, external	Exogeneous demand for goods and services	External	CA growth rule parameters	Markov

Table 2 Main approaches used for the analysis of drivers. Additionally, models may use tools to help understanding LUC and setting model´s parameters

	Pattern-based models (PBM)				Constraint CA-based models (CCAM)		
	CA_Markov TerrSet	LCM TerrSet	Dinamica EGO	CLUMondo	Metronamica and APoLUS	SLEUTH	LucSim
Data driven statistical approach		Logistic regression	Weight of evidence	Logistic regresion		Cellular automata	
Data driven machine learning		Multiplayer perceptron sim weight	Genetic algorithm	May be used to generate suitability map external to model		Brute force, Genetic algorithm	Decision trees
Knowledge driven approach	Multicriteria evaluation		Weight edition	Expert based parameterization several parameters	Empirical, trial and error tested against benchmarks.	CA rules are hard coded, but adjust	No

In CA constrained-based models, calibration involves parameter values for the neighborhood (van Vliet et al. 2013). For instance, models such as Metronamica and APoLUS compute a total transition potential combining the neighborhood effect, accessibility, suitability and zoning (see the technical note about the NASZ model in Part IV of this book). Usually, calibration proceeds systematically by fitting the simulated and the real maps for the calibration period as well as possible for each of the parameters. The procedure for evaluating goodness of fit involves visual inspection, cell-by-cell comparison measures such as Ksim (van Vliet et al. 2011) and spatial pattern indices such as fractal dimension or clumpiness.

The LucSim model uses a decision tree algorithm to determine a set of transition rules. Calibration data is split into training and testing data to avoid overfitting (see the short presentation about LucSim in Part V of this book). When calibrating the SLEUTH model, the model simulates a map of land-use at the end of the calibration period, carrying out a large number of simulations to assess its consistency. Thirteen performance metrics are used to assess the coefficient values. The best five coefficients are selected using brute force or a genetic algorithm (Silva and Clarke 2002, Clarke in this book) (Table 2).

3.3 Spatial Patterns

Spatial patterns involve the distribution, shape and size of the change patches in the landscape. Cellular automata (CA) are often used to enable the creation of small groups of cells which underwent change, simulating spatial patterns such as agriculture extension and urban growth (see Technical Notes in Part IV of this book). CA is a popular spatial simulation tool due to its simplicity, its ability to reproduce complex emergent dynamics and its affinity with raster GIS format (Torrens and O'Sullivan 2001). Calibration involves identifying the parameters which control the CA behavior based on training data. Torrens and O'Sullivan (2001) pointed out the need for stronger calibration techniques for CA because they are often calibrated by manual tuning.

A few studies incorporate landscape pattern metrics into the calibration procedure to establish the parameters for CA. For instance, Silva and Clarke (2002) determined the parameters of SLEUTH model CA by brute force, trying many combinations of the control parameters and computing measures of the goodness of fit between the simulated pattern and the real one. Soares-Filho et al. (2002) used a trial-and-error method to calibrate CA using landscape indices. Due to the large number of simulations involved, these methods are computation intensive. Li et al. (2013) proposed a pattern-calibrated method based on landscape metrics for calibrating CA using genetic algorithms. Liu et al. (2014) proposed an index called landscape expansion index (LEI) to calibrate a CA able to simulate infilling, edge-expansion and outlying urban growth patterns. Certain models such as Dinamica EGO have a mechanism for controlling the distribution of change with respect to the change potential and avoid restricting the simulated change to the highest change potential cells. This mechanism is controlled by a parameter which should be determined during calibration. Mas et al. (2015) used a genetic algorithm to calibrate this parameter along with CA parameters.

Finally, some models are based on objects rather than on cells. For example, Houet et al. (2014) carried out landscape simulations at fine resolution, based on elementary units (agricultural parcels) represented by vector-based objects.

Spatial patterns also involve the identification of zoning effects related with incentives or constraints in land-use regulation policies such as subsidies for cattle ranching or conservation. The zoning effect is often controlled by a coefficient to adjust the change potential in these areas. Highly restrictive zoning may result in a deterministic and unrealistic model. These patterns are also identified and quantified during calibration.

At another level, the spatial pattern may involve the identification of sub-regions, which present different processes and patterns of change. For instance, when a study area includes both mountainous and flood plain areas, different rates and patterns of change can be expected even for the same transition. In such cases, it may be useful to split the study area into various sub-regions in which independent calibration processes are carried out. For instance, Mas (2016) used a Geographically Weighted Regression model to identify sub-regions with different patterns of deforestation and carried out an independent calibration in each region.

3.4 *Temporal Patterns*

Temporal patterns include the sequence of land-use observed in the landscape. For example, when Houet et al. (2014) simulated LUCC they took into account farming practices such as crop successions. Chang and Mas (2017) develop a model of a slash and burn agriculture landscape in which a fallow period is necessary after a few years of cropping. This temporal behavior is generally calibrated using information from the literature or interviews, as a multi-date database with a high temporal resolution (e.g. yearly map) is often not available.

4 Calibration Evaluation

Calibration can be evaluated using the same methods as used to validate the model (see Chap. 4 about validation). For instance, for the past-trend pattern model, the change potential map can be compared with the changes that took place during the calibration period. Change can also be simulated from the beginning of the calibration period to create a simulated map for the end of this period. The simulated and observed (true) maps can then be compared. However, this evaluation only provides information about the goodness-of-fit of the calibration procedure. As we will see in the next section, this goodness-of-fit is not always a good indication of the predictive power of the model.

5 Source of Uncertainty

There are many sources of uncertainty that can obstruct the calibration of the model.

First, difficulties may arise in identifying the causal relationships between the land change processes and the explanatory variables used during calibration. In certain cases, the true drivers of land change are not identified or are not available. However, it is often impossible to establish a strong relationship between the land change and a particular set of variables due to the complexity of land change. Land change is related with environmental, socio-economic, historical and cultural drivers and acts as a complex system. Brown et al. (2004) argue that the failure to incorporate detailed information (e.g. survey-based) about household or community structures can create a specification bias because LUCC processes may be different for different types of households or communities. Additionally, in a large or heterogeneous area, different drivers may be active in different places, which makes finding causal explanations difficult (Walker et al. 2000).

Data inputs can also be a serious source of uncertainty. Land-use changes are often obtained from remote sensing data and accuracy assessments show that image processing often produces a large number of errors due to spectral confusion and other limitations. Consequently, the estimated rate of change and its spatial distribution can show a large amount of error that will propagate in subsequent processing. Change obtained from other sources such as interviews or volunteered geographical information can also show many errors or bias (Foody et al. 2013). Similarly, the explanatory variables used in the model may also have errors or be outdated. When using aggregated data such as census data, models can suffer from the modifiable areal unit problem where the shape and size of data aggregation (e.g. municipalities) affects the results of statistical analysis (Openshaw 1984).

Another source of uncertainty is the non-stationarity of the land change processes. As shown in Fig. 2, the rate of change can present large fluctuations over time. This lack of consistency can make the change patterns during the calibration and simulation periods very different. The non-stationarity of the land change process involves not only the rate of change but also the nature and the spatial distribution of the changes. For instance, agriculture can undergo drastic changes in response to demand for new crops. It is possible that the new crops may be grown on land with adverse environmental conditions where previously no crops could be planted, so rendering the change potential map obsolete. For instance, Mas et al. (2004) reported that the variation in the relative importance of the explanatory variables of deforestation in a tropical region of Mexico between the calibration period, dominated by cattle ranching, and the simulation period, when rice cultivation was introduced, led to errors in predicting the location of deforestation.

Finally, uncertainty can be the result of the design of the model itself. The model ignores important exogenous dynamics (e.g. price fluctuations, new market emergency) and oversimplifies certain relationships. For example, logistic regression can only model an S-shaped relationship between land change occurrence and an explanatory variable, when the true relationship may be an optimal range.

6 Concluding Remarks

Calibration enables modelers to set the model parameters that will control the behavior of the simulation with respect to aspects such as the quantity of change, its spatial distribution and spatio-temporal patterns such as the size and shape of patches and the succession of land-use categories over time. Many approaches are used to calibrate land change models including statistical analysis (mainly regression models and weights of evidence), machine learning (neural networks and genetic algorithms) and expert knowledge. Van Vliet et al. (2016) carried out a review of calibration approaches reported in recently published applications of land change models. They found that statistical analyses and automated procedures are the most common approaches, while expert knowledge and manual calibration are less frequently used.

Houet et al. (2016) distinguish two contrasting modeling approaches: (1) a path-dependent approach aimed at mimicking past changes into the future by applying the calibration procedure to a past period. In this first approach, the amount of change can be modified and incentives or constraints maps can be incorporated to produce different scenarios. These models enable researchers to simulate trend-based scenarios and explore various alternative land management scenarios when the quantity and the processes of change do not differ significantly from observed past changes. (2) A non path-dependent approach which assumes that LUCC models are used to spatially represent pre-defined contrasted scenarios. In this case, the parameterization of the future quantity of change does not depend on input maps which represent past changes. However, the parameterization of the allocation of future changes is usually defined by change potential maps obtained by observing past changes. In both modeling approaches, calibration is therefore a critical step. Success in calibrating the model will depend on the stationarity of change, especially in the path-dependent approach.

New applications of land change models involving the evaluation of land-based policies will require increasingly process-based models, able to model complex processes with feedbacks within and between the socioeconomic and biophysical systems across scales (National Research Council 2014). The improvement of land change models is likely to draw on multidisciplinary and interdisciplinary developments and drastically change the way models are calibrated.

Acknowledgements This study was supported by the Consejo Nacional de Ciencia y Tecnología (CONACYT) and the Secretaría de Educación Pública through the project entitled "¿Puede la modelación espacial ayudarnos a entender los procesos de cambio de cobertura/uso del suelo y de degradación ambiental?—Fondos SEP-CONACyT 178816". This work was also supported by the BIA2013-43462-P project funded by the Spanish Ministry of Economy and Competitiveness and by the FEDER European Regional Fund.

References

Aldwaik S, Pontius R (2012) Intensity analysis to unify measurements of size and stationarity of land changes by interval, category, and transition. Landsc Urban Plann 106:103–114

Almeida CM, Monteiro AMV, Soares-Filho BS, Cerqueira GC, Pennachin CL, Batty M (2005) GIS and remote sensing as tools for the simulation of urban land-use change. Int J Remote Sens 26(4):759–774

Barni PE, Fearnside PM, de Alencastro Lima, Graça PM (2015) simulating deforestation and carbon loss in Amazonia: impacts in Brazil's Roraima State from reconstructing highway BR-319 (Manaus-Porto Velho). Environ Manage 55:259–278

Bonham-Carter GF (1994) Geographic information systems for geoscientists: modelling with GIS. Pergamon, Ontario

Brown DG, Walker R, Manson S, Seto K (2004) Modeling land use and land cover change in land change science. Remote Sens Digit Image Process 6:395–409 (Springer)

Chang LA, Mas JF (2017) Modelación espacio temporal de un sistema roza-tuba-quema. In: Mas JF (ed) Análisis y modelación de patrones y procesos de cambio. CIGA-UNAM, Mexico

Chen H, Pontius RG (2011) Sensitivity of a land change model to pixel resolution and precision of the independent variable. Environ Model Assess 16:37–52

Collins L, Drewett R, Ferguson R (1974) Markov models in geography. The. Statistician 23: 179–209

Foody GM, See L, Fritz S, Van der Velde M, Perger C, Schill C, Boyd DS (2013) Assessing the accuracy of volunteered geographic information arising from multiple contributors to an internet based collaborative project. Trans GIS 17:847–860. doi:10.1111/tgis.12033

Goodchild MF (2007) Citizens as sensors: the world of volunteered geography. GeoJournal 69(4):211–221. doi:10.1007/s10708-007-9111-y

Hewitt R (2015) APoLUS model full system documentation. Project Report for EU FP7 Programme COMPLEX Project (deliverable 3.5) https://www.researchgate.net/publication/292047909_APoLUS_model_full_system_documentation

Houet T, Aguejdad R, Doukari O, Battaia G, Clarke K (2016) Description and validation of a "non path-dependent" model for projecting contrasting urban growth futures. Cybergeo: European Journal of Geography 759 doi:10.4000/cybergeo.27397

Houet T, Schaller N, Castets M, Gaucherel C (2014) Improving the simulation of fine-resolution landscape changes by coupling top-down and bottom-up land use and cover changes rules. Int J Geogr Inf Sci 28(9):1848–1876. doi:10.1080/13658816.2014.900775

Jokar Arsanjani J, Helbich M, Bakillah M, Hagenauer J, Zipf A (2013) Toward mapping land-use patterns from volunteered geographic information. Int J Geogr Inf Sci 27(12):2264–2278. doi:10.1080/13658816.2013.800871

Li X, Lin J, Chen Y, Liu X, Ai B (2013) Calibrating cellular automata based on landscape metrics by using genetic algorithms. Int J Geogr Inf Sci 27(3):594–613. doi:10.1080/13658816.2012.698391

Liu XL, Ma X, Li B, Ai S, Li He Z (2014) Simulating urban growth by integrating landscape expansion index (LEI) and cellular automata. Int J Geogr Inf Sci 28(1):148–163. doi:10.1080/13658816.2013.831097

Mas JF (2016) Combining Geographically Weighted and pattern-based models to simulate deforestation processes. In: Sauvage S, Sánchez-Pérez JM, Rizzoli AE (eds) Proceedings of the 8th international congress on environmental modelling and software July 10–14, Toulouse, France, pp 1321–1327. ISBN: 978-88-9035-745-9

Mas JF, Kolb M, Paegelow M, Camacho Olmedo MT, Houet T (2014) Inductive pattern-based land use/cover change models: a comparison of four software packages. Environ Model Softw 51(1):94–111. doi:10.1016/j.envsoft.2013.09.010

Mas JF, Puig H, Palacio JL, Sosa AA (2004) Modelling deforestation using GIS and artificial neural networks. Environ Model Softw 19(5):461–471

Mas JF, Soares-Filho B, Rodrigues H (2015) Calibrating cellular automata of land use/cover change models using a genetic algorithm, Int. ISPRS Geospatial Week 2015, La Grande Motte, France, 28th September - 2nd October 2015. International Archives of the Photogrammetry-Remote Sensing and Spatial Information Sciences XL-3/W3 67–70

Mendoza Ponce AV, Galicia Sarmiento L, Corona Núñez RO (2017) Cambios de usos y cobertura del suelo bajo diferentes escenarios socioeconómicos y climáticos en México. In Mas JF (ed) Análisis y modelación de patrones y procesos de cambio. CIGA-UNAM, Mexico

National Research Council (2014) Advancing land change modeling: opportunities and research requirements. The National Academies Press, Washington DC

Openshaw S (1984) The modifiable areal unit problem concepts and techniques in modern geography, vol 28. Geo Books, Norwich

Overmars KP, Verburg PH, Veldkamp TA (2007) Comparison of a deductive and an inductive approach to specify land suitability in a spatially explicit land use model. Land Use Policy 24:584–599

Paegelow M, Camacho Olmedo MT, Mas JF, Houet T (2014) Benchmarking of LUCC modelling tools by various validation techniques and error analysis. Cybergeo document 701 http://cybergeo.revues.org

Petit C, Scudder T, Lambin E (2001) Quantifying processes of land-cover change by remote sensing: resettlement and rapid land-cover changes in south-eastern Zambia. Int J Remote Sens 22(17):3435–3456

Pontius RG, Malanson J (2005) Comparison of the structure and accuracy of two land change models. Int J Geogr Inf Sci 19(2):243–265

Pontius RG, Shusas E, McEachern M (2004) Detecting important categorical land changes while accounting for persistence. Agr Ecosyst Environ 101:251–268

Runfola DM, Pontius RG (2013) Measuring the Temporal Instability of Land Change using the Flow Matrix. Int J Geogr Inf Sci 27(9):1696–1716

Sangermano F, Toledano J, Eastman J (2012) Land cover change in the Bolivian Amazon and its implications for REDD+ and endemic biodiversity. Landsc Ecol 27(4):571–584 https://doi.org/10.1007/s10980-012-9710-y

Shryock HS, Siegel JS (1976) The methods and materials of demography, studies in population. Academic Press, San Diego. ISBN 9780126411508 http://dx.doi.org/10.1016/B978-0-12-641150-8.50001-9

Silva EA, Clarke KC (2002) Calibration of the SLEUTH urban growth model for Lisbon and Porto. Portugal Comput Environ Urban Syst 26(6):525–552. doi:10.1016/S0198-9715(01)00014

Soares-Filho BS, Coutinho Cerqueira G, Lopes Pennachin C (2002) DINAMICA – a stochastic cellular automata model designed to simulate the landscape dynamics in an Amazonian colonization frontier. Ecol Model 154:217–235

Soares-Filho B, Rodrigues H, Follador M (2013) A hybrid analytical-heuristic method for calibrating land-use change models. Environ Model Softw 43:80–87. doi:10.1016/j.envsoft.2013.01.010

Takada T, Miyamoto A, Hasegawa SF (2010) Derivation of a yearly transition probability matrix for land-use dynamics and its applications. Landsc Ecol 25(4):561–572

Torrens PM, O'Sullivan D (2001) Cellular automata and urban simulation: where do we go from here? Environ Plan 28:163–168

van Asselen S, Verburg PH (2013) Land cover change or land-use intensification: simulating land system change with a global-scale land change model. Glob Chang Biol 19:3648–3667. doi:10.1111/gcb.12331

van Vliet J, Bregt AK, Brown DG, van Delden H, Heckbert S, Verburg PH (2016) A review of current calibration and validation practices in land-change modeling. Environ Model Softw 82:174–182. doi:10.1016/j.envsoft.2016.04.017

van Vliet J, Naus N, van Lammeren RJ, Bregt AK, Hurkens J, van Delden H (2013) Measuring the neighbourhood effect to calibrate land use models. Comput Environ Urban Syst 41:55–64

van Vliet J, Bregt AK, Hagen-Zanker A (2011) Revisiting Kappa to account for change in the accuracy assessment of land-use change models. Ecol Model 222(8):1367–1375

Vieilledent G, Grinand C, Vaudry R (2013) Forecasting deforestation and carbon emissions in tropical developing countries facing demographic expansion: a case study in Madagascar. Ecol Evol 3(6):1702–1716

Voinov A, Kolagani N, McCall MK, Glynn PD, Kragt ME, Ostermann FO, Pierce SA, Ramu P (2016) Modelling with stakeholders—Next generation. Environ Modell Softw 77:196–220. ISSN 1364-8152 http://dx.doi.org/10.1016/j.envsoft.2015.11.016

Walker RT, Moran E, Anselin L (2000) Deforestation and cattle ranching in the Brazilian Amazon: External capital and household processes. World Dev 28(4):683–699

Chapter 3
The Simulation Stage in LUCC Modeling

M.T. Camacho Olmedo, J.F. Mas and M. Paegelow

Abstract In land change modeling, the simulation stage uses parameters and processes to allocate changes by resolving competition between transitions. They are also used to reproduce spatiotemporal patterns of modeled change. There are also several advanced options that try to improve the simulation outputs. This chapter focuses on these simulation steps and on the different types of simulated maps (soft and hard outputs). A theoretical presentation of concepts and methods for each simulation step and simulation output is followed by a comparative analysis of the different approaches for estimating the parameters for the most common pattern-based models (PBM) and constraint cellular automata-based models (CCAM).

Keywords Simulation · Land use and cover changes · Land change models · Soft outputs · Hard outputs

---

M.T. Camacho Olmedo (✉)
Departamento de Análisis Geográfico Regional y Geografía Física,
Universidad de Granada, Granada, Spain
e-mail: camacho@ugr.es

J.F. Mas
Centro de Investigaciones en Geografía Ambiental, Universidad
Nacional Autónoma de México (UNAM), Morelia, Michoacán, Mexico
e-mail: jfmas@ciga.unam.mx

M. Paegelow
GEODE UMR 5602 CNRS, Université de Toulouse Jean Jaurès,
Toulouse, France
e-mail: paegelow@univ-tlse2.fr

M.T. Camacho Olmedo et al. (eds.), *Geomatic Approaches for Modeling Land Change Scenarios*, Lecture Notes in Geoinformation and Cartography,
https://doi.org/10.1007/978-3-319-60801-3_3

1 Introduction

One of the greatest challenges in land change models is to find effective ways of simulating spatiotemporal changes. Simulation outputs may reflect the real situation quite well, but the objective is more to design possible lines of development, than to forecast real future development. Several terms are used to identify the different kinds of land change model outputs, such as simulation, projection, prediction, forecast or scenario. Sometimes the difference between these terms is not clear (Paegelow and Camacho Olmedo 2008; Brown et al. 2013; Kelly et al. 2013).

The most generalized term used to designate the result of a time projection model is simulation, which is also used to describe the process involved. "Simulation consists to make evolving a system abstraction over time so as to understand the functioning and the behavior of the system and to grasp some of its dynamic characteristics with the aim to evaluate different decisions" (Coquillard and Hill 1997). Simulations can be obtained for a current situation (in order to compare with the real situation and to validate the model), a past situation (to understand how land-use has evolved) (Gonçalves and Dentinho 2007; Camacho et al. 2008; Chang-Martínez et al. 2015) or, and most frequently, to assess future change. This kind of simulation is known as a projection. A projection is a description of a future land system and the pathway leading to it.

Prediction and forecast are time extrapolations and the—predicted—result shows what is most likely to happen at an unknown moment in the future (prospective simulation). Kelly et al. (2013) differentiate between prediction, when estimating the value "… of a system variable in a specified time period given knowledge of other system variables in the same time period", and forecasting, which "… refers to predicting the value of a system variable in future time periods… without knowledge of the values of other system variables in those periods…". Prediction therefore seeks to achieve some degree of accuracy and needs data for calibration and validation. Forecasting is more uncertain and can include likely scenarios. Scenario is "a coherent, internally consistent, and plausible description of a possible future state of the world" (Houghton 1995). It shows us what could happen. Modelers frequently apply different underlying conditions (such as macroeconomic parameters) or dynamic variables (that are changing during the simulation) so that the simulations produce different results that describe a framework of possibilities providing predictive answers (see Chap. 5 about scenarios).

In order to produce the most appropriate land change model output, the calibration stage (see Chap. 2 about calibration) integrates all the necessary parameters and processes required to enable the model to evaluate potential change and estimate the quantities of change that will take place. Later, although the two stages may sometimes overlap, we embark on the simulation stage, which provides parameters and processes to allocate changes and reproduce spatiotemporal patterns, in addition to several advanced options (Fig. 1).

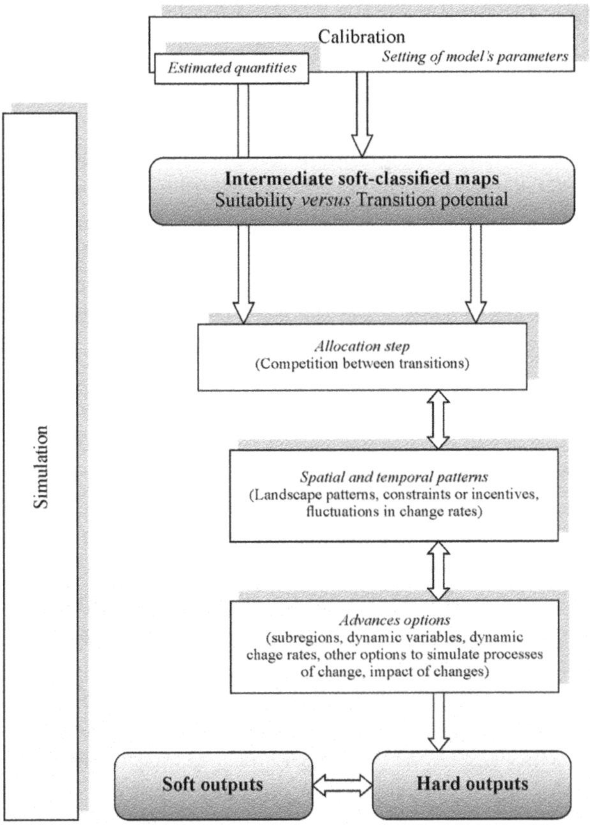

Fig. 1 Flowchart of the general procedure in the simulation stage in LUCC modeling approaches

In this chapter we will focus on these simulation steps. We will begin by presenting the different types of simulation maps, which must later be validated using the relevant methods and tools (see Chap. 4 about validation, Paegelow et al. 2014). A theoretical presentation of concepts and methods for each simulation step and simulation output is followed by a comparative analysis of the most common land use and cover change (LUCC) models.[1] Some of them, such as CA_MARKOV, Land Change Modeler (LCM) (both in TerrSet), Dinamica EGO and CLUMondo, are pattern-based models (PBM), while Metronamica, APoLUS, SLEUTH and LucSim are constraint cellular automata-based models (CCAM).

[1]See the short presentations in Part V of this book about (in alphabetical order) APoLUS, CA_MARKOV, CLUMondo, Dinamica EGO, Land Change Modeler (LCM), LucSim, Metronamica and SLEUTH. The authors are also grateful to all the contributors who helped us understand the different software packages.

2 Simulation Outputs

2.1 Soft Outputs

The first product of an LUCC modeling process is a soft output (soft-classified map) (Mas et al. 2014), where each pixel shows the potential or the likelihood of a specific transition or category occurring, or the probability of belonging to one or more classes. Several expressions have been used to describe soft outputs in the literature: change potential (Mas et al. 2011, 2014; Pérez-Vega et al. 2012), susceptibility to land change (Silva and Tagliani 2012), propensity to change (Mas et al. 2014), transition potential (Bonham-Carter et al. 1989), transition probability (Soares-Filho et al. 2001) and suitability (Steiner et al. 2000). While several of these terms are intended to be generic (such as change potential), others describe a specific type of soft result.

One of the most important characteristics that define and classify soft outputs is their temporal reference. Soft outputs are obtained from a calibration period (from two time points t_0-t_1, or from just one time point t_1) and, while some soft outputs are not related to a specific subsequent date, i.e. they have no temporal reference, other soft outputs are related to a subsequent date (t_2), attempting to estimate the quantity of changes/demands and their allocation in space. The former are called *intermediate soft-classified maps* (Camacho Olmedo et al. 2013) and the latter *soft-classified maps.*

In the wide body of research on output validation, only a few contributions focus exclusively on soft simulation results (Pontius and Cheuk 2006; Conway and Wellen 2011; Wang and Mountrakis 2011; Pérez-Vega et al. 2012; Kolb et al. 2013).

2.1.1 Intermediate Soft-Classified Maps

Intermediate soft-classified maps are used as rank-order indices and do not refer to a specific date in the future. They do not assess the probability of change, because they do not include the information about how many pixels will be affected by estimated transitions. Mas et al. (2014) argued that, when using neural networks or other machine learning tools, the values cannot be considered as probabilities in a strict sense, even if they are interpreted in the same manner, i.e. as values ranking the potential for change. This means that these maps are difficult to compare because a higher value (ranking) does not necessarily imply a higher anticipated quantity of change than a lower value. The simulated amount of changes will depend on how the model estimates these amounts, how it allocates changes after conflict resolution, and how model parameters produce variations in both quantity and allocation (Camacho Olmedo et al. 2015).

We identified two main, slightly different approaches for *intermediate soft-classified maps*: *suitability* maps and *transition potential* maps, that is, the suitability of a location for a given land use and cover (LUC) versus the potential of this transition occurring (Camacho Olmedo et al. 2013; Kolb et al. 2013). The distinction between suitability (i.e. the LUC state) and transition potential (i.e. the LUC transitions) is a major issue when comparing the application of modeling tools because it lies at their conceptual core. Using different terms, Koomen and Stilwell (2007) discuss allocation models, which allocate land use to a particular location based on its characteristics, or transformation models, which begin with one type of land use and simulate its possible conversion into a different type.

A transition potential map is "…an index on a scale from 0 to 1, where higher numbers indicate pixels that have a combination of explanatory values that are more similar to places where the particular transition occurred during the calibration interval compared to places where the transition did not occur…" (NRC 2014). We can therefore assume that a suitability map is also an index on a scale from 0 to 1, where the higher numbers indicate pixels that have a combination of explanatory values that are more similar to places where the particular LUC category is located in the calibration period/date than to the places where the LUC category is not located (Fig. 2). The combination of values corresponding to explanatory variables is determined in different ways and is one of the most important challenges in land change models (see Chap. 2 about calibration).

One of the main differences between the suitability-based approach and the transition potential maps approach is in how changes over time are considered. The suitability-based approach does not explicitly consider past changes. It does not necessarily pay attention to past history. Moreover, a suitability map may be considered either a static map or a global evaluation of the state of each land use and cover (LUC). A suitability map expresses the most appropriate use of a parcel of land according to a subjective decision based on knowledge or opinion, in this way determining to what extent a given piece of land is suitable for a specific use (Steiner et al. 2000), or assessing its potential for a specific use (Littleboy et al. 1996). Suitability is not designed necessarily for prediction, since humans frequently use land for purposes for which it is not suitable. In cellular models based on von Thünen (1966) and Alonso (1964), land suitability depends on the cost of renting land for different uses. Starting from the premise that land users aim to maximize profit, each parcel is converted to the use with the highest land rent at that location (NRC 2014). In some constraint cellular automata-based models suitability has a more limited meaning and is related specifically to the biophysical conditions (e.g. slope, etc.) that can influence the likely LUC category (Van Delden et al. 2007).

Suitability maps are obtained more frequently in up to one time step (generally, the last step t_1); however, several land change models based on suitability maps (Conway and Wellen 2011; Yu et al. 2011, 2015) incorporate information from a previous time period as past changes (e.g., gains, losses or no change) that can be introduced as a factor or as a constraint to enhance the approach (Paegelow and Camacho Olmedo 2005; Villa et al. 2007). The inclusion of this information from

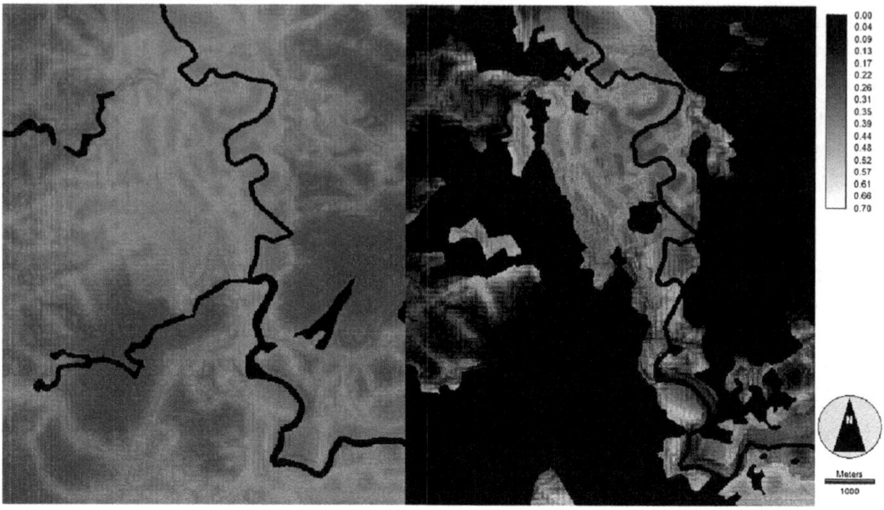

Fig. 2 Examples of intermediate soft-classified maps: suitability for urban areas (*left*) and transition potential maps from irrigated crops to urban areas (*right*)

the calibration period or from the last date does not produce a pure suitability map but one that pays partial attention to previously implemented human land-uses that are not suitable but are common.

In contrast, the modeling tool based on transition potential evaluates the change potential for each possible transition, where the future potential of space is split into specific transitions across a finite number of LUC categories. This approach indicates that the transition potential maps are derived explicitly from past changes (from two known LUC time steps t_0– t_1 from the calibration period), and thus, the reference map for creating the factor must be a real transition from t_0 to t_1. A transition potential map expresses a researcher's knowledge of the relative likelihood of transition of one parcel relative to another (Bonham-Carter et al. 1989; Sklar and Costanza 1991; Eastman et al. 2005; Sangermano et al. 2010; Wang and Mountrakis 2011; Mozumder et al. 2016), that is, it communicates the likely future based on an extrapolation from the observed past period.

Comparing both approaches, suitability-based models are more stable and provide better results for simulations over a long period with non-stationarity change patterns, while models based on transition potential are more appropriate for simulations over a short period and stationarity change patterns. Suitability maps reflect the changes accumulated throughout the human history of a region and do not specify the candidates for LUC change. This means they are made up of all the different processes that can occur to and from the different LUC categories.

However, transition potential maps focus on changes that occur during the calibration period and are specific to each LUC candidate transition. Suitability maps could offer larger statistical significance, because they cover all historical changes and stability, while transition potential maps could be less representative because they refer to a short, well-defined, recent period. These two types of intermediate soft-classified maps are compared in other papers (Camacho Olmedo et al. 2013; Kolb et al. 2013; Mas et al. 2014; Ozturk 2015).

2.1.2 Soft-Classified Maps

Depending on the models, *soft-classified maps* are maps where each pixel has a partial membership of several categories simultaneously. They can be considered as probabilities because they are projected to a specific date taking into account the estimated quantities during the simulation interval (Hsieh and Juang 2009; NRC 2014). These maps can therefore be effectively compared because a higher value in these maps does imply a higher anticipated quantity of change than a lower value. They refer to a specific future date, estimating the quantity of land changes/ demands and their allocation in space. Nevertheless, the soft output does not show the areas that will change, but rather their vulnerability to change or their likelihood to precipitate change (Eastman 2015) (Fig. 3).

The soft-classified map or set of soft-classified maps (one for each category or class or for each transition between them) can have values in an interval of 0-1, indicating the probability that the class exists within the pixel, which is derived from the uncertainty. As Pontius and Cheuk (2006) state "…many classes may exist within the pixel, but the scientist may be uncertain concerning where the specific classes are located within the pixel; therefore, the scientist assumes a random spatial distribution of the classes within the pixel. The membership to each class is the probability that the class exists at a randomly selected point within the pixel. This probability equals the proportion of the pixel that the class constitutes…".

In recent research, there is an increasing interest in these types of map, which are not limited to a single discrete LUC per pixel, but provide continuous fields that can consider LUC as quantitative data or fractional cover of different categories (NRC 2014). Although until recently soft-classified maps tended to be disregarded in the results, they are now drawing more and more attention in the literature because they can offer a better picture of the true proportion of each category within the study area (Fisher and Pathirana 1990; Settle and Drake 1993; Foody and Cox 1994; Zhang and Foody 1998; Pontius and Cheuk 2006).

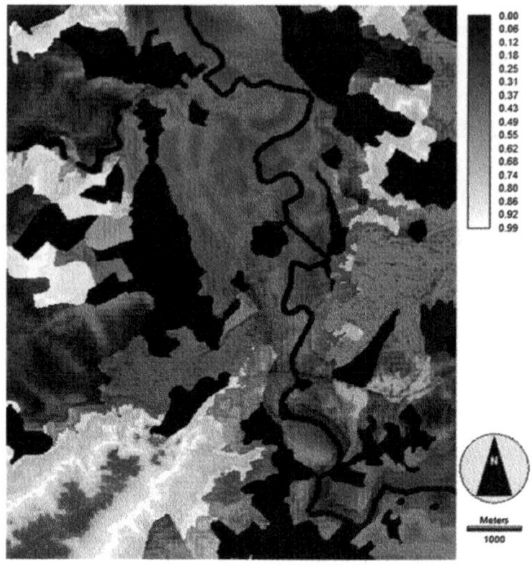

Fig. 3 Example of a soft-classified map created by combining maps of areas with transition potential to urban areas, irrigated crops and natural vegetation

2.2 *Hard Outputs*

The hard output (hard-classified map), also known as an allocation map, is generally accepted as one of the most important results in the simulation stage. It is viewed as simulation sensu stricto. In a hard-classified map, each pixel is allocated the same categories as those used in the calibration step and each pixel belongs to exactly one category. This form of simulation is subject to two main conditions. Firstly that the modeled variable has a discrete number of states (categories or classes), which are the same at any time step, unlike soft outputs, and secondly that the state can only change at discrete time steps (Legendre and Legendre 1984).

In this last sense, hard outputs are usually referred to as static representation, even if some models can produce successive hard outputs relating to each time step. Therefore, a descriptive model alone cannot cover the system's dynamics and complex processes (Batty 2003). As Bregt et al. (2002) explain "…Combinations of data, representing the initial status, and some rules or models describing the change of the environment over time, are needed. These rules range from relatively simple expert tables describing change in discrete intervals over time to complex dynamic simulation models describing change at continuous time intervals…".

Hard outputs are commonly obtained in prospective modeling, i.e. for future dates. Nevertheless, examples of hard outputs from retrospective simulations can also be found in Camacho Olmedo et al. (2007, 2008), Fuchs et al. (2015), Gonçalves and Dentinho (2007) and Chang-Martínez et al. (2015).

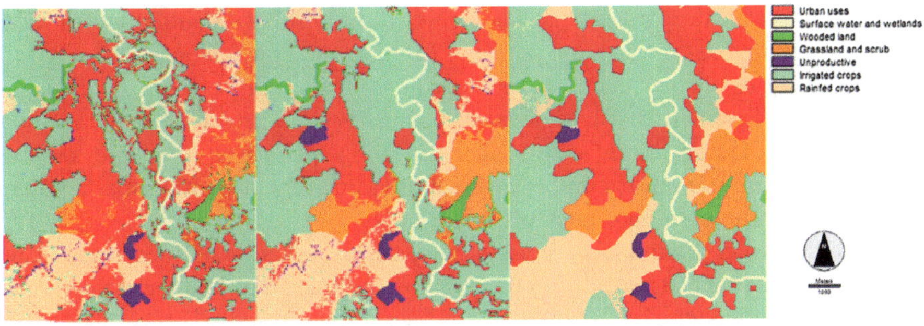

Fig. 4 Examples of hard outputs using different simulation parameters and models

A hard-classified map varies depending on the decisions taken during simulation, regarding allocation type, the spatial and temporal patterns and procedures and the advanced options chosen for each model approach (Fig. 4) (cf. Sect. 3). As Pontius and Malanson (2005) demonstrate, the choice of the parameters within a single model produces greater variation in accuracy than the choice between alternative models. Most hard output assessment methods try to gauge their accuracy with respect to real data. The comparison and assessment of hard outputs using different simulation parameters and models is of increasing interest in land change modeling literature (Villa et al. 2007; Pontius et al. 2008; Fuller et al. 2011; Mas et al. 2011, 2014; Wang and Mountrakis 2011; Arsanjani 2012; Sinha and Kumar 2013; Camacho Olmedo et al. 2015).

2.3 Comparing the Simulation Outputs of Different LUCC Models

If we compare the simulation outputs of the different LUCC models (Table 1) with regard to intermediate soft-classified maps, CA_MARKOV and CLUMondo use maps of the suitability of a location for each of the LUC categories, while Dinamica EGO, LCM, Metronamica and LucSim use transition potential maps. In APoLUS, by default the outputs only allocate land use, but the model can be modified to output the intermediate map, i.e. transition potential. In SLEUTH, the intermediate soft-classified maps are expressed as land use uncertainty and the probability of urban growth.

LCM TerrSet, Metronamica, APoLUS and SLEUTH provide soft-classified maps. LCM TerrSet can obtain the soft-classified map for each stage in the simulation period. Users can choose specific transitions to aggregate their corresponding transition potentials. The aggregation type by default is logical OR, where if a pixel presents transition potential for more than one claimant class, it will be more likely to change than a pixel that shows transition potential for a single claimant class (Eastman 2015). Metronamica and APoLUS provide a new transition potential map at any time within the modeling period, a dynamic output. In SLEUTH, land use uncertainty is given at the same time as hard output, and the urban class can be both hard and soft.

All models offer a simulation sensu stricto, that is, a hard-classified map. LCM TerrSet, Metronamica and APoLUS produce yearly time steps and these time steps can be saved as independent LUC maps.

3 Simulation Steps

3.1 Allocation Step

The allocation step is a decision process that selects from the *intermediate soft-classified maps* (transition potential maps, suitability maps, etc.) the pixels that are most likely to change from one category to another or the pixels that are most likely to be a category. These pixels are likely to be those with the highest potential for change or the highest suitability for the "destination" LUC category. The allocation of these changes in space finally gives rise to *soft-classified maps*, and, more usually, to *hard classified maps*. Both are related to estimated quantities in the calibration stage. In this section we will be referring exclusively to the hard output maps.

This procedure can vary greatly depending on several parameters. For example, the modeling of binary maps, the most basic procedure, is based on a simple cutoff of the change potential map. The cutoff value can be determined when the quantity of each land use to be allocated is reached (Pontius et al. 2001). Values over the cutoff value are then assigned to the desired class.

However, multiple land use transitions in the same model create what is known as the Multiple Classifications (MC) problem (Ho 2000; Tayyebi and Pijanowski 2014). In these cases, models use maps with several classes and transitions and MC can be solved by a simple hierarchical approach, in which for example urban use is considered the top priority use (Letourneau et al. 2012; NRC 2014), or by a competitive approach in land allocation.

Competition and conflicts between different transitions are resolved by iterative processes which take into account the change potential and the quantity of change during the simulation, according to the estimated quantities based on a Markov chain (see Technical Notes in Part IV of this book) or an external chain, or by normalizing the probabilities of the possible transitions during the calibration stage.

Table 1 Comparing the simulation outputs of LUCC models

	Pattern-based models (PBM)				Constraint CA-based models (CCAM)		
	CA_Markov TerrSet	LCM TerrSet	Dinamica EGO	CLUMondo	Metronamica and APoLUS	SLEUTH	LucSim
Intermediate soft-classified	Suitability	Transition potential	Transition potential	Suitability	Transition potential	Land use uncertainty/ probability of urban growth	Transition potential
Soft-classified	No	Yes	No	No	Yes	Yes	No
Hard-classified	Yes	Yes	Yes	Yes	Yes	Yes	Yes

Claims for land uses in bottom-up approaches are determined by allocation rules and in top-down approaches by driving factors, while hybrid approaches use feedback between the two (Verburg et al. 2006). Tayyebi and Pijanowski (2014) propose a simple method for solving the conflicts for multiple land-use transitions by eliminating ambiguous predictions using non-linear tools. A stochastic selection algorithm can also resolve incompatibilities.

Users' choices can also affect the general procedure in the allocation step. Users generally prefer to model just some of the transitions rather than all of them. This is due to various factors i.e. the rigid method and other reasons relating to erroneous data, non-probable changes, very small transitions and their corresponding very small reference maps for obtaining factors and the specific objectives of each user.

The set of algorithms and their parameters in the allocation step have important consequences in simulations. As Prestele et al. (2016) reveal, in land change models, a high uncertainty is related to "…allocation of projected changes, which can severely impact the results of environmental assessments…" Camacho Olmedo et al. (2015) demonstrated that the parameters used in the allocation procedure can affect the simulated quantities of change, because some parameters force the model to disregard the estimated quantities. The assessment of variation in the allocation of land change and persistence can throw up important conclusions as to how models work.

3.1.1 Comparing the Allocation Process in the Different LUCC Models

The allocation process involves various steps. In the first step, all models are based on ranking the intermediate maps. For both CA_MARKOV and LCM TerrSet models (Table 2), concurrences between different uses or transitions are solved using a Multi-Objective Land Allocation (MOLA) algorithm, which uses an approximation procedure based on a minimum-distance-to-ideal-point rule using the weighted ranks from the change potential maps (Eastman et al. 1995). In

Table 2 Comparing the allocation process in LUCC models

	Pattern-based models (PBM)				Constraint CA-based models (CCAM)		
	CA_Markov TerrSet	LCM TerrSet	Dinamica EGO	CLUMondo	Metronamica and APoLUS	SLEUTH	LucSim
Competition between transitions	Ranking/ multi-objective land allocation	Ranking/ multi-objective land allocation	Ranking/ stochastic selection algorithm	Ranking/dynamic (iterative), based on competitive advantage	Dynamic ranking/ stochastic	Deltatron model, based on slope	Ranking/ stochastic

CA_MARKOV the user must also incorporate the number of iterations, i.e., the number of time units in the simulation period, the last iteration being the prediction for the later date (see the technical note about Multi-Objective Land Allocation in Part IV of this book).

In Dinamica EGO, a stochastic selection algorithm resolves the competition between transitions ranking the pixels with respect to their change potential maps, which are previously normalized. To avoid restricting the simulated change to the highest change potential cells, Dinamica EGO selects a large amount of candidate cells and carries out a lottery-type process, which enables a small number of cells with lower change potential to present the transition. This behavior is controlled by a parameter called the pruning factor. If the chosen pruning factor is one, it produces a deterministic result, while a high pruning factor allows simulated changes to occur in less likely areas (Mas et al. 2014). In LucSim, the allocation step is also based on ranking the pixels with greatest potential and a stochastic algorithm to resolve incompatibilities.

The CLUMondo model consists of a non-spatial demand module and a spatially explicit allocation module (Liu and Yang 2015). CLUMondo ranks the most suitable land use, and uses an allocation process based on competitive advantage with respect to goods and services. Competitiveness is updated during a dynamic iterative process to match the allocation and land demand.

In Metronamica and APoLUS, the allocation step is a dynamic process based on ranking the intermediate maps produced for each time unit. In both models a random factor is added to represent the stochastic uncertainty of the land allocation process. In SLEUTH the competition between transitions is resolved in the Deltatron model, based on slope (Candau et al. 2000).

3.2 *Spatial and Temporal Patterns*

In land change models there is a growing interest in producing realistic simulations. The simulation of spatial pattern involves a realistic distribution of the simulated change in the landscape and should consider the competition between different transitions for the same locations. Ideally the model will be able to simulate patches of land change with realistic shape and size. Depending on the objective of the model, this aspect can be crucial or merely cosmetic. For instance, a model that can predict a deforestation pattern in which small agricultural plots perforate forest can be very useful for biodiversity conservation issues.

Spatially explicit land use/land cover change (LUCC) models aim to simulate the patterns of change—spatial and temporal patterns—on the landscape (Paegelow et al. 2013). This interest is evident in models such as machine learning, data mining, statistical methods, cellular automata (CA), pattern-based models (PBM) and other spatially explicit models. These kinds of models are focused on identifying patterns, as "descriptions of observed phenomena over some time interval or spatial area", as opposed to models focusing on processes, "the mechanisms that generate observed patterns" (NRC 2014). Other concepts such as equifinality, i.e. when two processes produce the same pattern, and multifinality, when one process produces different patterns, are also of interest for assessment methods (Brown et al. 2005, 2006).

Spatial and temporal patterns are closely related in land change models; nevertheless there are some processes that focus more on one than the other.

In previous research, models tried to generate realistic spatial patterns using different spatial units. Koomen and Stillwell (2007) differentiate, in relation to the spatial level of detail, between models based on grids—regular raster patterns, cell, element or pixel-based analysis—which is the basic unit in land change modeling, and models based on zones—relatively homogeneous, irregularly shaped areas or vector polygons. Still few in number, some researchers prefer vector-based representation (Schaldach and Alcamo 2006; Rasmussen and Hamilton 2012) as referenced by Kelly et al. (2013).

Landscape metrics or spatial metrics are related to a group of pixels or to an object-based analysis (Murayama and Thappa 2011; Chen et al. 2012) and validation methods are increasing focused on using landscape metrics instead of spatial coincidence (Mas et al. 2010; Aguilera et al. 2011; Bradley et al. 2016). Several characteristics must be considered: size, shape, neighborhood, distribution, connectivity or continuity of LUC, which can identify significant impacts on landscape and human-environment systems (McGarigal et al. 2012). Neighborhood analysis is one of the most highly developed parameters for spatial patterns modeling. It can be a simple algorithm of surrounding cells, i.e. it simulates that the pixels that are contiguous to existing LUC pixels in a particular category are likely to belong to or to change to this category. This kind of algorithm is included in some PBM models.

Neighborhood analysis can also be conducted using complex algorithms based on network analysis or predefined regions or in the form of cellular automata (CA) (Gardner 1970; Couclelis 1985; Batty and Xie 1994; White et al. 1997; Zhao and Murayama 2011; van Vliet et al. 2013). The success of CA models can be partly attributed to their simple data format, discrete spatial units—pixels, that match the format of the LUC data derived from remote sensing. Nevertheless, the algorithm affects spatial units that are more like real objects such as parcels, regions or other land units, so creating a much more complex structure (Lazrak et al. 2010).

In brief, a cellular automata approach is made up of the following elements: cell space, cell states, time steps and transition rules (White and Engelen 2000). Some constraint cellular automata-based models also integrate four parameters: Neighborhood (N), Accessibility (A), Suitability (S) and Zoning (Z) (see the technical note about the NASZ model in Part IV of this book).

Linked to spatial patterns, time analysis and reproducing temporal patterns are amongst the most important challenges in land change models. There is an increasing interest in temporal patterns in the literature (Liu and Anderson 2004; Mas et al. 2011; Runfola and Pontius 2013), despite which they are still rarely implemented (NRC 2014). Changes in LUC are often non-linear and LUCC often shows high degrees of temporal complexity and feedback mechanisms (Verburg 2006; Verburg et al. 2006). Moreover, as Paegelow and Camacho (2008) pointed out, LUC dynamics can be progressive or regressive, slow or fast. Time can be short, medium or long, continuous or split into discrete time steps and temporal resolution depends on databases and modeling methods (some models need more dates to work correctly than others).

Concepts such as non-stationarity, the opposite of stationarity, or stability in two subsequent periods lie at the heart of this research. The relation between non-stationary processes and predictability needs to be explored (Müller et al. 2014) and uncertainty in land change models can increase with the length of simulation (Chaudhuri and Clarke 2013; van Kliet et al. 2016), which can be an important limitation. Most applied models produce reasonable projections for short simulation periods (around 20 years), because temporal fluctuations and feedbacks are easy to implement. However, projections for longer periods are not really capable of including complex change rates and feedbacks, because of modifications to the conversion rules and changing decision-making strategies by the agents involved (NRC 2014; Meyfroidt 2013).

With respect to types of models, PBM are more useful for stationarity and short simulation periods extrapolating past patterns, but are of limited use for proposing large projections involving structural changes and non-stationarity processes, where process-based models seem to be more appropriate. Because the assumption is that past and present trends will continue into the future, PBM, machine learning and statistical models "…tend to oversimplify the temporal complexity of land change processes…" (Liu and Yang 2015) With respect to CA models, they forecast land cover patterns by evaluating "changes in spatial controls without market feedbacks" (NRC 2014).

Some parameters try to reproduce temporal patterns in order to achieve realistic simulations in land change models. Some researchers introduce spatial and temporal non-stationarity into the transition probabilities (Brown et al. 2000). Rosa et al. (2014) suggest that "…the next generation of LCC models may need to incorporate temporal variability in the parameters associated to the drivers of changes in order to allow the processes determining LCC to change through time and exert their influence on model predictions…" Intensity analysis (Aldwaik and Pontius 2012; Pontius et al. 2013) quantifies the behavior of a categorical variable across several time intervals to measure the degree to which changes are non-uniform.

In some models, spatial and temporal parameters such as fluctuations in change rates can be applied in the simulation stage. They refer to the sojourn time for certain transitions that are deterministic and depend on particular processes, and to saturation effect, the fact that certain transitions stop when the amount of change has reached a given level. For instance, a deforestation front will move forward, resulting in a certain number of remaining forest fragments (Mas et al. 2014).

3.2.1 Comparing LUCC Models in Spatial and Temporal Patterns

Inductive pattern-based models (PBM) and constraint cellular automata-based models (CCAM) work very differently in the simulation step for reproducing spatial and temporal patterns (Table 3).

The parameters in several models try to reproduce spatial patterns in hard-classified maps. These include for example the use of cellular automata and emergent patterns for *landscape patterns* simulation and the inclusion of areas of exclusion (Pas) *constraints or incentives.*

To simulate *landscape patterns*, the PBM CA_MARKOV and Dinamica EGO use a cellular automata (CA) approach to obtain a proximity effect that causes changes to occur in the form of patches. Nevertheless, the procedures are quite different in these two models. The Cellular Automata (CA) incorporated by default in CA_MARKOV, reduces the suitability away from existing areas of each LUC, giving more chance to pixels that are both suitable and close to the existing LUC, producing a dilation effect around existing patches and partially avoiding the "salt-and-pepper" effect (see technical note about Cellular Automata in CA_MARKOV in Part IV of this book). Dinamica EGO uses two complementary CA: the Expander, that expands or contracts previous patches, and the Patcher, that

Table 3 Comparing LUCC models in spatial and temporal patterns

	Pattern-based models (PBM)				Constraint CA-based models (CCAM)		
	CA_Markov TerrSet	LCM TerrSet	Dinamica EGO	CLUMondo	Metronamica and APoLUS	SLEUTH	LucSim
Landscape patterns	CA: Contiguity 5 × 5 filter	No	CA: Expander & Patcher	Optional CA: Neighborhood	Implicit CA: NASZ In APoLUS: NASZD	Two implicit CA	Implicit CA: Neighborhood
Constraints or incentives	In suitability maps	Yes	Yes	Yes	Yes	Yes	In transition potential maps
Fluctuations in change rates (Sojourn time, saturation)	Could be implemented	No	Easily implemented	Sojourn time; saturation easily implemented	Could be implemented	Integral part of model (self-modification)	Saturation easily implemented

generates new patches through a seeding mechanism. Users can set several parameters such as patch size, variance, etc. (see technical note about Cellular Automaton in Part IV of this book). CLUMondo can include neighborhood interaction and influences the suitability maps through spatial filters (Verburg et al. 2004). It allows a user-defined neighborhood with positive or negative values that can produce attraction and repulsion between land uses. LCM does not apply a CA procedure.

Constraint cellular automata-based models, such as Metronamica, APoLUS, SLEUTH and LucSIM, apply an implicit CA based on neighborhood interactions. Metronamica and APolUS are based on the NASZ model and APoLUS adds Actor Dynamics (D). SLEUTH consists of two coupled cellular automata models: the Urban Growth Model and the Deltatron land use change model. LucSim is a cellular automata model based on Torrens's definition (Torrens 2011) and one of the two models it integrates, the potential model, is based on spatial interaction and neighborhood parameters.

Constraints or incentives limit the simulated maps to particular areas or to apply spatial policies. LCM and Dinamica EGO use them at a certain time step of simulation. They also include an implicit constraint because only modeled transitions such as transition potential maps are simulated. Therefore, LCM and Dinamica EGO simulate transitions by excluding some origin LUC that users in the study areas consider illogical or unlikely to be changed to another LUC. In CLUMondo, a hard constraint can be implemented by supplying a map with pixel values between 0 and 1, and these maps can be considered as a soft incentive or constraint.

In CA_MARKOV and LucSim, constraint and incentive areas can only be considered through the suitability or transition potential maps, respectively. In CA_MARKOV, a script can be written to simulate an effect that will occur at a certain time, and constraints with specific LUC that users consider inappropriate for the destination LUC can be included in the multi-criteria evaluation (MCE).

In Metronamica and APoLUS constraints and incentives are possible in suitability, accessibility and more specifically in zoning (NASZ), which is applied at each simulation step. In SLEUTH, a separate excluded layer can be included, acting as a constraint during simulation.

SLEUTH incorporates *fluctuations in change rates* for reproducing spatial and temporal patterns as part of the model (self-modification). CLUMondo is the only model that explicitly includes a sojourn time for each transition. In this model, the saturation effect can be included and it also uses the elasticity values to manage the amount of change from one LUC to another necessary to fulfill the established change rules (Mas et al. 2014). This elasticity allows some categories to be more resistant to change than others (Verburg et al. 2006). In LucSim, the saturation effect is easy to control. In Dinamica EGO, both saturation effect and sojourn time are easy to implement. In order to simulate the saturation effect in deforestation processes and establish a minimum remaining forest area, the probability of a transition can be reduced taking into account the cell neighborhood by means of a kernel window. In CA_MARKOV, Metronamica and APoLUS these options could be implemented. In LCM it is not possible to include any kind of fluctuations in change rates.

3.3 Comparing LUCC Models in Advanced Options

Region-based, compartmental spatial models (Kelly et al. 2013) are one of the advanced options in modeling for integrated environmental management. Complex models include splitting the study areas into *subregions* (Mas et al. 2014). Outputs are generated for homogeneous subareas of the total area. Different dynamics can be produced by different rates of change, transitions, explanatory variables and their effects. As an advanced option, *subregions* analysis can be incorporated into the models. Dinamica EGO and LucSim (Table 4) can divide the study area into regions using certain particular parameters. For example, interactions between the subregions enable certain variables to have an effect on certain subregions, and variables based on distance to affect the entire study. In Metronamica, subregion analysis is based on interconnected models and APoLUS includes the regional layer, but only related to actor behavior. SLEUTH does not incorporate regions in simulation, as it is a pixel-based model at any extent or scale. In CLUMondo, LCM and CA_MARKOV, it is possible to run independent models to obtain subregions and, then to mosaic the simulated maps, but these results will not interact. Consequently, incompatibilities could happen on the boundaries between regions.

Dynamic variables are another advanced option in simulation as they are updated at each time step. This enables events in previous steps to have an effect on subsequent steps. Some models can include a temporally lagged variable to take into account the fact that there are lags between causes and consequences in land change processes (Agarwal et al. 2002; Liu and Yang 2015). Almost all models can handle *dynamic variables* in the simulation step. In SLEUTH, most of the variables are dynamic. In Metronamica and APoLUS, the neighborhood parameter changes at any time step, and accessibility, suitability and zoning can change if required. In CA_MARKOV, users decide the number of CA iterations (time steps) and therefore their contiguity effect. CA_MARKOV does not consider more dynamic variables, although TerrSet's macro modeler can develop this option. For its part, LCM allows us to change some variables such as infrastructure and spatial constraints/incentives in the simulation stage. Dinamica EGO allows explanatory variables to be substituted at certain time-steps in the simulation. In CLUMondo, the explanatory variables can be changed in given time steps. LucSim also allows some dynamic variables.

Dynamic change rates show generic variations or trends over time that must be included in the model. These options complement a more linear or more stationary estimation of quantities, the most commonly used matrix in the calibration stage (see Chap. 2 about calibration). Feedback mechanisms are also included in several dynamic models, for example, via the interaction between different processes that can produce different results in each run of the model (Claessens et al. 2009; van

Table 4 Comparing LUCC models in advanced options

	Pattern-based models (PBM)				Constraint CA-based models			
	CA_Markov TerrSet	LCM TerrSet	Dinamica EGO	CLUMondo	Metronamica	APoLUS	SLEUTH	LucSim
Subregions	No	No	Region operators	No	Interconnected models	Regional layer. Only related to actor behavior	No	Region operators
Dynamic variables	Only CA	Yes	Yes	Yes	Yes	Yes	Yes	Yes
Dynamic change rates	Concatenating models	From external transition matrix or concatenating models	From external or internal sub models	Demand from external models. Intensification and expansion is endogenous	From external/ internal sub models	Exogenous but affected by actors.	Self-generated from bottom-up change	No
Other options for simulating change processes	No	Road constructor, expected infrastructure changes	Road constructor	No	Transport model, population model (additional components)	Dynamic actor behavior	No	No
Impacts of change	No	Yes	Yes	No	Indicators	No	No	No

Vliet et al. 2016). Other models can represent time lags in land change decisions (Irwin and Bockstael 2002).

Dynamic change rates or different amounts of change over time are possible in the PBM LCM, CLUMondo and Dinamica EGO from internal or external transitions and sub-models. All these models allow us to use matrices other than the Markov matrix. LCM allows us to use an external transition matrix. In CLUMondo, demand can be incorporated from external models, but the trade-off between expansion and intensification is endogenous. Dinamica EGO is the most flexible: it can replace the Markov matrix at specific steps of the simulation and it can be coupled with an external model that calculates dynamic transition rates and passes them on to the model. Moreover, simulation can be performed if conditional execution functions ("if then", "while") are used. In CA_MARKOV, dynamic change rates are not included, although some transition dynamics and concatenate models could run if TerrSet's macro modeler were used.

With respect to Cellular automata approaches, Metronamica incorporates dynamic change rates from external or internal sub-models, while in APoLUS the exogenous rates can be affected by the stakeholders. In SLEUTH, dynamic change rates are self-generated from bottom-up change. The SLEUTH model simulates change and changes feedback at each time increment. LucSim does not include dynamic change rates.

Some models can incorporate dynamic change rates by concatenating models, so that one model's output is the next model's input, and splitting the simulation horizon into several periods (Eastman 2016).

Other options for simulating change processes are for example the inclusion of road network as a predictor of the LUCC spatial patterns, because it is easy to simulate the effect of a new road in the simulation period, as in Dinamica EGO and LCM. New road end-points are stochastically selected in the areas with the highest change potential. They are then connected to the existing road network using friction maps (e.g. related to topography) in order to create the least-cost path and/or to link various areas with high change potential. Expected infrastructure changes can also be included in LCM simulations. Metronamica also allows for additional components such as a transport model and a population model. In a future version of APoLUS (APoLUS-SD) dynamic actor behavior will be available. CA_MARKOV, SLEUTH, CLUMondo and LucSim do not incorporate other options to simulate processes of change, except for those integrated into the intermediate soft-classified maps by Multicriteria Evaluation in the two first models, logistic regression in CLUMondo and in the potential model in LucSim.

LCM provides tools to assess the *impact of change* for ecological sustainability and conservation planning. These include tools for species-specific habitat assessment and change studies, gap analysis, landscape pattern evaluation, biodiversity analysis and CO2 emission assessment. Dinamica EGO provides some additional tools to model wood harvest volumes processed by sawmills and carbon pool mapping. Metronamica produces indicators of the impact of change such as soil sealing. CA_MARKOV, CLUMondo, LucSim, SLEUTH and APoLUS do not incorporate additional tools for assessing the impact of change, but some of them, such as APoLUS, are flexible enough to incorporate them easily.

4 Concluding Remarks

Land change modeling processes integrate different, albeit interwoven steps. After calibrating the models, setting all the necessary parameters for the evaluation of change potential and the estimated quantities of change, the simulation stage offers a large variety of parameters for modelers to choose from. Some of these reproduce the spatiotemporal patterns and other advanced options are shared between the calibration and simulation stages or act as a dynamic and iterative process. Other parameters, such as the allocation of changes, belong exclusively to the simulation stage. As Pontius and Malanson (2005) make clear choosing which parameters to use in a model from the large panoply available is one of the most important and critical decisions, because it produces greater variations in accuracy than the choices between alternative models. Variety is also related to simulation outputs. Even though hard output is considered as simulation sensu stricto, and is the principal focus of the validation methods and tools, there is increasing interest in the intermediate and soft outputs that represent the change potential and the true proportion of change. As a result various specific assessment methods have been developed to contribute to their evaluation. Moreover, the LUCC models we analyzed work differently depending on their particular nature as either pattern-based models (PBM) or constraint cellular automata-based models (CCAM) and consequently their parameters and outputs must be contextualized. Current and future lines of development, incorporating process-based models and agent-based models, will help expand the knowledge base to allow modelers to improve land change modeling results.

Acknowledgements This work was supported by the BIA2013-43462-P project funded by the Spanish Ministry of Economy and Competitiveness and by the FEDER European Regional Fund. This study was also supported by the Consejo Nacional de Ciencia y Tecnología (CONACYT) and the Secretaría de Educación Pública through the project ¿Puede la modelación espacial ayudarnos a entender los procesos de cambio de cobertura/uso del suelo y de degradación ambiental? Fondos SEP-CONACyT 178816.

References

Agarwal C, Green GM, Grove JM et al. (2002) A review and assessment of land-use change models: dynamics of space, time, and human choice. United States Department of Agriculture

Aguilera F, Valenzuela LM, Botequilha A (2011) Landscape metrics in the analysis of urban land use patterns: A case study in a Spanish metropolitan area. Landsc Urban Plann 99(2011): 226–238

Aldwaik SZ, Pontius RG Jr (2012) Intensity analysis to unify measurements of size and stationarity of land changes by interval, category, and transition. Landsc Urban Plann 106: 103–114
Alonso W (1964) Location and land use. Toward a general theory of land rent. Harvard University Press
Arsanjani JJ (2012) Dynamic land-use/cover change simulation: geosimulation and multi agent-based modelling. Springer
Batty M (2003) New developments in urban modeling: simulation, representation, and visualization. In: Guhathakurta S (ed) Integrated land use and environmental models: a survey of current applications and research. Springer, Berlin
Batty M, Xie Y (1994) From cells to cities. Environ Plan 21(Suppl):s31–s48
Bradley AV et al (2016) SimiVal, a multi-criteria map comparison tool for land-change model projections. Environ Model Softw 82:229–240
Bregt AK, Skidmore AK, Nieuwenhuis G (2002) Environmental modeling: issues and discussion. In: Skidmore A (ed) Environmental modelling with GIS and remote sensing. Taylor and Francis, London
Brown DG, Pijanowski BC, Duh JD (2000) Modeling the relationships between land-use and land-cover on private lands in the Upper Midwest, USA. J Environ Manage 59:247–263
Brown DG, Page S, Riolo R, Zellner M, Rand W (2005) Path dependence and the validation of agent-based spatial models of land use. Int J Geogr Inf Sci 19:153–174
Brown DG, Aspinall R, Bennett DA (2006) Landscape models and explanation in landscape ecology: A space for generative landscape science? Prof Geogr 58(4):369–382
Brown DG, Verburg PH, Pontius RG, Lange MD (2013) Opportunities to improve impact, integration, and evaluation of land change models. Curr Opin Environ Sustain 5(5):452–457
Bonham-Carter GF, Agterberg FP, Wright DF (1989) Weights of evidence modelling: a new approach to mapping mineral potential. In: Agterberg FP, Bonham-Carter GF (eds) Statistical applications in the Earth sciences. Geol Surv Canada 89(9):171–183
Camacho Olmedo MT, Paegelow M, García Martínez P (2007) Modélisation géomatique rétrospective des paysages par évaluation multicritères et multiobjectifs. Cybergéo (European Journal of Geography). 09–03-2007. Systèmes, modélisation & géostatistiques, N° 365, 24 pp. ISSN: 1278-3366
Camacho Olmedo MT, Paegelow M, García Martínez P (2008) Retrospective geomatic landscape modeling. A probabilistic approach. In: Paegelow M, Camacho Olmedo MT (eds) Modelling environmental dynamics. Advances in geomatic simulations, pp 247–268. Series Environmental science. Springer, Heidelberg
Camacho Olmedo MT, Paegelow M, Mas JF (2013) Interest in intermediate soft-classified maps in land change model validation: suitability versus transition potential. Int J Geogr Inf Sci 27 (12):2343–2361 (Published by Taylor & Francis)
Camacho Olmedo MT, Pontius RG Jr, Paegelow M, Mas JF (2015) Comparison of simulation models in terms of quantity and allocation of land change. Environ Modell Softw 69:214–221 (Publisher By: Elsevier)
Candau J, Rasmussen S, Clarke KC (2000) A coupled cellular automaton model for land use/land cover dynamics. In: 4th International conference on integrating gis and environmental modeling (GIS/EM4): problems, prospects and research needs. Banff, Alberta, Canada
Chang-Martínez LA, Mas JF, Valle NT et al (2015) Modeling historical land cover and land use: a review from contemporary modeling. ISPRS Int J Geo-Inf 4(4):1791–1812
Chaudhuri G, Clarke KC (2013) Temporal accuracy in urban growth forecasting: a study using the SLEUTH model. Trans GIS 18:302–320
Chen G, Hay GJ, Carvalho LMT, Wulder MA (2012) Object-based change detection. Int J Remote Sens 33(14):4434–4457
Claessens L, Schoorl JM, Verburg PH, Geraedts L, Veldkamp A (2009) Modelling interactions and feedback mechanisms between land use change and landscape processes. Agric Ecosyst Environ 129:157–170

Conway TM, Wellen CC (2011) Not developed yet? Alternative ways to include locations without changes in land use change models. Int J Geogr Inf Sci 25(10):1613–1631
Coquillard P, Hill DRC (1997) Modélisation et simulation d'écosystèmes. Des modèles déterministes aux simulations à événements discrets, Paris Masson
Couclelis H (1985) Cellular worlds: A framework for modeling micro-macro dynamics. Environ Plann A 17(5):585–596
Eastman JR (2015) TerrSet, Manual. Clark University 394 p. © 1987–2015. http://www.clarklabs.org/
Eastman JR, Jin W, Kyem PAK, Toledano J (1995) Raster procedures for multi-criteria/multi-objective decisions. Photogramm Eng Remote Sens 61(5):539–547
Eastman JR, Solorzano LA, Van Fossen ME (2005) Transition potential modeling for landcover change. In: Maguire DJ, Batty M, Goodchild MF (eds) GIS, spatial analysis, and modeling. ESRI, Redland, CA, pp 357–385
Fisher PF, Pathirana (1990) The evaluation of fuzzy membership of land cover classes in the suburban zone. S Remote Sens Environ 34(2):121–132
Foody GM, Cox DP (1994) Sub-pixel land cover composition estimation using a linear mixture model and fuzzy membership functions. Int J Remote Sens 15(3):619–631
Fuchs R, Herold M, Verburg PH, Clevers JGPW, Eberle J (2015) Gross changes in reconstructions of historic land cover/use for Europe between 1900 and 2010. Glob Change Biol 21:299–313
Fuller DO, Hardiono M, Meijaard E (2011) Deforestation projections for carbon-rich peat swamp forests of central Kalimantan, Indonesia. Environ Manag 48(3):436–447
Gardner M (1970) Mathematical games: The fantastic combinations of John Conway's new solitaire game 'life'. Sci Am 223:120–123
Gonçalves J, Dentinho T (2007) A spatial interaction model for agricultural uses. In: Koomen E et al. (eds) Modelling land-use change. Springer, pp 133–144
Ho TK (2000) Complexity of classification problems and comparative advantages of combined classifiers. Lect Notes Comput Sci 1857:97–106
Houghton JT (1995) Climate Change 1994: Radiative forcing of climate change and an evaluation of the IPCC 1992 IS92 Emission Scenarios. Intergovernmental Panel on Climate Change. Working Group I, Intergovernmental Panel on Climate Change. Working Group III. Cambridge University Press
Hsieh CC, Juang HH (2009) The Niche Characteristic of Urban Land Use Strength in Taichung City-A Case Study of TDR. J Geogr Res 51:23–43
Irwin EG, Bockstael NE (2002) Interacting agents, spatial externalities and the evolution of residential land use patterns. J Econ Geogr 2(1):31–54
Kelly RAB, Jakeman AJ, Barreteau O et al (2013) Selecting among five common modelling approaches for integrated environmental assessment and management. Environ Model Softw 47:159–181
Kolb M, Mas JF, Galicia L (2013) Evaluating drivers of land-use change and transition potential models in a complex landscape in Southern Mexico. Int J Geogr Inf Sci 27(9):1804–1827
Koomen E, Stilwell J (2007) Modelling land-use change. In: Koomen E et al (eds) Modelling land-use change. Progress and applications. Springer, pp 1–24
Lazrak E, Mari JF, Benot M (2010) Landscape regularity modelling for environmental challenges in agriculture. Landsc Ecol 25(2):169–183
Legendre L, Legendre P (1984) Écologie numérique, 2ième éd. Tome 2: La structure des données écologiques. Masson, Paris. Presses de l'Université du Québec
Letourneau A, Verburg PH, Stehfest E (2012) A land-use systems approach to represent land-use dynamics at continental and global scales. Environ Model Softw 33:61–79
Littleboy M, Smith DM, Bryant MJ (1996) Simulation modelling to determine suitability of agricultural land. Ecol Model 86(2):219–225
Liu X, Anderson C (2004) Assessing the impact of temporal dynamics on land-use change modeling. Comput Environ Urban Syst 28(1–2):107–124
Liu T, Yang X (2015) Land Change Modeling: Status and Challenges. In: Li J, Yang X (eds) A geomatics perspective. Monitoring and modeling of global changes. Springer, Berlin

Mas JF, Pérez Vega A, Clarke K (2010) Assessing simulated land use/cover maps using similarity and fragmentation indices. In: ASPRS annual conference, April 26–30, San Diego, CA (CD)
Mas JF, Kolb M, Houet T, Paegelow M, Camacho Olmedo MT (2011) Eclairer le choix des outils de simulation des changements des modes d'occupation et d'usages des sols. Une approche comparative. Revue Internationale de Géomatique 3:405–430
Mas JF, Kolb M, Paegelow M, Camacho Olmedo MT, Houet T (2014) Inductive pattern-based land use/ cover change models: a comparison of four software packages. Environ Modell Softw 51:94–111 (Publisher By: Elsevier)
McGarigal K, Cushman SA, Ene E (2012) FRAGSTATS v4: Spatial Pattern Analysis Program for Categorical and Continuous Maps. Computer software program produced by the authors at the University of Massachusetts, Amherst. http://www.umass.edu/landeco/research/fragstats/fragstats.html
Meyfroidt P (2013) Environmental cognitions, land change, and social–ecological feedbacks: an overview. J Land Use Sci 8(3):341–367
Mozumder C, Tripathi NK, Losiri C (2016) Comparing three transition potential models: a case study of built-up transitions in North-East India. Comput Environ Urban Syst 59:38–49
Müller D, Sun Z, Vongvisouk T, Pflugmacher D, Xu J, Mertz O (2014) Regime shifts limit the predictability of land-system change. Glob Environ Change 28:75–83
Murayama Y, Thapa RB (2011) Spatial analysis: evolution, methods and applications. In: Murayama Y, Thapa RB (eds) Spatial analysis and modeling in geographical transformation process: GIS-based applications. Springer, New York, pp 1–26
National Research Council (NRC) (2014) Advancing land change modeling: opportunities and research requirements. The National Academies Press, Washington DC
Ozturk D (2015) Urban growth simulation of Atakum (Samsun, Turkey) using cellular automata-Markov chain and multi-layer perceptron-Markov chain models. Remote Sens 7:5918–5950
Paegelow M, Camacho Olmedo MT (2005) Possibilities and limits of prospective GIS land cover modeling - a compared case study: Garrotxes (France) and Alta Alpujarra Granadina (Spain). Int J Geogr Inf Sci 19(6):697–722
Paegelow M, Camacho Olmedo MT (eds) (2008) Modelling environmental dynamics. Advances in geomatic simulations. Series Environmental Science. Springer, Berlin
Paegelow M, Camacho Olmedo MT, Mas JF, Houet T, Pontius RG Jr (2013) Land Change Modelling: moving beyond projections. Int J Geogr Inf Sci 27(9):1691–1695 (Published By: Taylor & Francis)
Paegelow M, Camacho Olmedo MT, Mas JF, Houet T (2014) Benchmarking of LUCC modelling tools by various validation techniques and error analysis. Cybergeo, document 701, mis en ligne le 22 décembre 2014. http://cybergeo.revues.org
Pérez-Vega A, Mas JF, Ligmann-Zielinska A (2012) Comparing two approaches to land use/cover change modeling and their implications for the assessment of biodiversity loss in a deciduous tropical forest. Environ Model Softw 29(1):11–23
Pontius RG Jr, Cheuk ML (2006) A generalized cross-tabulation matrix to compare soft-classified maps at multiple resolutions. Int J Geogr Inf Sci 20(1):1–30
Pontius RG Jr, Malanson J (2005) Comparison of the structure and accuracy of two land change models. Int J Geogr Inf Sci 19(2):243–265
Pontius RG Jr, Cornell and Hall (2001) Modeling the spatial pattern of landuse change with GEOMOD2: Application and validation for Costa Rica. Agric Ecosyst Environ 85(1/3)
Pontius RG Jr et al (2008) Comparing input, output, and validation maps for several models of land change. Ann Region Sci 42(1):11–47
Pontius RG Jr, Gao Y, Giner NM, Kohyama T, Osaki M, Hirose K (2013) Design and interpretation of intensity analysis illustrated by land change in Central Kalimantan, Indonesia. Land 2013(2):351–369
Prestele R, Alexander P, Rounsevell MDA et al (2016) Hotspots of uncertainty in land-use and land-cover change projections: a global-scale model comparison. Global change Biology (Wiley Online Library)

Rasmussen R, Hamilton G (2012) An approximate Bayesian computation approach for estimating parameters of complex environmental processes in a cellular automata. Environ Model Softw 29(1):1–10

Rosa IMD, Purves D, Carreiras JMB, Ewers RM (2014) Modelling land cover change in the Brazilian Amazon: temporal changes in drivers and calibration issues. Reg Environ Change 15:123–137

Runfola D, Pontius RG Jr (2013) Measuring the temporal instability of land change using the flow matrix. Int J Geogr Inf Sci 27(9):1696–1716

Sangermano F, Eastman JR, Zhu H (2010) Similarity weighted instance based learning for the generation of transition potentials in land change modeling. Trans GIS 14(5):569–580

Schaldach R, Alcamo J (2006) Coupled simulation of regional land use change and soil carbon sequestration: a case study for the state of Hesse in Germany. Environ Model Softw 21: 1430–1446

Settle JJ, Drake A (1993) Linear mixing and the estimation of ground cover proportions. Int J Remote Sens 14(6):1159–1177

Silva TS, Tagliani PRA (2012) Environmental planning in the medium littoral of the Rio Grande do Sul coastal plain—Southern Brazil: Elements for coastal management. Ocean Coast Manag 59:20–30

Sinha P, Kumar L (2013) Markov land cover change modeling using pairs of time-series satellite images. Photogramm Eng Remote Sens 79(11):1037–1051

Sklar FH, Costanza R (1991) The development of dynamic spatial models for landscape ecology: a review and prognosis. In: Turner MG, Gardner RH (eds) Quantitative methods in landscape ecology. Springer, New York, pp 239–288

Soares-Filho BS, Assunção RM, Pantuzzo AE (2001) Modeling the spatial transition probabilities of landscape dynamics in an amazonian colonization frontier: Transition probability maps indicate where changes may occur in the landscape, thus enabling better evaluation of the ecological consequences of landscape evolution. Bioscience 51(12):1059–1067

Steiner F, McSherry L, Cohen J (2000) Land suitability analysis for the Upper Gila River watershed. Landsc Urban Plann 50(4):199–214

Tayyebi A, Pijanowski BC (2014) Modeling multiple land use changes using ANN, CART and MARS: Comparing tradeoffs in goodness of fit and explanatory power of data mining tools. Int J Appl Earth Obs Geoinf 28:102–116

Torrens PM (2011) Calibrating and validating cellular automata models of urbanization. In: Yang X (ed) Urban remote sensing. Synthesis and modeling in the urban environment. Wiley, Monitoring, pp 335–345

Van Delden H, Luja P, Engelen G (2007) Integration of multi-scale dynamic spatial models of socio-economic and physical processes for river basin management. Environ Model Softw 22 (2):223–238

Van Vliet J, Naus N, van Lammeren RJA, Bregt AK, Hurkens J, van Delden H (2013) Measuring the neighbourhood effect to calibrate land use models. Comput Environ Urban Syst 41:55–64

Van Vliet J, Bregt AK, Brown DG, van Delden H, Heckbert S, Verburg PH (2016) A review of current calibration and validation practices in land-change modeling. Environ Model Softw 82:174–182

Verburg PH (2006) Simulating feedbacks in land use and land cover change models. Landsc Ecol 21:1171–1183

Verburg PH, Schot PP, Dijst MJ, Veldkamp A (2004) Land use change modeling: current practice and research priorities. GeoJournal 61:309–324

Verburg PH et al (2006) Modeling land-use and land-cover change. In: Lambin EF, Geist H (2006) Land-use and land-cover change: local processes and global impacts. Springer, pp 117–135

Villa N, Paegelow M, Camacho MT, Cornez L, Ferraty F, Ferré L, Sarda P (2007) Various approaches for predicting land cover in mountain areas. Commun Stat Simul Comput 36 (1):73–86

von Thünen JH (1966) Der Isolierte Staat in Beziehung der Landwirtschaft und Nationalökonomie. P Hall. Pergamon, Oxford

Wang J, Mountrakis G (2011) Developing a multi-network urbanization model: a case study of urban growth in Denver, Colorado. Int J Geogr Inf Sci 25(2):229–253
White R, Engelen G (2000) High-resolution integrated modelling of the spatial dynamics of urban and regional systems. Comput Environ Urban Syst 24(5):383–400
White R, Engelen G, Uijee I (1997) The use of constrained cellular automata for high resolution modelling of urban land-use dynamics. Environ Plann B 24(3):323–343
Yu J, Chen Y, Wu J, Khan S (2011) Cellular automata-based spatial multi-criteria land suitability simulation for irrigated agriculture. Int J Geogr Inf Sci 25(1):131–148
Yu Z, Zhang W, Liang J et al (2015) Progress in evaluating suitability of spatial development and construction land. Progr Geogr 34(9):1107–1122
Zhang J, Foody GM (1998) A fuzzy classification of sub-urban land cover from remotely sensed imagery. Int J Remote Sens 19(14):2721–2738
Zhao Y, Murayama Y (2011) Modeling neighborhood interaction in cellular automata-based urban geosimulation. In: Murayama Y, Thapa RB (eds) Spatial analysis and modeling in geographical transformation process: GIS-based applications. Springer, New York, pp 75–83

Chapter 4
Techniques for the Validation of LUCC Modeling Outputs

M. Paegelow, M.T. Camacho Olmedo and J.F. Mas

Abstract Validation is the third stage in the modeling process, after calibration and simulation, and also applies to scenarios. It is an essential part of the process in that the credibility of a model depends on the accuracy of its output. A large range of validation approaches and tools exist, many of which can also be used during the calibration stage. In this chapter we distinguish between purely quantitative validation techniques and those that also consider the spatial allocation of simulated land use/cover changes (LUCC). According to model outputs and objectives, simulation maps can be either hard-classified or soft-classified. While some validation techniques apply to both types of map (cross tabulation matrices and indices, congruence of model outputs), others are specific to one. Techniques such as LUCC indicators, feature and pattern recognition and error analysis are used to validate hard-classified simulation maps, while ROC is used to test soft-classified maps. We then look at a second validation approach based on LUCC dynamics such as LUCC components, intensity analysis, data uncertainty and the impact of spatial and temporal scales. Finally, we compare a group of the most common model software programs (those used by the contributors to parts II and III of this book), in order to list their validation capabilities.

Keywords Validation · Error analysis · Land use and cover changes · Modeling · Simulation assessment

---

M. Paegelow (✉)
Département de Géographie, Aménagement et Environnement,
Université de Toulouse Jean Jaurès, Toulouse, France
e-mail: paegelow@univ-tlse2.fr

M.T. Camacho Olmedo
Departamento de Análisis Geográfico Regional y Geografía Física,
Universidad de Granada, Granada, Spain
e-mail: camacho@ugr.es

J.F. Mas
Centro de Investigaciones en Geografía Ambiental, Universidad Nacional
Autónoma de México (UNAM), Morelia, Michoacán, Mexico
e-mail: jfmas@ciga.unam.mx

M.T. Camacho Olmedo et al. (eds.), *Geomatic Approaches for Modeling Land Change Scenarios*, Lecture Notes in Geoinformation and Cartography,
https://doi.org/10.1007/978-3-319-60801-3_4

1 Introduction

"Despite its apparently scientific nature, modeling is a matter of judgement" (Abdou et al. 2012). "However, the validity of a model should not be thought of as a binary event… model can have a certain degree of validity" (Croks and Heppenstall 2012). "Until more guidance is provided in the literature, calibration and validation will remain a key challenge" (Ngo and See 2012). Rykiel (1996) noted that there is no universal agreement as to how to evaluate the goodness-of-fit of validation. "Depending on their position on this spectrum, models may have different calibration and validation requirements… Models can be calibrated with vast quantities of detailed data, and using sophisticated procedures. They can be validated for historical time periods with high degrees of success. However, a model is only as good as the rules that drive its behavior. Good rules require good theory" (Torrens 2011). Spatial models cannot be validated in a rigorous way (Oreskes et al. 1994).

These quotations from the literature give us some idea both of the difficulty of designing a model that closely reflects future reality and the ambiguity or debate as to what validation actually means. Model validation becomes crucial in a world that produces an ever-increasing number of simulations and scenarios over a large thematic range. In order to give credit to the output of a model, we need information about its robustness and accuracy.

1.1 What Is Validated in Land Change Models?

In this chapter we begin by outlining that the validation techniques discussed here focus on path-dependent models, although there are others that are not path-dependent. Also known as the SAS (story and simulation) approach (Alcamo 2008), these models try to render contrasted, spatially explicit scenarios defined by experts or in a participatory manner: narratives which are then translated into quantitative scenarios (Houet et al. 2016). For their part, the path-dependent models produce scenarios known as trend scenarios or BAU (business as usual) scenarios.

Over the last decade, there has been an important and increasing interest in the validation of simulation models that predict changes over time, particularly from an initial time in the past to a subsequent time in the future (Pontius and Petrova 2010), with a focus on land use and cover change (LUCC), often simplified as land change (Jansen and Veldkamp 2011).

A model's credibility depends on its validation, and this general concept includes three stages, which have been widely endorsed: Verification, Calibration and Validation (Coquillard and Hill 1997; Torrens 2011; Croks and Heppenstall 2012; Ngo and See 2012). Verification refers to the entire process of certifying the correct internal operation of a model (including Face Validation and Sensitivity Analysis); during calibration (see Chap. 2 about calibration), the model is tested using several specific parameters and context-like training periods or dates; while validation involves evaluating the accuracy of the results produced by the model during the simulation stage (see Chap. 3 about simulation), including scenarios (see Chap. 5 about scenarios). North and Macal (2007) also state that "Verification is the process of making sure that an implemented model matches its design, validation is the process of making sure than an implemented model matches the real-world" (cited by Croks and Heppenstall 2012).

Calibration and validation are individually and separately defined, and the period used for calibration purposes may be different from or unknown in the validation period. While the first step refers to a date (t1) and/or a period prior to it (t0−t1), the second step is focused on simulations after t1, which is the point in time for which the predictive extrapolation with the horizon T (t1–T) begins. Pontius and Malanson (2005) highlight this difference, referring specifically to the confusion detected in several studies regarding the goodness-of-fit of the calibration stage for quantifying the predictive power of a model rather than using the goodness-of-fit of the validation stage. In fact, a good fit for calibration does not necessarily imply a good fit for validation or that the latter is an appropriate indicator of a model's predictive power (Pontius and Pacheco 2004). Following White et al. (2012), the time periods for calibration and validation must be sufficiently long to minimize the impact of unrepresentative details during the training period. Calibration and validation over short time periods are notoriously unreliable. Even an empirically excellent calibration may be fundamentally in error either because over-calibration tunes the model to idiosyncratic details of the particular data set or more fundamentally because the data set may be unrepresentative of the range of possibilities present in the system being modeled (Brown et al. 2005; Engelen and White 2007).

1.2 How to Validate Land Change Models?

Modeling land use/land cover changes (LUCC) can help us understand complex social and ecological interactions and provides useful information for decision-makers such as planners (Paegelow et al. 2013). The usefulness of LUCC models can be measured by the accuracy of their output.

According to Torrens (2011), validation evaluates the correctness of a model while Croks and Heppenstall (2012) described it this way: "Verification is the process of making sure that an implemented model matches its design, validation is the process of making sure that an implemented model matches the real world". Coquillard and Hill (1997) proposed that model validation should consist of three progressive steps: verification, (Does the model run correctly?), calibration (Does the model correctly simulate a known state?) and validation (Does the model correctly predict an unknown state?). "To improve the robustness and the acceptance of a model, the data at the validation date must be model unknown, in other words data that has not been used in the building and calibration of the model"

(Paegelow and Camacho Olmedo 2008). If not, simulation must be considered as a step in the calibration process.

Rykiel (1996) distinguishes between "conceptual" and "operational" validation. Conceptual validation warrants that the assumptions underlying the conceptual model are correct or plausible. Operational validations measure the accuracy of model output. When modeling the future, the model can be partially validated by comparing the results with expert knowledge, by assessing its robustness by measuring the constancy of model outputs during iterative model runs. A complementary technique is gauging the degree of congruence between the outputs of different software programs that use the same data set and parameters. Gómez Delgado and Tarantola (2006) tested model stability using sensitivity analysis. To this end they developed several indices to measure the variability of model outputs when input parameters are changed. In this context, Gomez Delgado and Barredo (2005) describe techniques to assess risk when using model outputs and Jokar Arsanjani (2012) focuses on model data and drivers of uncertainty.

There is a large range of statistical tools for measuring the accuracy of hard and soft predictions. Hard predictions can be validated by comparing between simulated and observed LUCs. However, a soft prediction is evaluated by comparing potential changes or LUC suitability with observed LUC or LUCC. This is often done by measuring the area under the ROC (Relative Operating Characteristics) curve (Pontius and Schneider 2001). Eastman et al. (2005) and Pérez-Vega et al. (2012) focused on the potential for change. With this in mind, they compared dynamic areas relative to persistent ones and developed a measure called DiP (Difference in Change Potential). Of the two forms of model output—hard or soft prediction—the validation of hard maps is more common and there is a larger spectrum of statistical tools. These tools focus on different aspects: accuracy of quantity and allocation, correctness of LUCC components, similarity of the landscape pattern, model congruence and error analysis.

As regards quantitative agreement, modelers distinguish between matching the sum total of the LUC area and the pixel-by-pixel comparison, which also evaluates matching in allocation (Torrens 2011). As a first step, an overall agreement may be obtained by calculating statistical indices, such as Chi-square or Kappa (Pontius 2002). However, Pontius and Millones (2011) indicate that the KIA (Kappa Index of Agreement) is not suitable for LUCC model validation because it assumes randomness. The sample matrix must therefore be converted into an estimated population matrix. The Chi-square index has the same drawback, as pixels cannot be considered as independent observations. For map comparison we recommend easier indices such as quantity and allocation disagreement. Various validation techniques that consider changes have been developed. For example, Pontius (2000) and Pontius et al. (2004a, b, 2008) propose a technique that splits the LUCC-budget into gain, loss, net change and swap (see Technical Notes in Part IV of this book). Pontius et al. (2008) also developed several statistical LUCC indices for determining accuracy, including a figure of merit (see Technical Notes in Part IV of this book), a ratio between correct predicted changes and the sum total of observed and predicted changes.

Further validation techniques focus on fuzzy allocation agreement, with indices (Hagen 2003; Hagen-Zanker et al. 2005; Rodrigues et al. 2007) that measure the relative allocation agreement and overcome the limitations produced by exclusive cell state and exact allocation (see Technical Notes in Part IV of this book). In the same way Procrustes analysis (Jackson 1995) performs pixel-by-pixel comparison by linearly transforming one grid as rotation, translation or scaling to achieve the best fit with the reference grid. Furthermore, Kuhnert et al. (2005) describe algorithms that test the similarity of raster matrices by using different weights and by varying the window size.

Spatial analysis measurements consider the distribution and shapes of land patterns (White et al. 1997) at multiple scales (Gaucherel 2007; Gaucherel et al. 2008) and are mainly inspired by landscape ecology metrics (Forman 1995; McGarigal and Marks 1995; Botequilha et al. 2006). In addition, error analysis highlights conceptual and model parameter inaccuracy by measuring errors in simulated LUC categories or transitions and their allocation (Pontius 2000; Pontius and Petrova 2010).

There are several studies that provide a comprehensive review of the validation techniques designed for spatial models (Turner et al. 1989; Pontius et al. 2004a; Paegelow and Camacho Olmedo 2008; Shirley and Battaglia 2008; Sargent 2009), while van Vliet et al. (2016) provide the results of a large study about calibration and validation techniques applied in recent land change modeling papers.

These few lines of introduction are intended to outline the importance of setting objectives for LUCC modeling. Do we care about the entire space or should we focus only on changing land? Do we want to achieve quantitative accuracy or a realistic landscape or urban pattern? Evaluating the accuracy of a model is clearly a matter of assessing its true purpose: do we want a model that makes predictions or one that presents a range of plausible futures?

In this chapter, we will be focusing on three aspects of validation. We will begin by presenting validation methods and tools according to model outputs and objectives (Fig. 1). Model outputs may be *hard* (maps with the same legend as training LUC maps), or *soft* (simulation maps expressing the potential of places to become a particular land cover or land use). Modeling objectives may be different: focusing on accuracy in terms of quantity, of allocation, of realistic landscape patterns. A second aspect is that validation depends on LUCC dynamics, as manifested in the intensity or rate of land change and also in the impact of the particular spatial and temporal scales used. Thirdly we describe validation according to LUCC models.[1] A presentation of selected software validation tools is completed with a table comparing them.

[1]See the short presentations in Part V of this book about (in alphabetical order) APoLUS, CA_MARKOV, CLUMondo, Dinamica EGO, Land Change Modeler (LCM), LucSim, Metronamica and SLEUTH. The authors are also grateful to all contributors who helped us understand the different software packages.

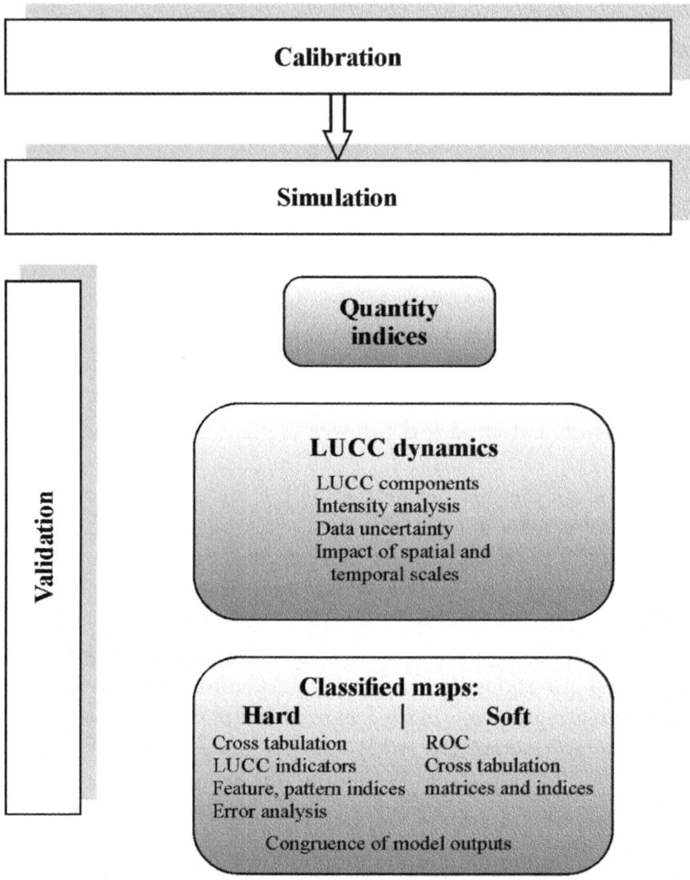

Fig. 1 General overview of validation techniques

2 Validation in Terms of Model Outputs and Objectives

Simulation Output Forms: Hard Versus Soft

As mentioned in the chapter in this book about simulation (see Chap. 3), model outputs can be split into two categories: hard outputs, in which each pixel in a raster map is assigned to exactly one category of land use or cover (LUC) (hard-classified map) and soft outputs, in which each pixel has a partial membership of several classes simultaneously (soft-classified maps). During the validation step, soft simulation results show the partial membership of a specific land use category or land transition and the level of membership indicates the degree of uncertainty. Most spatial land-change models focus on hard simulation results and their validation. In several cases, a quick reference to soft simulation is made, but only a few contributions focus exclusively on soft simulation results and their validation (Pérez-Vega et al. 2012; Wang and Mountrakis 2011; Conway and Wellen 2011).

2.1 Validation in Terms of Quantity Estimation

Modeling over time and space typically produces results about the quantity of land-use change (quantity) and where it takes place (allocation). Validation can focus on one or both of these output components (Fig. 2). Generally, both components are evaluated together. This is the domain of map comparison techniques using matrices to compute correct predictions as quantities correctly allocated. The spatial component does not only refer to prediction at the correct place. Validation focusing on allocation can also evaluate spatial shapes and patterns.

Evaluating only predicted quantities (cumulated area) without considering correct allocation is much easier than predicting the correct amount of land change at the correct place (Paegelow and Camacho Olmedo 2005; Paegelow et al. 2014).

The amount of expected land change may be predicted or given. The latter choice is made by "what happens if" scenarios that design a range of plausible futures. Quantitative prediction often uses a probabilistic approach such as Markov chains (see Technical Notes in part IV of this book). In this context, we will be specifically focusing on Markov chains and their implications on accurate

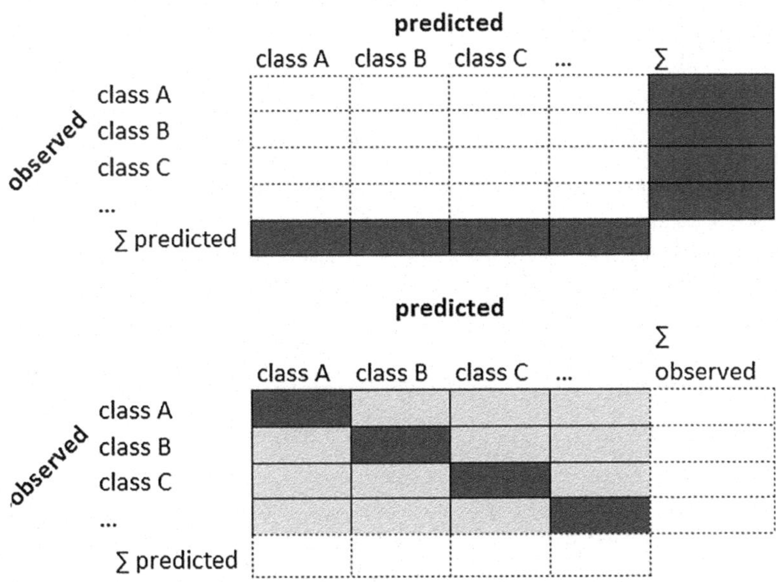

Fig. 2 Validation of cumulated surface (*above*) versus pixel-by-pixel matrix validation of quantity and allocation (*below*)

prediction. Two important aspects will therefore be analyzed: the impact of the software (Mas et al. 2011, 2014) and its algorithms and the assumed or specified level of confidence in training data.

When focusing exclusively on the quantitative aspect of model output, it is important to put the comparison between observed and simulated LUC at t_2 into perspective by also indicating former LUC quantities at t_1 (end of calibration—model known—period). This enables us to compare observed and modeled land change. We will come back to this point in more detail when discussing map comparison techniques in the next paragraph. As for integrating dynamics into quantitative validation, error analysis will be discussed further by taking into account the allocation aspect too.

2.2 *Hard Classified Maps*

The initial validation may be visual or qualitative (Torrens 2011), a more intuitive means of assessing the resemblance between model output and the validation data, e.g. simulated land use and observed land use. However, visual inspection only provides an initial impression and model accuracy has to be tested in other ways, generally statistically.

2.2.1 Pixel-by-Pixel Matrices and Comparison with the Null Model

For hard-classified maps, a full validation is the most common method, where comparisons between simulated and observed LUC referring to the same data are possible, i.e. both documents have the same nomenclature and temporal reference. The model's accuracy is evaluated by comparing simulated LUC with its reference image to a null, no change model (Pontius and Malanson 2005). In a relative minority of cases, researchers have compared different models or individual runs of the same model in different places and times (Pontius Jr. et al. 2008, cited by Torrens 2011). A large range of statistical tools may be used to assess the correctness of model output. The range of tools for comparing observed and simulated results or various different simulations, include the following pixel-matching techniques (performed on a pixel-by-pixel basis):

LUCC Indicators

Sohl et al. (2012) used this pixel-by-pixel technique (Fig. 3) to compare various LUCC scenarios by measuring the disagreement in quantity and allocation.

Prediction errors may be split into omission errors and commission errors for each class (Fig. 4). Omission refers to areas observed as change but not predicted as such. Commission error means the part of predicted change that, in fact, did not

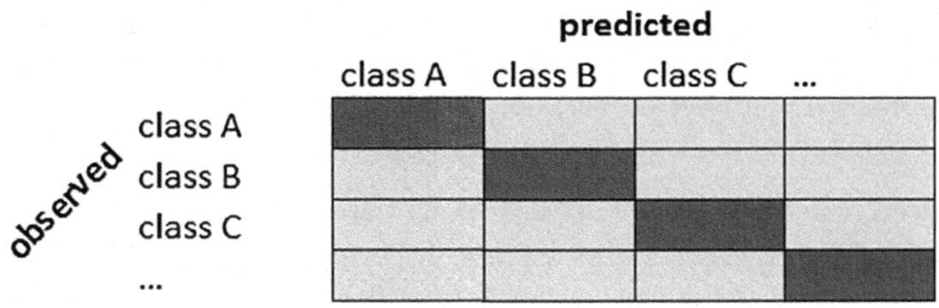

Fig. 3 LUC matrix comparing observed and predicted LUC. Accurate prediction (*hits*) are located on the matrix diagonal (*dark cells*), errors in the rest of the matrix (*light cells*)

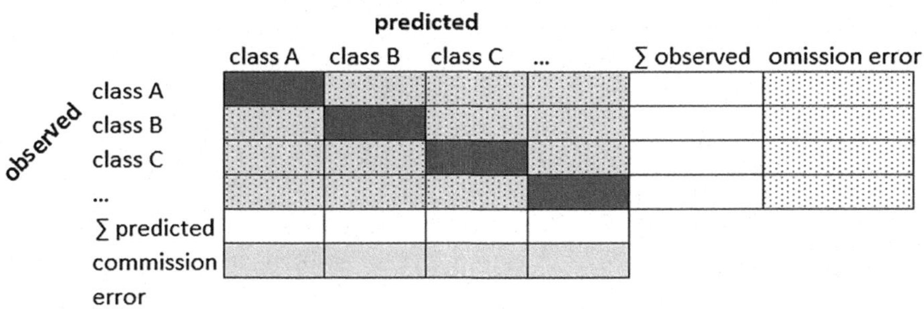

Fig. 4 Omission and commission errors

change. Commission is sometimes also referred to as consumer's accuracy and omission as producer's accuracy such as in cross tabulation techniques in remote sensing. In Fig. 4 omission is the total per line minus correct predicted (diagonal matrix cell), while commission is equal to the total per column minus correct predicted (diagonal matrix cell).

When introducing a third map into the comparison, e.g. observed LUC at the beginning of the simulation period (generally the last known date for the model is the end of the calibration period), it will be possible to compare observed and predicted change and to distinguish between hits (observed persistence or change predicted as such) and errors due to observed change predicted as persistence (omission), observed persistence predicted as change (commission) and observed change predicted as such, but with incorrect LUC categories.

Some software programs provide tools for cross validation between t_1 observed, t_2 observed and t_2 predicted by differentiating between 'Hits' (correctly predicted changes), 'misses' (omission errors) and 'false alarms' (commission errors).

These validation techniques rely on a technique of land change analysis. Pontius (2000) and Pontius et al. (2004a, b) established a comprehensive way of analyzing LUCC and measuring the accuracy of the model outputs based on LUC persistence and changes. They called this technique LUCC- budget (see the technical note about LUCC budget in Part IV of this book).

On the basis of previous research by Klug et al. (1992) and Perica and Foufoula-Georgiou (1996), Pontius et al. (2008) calculated various LUCC indices by splitting map comparison between the observed and predicted LUCs into percent correct and percent error distinguishing the following components:

A = Observed change predicted as persistence: error
B = Observed change predicted as such with correct LUC categories: correct
C = Observed change predicted as such but with incorrect LUC categories: error
D = Observed persistence predicted as change: error

These components allowed the following three derived measurements to be calculated:

- Figure of Merit—the ratio of B/(A + B + C + D) which expresses the overlap between observed and predicted change. This value ranges from 0 (no overlap) to 100% (perfect overlap).
- Producer's Accuracy—the ratio of B/(A + B + C) which expresses "the proportion of pixels that the model predicts accurately as change, given that the reference maps indicate observed change" (Pontius et al. 2008).
- User's Accuracy—the ratio of B/(B + C + D) which expresses the part of the pixels accurately predicted as change compared to all model-predicted changes.

2.2.2 Disagreement Indices Based on Cross Tabulation

Krüger and Lakes (2015) present an innovative method for quantifying disagreement between different simulations using cross-tabulation techniques applied to binary maps (e.g. deforestation or not). Their disagreement index also includes quantity as allocation matching and may be used for hard classified maps as continuous probability simulations. They started with a well-known cross-tabulation matrix (Hagen-Zanker 2009; Mas et al. 2013) as shown in Fig. 5. "The diagonal from upper left to lower right represents agreement while the diagonal from lower left to upper right represents disagreement" (Krüger and Lakes op. cit.). By considering soft-classified maps as original simulation output and following Pontius and Milliones (2011), Krüger and Lakes considered the two disagreement cells of

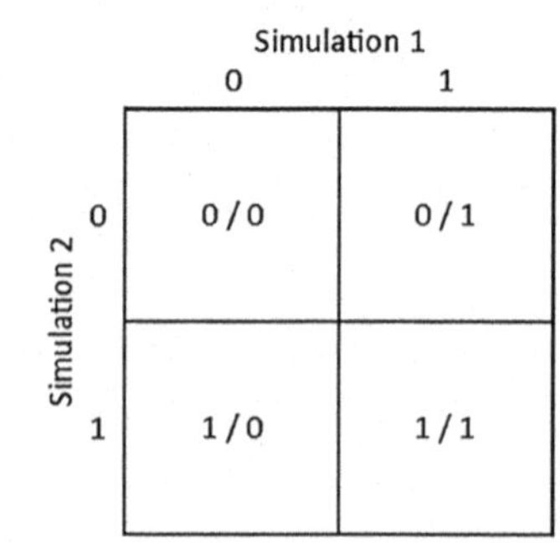

Fig. 5 Cross-tabulation between two binary simulation maps showing the four possible combinations

the matrix as a base to split the disagreement between different simulations into their quantity and allocation components. Their method allowed us to quantify the distance between two maps from the diagonal (perfect fit) in an orthogonal diagram whose two axes express quantity and allocation.

As the authors themselves make clear, their method is established for comparison between binary maps but can be extended to multi-categorical maps by splitting them into monothematic maps. However, we must bear in mind that when doing so, we lose the relations between LUC categories. This means for example that we cannot measure how wrong a simulation is by comparing simulated and observed LUC. Some errors could be considered more important than others, e.g. simulating woodland instead of shrubs could be a more important disagreement than simulating urban.

2.2.3 Fuzzy Logic Indices

There are various alternative techniques to hard pixel-by-pixel comparison. Indices based on fuzzy logic (Hagen 2003; Hagen-Zanker et al. 2005, 2009) (see Technical Notes in Part IV of this book) measure the agreement of location and overcome the limitations due to exclusive cell state and exact allocation. Some popular modeling software programs incorporate vicinity-based comparison tools measuring the fuzziness of location (Rodrigues et al. 2007), allowing a more gradual and flexible method than the classic cell-to-cell comparisons.

2.2.4 Procrustes Analysis

Jackson (1995) described the usefulness of Procrustes analysis. He compared the fit between different matrices by linear transformation (rotation, translation, scaling) of one grid to achieve the best fit with the reference grid. Pontius et al. (2004b) chose multiple resolutions to analyze the nature of allocation errors (cf. Sect. 2). More recently, Pontius et al. (2007) proposed a validation method that considered a nested stratification structure.

2.2.5 Feature and Pattern Recognition

Spatial analysis measurements take into account spatial pattern, its distribution and shapes (White el al. 1997). Many metrics were derived from landscape ecology such as shape, compactness, diversity and fragmentation (Forman 1995; McGarigal and Marks 1995; Botequilha et al. 2006). White et al. (2012) analyzed cluster size-frequency distributions. In addition to quantitative accuracy measurements, landscape pattern agreement offers a useful, supplementary validation approach. The simplest indicators are the size and shape of the patches. Dinamica EGO software allows us to model these parameters by average and standard deviation of patch size and the degree of compactness as a ratio between surface area and perimeter. Validation may be done by map comparison techniques that focus on the number, size and compactness of observed and simulated patches.

2.2.6 Error Analysis

Error analysis provides useful information about model logic and underlying conceptual approaches, so giving the modeler a better understanding of the model. In addition, the previously presented techniques can be completed by analyzing the possible origins of error. Seen from this point of view, LUCC analysis and Figure of Merit (see Technical Notes in part IV of this book) can be considered alongside validation techniques such as error analysis. Error analysis tries to answer the question 'how wrong is the prediction?' To do so, it generally focusses on two components: categorical or transitional errors and error in allocation.

LUC Category Errors

Various techniques measure disagreement between observed and simulated LUC. While quantitative data (e.g., percent of tree cover) enable us to measure the magnitude of inaccuracy, categorical data generally needs to be transformed into quantitative data or ordered on a scale before being analyzed. Ahlqvist (2008) offers a technique of fuzzy change estimation about the closeness between observed and simulated LUC categories. Paegelow et al. (2014) measured the magnitude of error between simulated and observed LUC expressed as categories. However, if LUC legends form a ranking order that reflects spontaneous vegetation succession from bare soil to woodland, land use intensity or other criteria that enable us to place LUC categories in an ordered scale, we can measure the parametric distance between observed and simulated LUC. Prediction error is measured by the absolute categorical distance between observed and simulated LUC. In many situations, modelers will probably have difficulties quantifying the exact distance between different LUC on an ordered scale. A possible coarse approach is to use equal distance between original categories. Paegelow et al. (2014) did so to rank LUC by the covering rate from bare soil to woodland.

Allocation Error

A large number of metrics can be calculated. Paegelow et al. (2014) created a distance map for each LUC category for which the considered LUC was the origin. The distance map was then crossed with simulation errors (omissions, commissions and prediction of false gaining categories). For each wrongly predicted patch of a given LUC category, these authors measured the minimum distance to the nearest correct location and then calculated the average for each LUC category.

2.2.7 Congruence of Model Outputs

Another form of validation consists of using the same data set to simulate LUCC with different models (Figs. 6 and 7). The closeness of the resulting simulation maps is measured and the degree of congruence is considered as an indicator of the stability of the model and the plausibility of the simulations (Paegelow et al. 2014). The same procedure also provides useful information about model robustness (Camacho Olmedo et al. 2015). Sohl et al. (2012) applied the same approach to multiple LUCC scenarios computed for the Great Plains in the United States, a procedure they described as "scenario diversity".

2.2.8 Other Approaches

Torrens (2011) proposed running models exhaustively (specifically in stochastic or probabilistic models). Several other authors use histograms (Conway and Wellen 2011) with several choices (equal weights, difference...) (Bone et al. 2011; Kamusoko et al. 2009), while Li et al. (2011) proposed a geographical simulation

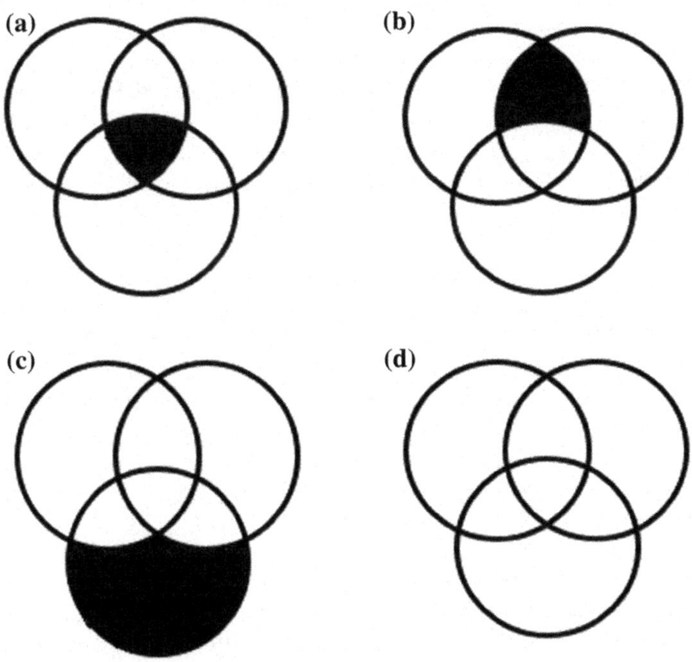

Fig. 6 Different congruence levels of simulation maps computed by three different models: **a** perfect intersection, which means total congruence of correctly predicted land use, **b** congruence of two models, **c** only one model gives correct prediction, **d** no model predicts correctly

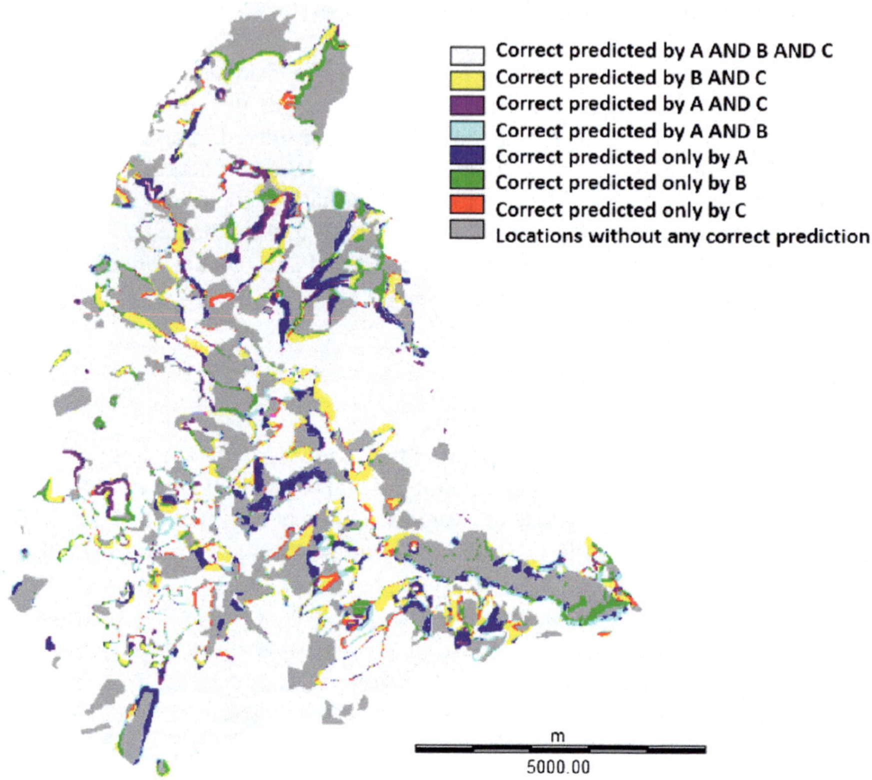

Fig. 7 Congruence of three simulations computed by *(A)* CA_MARKOV model, *(B)* Multi Layer Perceptron, *(C)* Statistical regression model, applied to Garrotxes catchment (Eastern Pyrenees, France)

and optimization system to model the reciprocal relationships between simulation and spatial optimization, including future simulations.

2.3 *Soft-Classified Maps*

2.3.1 Soft-Classified Maps

Of the various methods for assessing the accuracy of simulation maps, the first, most intuitive comparison method is usually visual or qualitative validation. This is also used in soft results (Torrens 2011) and in different types of superposition between soft-classified and real maps (observed and non-observed transition or land use) and in the analysis of frequency distributions (Yu et al. 2011; Alcamo et al. 2011; Camacho et al. 2013; Wang and Mountrakis 2011). Paegelow and Camacho

Olmedo (2005) compared the performance average and the standard deviation suitability scores for each candidate land cover with all of the other categories.

2.3.2 ROC

While hard prediction leads to cells being classified within one specific LUC category, some modeling programs provide soft prediction maps expressing the vulnerability of the land to change or suitability maps for each LUC category, which are computed by multi-criteria evaluation (MCE) (Eastman et al. 1995) (see Technical Notes in Part IV of this book).

In this context, Relative Operating Characteristic (ROC) (Hanley and McNeil 1982; Pontius and Schneider 2001) (see Technical Notes in Part IV of this book) is a measure of the spatial likelihood between a reference map and a suitability map. The reference is binary and shows the spatial distribution of a specific LUC category or transition, while the suitability map expresses the potential for this category or the propensity to change in the case of analyzing transitions. The procedure consists in ranking these suitability or vulnerability-to-change scores into *n* classes and computes the proportion of true (presence on reference map) and false (absence) positives. ROC assumes that the high scores in the comparison map are more likely to be truly positive. Pontius and Schneider (2001) provide a graphic illustration for this technique. Various other researchers have applied ROC in land change models (Wang and Mountrakis 2011; Alcamo et al. 2011; Lin et al. 2011; Jokar Arsanjani 2012; Ngo and See 2012), comparing different study areas (Paegelow and Camacho Olmedo 2005), calibration and validation periods (Conway and Wellen 2011) or different results after a number of drivers had been considered (Huang et al. 2012). Eastman et al. (2005) and Pérez-Vega et al. (2012) applied ROC and DiP to compare modeling approaches. Conway and Wellen (2011) compared ROC between the calibration and validation period. Pontius and Si (2013) introduced a variant of ROC: TOC—the Total Operating Characteristic, which enables the user to calculate the AUC, while also showing all the information in the contingency table for each threshold.

2.3.3 Cross Tabulation Matrices and Indices

This type of validation compares two or more types of soft-classified maps. All of the maps are likelihood maps. Nevertheless, overlay maps based on pixel matching (performed on a pixel-by-pixel basis) can be applied after reducing soft maps to several classes or binary maps, and this method can reach conclusions regarding the convergence of the results. This transformation makes it possible to use the most common validation techniques (Paegelow and Camacho Olmedo 2008). For example, Syphard et al. (2011) overlaid binary maps of urban predictions (only including land with a high-probability of development) for several future scenarios, in order to map and quantify where urban growth predictions converged over time.

They also carried out a data reduction by placing the probability images in classes. Another technique known as soft cross-tabulation involves a process of cross-tabulation on soft-classified maps, which preserves continuous values without reducing them into classes, performing a pixel-by-pixel comparison between two maps in which the pixel values have simultaneous memberships of more than one category (also called fuzzy classification). Pontius and Cheuk (2006) compared this method to existing techniques, and proposed that it should be applied to both hard-classified and soft-classified data at any scale. A cross-tabulation tool of this kind for soft-classified maps in which the spatial resolution can be varied is implemented in TerrSet software. The potential of ROC statistics within the framework of land change modeling is analyzed in detail in Mas et al. (2013).

Assessment methods developed for hard-classified maps that focus on the similarity or correspondence between them can also be used for soft-classified maps. The most commonly used tools are Spearman and Pearson correlation indices: similarity can be tested at ordinal and quantitative data level. Using the Spearman rank correlation, Conway and Wellen (2011) evaluated two suitability maps using histograms showing the degree of similarity between the two maps.

2.3.4 DIP—Difference in Change Potential

Difference in Change Potential (DiP) is an assessment technique measuring the difference between the mean potential in the areas of change and the mean potential in the areas of no change, as manifested in the form of hits (correct forecast of change) and false alarms (incorrect forecast of change) (Eastman et al. 2005; Pérez-Vega et al. 2012).

DiP is based on the Peirce Skill Score (PSS) defined as:

$$PSS = H - F$$

where H is the mean potential in the areas of change and F is the mean potential in the areas of no change respectively, and PSS is the difference between them. A value of 1.0 indicates perfect agreement, while a value close to 0 shows random behavior (Pérez-Vega et al. 2012).

2.3.5 Other Validation Techniques/Crossing Techniques

A large number of studies combine various validation techniques. Wang and Mountrakis (2011) compared three models at both per-pixel and neighborhood levels. In the first, they included the confusion matrix, KIA, the receiver operating characteristic (ROC) curve, and multi-scale summary accuracy. The same authors recommended that the results obtained by binary comparison (accurate or not), the probability of change and the spatial accuracy of predicted change be compared.

Lin et al. (2011) use ROC, KIA, multiple resolution validation and landscape metrics to analyze the accuracy of model outputs.

3 Validation According to LUCC Dynamics

The relative importance of the validation techniques presented here also depends on the objective of the model. If the model aims to predict land change, the accuracy of the estimated amount of change is just as important as its allocation. By contrast, if the objective is to design plausible, contrasted, scenarios, the modeler implements quantitative targets with regard to the expected LUC area or changes. From this perspective, validation techniques focus more on spatial pattern and error analysis. Furthermore, the map comparison techniques presented above (particularly the Figure of Merit when computed for outputs of various models) provide useful information about the performance of the models in predicting persistence and change, change components such as net change and swap, and the realism of the landscape. They also allow the modeler to choose the most appropriate model according to the objectives.

Tests with various training dates used for Markov chains show that quantitative accuracy depends on the choice of these dates (see Chap. 7 in part II of this book). This finding shows why it is so important for the modeler to have the key dates at his/her disposal because the Markov chain is strongly dependent on previous trends. If relatively few LUC dates are available, this increases random chance because the Markov chain determines the overall accuracy of the model. If available LUC maps do not allow us to trace past trends or if these trends are not informative for future evolution, it is advisable to support trend-based simulation, also known as the baseline scenario, with various scenarios that deliberately break with Markovian conditional transitions calculated on a basis that is incomplete or becoming obsolete. By varying quantitative assumptions, this geoprospective model (Houet and Gourmelin 2014) implements the allocation of these hypotheses and designs plausible futures.

3.1 Intensity of Dynamics

3.1.1 Splitting Dynamics into Components of Interest: Persistence, Net Change and Swap

LUCC allows us to analyze observed and simulated land change at different levels. The first level is obtained by cross-tabulation of the whole area (Fig. 8). An example from a study of dynamics in a typical European mountain region first shows: persistence (sum of diagonal cells) which amounts to about 97.09%. This means that land use has changed in less than 3% of the study area. Having said this,

		from: 2000						
		Coniferous forest	Deciduous forest	od recolonization	Broom land	Grassland	Crops	total 2009
to 2009	Coniferous forest	37.37	0.00	0.03	0.20	0.01	0.00	37.61
	Deciduous forest	0.00	8.04	0.00	0.01	0.05	0.00	8.10
	od recolonization	0.19	0.00	18.91	1.56	0.83	0.00	21.49
	Broom land	0.01	0.00	0.00	25.68	0.01	0.00	25.71
	Grassland	0.01	0.00	0.00	0.00	7.09	0.00	7.10
	Crops	0.00	0.00	0.00	0.00	0.00	0.00	0.00
	total 2000	37.58	8.05	18.94	27.45	7.98	0.00	100.00

Fig. 8 LUCC 2000–2009 in Garrotxes (French Pyrenees); data in percent of study area (8750 ha)

some LUC categories underwent important changes. Expressed as a percentage of the surface area in 2000 (start date), most LUC components remained stable. For example, all land changes from and to coniferous forest (gains and losses) totaled only 1.2% of its extent in 2000. At the other end of the scale, changes in the "wood recolonization" category amounted to 13.9% of the land it occupied in 2000. This means that global or dominant persistence can mask important individual transitions.

Validations may be performed at individual LUC category or transition level or at a global level by considering the overall change potential map (superposition of all the maps) (Pérez-Vega et al. 2012). If we set persistence aside to focus exclusively on change, the accuracy of predicted land change is considerably lower (Brown et al. 2005).

3.1.2 Intensity Analysis

Another means of analyzing model accuracy is to put it into perspective with the intensity of land change (Pontius et al. 2013). *Intensity* is the amount of land change per time unit (e.g. the annual rate). Land change intensity may be analyzed by comparing the amount of change over the study period in several LUC categories (Fig. 9) or by comparing their rate of change over different time periods (Fig. 10). Figure 9 shows that three LUC categories (coniferous and deciduous forest, crops) were more persistent, while broom land and wood recolonization underwent more significant changes over the study period.

For the extent 1942–2009, Fig. 10 shows two intervals in which there was a slow rate of change (1980–1989; 2000–2009), one interval that was close to the average (1942–1962) and two intervals characterized by fast dynamics (1962–1980; 1989–2000). This shows that model accuracy is highly dependent on the comparison interval selected.

Runfola and Pontius (2013) proposed a number of indices based on the difference between individual change rates and the average annual rate of change. For their part, Aldwaik and Pontius (2012) developed tools to measure the intensity of land change at three levels: interval, category and transition. They created indices

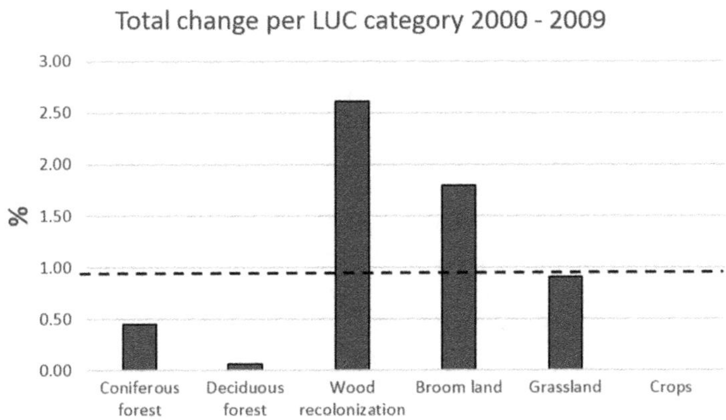

Fig. 9 Total change (expressed in % of the entire study area) per LUC category, Garrotxes 2000–2009. The *dotted line* shows the average LUCC rate

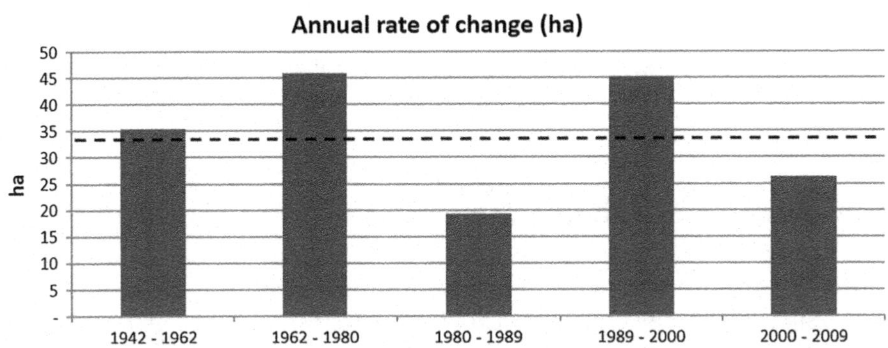

Fig. 10 Annual rate of change in ha (all categories) over the different time periods, Garrotxes. The *dotted line* is the average rate of change over the extent 1942–2009

based on cross-tabulation matrices and distinguished between slow and fast intensities of change with respect to average annual change over different intervals within the whole time extent. They also explored the relative importance of changes (Fig. 8) by unraveling the annual rate of change, expressed in area units or percent, as a proportion of the study area and the amount of annual change expressed in percent of the total area covered by each LUC category. This is important for measuring changes affecting small areas involving relatively less significant (in terms of area) LUC categories. Huang et al. (2012) applied these intensity measures to a coastal watershed in south-eastern China and qualified the categories in which total change was below or above the average as "dormant" versus "active" categories, which respectively "avoid" or "target" transitions.

3.1.3 Data Uncertainty

Pontius et al. (2006); Pontius and Lippitt (2006) proposed a way of using model accuracy measurements to extrapolate predictive uncertainty. Pontius and Petrova (2010) considered the question of whether map error can explain the differences between LUC maps from two points in time. This paper is unique in that it was the first in this series to consider how the level of accuracy in the reference maps influences the interpretation of model validation, and it examines the results for each entry in several cross-tabulation matrices, rather than just overall agreement (Pontius and Millones 2011). This alternative approach had a major impact because most LUCC simulations rely on category data to calibrate and validate the model, and these data often do not have a clear level of accuracy or error structure. The issue of data misclassification within LUCC models has only recently been explored, as have the procedures to follow when the available error information is incomplete. For example, Pontius and Petrova (2010) developed a method for evaluating predicted results when the level of accuracy of the reference data is unknown (Conway and Wellen 2011). Uncertainty in the data is often related to the statistical level of LUC data. This is because most studies are based on qualitative data, which means that LUC is described by categories. The coarser the legend, the higher the uncertainty of the data due to intra-category variance (Paegelow et al. 2014).

3.2 Impact of Spatial and Temporal Scales (Resolution)

Jansen (2006) distinguish three dimensions of scale: (1) space, (2) time and (3) the organizational hierarchy as constructed by the observer. This organizational hierarchy is synonymous with the variation in the semantic contents of data expressed as differences in categorization (Feng and Flewelling 2004). Of these three dimensions, scientists paid little attention to the latter. In fact, so little that this dimension was not even included in the definition of scale cited above (Jansen and Veldkamp 2011). The organization of the data which is finally expressed in the legends of LUC maps is also a critical point, as mentioned above, about which we feel we must insist.

3.2.1 Impact of Spatial Resolution

The concept of scale and resolution is closely linked to the level of detail available in geographic data. Scale refers to printed maps and the level of detail for a given scale is expressed by the minimum mapping unit. The notion of resolution is closely linked to numerical data, especially in raster format and is expressed by the pixel size.

Pontius et al. (2004b) showed that spatial resolution impacts on LUCC components as net change and swap. Using an example of LUC maps for several towns

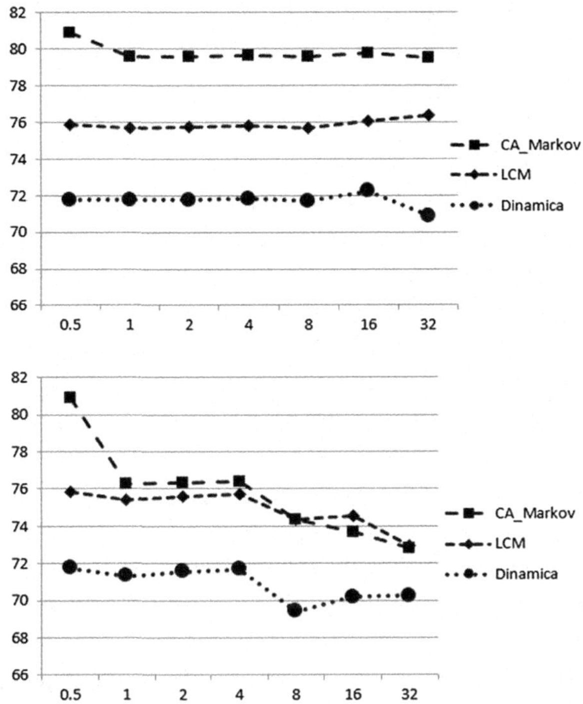

Fig. 11 Varying spatial resolution (geometric sequence) in cross-tabulation between observed LUC and LUC simulated by three land change model tools: CA_MARKOV, LCM and Dinamica Ego applied to pasture land in Eastern Pyrenees. The abscissa shows the spatial resolution in meters while the Y-axis is the percentage of correctly simulated LUC. The *top* figure shows the accuracy rate by pixel thinning and the *bottom* one shows the impact of applying the majority rule

in central Massachusetts, they discovered that the swap component in LUCC budgets is related to spatial scale. The coarser the spatial resolution, the lower the swap. Varying resolution may have different effects when it comes to validating hard-classified land change simulation. We performed pixel-by-pixel cross tabulation between LUC simulated by three models and observed (model unknown) LUC on pasture land in the Eastern Pyrenees by varying the spatial resolution (geometric sequence) and the method of calculating pixel values (pixel thinning and majority). As Fig. 11 shows, the prediction score remains almost stable with coarser resolution when the pixel-thinning technique is applied, while it falls with coarser resolution when the majority rule is applied.

3.2.2 Impact of Temporal Resolution

The influence of scale or resolution—in our case the duration of the time interval—is well known in various disciplines (Allen and Starr 1982; Kim 2013). Several recent studies have formalized the impact of time intervals on the amount of change (Burnicki et al. 2007; Lee et al. 2009; Liu and Deng 2010).

Using various data sources and resolutions, Colas (2016) observed that, as in the case of spatial resolution, short time intervals generate a high rate of change while change intensity decreases with longer intervals. Figure 12 underlines this finding

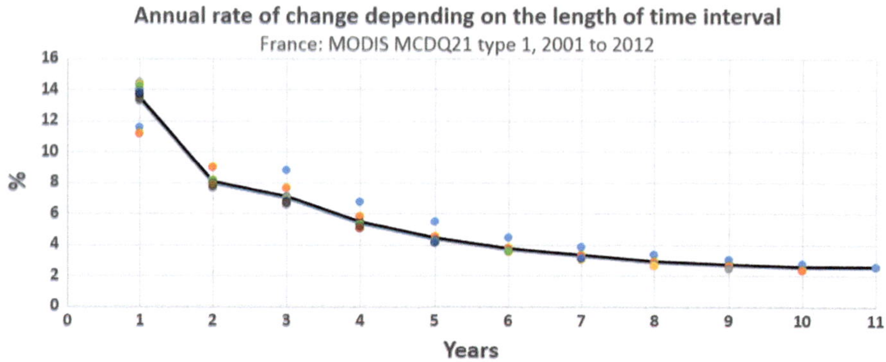

Fig. 12 Annual rate of change (%) depending on the length of the time interval (years). Applied to MODIS MCDQ21 type 1 data for France, 2001–2012

by using MODIS 250 m MCDQ21 data with a type-1 legend for France. The available data are for 2001–2012. The figure shows that the intensity of change decreases exponentially with increasing length of time intervals.

4 Validation According to LUCC Models

The modeling software packages discussed here use either internal validation tools implemented within the modeling program, or external techniques such as parent software, GIS or specific raster tools such as Map Comparison Kit (Visser and de Nijs 2006), especially recommended for CLUMondo (Table 1).

As regards those with built-in validation techniques, all the software packages we considered except for CLUMondo offer cross tabulation to compare hard predictions to observed data. The majority of programs also do this for soft prediction maps, while only TerrSet and Dinamica EGO allow a validation of this kind with multiple resolutions. For their part, Dinamica EGO and APoLUS allow a spatial validation by fuzzy allocation.

With the exception of LucSim and CLUMondo all programs offer various similarity indices for comparing maps. The situation varies more with regard to comparison tools, in that they do not all have tools that offer omissions and commissions and pattern analysis. All of the software packages we considered do however perform a quantitative validation and most of them use ROC statistics.

Table 1 Comparing LUCC models in the validation stage

	Pattern-based models (PBM)				Constraint CA-based models (CCAM)			
	CA_Markov TerrSet	LCM TerrSet	Dinamica EGO	CLU-Mondo	Metronamica	APoLUS	SLEUTH	LucSim
Cross-tabulation for hard classified maps	Yes	Yes	Yes	No, external	Yes	Implemented in parent software(R)	Model creates transition and contingency matrices	Yes
Cross tabulation for soft classified maps	Yes	Yes	Yes: DIP	No	Yes	Implemented in parent software(R)	No	No
Cross tabulation for multiple resolutions	Yes	Yes	YES	No, external	No	Implemented in parent software(R)	Multi-resolution can be used in calibration	No
Fuzzy coincidence	No	No	Yes	No, external	No although available in MCK1	Implemented in parent software(R)	No	No
Map comparison similarity indexes	Cramer's V, KIA, KIA multiple resolutions	Yes	Yes	No, external	Yes, mainly through accom-panying MCK	Ksim, KsimF (MCK, currently working on native R solution)	Model uses 13 statistics based on data matching. Post comparison must be performed independently	No
Map comparison showing correctly predicted changes, omissions, commissions	Yes	Yes	Yes	No, external	Yes, mainly through accom-panying MCK	No. Per category map or Ksim	Post comparison must be performed independently	Confusion matrix
Pattern analysis	Compactness ratio, landscape metrics	Yes	Yes, various	No, external	Yes, mainly through accom-panying MCK	Various pattern based (SDMTools, Fragstats, MCK)	Post comparison must be performed independently.	No
Quantity	Yes	Yes	Yes	Yes	Yes	Yes	YES	Yes
ROC statistics	Yes	Yes	Yes	Yes	Yes	No	No	No

5 Concluding Remarks

Everyone agrees on the importance of model validation. The credibility of the model depends on it. However, the specific nature of each land change model software program and its various options make detailed comparisons impossible. On the other hand, the efforts undertaken by the scientific community in recent years are beginning to bear fruit. Modelers—and critical model users—have never had as many tools at their disposal for assessing the credibility of a simulation or that enable them to focus on particular aspects such as quantitative accuracy, in particular the accuracy of LUCC components, or to pay more attention to landscape pattern similarity. Nevertheless, the impressive array of techniques for calculating validation indices should not make us forget certain limitations. Firstly, the fact that in this chapter we have focused on path-dependent modeling approaches (Houet et al. 2016), while the validation of non trend-based scenarios (also known as contrasted scenarios) is even more difficult. Secondly, we centered on pattern-based models (PBM) while the large panoply of agent-based models (ABM) require their own particular tools, especially when we go beyond purely operational validation to consider conceptual realism as well (Rykiel 1996). Finally, model output accuracy depends above all on the quality of data and its conscientious use, as countless studies have proved.

Acknowledgements This work was supported by the BIA2013-43462-P project funded by the Spanish Ministry of Economy and Competitiveness and by the Regional European Fund FEDER. This study was also supported by the Consejo Nacional de Ciencia y Tecnología (CONACYT) and the Secretaría de Educación Pública through the project ¿Puede la modelación espacial ayudarnos a entender los procesos de cambio de cobertura/uso del suelo y de degradación ambiental? Fondos SEP-CONACyT 178816.

References

Abdou M, Hamill L, Gilbert N (2012) Designing and Building an Agent-Based Model. In: Heppenstall et al (eds) Agent-based models of geographical systems. Springer, pp 141–166

Ahlqvist O (2008) Extending post-classification change detection using semantic similarity metrics to overcome class heterogeneity: a study of 1992 and 2001 US National Land Cover Database changes. Remote Sens Environ 112(3):1226–1241

Aldwaik SZ, Pontius RG Jr (2012) Intensity analysis to unify measurements of size and stationaity of land changes by interval, category, and transition. Landsc Urban Plann 106:103–114

Alcamo J (2008) Environmental futures: the practice of environmental scenario analysis. Developments in integrated environmental assessment, vol 2. Elsevier, Amsterdam

Alcamo J, Schaldach R, Koch J, Kölking C, Lapola D, Priess J (2011) Evaluation of an integrated land use change model including a scenario analysis of land use change for continental Africa. Environ Model Softw 26(8):1017–1027. doi:10.1016/j.envsoft.2011.03.002

Allen THF, Starr B (1982) Hierarchy: Perspectives for ecological complexity. University of Chicago Press, Chicago 310 p

Burnicki AC, Brown DG, Goovaerts P (2007) Simulating error propagation in land-cover change analysis: The implications of temporal dependence. Comput Environ Urban Syst 31:282–302

Bone C, Dragicevic S, White R (2011) Modeling-in-the middle: bridging the gap between agent-based modeling and multi-objective decision-making for land use change. Int J Geogr Inf Sci 25(5):717–737. doi:10.1080/13658816.2010.495076

Botequilha A, Miller J, Ahern J, McGarigal K (2006) Measuring landscapes. A planner´s handbook. Island Press, Washington

Brown D et al (2005) Path dependence and the validation of agent-based spatial models of land use. Int J Geogr Inf Sci 19:153–174

Camacho Olmedo MT, Paegelow M, Mas JF (2013) Interest in intermediate soft-classified maps in land change model validation: suitability versus transition potential. Int J Geogr Inf Sci 27(12):2343–2361

Camacho Olmedo MT, Pontius RG Jr, Paegelow M, Mas JF (2015) Comparison of simulation models in terms of quantity and allocation of land change. Environ Modell Softw 69:214–221 (Elsevier)

Colas R (2016) Comprendre les changements d'utilisation des terres en France pour mieux estimer leurs impacts sur les émissions à effet de serre. De l'observation à la modélisation. PhD thesis, University Paris 7, 530 p
Conway TM, Wellen CC (2011) Not developed yet? Alternative ways to include locations without changes in land use change models. Int J Geogr Inf Sci 25(10):1613–1631. doi:10.1080/13658816.2010.534738
Coquillard P, Hill DRC (1997) Modélisation et simulation d'écosystèmes. Des modèles déterministes aux simulations à événements discrets, Paris, Masson
Croks AT, Heppenstall AJ (2012) Introduction to Agent-based modelling. In: Heppenstall et al (eds) Agent-based models of geographical systems. Springer, pp 85–108
Eastman JR, Jin W, Kyem PAK, Toleando J (1995) Raster procedures for multi-criteria/multi-objective decisions. Photogramm Eng Remote Sens 61(5):539–547
Eastman JR, Solorzano LA, Van Fossen ME (2005) Transition potential modeling for landcover change. In: Maguire DJ, Batty M, Goodchild MF (eds) GIS, spatial analysis, and modeling. ESRI, Redland CA, pp 357–385
Eastman JR (2009) Idrisi Taiga, Guide to GIS and Image Processing, manual version 16.02, Clark University 342 pp
Engelen G, White R (2007) Validating and calibrating integrated cellular automata based models of land use change. In: Albeverio S et al (eds) The dynamics of complex urban systems. An interdisciplinary approach, Physica, pp 185–211
Forman RTT (1995) Land Mosaics: the ecology of landscapes and regions. EEUU, Cambridge
Gaucherel C (2007) Multiscale heterogeneity map and associated scaling profile for landscape analysis. Landsc Urban Plann 82(3):95–102
Gaucherel C, Alleaume S, Hély C (2008) The Comparison Map Profile method: a strategy for multiscale comparison of quantitative and qualitative images. IEEE Trans Geosci Remote Sens 46(9):2708–2719
Gómez Delgado M, Barredo JI (2005) Sistemas de Información Geográfica y evaluación multicriterio en la ordenación del territorio (GIS and multicriteria evaluation for urban and regional planning). Ra-Ma, Madrid
Gómez Delgado M, Tarantola S (2006) Global sensitivity analysis, GIS and multi-criteria evaluation for a sustainable planning of hazardous waste disposal site in Spain. Int J Geogr Inf Sci 20:449–466
Hagen A (2003) Fuzzy set approach to assessing similarity of categorical maps. Int J Geogr Inf Sci 17(3):235–249
Hagen-Zanker A, Straatman B, Uljee I (2005) Further developments of a fuzzy set map comparison approach. Int J Geogr Inf Sci 19:769–785
Hagen-Zanker A (2009) An improved Fuzzy Kappa statistic that accounts for spatial autocorrelation. Int J Geogr Inf Sci 23(1):61–73. doi:10.1080/13658810802570317
Houet T, Gourmelon F (2014) La géoprospective—Apport de la dimension spatiale aux approches prospectives. Cybergéo 8 pp. http://cybergeo.revues.org/26194
Houet T, Aguejdad R, Doukari O, Battaia G, Clarke K (2016) Description and validation of a "non path-dependent" model for projecting contrasting urban growth futures. Cybergeo, document 759, 6 Jan 2016. http://cybergeo.revues.org/27397
Huang J, Pontius RG Jr, Li Q, Zhang Y (2012) Use of intensity analysis to link patterns with process of land change from 1986 to 2007 in a coastal watershed of southeast China. Appl Geogr 34:371–384
Jansen LJM (2006) Harmonisation of land-use class sets to facilitate compatibility and comparability of data across space and time. In: 12th CEReS international symposium, 13–14 Dec 2005, Japan 29 pp
Jackson DA (1995) PROTEST: a procrustean randomization test of community environment concordance. Ecoscience 2:297–303
Jokar Arsanjani J (2012) Dynamic land-use/cover change simulation: geosimulation and multi agent-based modelling. Springer, Springer Theses

Jansen LJM, Veldkamp TA (2011) Evaluation of the variation in semantic contents of class sets on modelling dynamics of land-use changes. Int J Geogr Inf Sci. doi:10.1080/13658816.2011.609989

Kamusoko C, Aniya M, Adi B, Manjoro M (2009) Rural sustainability under threat in Zimbabwe-Simulation of future land use/cover changes in the Bindura district based on the Markov-cellular automata model. Appl Geogr 29:435–447. doi:10.1016/j.apgeog.2008.10.002

Kim JH (2013) Spatiotemporal scale dependency and other sensitivities in dynamic land-use change simulations. Int J Geogr Inf Sci 27:1782–1803

Klug W, Graziani G, Grippa G, Pierce D, Tassone C (eds) (1992) Evaluation of long range atmospheric transport models using environmental radioactivity data from the Chernobyl accident: the ATMES report. Elsevier, London

Krüger C, Lakes T (2015) Revealing uncertainties in land change modeling using probabilities. Trans GIS 20(4):526–546

Kuhnert M, Voinov A, Seppelt R (2005) Comparing raster map comparison algorithms for spatial modelling and analysis. Photogramm Eng Remote Sens 71(8):975–984

Lee YJ, Lee JW, Chai DJ, Hwang BH, Ryu KH (2009) Mining temporal interval relational rules from temporal data. J Syst Softw 82:155–167

Li X, Chen Y, Liu X, Li D, He J (2011) Concepts, methodologies and tools of an integrated geographical simulation and optimization system. Int J GIS 25(4):633–655

Lin YP, Chu HJ, Wu CF, Verburg PH (2011) Predictive ability of logistic regression, auto-logistic regression and neural network models in empirical land-use change modeling—a case study. Int J Geogr Inf Sci 25(1):65–87. doi:10.1080/13658811003752332

Liu J, Deng X (2010) Progress of the research methodologies on the temporal and spatial process of LUCC. Chin Sci Bull 55:1354–1362

Mas JF, Kolb M, Houet T, Paegelow M, Camacho Olmedo MT (2011) Eclairer le choix de modèles de simulation des changements des modes d'occupation et d'usages des sols. Revue Internationale de Géomatique 21(3):405–430

Mas JF, Soares Filho B, Pontius RG Jr, Farfan Gutierrez M, Rodrigues H (2013) A suite of tools for ROC analysis of spatial models. ISPRS Int J Geo-Inf 2(3):869–887. doi:10.3390/ijgi2030869

Mas JF, Kolb M, Paegelow M, Camacho Olmedo MT, Houet T (2014) Inductive pattern-based land use/cover change models: a comparison of four software packages. Environ Model Softw 51:94–111

McGarigal K, Marks BJ (1995) FRAGSTATS: spatial pattern analysis program for quantifying landscape structure. USDA For Serv Gen Tech Rep

Mobaied S, Riera B, Lalanne A, Baguette M, Machon N (2011) The use of diachronic spatial approaches and predictive modelling to study the vegetation dynamics of a managed heathland. Biodivers Conserv 20:73–88. doi:10.1007/s10531-010-9947-1

Ngo TA, See L (2012) Calibration and validation of agent-based models of land cover change. In: Heppenstall et al (eds) Agent-based models of geographical systems. Springer, pp 181–198

North MJ, Macal CM (2007) Managing business complexity: Discovering strategic solutions with agent-based modelling and simulation. Oxford University Press, New York

Oreskes N, Shrader-Frechette K, Belitz K (1994) Verification, validation and confirmation of numerical-models in the earth-sciences. Science 263:641–646

Paegelow M, Camacho Olmedo MT (2005) Possibilities and limits of prospective GIS land cover modelling—a compared case study: Garrotxes (France) and Alta Alpujarra Granadina (Spain). Int J Geogr Inf Sci 19(6):697–722

Paegelow M, Camacho Olmedo MT (eds) (2008) Modelling environmental dynamics. Advances in geomatic solutions. Springer, series: Environmental science and engineering

Paegelow M, Camacho Olmedo MT, Mas JF, Houet T, Pontius RG Jr (2013) Land change modelling: moving beyond projections. Int J Geogr Inf Sci 27(9):1691–1695

Paegelow M, Camacho Olmedo MT, Mas JF, Houet T (2014) Benchmarking of LUCC modelling tools by various validation techniques and error analysis. Cybergeo 701, 22 Dec 2014. http://cybergeo.revues.org

Perica S, Foufoula-Georgiou E (1996) Model for multiscale disaggregation of spatial rainfall based on coupling meteorological and scaling descriptions. J Geophys Res 101(D21):26347–26361
Pérez-Vega A, Mas JF, Ligmann-Zielinska A (2012) Comparing two approaches to land use/cover change modelling and their implications for the assessment of biodiversity loss in a deciduous tropical forest. Environ Model Softw 29(1):11–23
Pontius RG Jr (2000) Quantification error versus location error in comparison of categorical maps. Photogramm Eng Remote Sens 66(8):1011–1016
Pontius RG Jr (2002) Statistical methods to partition effects of quantity and location during comparison of categorical maps at multiple resolutions. Photogramm Eng Remote Sens 68:1041–1049
Pontius RG Jr, Pacheco P (2004) Calibration and validation of a model of forest disturbance in the Western Ghats, India 1920–1990. GeoJournal 61:325–334
Pontius RG Jr, Schneider LC (2001) Land-cover change model validation by an ROC method for the Ipswich watershed, Massachusetts, USA. Agric Ecosyst Environ 85:239–248
Pontius RG Jr, Huffaker D, Denman K (2004a) Useful techniques of validation for spatially explicit land-change models. Ecol Model 179(4):445–461
Pontius RG Jr, Shusas E, McEachern M (2004b) Detecting important categorical land changes while accounting for persistence. Agric Ecosyst Environ 101:251–268
Pontius RG Jr, Malanson J (2005) Comparison of the structure and accuracy of two land change models. Int J Geogr Inf Sci 19:243–265
Pontius RG Jr, Cheuk ML (2006) A generalized cross-tabulation matrix to compare soft-classified maps at multiple resolutions. Int J Geogr Inf Sci 20(1):1–30. doi:10.1080/13658810500391024
Pontius RG Jr, Lippitt CD (2006) Can error explain map differences over time? Cartogr Geogr Inf Sci 33(2):159–171
Pontius RG Jr, Verluis AJ, Malizia NR (2006) Visualizing certainty of extrapolations from models of land change. Landsc Ecol 21(7):1151–1166
Pontius RG Jr, Walker RT, Yao-Kumar R, Arima E, Aldrich S, Caldas M, Vergara D (2007) Accuracy assessment for a simulation model of Amazonian deforestation. Ann Assoc Am Geogr 97:677–695
Pontius Jr RG, LI X (2010) Land transition estimates from erroneous maps. J Land Use Sci 5 (1):31–44
Pontius RG Jr, Petrova SH (2010) Assessing a predictive model on land change using uncertain data. Environ Model Softw 25:299–309
Pontius RG Jr, Boersma W, Castella JC, Clarke K, de Nijs T, Dietzel C, Duan Z, Fotsing E, Goldstein N, Kok K, Koomen E, Lippitt CD, McConnell W, Sood AM, Pijankowski B, Pidhadia S, Sweeney S, Trung TN, Veldkamp AT, Verburg PH (2008) Comparing the input, output, and validation maps for several models of land change. Anna Region Sci 42(1):11–27
Pontius RG Jr, Millones M (2011) Death to kappa and to some of my previous work: a better alternative. Int J Remote Sens 32(15):4407–4429
Pontius RG Jr, Gao Y, Giner NM, Kohyama T, Osaki M, Hirose K (2013) Design and interpretation of intensity analysis illustrated by land change in Central Kalimantan, Indonesia. Land 2(3):351–369. doi:10.3390/land2030351
Pontius Jr RG, Si K (2013) The total operating characteristic to measure diagnostic ability for multiple thresholds. Int J Geogr Inf Sci
Rodrigues HO, Soares Filho BS, de Souza Costa WL (2007) Dinamica EGO, uma plataforma para modelagem de sistemas ambientais. Anais XIII Simposio Brasileiro de Sensoriamento Remoto, INPE 3089–3096
Runfola DM, Pontius RG Jr (2013) Measuring the temporal instability of land change using the flow matrix. Int J Geogr Inf Sci 26(9):1696–1716
Rykiel EJJ (1996) Testing ecological models: the meaning of validation. Ecol Model 90:229–244
Sargent RG (2009) Verification and validation of simulation models. In: Rosetti MDD, Hill RR, Johansson B, Dunkin A, Ingalls RG (eds) Proceedings of the 2009 winter simulation conference. IEEE, Syracuse, NY, pp 162–176

Schneider LC, Pontius RG Jr (2001) Modeling land-use change in the Ipswich watershed, Massachusetts, USA. Agric Ecosyst Environ 85:83–94

Shirley LJ, Battaglia LL (2008) Projecting fine resolution land cover dynamics for a rapidly changing terrestrial-aquatic transition in Terrebone basin, Louisiana, USA. J Coastal Res 246:1545–1554

Silva TS, Tagliani PRA (2012) Environmental planning in the medium littoral of the Rio Grande do Sul coastal plain—Southern Brazil: elements for coastal management. Ocean Coast Manag 59:20–30

Soares-Filho BS, Pennachin CL, Cerqueira G (2002) DINAMICA—a stochastic cellular automata model designed to simulate the landscape dynamics in an Amazonian colonization frontier. Ecol Model 154(3):217–235

Sohl TL et al (2012) Spatially explicit land-use and land-cover scenarios for the Great Plains of the United States. Agric Ecosyst Environ 153:1–15. doi:10.1016/j.agee.2012.02.019

Steiner F, McSherry L, Cohen J (2000) Land suitability analysis for the upper Gila River watershed. Landsc Urban Plann 50(4):199–214

Syphard AD, Clarke KC, Franklin J, Regan HM, Mcginnis M (2011) Forecasts of habitat loss and fragmentation due to urban growth are sensitive to source of input data. J Environ Manage 92:1882–1893

Takada T, Miyamoto A, Hasegawa SF (2010) Derivation of a yearly transition probability matrix for land-use dynamics and its applications. Landsc Ecol 25(4):561–572

Torrens PM (2011) Calibrating and validating cellular automata models of urbanization. In: Yang X (ed) Urban remote sensing. Monitoring, synthesis and modelling in the urban environment. Wiley, New York, pp 335–345

Turner MG, Constanza R, Sklar FH (1989) Methods to evaluate the performance of spatial simulation models. Ecol Model 48:1–18

Villa N, Paegelow M, Camacho Olmedo MT, Cornez L, Ferraty F, Ferré L, Sarda P (2006) Various approaches for predicting land cover in Mediterranean mountains. Commun Stat 36 (1):73–86

Visser H, de Nijs T (2006) The map comparison kit. Environ Model Softw 21:346–358

van Vliet J, Bregt AK, Brown DG, van Delden H, Heckbert S, Verburg PH (2016) A review of current calibration and validation practices in land change modelling. EMS 82:174–182

Wang J, Mountrakis G (2011) Developing a multi-network urbanization model: A case study of urban growth in Denver, Colorado. Int J Geogr Inf Sci 25(2):229–253. doi:10.1080/13658810903473213

White R, Uljee I, Engelen G (2012) Integrated modelling of population, employment, and land use change with a multiple activity based variable grid cellular automaton. Int J GIS 26:1251–1280

White R, Engelen G, Uljee I (1997) The use of constrained cellular automata for high resolution modelling of urban land use dynamics. Environ Plan 24:323–343

Yu J, Chen Y, Wu J, Khan S (2011) Cellular automata-based spatial multi-criteria land suitability simulation for irrigated agriculture. Int J Geogr Inf Sci 25(1):131–148. doi:10.1080/13658811003785571

Chapter 5
LUCC Scenarios

F. Escobar, H. van Delden and R. Hewitt

Abstract Since ancient times people have been curious to know more about how the future could unfold, and have proposed different scenarios as a tool for exploring the future of their societies. Examples abound, from Plato's description of his ideal Republic to Orwell's vision of 1984 in 1948. As a strategic planning tool, scenario techniques originated as a means of enhancing military strategies, first appearing in the form of war games. Today's scenario techniques emerged after World War 2 and have a wide range of industrial and government applications. Concerns about the possible impacts of climate and global change have boosted studies in which scenarios play a key role as an analytical technique. Current development of modeling techniques within Geographic Information Systems (GIS) and the increasing availability of geospatial information have enabled the implementation of spatially-explicit scenarios of various kinds, including those on land-use cover change (LUCC) studied in this book. Such is the current popularity of scenario techniques in terms of the number of applications and users that the relevant literature reveals a wide array of different and often contradictory definitions and ideas about scenarios. These are accompanied by a large number of scenario planning techniques and models, leading some authors to describe the situation as "methodological chaos". This chapter has two main objectives: firstly,

F. Escobar (✉)
Department of Geology, Geography and Environmental Sciences, University of Alcalá, Alcalá de Henares, Spain
e-mail: francisco.escobar@uah.es

H. van Delden
Research Institute for Knowledge Systems (RIKS), Maastricht, The Netherlands
e-mail: hvdelden@riks.nl

H. van Delden
The University of Adelaide, Adelaide, Australia

R. Hewitt
Observatorio para una Cultura del Territorio, Madrid, Spain
e-mail: richard.hewitt@observatorioculturayterritorio.org

R. Hewitt
James Hutton Institute, Dundee, Scotland, UK

M.T. Camacho Olmedo et al. (eds.), *Geomatic Approaches for Modeling Land Change Scenarios*, Lecture Notes in Geoinformation and Cartography,
https://doi.org/10.1007/978-3-319-60801-3_5

to offer insights into the topic and to clarify some definitions of scenario-related terms and techniques, and secondly to serve as a guide for LUCC scenario planning and modeling.

Keywords Scenario · LUCC · Modeling · Validation

1 Introduction

The Oxford English Dictionary defines scenario as "a postulated or projected situation or sequence of potential future events; (also) a hypothetical course of events in the past, intended to account for an existing situation, set of facts, etc. Also more generally: a set of circumstances; a pattern of events" (OED 2017). Although these definitions seem clear and easy to understand, literature on scenario planning and applications, particularly in land-use studies, shows a confusing array of definitions and interrelations with similar, albeit not identical, terms. For instance, in land-use modeling literature, scenarios refer sometimes to the storylines that the modeling exercise attempts to reproduce in the form of a hypothetical future land-use map. On other occasions, however, the scenario is the future land-use map itself. And what is more worrying, scenarios are often considered as predictions. The concept of 'scenario' is therefore one of the most basic, and at the same time one of the most disputed concepts in the field of futures studies (Börjeson et al. 2006). It therefore seems necessary to begin this chapter by defining the term scenario and positioning ourselves within the large variety of meanings and interpretations of the term.

Herman Kahn, who is considered the father of scenario planning, defined it as "a set of hypothetical events set in the future constructed to clarify a possible chain of causal events as well as their decision points" (Bradfield et al. 2005). In his definition, Kahn highlights the existing causal relationship between events leading to a hypothetical future and the key points in time (drivers) that make that future possible.

As early as 1994, the International Permanent Committee on Climate Change (IPCC) defined scenario as "a coherent, internally consistent, and plausible description of a possible future state of the world" (IPCC 1994) and added that scenarios are not forecasts. The IPCC views scenario as an alternative image of how the future could unfold. Later on, in 2000 the IPCC redefined scenarios as "alternative images of how the future might unfold" and "an appropriate tool with which to analyze how driving forces may influence future emission outcomes and to assess the associated uncertainties" (IPCC 2000). The interest of this definition lies in the fact that it has been adapted to climate change (future emissions) and, more importantly, scenario is considered not only as a hypothetical future but as a valid tool for assessing the uncertainties associated with it.

Also considering future impacts of climate change, Carter et al. opted for almost the same definition as that provided by the IPCC in 1994 and stated that a scenario is “a coherent, internally consistent, and plausible description of a possible future state of the World”. Scenarios are commonly required in climate change impact, adaptation, and vulnerability assessments to provide alternative views of future conditions considered likely to influence a given system or activity (Carter et al. 2001). Characteristics such as coherent, consistent and plausible are also typical conditions in the validation of scenarios, as we will see later on in this chapter.

Although the above definitions are slightly different, they are coherent and evoke a fairly clear image of what is commonly understood by the word “scenario”. However, serious confusion arises when other terms with different meanings are used as synonyms. For instance, the term “storyline” is often used in the literature as a synonym for scenario. Although they are closely related, our understanding is that a storyline provides a description of a scenario, in a similar way to other descriptive methods, such as figures, diagrams and maps. In other words, scenarios are images or descriptions of the future, while storylines are materializations of these images, in the form of literal explanations. Storylines may take the form of narrative descriptions which are later “realized” in the form of visual images of the future territory (see e.g. Volkery et al. 2008). Similarly, a map representing a future land-use scenario is not a scenario itself but its cartographic image or representation.

As made clear by the IPCC (1994), scenarios are not forecasts. Neither are they projections or predictions. A projection can be regarded as any description of the future and the pathway leading to it and can serve as basic information for building a scenario. Since projections are by nature uncertain, scenarios can be used to reflect the different implications of these uncertainties. For its part, a forecast is defined as the most likely projection, i.e. the projection with least uncertainties. Prediction and forecast are synonymous. Despite these differences it is not uncommon to find examples in the literature where land-use scenario modeling is presented as a predictive model, even in models where the level of uncertainty is very high as in Yuan et al. (2013) where the simulation period runs as far as 2100.

In line with the IPCC, Wack (1985a, 1985b) notes that a scenario does not predict the future, but “it explores multiple plausible future situations with the purpose of extending the sphere of thinking of the participants in the scenario development process”. Scenarios differ from forecasts as in the former, a range of possible outcomes resulting from uncertainty can be explored, whereas the purpose of forecasts is to identify the most likely pathway to that outcome and from there estimate the uncertainties.

Since scenarios do not predict the future one could ask why we need them. Amer et al. (2013) cite multiple previous works to conclude that “consideration of multiple possible future alternatives helps to conduct future planning in a holistic manner and significantly enhance the ability to deal with uncertainty”. The scenario planning process helps to make the desirable future real, stimulates strategic thinking and helps to overcome thinking limitations by creating multiple futures (Saliba 2009). From the decision-making point of view, scenarios are considered a valuable tool that helps organizations to prepare for possible eventualities (Amer et al. 2013).

ESPON (2015) states that scenarios can be used to communicate insights and discuss potential territorial developments, the effects of policies with territorial impact, and the political choices to be made.

In view of these definitions and the fact that, according to the IPCC, scenarios are considered as a tool for assessing uncertainties about the future, it is necessary to differentiate between its two meanings. Scenarios offer images of a plausible future, whether in the form of storylines, maps, graphs or figures, and they are also considered as tools that can reveal peculiarities and uncertainties, and make discoveries, about a possible future. This is of particular interest in spatially-explicit scenarios such as land-use cover change (LUCC) scenarios where the models (the maps) representing the scenarios also provide information about where change is taking place.

After clarifying the definition of this concept, in the following sections we provide an overview of the various applications of scenarios, their history and the techniques used to develop them. This is followed by a review of the scenarios used in LUCC and their validation techniques.

2 The Advent of LUCC Scenario Planning

As pointed out by Bradfield et al. (2005), the concept of scenarios has been around since ancient times and has been used as a way of exploring the future of societies and institutions. However, as a strategic planning tool, scenario techniques first appeared within the military in the form of war game simulations. The first documents describing what today we regard as scenarios, appeared in the 19th century in the writings of two Prussian military strategists also credited with having 'first formulated the principles of strategic planning' (Bradfield et al. 2005). Modern day scenario techniques emerged in the 1950s in the US at the RAND (from "Research and Development") Corporation. After being used by the US Department of Defense as a method for military planning, the same scenario methodology was extensively used for social forecasting, public policy analysis and decision making in the 1960s.

Despite these developments, scenario planning was not widely used until the 1973 oil crisis, after which the number of users of scenario planning almost doubled. Given that interest in scenario studies seemed to increase in times of social and economic crisis, Malaska et al. concluded that the adoption of scenario planning was associated with the increasing unpredictability of the corporate environment in the 1970s (Malaska et al. 1984).

Linneman and Klein (1983) estimate that in the early 1980s, almost half of the top 1000 industrial firms in the US, half of the top 300 non-industrial firms and half of the top 500 foreign industrial firms were actively using scenario planning.

The *Centre d'Etudes Prospectives*, also known as *La Prospective*, appeared around the same time in France. Founded by the philosopher Gaston Berger, it developed a scenario approach to long-term planning (Amer et al. 2013). The main idea behind *La Prospective* was that the future is not part of a "predetermined temporal continuity" but something which is to be created and which can be "consciously modeled to be humanly beneficial". *La Prospective* was materialized in the creation of DATAR (*Délégation interministérielle à l'Aménagement du Territoire et à l'Attractivité Régionale* or Delegation for Planning and Regional Action) in 1963. DATAR is still operational today and is controlled directly by the Prime Minister. It has funded numerous interesting studies on for example the implementation of the high speed train system in France and its impact on the territory (Cauvin et al. 1992).

During the first years of RAND Corporation and *La Prospective* the lack of appropriate tools made it very difficult, or virtually impossible, to develop spatially-explicit models to represent the scenarios they were constructing. The appearance and diffusion of Geographic Information Systems (GIS) in the 1970s and the later increasing availability of spatial information, enabled the production of the first static cartographic models on land-use change. Simultaneously, in the late 1970s, the geographer Waldo Tobler developed his "cellular geography" and proved that the cellular automata (CA) theory could be adopted in land-use modeling (Tobler 1979). However, it was not until the 1990s that the first CA-based models were run in a computing environment with real land-use data (White and Engelen 1993, 1994, 1997).

Once LUCC models had become dynamic, i.e. they could simulate changes occurring over time, a natural step forward was to use them in scenario planning studies. This together with increasing social concern about global warming and climate change gave rise to an increasing number of LUCC projects in which different scenarios were tested and implemented with the new CA-based models. Examples are the national/regional level scenarios of the PRELUDE study which are modelled with the Metronamica land use modeling framework (Van Delden and Hagen-Zanker 2009), and the MedAction and LUMOCAP scenarios (Kok and Van Delden 2009; Van Delden et al. 2010) modelled with an integrated policy support system that incorporates the Metronamica land use model (Van Delden and Hurkens 2011).

3 Scenario Typologies and Developing Techniques

3.1 Scenario Typologies

As noted by Börjeson and others (2006), there is no agreement as to what typologies of scenario can be established or as to which would be most useful to users. In Table 1 there is a summary of the typologies outlined by these authors.

Table 1 Scenario typologies and main characteristics

<table>
<tr><th>Proposed by:</th><th colspan="6">Typologies and focus of futures studies</th><th></th></tr>
<tr><td>Amara (1981)
Marien (2002)</td><td colspan="2">Possible
Forecasting
Foresight
Predicting</td><td colspan="2">Probable
Alternatives
Risk analysis
Scenario construction</td><td colspan="2">Preferable
Planning
Policy making
Visionary</td><td rowspan="8">User's needs</td></tr>
<tr><td>Masini (1993, 2006)</td><td colspan="2">Utopian
Positive or negative</td><td colspan="2">Extrapolation
Business as usual</td><td colspan="2">Vision
Desirable</td></tr>
<tr><td>Dreborg (2004)</td><td colspan="2">Predictive
Forecasting</td><td colspan="2">Eventualities
External scenarios</td><td colspan="2">Visionary
Backcasting</td></tr>
<tr><td rowspan="2">Börjeson et al. (2006)</td><td colspan="2">What will happen?
Predictive</td><td colspan="2">What can happen?
Exploratory</td><td colspan="2">How to target it?
Normative</td></tr>
<tr><td>Forecast</td><td>What-if</td><td>External</td><td>Strategic</td><td>Preserving</td><td>Trans-forming</td></tr>
<tr><td rowspan="3">Maier et al. (2016)</td><td colspan="2">What will happen?
Predictive</td><td colspan="2">What can happen?
Exploratory</td><td colspan="2">How to target it?
Normative</td></tr>
<tr><td>Forecast</td><td>What-if</td><td>Unframed</td><td>Framed</td><td>Preserving</td><td>Trans-forming</td></tr>
<tr><td colspan="4">Forward/problem focused approaches</td><td colspan="2">Inversed/solution fo-cused approaches</td></tr>
<tr><td>Inayatullah (1990)</td><td colspan="2">Predictive-empirical
Forecast</td><td colspan="2">Cultural-interpretative
Negotiation (among cultures) to achieve a desirable future</td><td colspan="2">Critical futures
Historical context</td><td rowspan="4">Use of the knowledge generated</td></tr>
<tr><td>Mannermaa (1991)</td><td colspan="2">Descriptive
Objective trends</td><td colspan="2">Scenario paradigm
Constructing futures and the paths to them</td><td colspan="2">Evolutionary
World-view of societies with combination of predictable and chaotic phases</td></tr>
<tr><td>Bell (1997)</td><td colspan="2">Critical realism
Desirable futures</td><td colspan="2">Positivism
Alternatives</td><td colspan="2">Post-positivism
Historical context</td></tr>
<tr><td>Habermas (2003)</td><td colspan="2">Technical
Objective trends</td><td colspan="2">Hermeneutic
Increasingly common understanding of social reality</td><td colspan="2">Emancipatory
Widening perceived scope of option</td></tr>
</table>

As reflected in Table 1, the typologies focus either on user's needs or on the function that the generated knowledge may play. The latter is a philosophical view commonly found in futures studies, while the former, although also part of futures studies, is more closely related to what this book is about; namely, exploring what the future may look like in order to increase preparedness and decision-making leading to a desirable future.

We are particularly interested in the typology proposed by Börjeson et al. in 2006 as it tackles the three different (and often intermixed) views of scenarios found in LUCC literature; predictive, exploratory and normative.

As outlined above, scenarios are not predictions. However, in LUCC modeling literature we often find the expression "what-if" referring to what these models are able to produce, i.e. hypothetical outcomes based on answering questions like "what might happen if we act in a certain way?". Börjeson et al. however relate "what-if" questions to predictive scenarios while LUCC modelers tend to relate them to an exploration of the future. Of course, this does not help clarify the already confusing terminology.

In line with Börjeson et al., we believe that LUCC scenarios, in the way they are currently being developed within the land use simulation community, are on a spectrum from rather predictive (mostly what-if) to exploratory, depending on their scope. However, we feel that LUCC depends on so many different factors (natural conditions, legislation, economy, demography…) and is so complex in nature (number of land-use categories, area occupied by each one, patterns of distribution among the categories, interrelations between them…) that predictive exercises have strong limitations and exploratory scenarios are therefore preferred. The latter helps to better understand the different ways the future might unfold and by obtaining a better appreciation for this assist in developing robust and/or adaptive plans to move towards a desired future.

3.2 *Scenario Developing Techniques*

Börjeson et al. (2006) state that there are three main tasks in scenario development: (1) generation of ideas and gathering of data, (2) integration of the data and (3) assessing the consistency of scenarios created. They outline a number of valid techniques for each of these tasks: surveys, workshops and Delphi methods for Task 1; time series analysis, explanatory modeling and optimizing modeling for Task 2; and morphological field analysis and cross impact analysis for Task 3. To a greater or lesser extent these techniques have been applied in most LUCC and other spatially-explicit scenario development techniques.

As acknowledged by Amer et al. (2013), there is such a large number of perspectives and techniques for developing scenarios and related models that some authors have dubbed it "methodological chaos".

One of the first questions to answer when attempting to design scenarios is how many of them to consider. This depends enormously on the type of scenario method: if we want to build normative scenarios then there should be as many scenarios as there are normative possibilities. If we are exploring future situations with limited alternatives or a narrow scope (i.e. alternative transport systems) then there should be as many scenarios as there are transport alternatives (see chapter on transport scenario planning for Bogota). The decision depends on how many uncertainties about the future are considered. Obviously, the shorter the period being simulated, the lesser the number of uncertainties and inversely, the longer the period considered, the greater the number of uncertainties to consider within the scenario.

Dator (2002) proposes four types; (1) continued growth, assuming that current conditions will improve, (2) collapse, current conditions do not improve and do not sustain growth, revealing deep contradictions, (3) steady state, where growth decelerates to find a better balance between economy and environment, envisaging a softer, fairer society; and (4) transformation, depicting a dramatic change in society caused by either technological or spiritual changes. All the papers reviewed in Amer et al. (2013) recommended between two and six scenarios and most of them suggested four. There was a simple explanation for this choice; most of them were based on a 2 × 2 matrix. For instance, Schwarth (2009) designed a matrix in which the number of businesses, higher versus lower, was contrasted with the speed of change, slow versus quick. In Piorr et al. (2011) the four scenarios analyzed come from relating on one side, private enterprise economic values versus public community and ecological values, and on the other global-macro top down dynamics versus regional-local bottom up dynamics. These four archetypes are found with slight variations in most scenario planning projects under different names. Thus, in the PLUREL project (Piorr et al. 2011) the four scenarios are referred to as "hypertech", "peak-oil", "extreme water" and "walls and enclaves", a direct adaptation of the A1, A2, B1 and B2 scenario families established by the IPCC in their special report on emissions scenarios (SRES) (Nakicenovik and Svart 2000).

Given the number of projects and modeling scenario exercises based on the IPCC SRES scenario families, we believe it is useful to summarize them here:

"The A1 family describes a world of very rapid economic growth, global population that peaks in mid-century and declines thereafter, and the rapid introduction of new and more efficient technologies. Major underlying themes are convergence among regions, capacity building, and increased cultural and social interactions.

A2 describes a very heterogeneous world. The underlying theme is self-reliance and preservation of local identities. Fertility patterns across regions converge very slowly, which results in continuously increasing global population. Economic development is primarily regionally oriented and per capita economic growth and technological change are more fragmented and slower than in other scenarios.

B1 describes a convergent world with the same global population that peaks in mid-century and declines thereafter, as in A, but with rapid changes in economic structures toward a service and information economy, with reductions in material intensity, and the introduction of clean and resource-efficient technologies. The emphasis is on global solutions to economic, social, and environmental sustainability, including improved equity, but without additional climate initiatives.

B2 describes a world in which the emphasis is on local solutions to economic, social, and environmental sustainability. It is a world with continuously increasing global population at a rate lower than A2, intermediate levels of economic development, and less rapid and more diverse technological change than in the B1 and A1. While the scenario is also oriented toward environmental protection and social equity, it focuses on local and regional levels" (IPCC 2000).

These four scenarios are frequently complemented by a fifth scenario named "more of the same" (Hicks et al. 1995) or "business as usual" (Masini 2006). This scenario assumes that trends observed in the past will continue along the same path towards the future, and that no dominant drivers will divert them from their course. Issues arise when deciding how far back in the past we should go to observe these trends. Since ups and downs have probably been observed in the past, the business as usual path will have to be described by averaging the whole period under consideration. In addition to this and as noted in the ESPON report (2014), in periods of economic and political crises, turbulence and uncertainty make defining consistent enough baseline trends difficult. In this report, the methodology used to define the baseline scenario was based on a mixture of qualitative and quantitative techniques and was divided into the following phases:

- Internal expert consultation and debates
- Analysis by sectors and macro-regions
- Analysis of the present state and trend
- Analysis of ongoing debates on policy reforms in Europe
- Identification of critical points of bifurcation (alternative pathways)
- Comparative analysis of existing baseline scenarios at both European and World levels
- Definition of baseline assumptions
- Quantitative analysis that provided relevant indicators at EU level
- Quantitative modeling using forecast models, and
- Analysis of territorial differences as a means of explaining territorial dynamics.

Within the phase in which critical points of bifurcation are identified, the ESPON report analyzed, among others the possibility of land-use becoming more hybrid due to ineffective planning.

Finally, the IPCC SRES scenarios themselves are in the process of replacement by a new generation of scenarios, the Representative Concentration Pathways (RCP) Scenarios (Moss et al. 2010). Since these have not yet been widely adopted by land use modelers, we do not discuss them here. However, we observe that the

RCPs do attempt to address many of the limitations found in the SRES scenarios, so their incorporation more broadly into land use change studies is to be encouraged.

Once the number of scenarios to be modeled has been established, it is time to choose which scenario planning method to use. A wide range of options are available.

Quantitative methods include INTERAX or Interactive Cross Impact Simulation (Enzer 1980) in which a detailed multidisciplinary database containing important information on a broad range of long-term strategic issues and future trends and events was developed through a Delphi study by a large group of experts. Using a similar approach, Interactive Future Simulations (IFS) identifies novel and diverse ideas, encourages contingency planning and provides an early warning system for major changes. Jetter and Schweinfort (2011) stand out for their use of the Fuzzy Cognitive Map (FCM) to represent social scientific information as an interconnected collection of causal graphs. The visual nature of these graphs facilitates understanding of the dependencies and contingencies between concepts.

For their part, qualitative methods focus on workshops and discussion group sessions, surveys, interviews and other participatory activities such as photovoice and brainstorming.

Ideally, scenario planning has to be supported by both quantitative and qualitative techniques. The latter facilitate the construction of storylines that help depict the scenario in a narrative and realistic manner, while the former provide figures to support all the assumptions included in the storyline.

4 LUCC Scenario Planning Examples

Within general scenarios, territory-based scenarios have become increasingly popular in the European context as a tool for investigating territorial cohesion issues and the impact of EU regional policies on member states.

One particular kind of territorial scenario is the land-use scenario. As stated by Veldkamp and Fresco (1997), land-use scenarios should be able to describe land-use as a result of changing biophysical and socioeconomic conditions, and should also trace the pathways for possible developments including feedbacks between land-use change and its drivers. Instead of exploratory scenarios, they present the CLUE model in which scenarios are made by changing, extrapolating and adjusting the relationship of land-use/cover drivers and related land use systems. They model six future scenarios: abolition of national parks, extension of national parks, urbanization, soil erosion and soil fertility depletion, crop disease in permanent crops below 300 m and a volcanic eruption.

Fischer et al. (2010) applied a "food first" paradigm in their estimations of the land potentially available for the production of biofuel feedstocks, without putting food supply or nature conservation at risk. Three land conversion scenarios were formulated: A base scenario that reflects developments under current policy settings and respects current trends in nature conservation and organic farming practices, by assuming moderate overall yield increases; an environment oriented scenario with higher emphasis on sustainable farming practices and maintenance of biodiversity; and an energy oriented scenario considering more substantial land use conversions including the use of pasture land.

Münier et al. (2004) combined ecological and economic modeling in four agricultural land use scenarios. Their research deals with the consequences for ecology, the environment and the economy of changes in agricultural production. They also seek to link vegetation ecology and farm economy.

In order to address various agricultural issues, Ewert et al. (2005) explored changes in crop productivity by means of scenarios that represent alternative economic and environmental pathways to future development. The scenarios modeled were based on changing conditions in the climate, atmospheric CO_2 concentration and technology development.

Eickhout et al. (2007) investigated the economic and ecological consequences of four European land-use scenarios, dealing with the complex interaction between agricultural trade, production, land-use change and the environment. In their scenarios, they focused on the major uncertainties likely to be experienced by regional trade blocks as a result of trade liberalization. They found that although these liberalizing scenarios did result in economic growth, the environmental threats posed by climate change and fertilizers for the sustainability of global agricultural practices produced new challenges for future food production.

These four scenarios were developed on the basis of four corresponding narratives or storylines. These narratives were an adaptation of those developed for the Special Report on Emission Scenarios (SRES) issued by the IPCC.

Also based on the IPCC report, Rounsevell et al. (2006) presented a range of future, spatially explicit, land-use change scenarios for the EU15, Norway and Switzerland. They discussed the technical and conceptual difficulties inherent in developing future land use change scenarios. These included the problems of the subjective nature of qualitative interpretations, the land-use change models used in scenario development, the problem of validating future change scenarios, the quality of the observed baseline, and statistical downscaling techniques.

In addition to these applications in agriculture, other studies have focused on the impact of LUCC on landscapes. Thus Hawkins and Selman (2002) modeled alternative land use scenarios based on landscape ecology.

Examples in developing countries are not as common as in Europe or North America. In this book we present a study of the city of Bogota in which two different transport scenarios were modelled (see chapter by Páez and Escobar). Some interesting work has also been done in Costa Rica where Stoorvogel (1995) integrated models and scenario tools to evaluate alternative land-use in agriculture, while Barredo et al. (2004) explored alternative scenarios for urban growth in Lagos, Nigeria.

In the European context the ESPON program has been proactive in developing spatially-explicit land-use scenarios and models (ESPON 2013, 2014). They have set three scenarios: (1) Europe of the Flows, which provides an image of European territory in which economic and population growth as well as public investments are mainly stimulated along the main corridors that structure the European continent; (2) Europe of the Cities where growth takes place in cities and (3) Europe of the Regions where the recipients of growth and investments are the regions. These scenarios have formed the basis of multiple modeling exercises conducted in Europe.

There are an impressive number of modeling exercises that offer a cartographic view of LUCC scenarios in different European settings. Numerous examples can be found in The Netherlands (De Nijs et al. 2004; Kok et al. 2001), probably the most modeled country in the world, in Spain (Hewitt et al. 2014; Escobar et al. 2015; Aguilera et al. 2011, Camacho et al. 2015; Gallardo and Martínez-Vega 2016; Plata et al. 2009), Portugal (Petrov et al. 2009), and a long list including the most important urban centers in Europe.

5 Scenario Validation

Since scenarios are by definition unrealized visions of the future, they cannot be validated by contrasting them with reality. Chermack et al. (2001) highlight the importance of establishing appropriate criteria for validation and state that scenarios must be checked for validity to ensure that they provide a solid reliable basis for making important decisions.

Amer et al. (2013) summarized scenario validation criteria found in the literature into seven groups; plausibility, consistency/coherence, creativity/novelty, relevance/pertinence, importance, transparency and completeness/correctness. Plausibility refers to the capacity of the scenarios to be capable of happening. Consistency guarantees that there is no built-in internal inconsistency or contradiction. Relevance indicates that the scenario should contribute specific insights into the future that help to make the decision. Novelty applies to the capacity of the scenario to challenge conventional thinking about the future and correctness to the credibility of the scenario.

Of all of these criteria, the only ones unanimously accepted by all authors are the first two. Importance is only mentioned by Durance and Godet (2010) and transparency by these authors and by De Brandere and Iny (2010) and Kosow and Gassner (2008).

Other authors propose different criteria for scenario validation. Van der Heijden (1996) states that at least two scenarios are needed to reflect uncertainty, each scenario must be plausible, all scenarios must be internally consistent, each scenario must be relevant to the client's concerns and all of them must produce a new and original perspective on the issues.

In addition to the criteria already mentioned, Durance and Godet (2010) argue that scenarios should also present a degree of likelihood. However this would partially or totally contradict the criteria of novelty and more visionary scenarios would not be proposed if they are considered likely to occur.

We believe that the additional criteria outlined by De Brandere and Iny (2010) are more useful. They suggest "easy to recount and illustrate" as a final criterion to complement those mentioned above. Indeed, a scenario that is not easy to recount or to illustrate is not worth being analyzed and will have difficulties in finding supporters. On similar lines, the research presented by Sheppard et al. (2011) successfully tested realistic visualization tools for community engagement and planning with scenarios.

Based on Amer et al. (2013) the methods commonly used to select and validate scenarios include: (1) A minimal approach in which scenarios are defined by two factors: the most important and the most uncertain; (2) The Wilson matrix that helps to evaluate and prioritize the scenario drivers against their potential impact and uncertainty (probability of developing into a significant issue in the future); (3) Morphological analysis which develops raw scenarios/input vectors and access plausibility; (4) Cross impact analysis to identify the strongest scenario drivers with the highest shaping potential; and (5) Consistency analysis to verify the internal consistency of scenarios.

In order to conduct consistency analysis, Pillkahn (2008, cited in Amer et al. 2013) suggested assigning a score on a scale of 1–5 to evaluate the consistency of the different scenarios. The scores are presented in a matrix containing all the scenario drivers in order. A score of 1 is assigned if there is total inconsistency (impossible combination) while a score of 5 is awarded if both factors/drivers are closely linked and positively impact on each other or are mutually dependent.

6 Concluding Remarks

Amer et al. (2013) insist on the importance of the internal consistency of the scenarios as emphasized in the literature on scenario planning. Consistency analysis is used to check the compatibility of combined variations of various scenario drivers and can also be used for reducing the number of scenarios to a manageable number. They conclude that the best scenario planning approaches offer a combination of qualitative and quantitative techniques through which they can generate robust scenarios, adding that the most appropriate number of scenarios to be modelled is between three and five.

Some researchers argue that preference for the scenario planning approach has declined slightly. They consider that scenario methods have evolved into a set of very complex sub-techniques which as they are difficult to implement discourage possible users from trying.

This undoubtedly applies to LUCC modeling software. On the positive side it has enabled researchers to put spatially-explicit scenarios into practice, to recreate them in the form of maps of the future. These maps are analyzed in the same way as we analyze maps of the present or the past and allow us to compute indicators and other metrics that help us characterize the future and reach conclusions about it. However, at the same time, their level of sophistication and the fact that they are difficult to use (substantial training is still needed) means that they tend to be most popular with experts or scientists, who often have little relation with real practitioners and users.

A vision for future land-use scenario modeling must include easy to use software, well-trained and experienced GIS land-use practitioners and up-to-date, high quality data.

Another possible reason for the decline of scenario-based research lies in the impact of fast changes in society and the environment which makes forecasting and visions for the future quickly obsolete. But perhaps for this very reason scenario analysis is more urgently required and more easily justified. Despite this decline, LUCC models remain popular in scientific literature probably due to social concern about climate change and the availability of geo-referenced databases, as outlined above in this chapter.

References

Aguilera F, Valenzuela LM, Botequilha-Leitão A (2011) Landscape metrics in the analysis of urban land use patterns: a case study in a Spanish metropolitan area. Landsc Urban Plann 99:226–238

Amara R (1981) The futures field: searching for definitions and boundaries. The Futurist 15(1): 25–29

Amer M, Daim T, Jetter A (2013) A review of scenario planning. Futures 46:23–40

Barredo J, Demicheli L, Lavalle C, Kasanko M, MacCormick N (2004) Modelling future urban scenarios in developing countries: an application case study in Lagos, Nigeria. Environ Plann B Plann Design 31(1):65–84

Bell W (1997) Foundations of futures studies: human science for a new era, vol. 1: history, purposes and knowledge. New Brunswick, Transaction Publishers

Börjeson L, Höjer M, Dreborg KH, Ekvall T, Finnveden G (2006) Scenario types and techniques: towards a user's guide. Futures 38:723–739

Brabandere L, Iny A (2010) Scenarios and creativity: thinking in new boxes. Technol Forecast Soc Chang 77:1506–1512

Bradfield R, Wright G, Burt G, Cairns G, Van Der Heijden K (2005) The origins and evolution of scenario techniques in long range business planning. Futures 37:795–812

Camacho Olmedo MT, Pontius RG Jr, Paegelow M, Mas JF (2015) Comparison of simulation models in terms of quantity and allocation of land change. Environ Model Softw 69:214–221

Carter TR, La Rovere EL, Jones RN, R. et al (2001) Developing and applying scenarios. In: McCarthy JJ, Canziani OF, Leary NA, Dokken DJ, White KS (eds) Climate Change 2001: impacts, adaptation, and vulnerability. Contribution of working group II to the third assessment report of the intergovernmental panel on climate change. Cambridge University Press, Cambridge and New York, pp 145–190 IPCC 2000

Cauvin C, Reymond H, Enaux C et al (1992) Pour une approche multiple de l'accessibilité: propositions méthodologiques. Etude réalisée à la demande de la DATAR, Strasbourg 135 pp
Chermack TJ, Lynham SA, Ruona WEA (2001) A review of scenario planning literature. Fut Res Q 17:7–31
Dator J (2002) Advancing futures: futures studies in higher education. Praeger, Westport, CT
De Nijs T, De Niet R, Crommentuijn L (2004) Constructing land-use maps of the Netherlands in 2030. J Environ Manage 72(1–2):35–42
Dreborg K (2004) Scenarios and structural uncertainty, Stockholm, department of infrastructure, Royal Institute of Technology
Durance P, Godet M (2010) Scenario building: uses and abuses. Technol Forecast Soc Chang 77:1488–1492
Eickhout B, van Meijl H, Tabeau A, van Rheenen T (2007) Economic and ecological consequences of four European land use scenarios. Land Use Policy 24:562–575
ESPON (2013) ESPON Atlas. Territorial Dimensions of the Europe 2020 Strategy. Luxembourg, ESPON Program.
ESPON (2014) ET2050 Territorial Scenarios and Visions for Europe. Final Report Volume 1—Approach to Scenario Building and Storylines. ESPON & MCRIT LTD
ESPON (2014) ET2050 Territorial Scenarios and Visions for Europe Volume 5 Land-use Trends and Scenarios. ESPON
ESPON (2015) Territorial scenarios for Europe, working paper. Annex to the ESPON Policy Brief "Territorial Scenarios for Europe towards 2050". ESPON 2020 Corporate Program
Enzer S (1980) INTERAX—An interactive model for studying future business environments: Part I. Technol Forecast Soc Chang 17(2):141–159
Escobar F, Hewitt R, Hernandez V (2015) Usos del suelo en los parques nacionales españoles. Evolución y modelado participativo, Proyectos de investigación en parques nacionales: 2010–2013. Madrid: Organismo Autónomo Parques Nacionales, Ministerio de Agricultura y Medio Ambiente pp 175–211
Ewert F, Rounsevell MDA, Reginster I, Metzger MJ, Leemans R (2005) Future scenarios of European agricultural land use I. Estimating changes in crop productivity. Agric Ecosyst Environ 107:101–116
Fischer G, Prieler S, van Velthuizen H, Berndes G, Faaij A, Londo M, Wit M (2010) Biofuel production potentials in Europe: sustainable use of cultivated land and pastures, Part II: Land use scenarios. Biomass Energy 34:173–187
Gallardo M, Martínez-Vega J (2016) Three decades of land-use changes in the region of Madrid and how they relate to territorial planning. Eur Plan Stud 24(5):1016–1033
Habermas J (2003) The future of human nature. Polity Press 136 pp
Hewitt R, van Delden H, Escobar F (2014) Participatory land use modelling, pathways to an integrated approach. Environ Model Softw 52:149–165
Hicks D, Holden C (1995) Visions of the future: why we need to teach for tomorrow: stoke-on-trent. Trentham Books, London, UK
Inayatullah S (1990) Deconstructing and reconstructing the future. Futures 22(2):115–141
IPCC (2000) Emissions Scenarios, A Special Report of IPCC Working Group III https://www.ipcc.ch/pdf/special-reports/spm/sres-en.pdf. Accessed 03/03/2017
Jetter A, Schweinfort W (2011) Building scenarios with fuzzy cognitive maps: an exploratory study of solar energy. Futures 43:52–66
Kok K, Farrow A, Veldkamp A, Verburg P (2001) A method and application of multi-scale validation in spatial land use models. Agric Ecosyst Environ 85:223–238
Kok K, Van Delden H (2009) Combining two approaches of integrated scenario development to combat desertification in the Guadalentín watershed, Spain. Environ Plan 36:49–66
Kosow H, Gassner R (2008) Methods of Future and Scenario Analysis, DIE Research Project. Development Policy: Questions for the Future. Bonn, Germany
Linneman R, Klein HE (1983) The use of multiple scenarios by US industrial companies: a comparison study, 1977–1981. Long Range Plan 16(6):94–101

Maier HR, Guillaume JHA, Van Delden H, Riddell GA, Haasnoot M, Kwakkel JH (2016) An uncertain future, deep uncertainty, scenarios, robustness and adaptation: How do they fit together? Environ Model Softw 81:154–164

Malaska P, Malmivirta M, Meristo T, Hanson S (1984) Scenarios in Europe: who uses them and why? Long Range Plann 17(5):45–49 cited in Bradfield et al

Mannermaa M (1991) In search of an evolutionary paradigm for futures research. Futures 23 (4):349–372

Marien M (2002) Futures studies in the 21st century: a reality-based view. Futures 34:261–281

Masini E (1993) Why Futures studies? London, Grey Seal

Masini E (2006) Rethinking futures studies. Futures 38(10):1158–1168

Moss RH, Edmonds JA, Hibbard KA, Manning MR, Rose SK et al (2010) The next generation of scenarios for climate change research and assessment. Nature 463(7282):747–756

Münier B, Birr-Pedersen K, Schou JS (2004) Combined ecological and economic modelling in agricultural land use scenarios. Ecol Model 174:5–18

Nakicenovic N, Swart R (2000) Special report on emissions scenarios. In: Nakicenovic N, Robert Swart (eds) Special report on emissions scenarios. Cambridge University Press, Cambridge, UK, 612 pp. ISBN 0521804930

Petrov L, Lavalle C, Kasanko M (2009) Urban land use scenarios for a tourist region in Europe: applying the MOLAND model to Algarve, Portugal. Landsc Urban Plann 92(1):10–23

Piorr A, Ravetz J, Tosics I (2011) Peri-urbanisation in Europe. Towards European Policies to Sustain Urban-Rural Futures. PLUREL Consortium, Berlin 143 pp

Plata Rocha W, Gomez Delgado M, Bosque Sendra J (2009) Cambios de usos del suelo y expansión urbana en la Comunidad de Madrid (1990-2000). Scripta Nova. Revista Electrónica de Geografía y Ciencias Sociales 13:281–309

Rounsevell MDA et al (2006) A coherent set of future land use change scenarios for Europe. Agric Ecosyst Environ 114:57–68

Saliba G (2009) Windows for the mind: the use of scenario planning for enhancing decision-making and managing uncertainty. J Fut Stud 13:123–128

Schwartz P (2009) Your future in 5 easy steps: wired guide to personal scenario planning., Wired Magazine 17(8)

Sheppard S, Shaw A, Flanders D, Burch S, Wiek A, Carmichael J, Robinson J, Cohen S (2011) Future visioning of local climate change: a framework for community engagement and planning with scenarios and visualization. Futures 43:400–412

Stoorvogel JJ (1995) Integration of computer-based models and tools to evaluate alternative land-use scenarios as part of an agricultural systems analysis. Agric Syst 49(4):353–367

Tobler WR (1979) Cellular Geography. http://www.geog.ucsb.edu/~tobler/publications/pdf_docs/Cellular-Geography.pdf, 6 pp. Accessed 1 Feb 2017

Van Delden H, Stuczynski T, Ciaian P, Paracchini ML, Hurkens J, Lopatka A, Shi Y, Gomez Prieto O, Calvo S, Van Vliet J, Vanhout R (2010) Integrated assessment of agricultural policies with dynamic land use change modelling. Ecol Model 221(18):2153–2166

Van Delden H, Hurkens J (2011) A generic integrated spatial decision support system for urban and regional planning. Keynote presented at MODSIM11 international congress on modelling and simulation, Perth, Australia

Van Delden H, Hagen-Zanker A (2009) New ways of supporting decision making: linking qualitative storylines with quantitative modelling. In: Geertman S, Stillwell JCH (eds) Best practice and new methods in planning support systems. Springer, Dordrecht, The Netherlands

Van Der Heijden K (1996) Scenarios: the art of strategic conversation. Wiley, Chichester, England

Veldkamp A, Fresco O (1997) Exploring land use scenarios, an alternative approach based on actual land use. Agric Syst 55(1):1–17

Volkery A, Ribeiro T, Henrichs T, Hoogeveen Y (2008) Your vision or my model? Lessons from participatory land use scenario development on a European scale. Syst Pract Action Res 21 (6):459–477

Wack P (1985a) Scenarios: uncharted waters ahead. Harvard Bus Rev 63 (September–October):73–89. https://hbr.org/1985/09/scenarios-uncharted-waters-ahead

Wack P (1985b) Scenarios: shooting the rapids. Harvard Bus Rev 63 (November–December):139–150. https://rjohnwilliams.files.wordpress.com/2016/02/wack-scenarios-hbr2-1985.pdf
White R, Engelen G (1993) Cellular automata and fractal urban form: a cellular modelling approach to the evolution of urban land use patterns. Environ Plann A 25:1175–1199
White R, Engelen G (1994) Cellular dynamics and GIS: modelling spatial complexity. Geogr Syst 1:237–253
White R, Engelen G (1997) Cellular automata as the basis of integrated dynamic regional modelling. Environ Plann B 24:235–246
Yuan Y, Zhao T, Wang W, Chen, S, Wu F (2013) Projection of the spatially explicit land use/cover changes in China, 2010–2100. Advances in Meteorology 2013, Article ID 908307, 9 pp

Part II
Methodological Developments and Case Studies: Methodological Developments

Chapter 6
Obtaining and Comparing Factors in Land Change Models Using One or Two Time Points Based Calibration

M.T. Camacho Olmedo

Abstract A land change model can be calibrated with the state at *one time point* or with the difference between *two time points*. For a case study in Spain we obtained the collections of factors for two calibration periods at *one time point* (dates 2000 and 2006) and the collections of factors for two calibration periods between *two time points* (periods 1990–2000 and 2000–2006). We used evidence likelihood to transform the explanatory variables into factors. We then compared these four collections of factors to show: how the choice of reference maps influences the factors, how these factors highlight the change patterns in two different calibration periods and how these factors highlight the change patterns in the calibration of two models. We ended by analyzing the detailed results for the different factors and LUC categories.

Keywords Land use and cover change · Land change models · Calibration · Factors

1 Introduction

Land Change Models are useful tools for environmental and geomatic research into land use and cover change (LUCC) (Turner et al. 1994; Paegelow et Camacho Olmedo 2008; Paegelow et al. 2013; NRC 2014). The simulation maps obtained from LUCC models help us to understand, forecast and anticipate the future evolution of a variety of applied environmental problems.

One of the most important challenges is to verify and clarify the validity of the model and its outputs (Paegelow et al. 2014). Pontius and Malanson (2005) demonstrate that output varies more as a result of the choice of model parameters than as a result of the choice of the model itself. One of these parameters relates to

M.T. Camacho Olmedo (✉)
Departamento de Análisis Geográfico Regional y Geografía Física,
Universidad de Granada, Granada, Spain
e-mail: camacho@ugr.es

M.T. Camacho Olmedo et al. (eds.), *Geomatic Approaches for Modeling Land Change Scenarios*, Lecture Notes in Geoinformation and Cartography,
https://doi.org/10.1007/978-3-319-60801-3_6

how changes over time and space are considered in the model, for the purpose of describing land use and cover (LUC) state patterns, i.e. *one time point* calibration, or LUC transition patterns, i.e. *two time points* calibration (Camacho Olmedo et al. 2013; Kolb et al. 2013).

A model that is calibrated with the state at *one time point* has certain advantages and disadvantages compared to a model that is calibrated with the difference between *two time points*. The first approach does not explicitly consider the distribution of land cover resulting from recent past changes and instead assesses the results of all the changes that have taken place since human activity in the area began (Paegelow and Camacho Olmedo 2005; Villa et al. 2007; Conway and Wellen 2011; Yu et al. 2011). By contrast, the second approach evaluates the change potential for each possible transition, where the future potential of the space is split into specific transitions across a finite number of LUC categories (Eastman et al. 2005; Sangermano et al. 2010; Wang and Mountrakis 2011).

When calibrating the model, the patterns of change (or change behavior) are analyzed by a collection of variables explaining LUC states and/or LUC transitions. From these variables, a collection of factors can be created with a large variety of methods and analyses, as described in previous research into land change modeling (Mas and Flamenco 2011; Pérez-Vega et al. 2012; Camacho Olmedo et al. 2013; Kolb et al. 2013; Soares-Filho et al. 2013; Mas et al. 2014; Osorio et al. 2015; Abuelaish and Camacho Olmedo 2016). Factors can be created without the need for specific data about LUC locations, either states or transitions, using several transformation methods (natural logarithm, fuzzy, etc.).

Alternatively, a collection of factors can be made on the basis of information about LUC locations. We chose this option because land change models describing LUC states or transitions must include LUC locations. This is possible if methods such as evidence likelihood are used to create the factors, using the LUC states as the reference areas in *one time point* calibration, and the LUC transitions in *two time points* calibration.

Several hypotheses can be proposed. First, in *one time point* calibration, the reference areas in one calibration period may be included in the reference areas in the next period (i.e. simulation period), so highlighting areas of land use persistence. However, in *two time points* calibration, the reference areas in the first calibration period are excluded from the reference areas in the next period, because areas affected by a specific transition in the past, e.g. natural to urban, cannot be affected by the same specific transition (although it may be affected by others) in the future in a discrete based model calibration. This could affect the similarity or dissimilarity of extracted factors.

Second, if change patterns are maintained in two consecutive periods, then modeling LUC transitions (*two time points*) could be an appropriate choice; however, if change patterns vary from one period to the next, then modeling the LUC state (*one time point*), i.e. the most recent period, could produce more realistic results. Previous authors found that if the changes during the calibration interval are not stationary with the changes during the validation interval, then an extrapolation from the calibration interval to the validation interval will probably have systematic errors (Pontius and Neeti 2010; Camacho Olmedo et al. 2015).

Third, if the patterns in the LUC state, i.e. destination category, are not the same as the patterns in the LUC transition, i.e. origin category to destination category, then the factors could be quite different.

Finally, if the sizes of the reference areas vary considerably, this can produce a different pattern in the extracted factors. On the contrary, if the transitions between land categories affect only a small portion of the study area, it will be difficult to predict changes accurately, especially when there are errors in the data (Pontius et al. 2008; Pontius and Petrova 2010).

Our goals are therefore to obtain and compare factors in order to show: (1) how the choice of LUC reference maps influences the factors, (2) how these factors represent the change patterns in two different calibration periods, (3) how these factors represent the change patterns in the two models calibrated in different ways, and, finally (4) the specific behavior of the different LUC categories and factors.

We illustrate the concepts using the TerrSet software (Clark Labs 2016). For a case study in Spain, we obtained the collections of factors for two calibration periods at *one time point* (dates 2000 and 2006) and the collections of factors for two calibration periods between *two time points* (periods 1990–2000 and 2000–2006). Evidence likelihood is used to transform the explanatory variables into factors. We then compared these four collections of factors so as to gain a better understanding of what we expected a priori to be different change patterns.

2 Test Area and Data Sets

The two types of calibration are based on land use and cover data for the different time periods and the related explanatory variables. Figure 1 shows the specific study area, which covers 2,300 km^2 in the province of Murcia (southern Spain). The maps of land use and cover (LUC) have four categories from the Corine Land Cover (CoORdination of INformation of the Environment, Instituto Geográfico Nacional, Spain) dataset: urban, industrial and transport uses; natural vegetation, unproductive land and water; irrigated crops; rainfed crops. In the rest of this article we refer to these categories as: urban, natural, irrigated and rainfed. Corine maps at 1990 (t0), 2000 (t1) and 2006 (t2) are used for model calibration. The explanatory variables are topographic variables, protected areas, territorial accessibility (roads diversity and quality), distance to roads and distance to hydrographic network (Gómez and Grindlay 2008).

The study area has undergone profound territorial and economic transformations in the recent past. The most important change has been the transition from rainfed crops to irrigated crops, due to the development of water-related infrastructures and the increase in the water supply (Gómez Espín et al. 2011). Urban growth is a secondary change driven by the development of transportation and communication infrastructures.

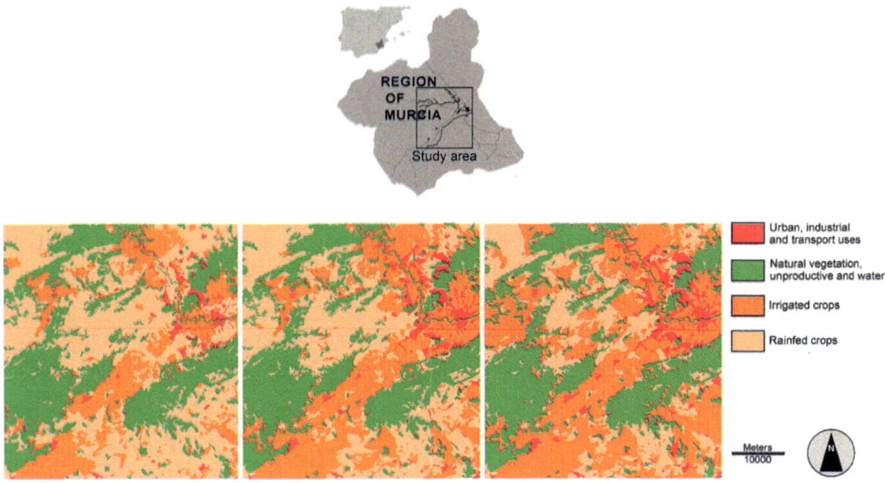

Fig. 1 LUC in 1990 (*left*), 2000 (*middle*) and 2006 (*right*) in the Murcia region in southern Spain. *Source* Corine Land Cover

3 Methodology and Practical Application to the Data Sets

3.1 Components of Models: Obtaining Factors

We used evidence likelihood to transform the explanatory variables into factors. This procedure analyzes the relative frequency of the different categories of a given variable within the areas of LUC states or LUC transitions. It is an efficient means of introducing categorical variables into the analysis, and it accepts continuous variables that have been binned into categories.

The reference areas represented in binary maps are therefore different for model calibration based on *one time point* or *two time points*. For *one time point*, the reference area is the most recent land use category, i.e. the LUC state. For *two time points*, the reference area is a map showing the changes that have taken place between two points in time, i.e. LUC transitions. This option aims to preserve the nature of the state of the categories and the nature of the changing categories. From now on, we refer to areas corresponding to an LUC state or an LUC transition as 'reference maps'.

We obtained four reference maps for each LUC category. In the first calibration period t0–t1, the reference map for *one time point* is a set of binary categorical LUC maps (one for each category) at t1, and for *two time points* is a set of binary categorical LUC maps (one for each transition) between t0–t1. In the second calibration period t1–t2, the reference map for *one time point* is every LUC state at t2 and for *two time points* is every LUC transition between t1–t2 (Table 1). Figure 2 shows the reference maps for irrigated crops as an example.

Table 1 Components of model: Reference maps for evidence likelihood in *one time point* and *two time points* based calibration in both calibration periods

	First calibration period	Second calibration period
One time point	2000 (t1)	2006 (t2)
	LUC state	LUC state
Two time points	1990 (t0)–2000 (t1)	2000 (t1)–2006 (t2)
	LUC transitions	LUC transitions

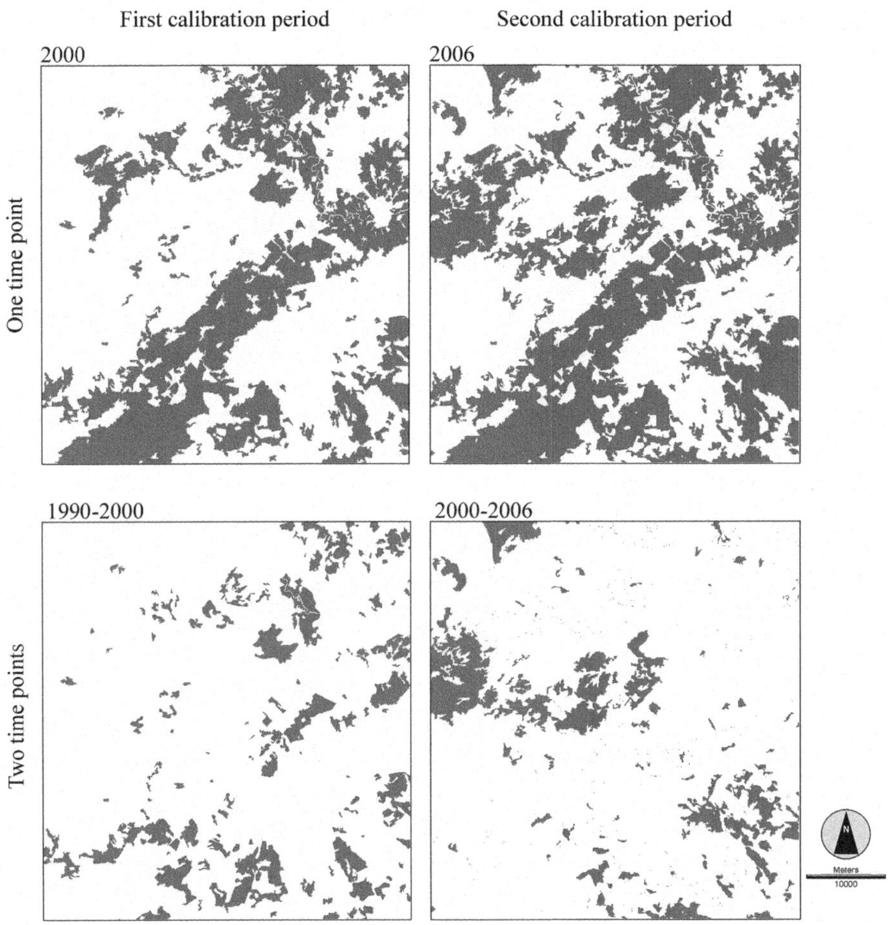

Fig. 2 Example of irrigated crops: Reference maps for evidence likelihood of the LUC state of irrigated crops in 2000 and in 2006 (*above*) and of the LUC transition to irrigated crops over the periods 1990–2000 and 2000–2006 (*below*)

Table 2 Equivalence between each LUC state in columns and the general dynamics to which the LUC transitions refer in rows

	Urban	Natural	Irrigated	Rainfed
Natural to urban Irrigated to urban Rainfed to urban	Urban gain			
Rainfed to natural		Natural gain		
Natural to irrigated Rainfed to irrigated			Irrigation gain	
Natural to rainfed				Rainfed gain

As is standard procedure in transition modeling, not all the transitions were used. We chose those affecting large surface areas and also on the basis of their homogeneity in dynamic behavior to the LUC destination. This procedure excludes very small transitions from the model due to the risk of errors in the data sets and/or strongly heterogeneous dynamics. For this reason, we established the degree of equivalence between each LUC state and the general dynamics to which the LUC transitions refer (Table 2). Various transitions involving the same explanatory variables were grouped together such that in the practical application only the following transitions are modeled: natural/irrigated/rainfed to urban; rainfed to natural; natural/rainfed to irrigated; natural to rainfed. By far the most important change in the area we studied is the transition to irrigated crops, which is followed some way behind by urban growth.

Using these reference maps we obtained four collections of factors for each LUC category: for *one time point* and for *two time points*, and both of these for two calibration periods. Figures 3 and 4 show the original elevation and slope values (variable) from the reference maps for irrigated crops. Evidence likelihood will compute the relative frequency in the elevation and slope categories within the LUC state and LUC transition areas.

3.2 Assessment Methods

A classical method for assessing the congruence of quantitative data, the Pearson correlation, was used for comparing factors. Instead of looking for a causality relationship between pairs of data, the Pearson correlation tries to establish whether there is a relationship between them. Values range from −1 to +1. High positive/negative Pearson values indicate a direct/indirect relationship between two data. Low positive/negative values indicate a lack of relationship.

The Pearson correlation was calculated between all pairs of factors for the *one* and *two time points* based models and for the two calibration periods. Factors are quantitative data from 0.0 to 1.0. The higher the Pearson coefficient, the stronger the correlation of factors. We consider values of over 0.8 to be very strong correlations.

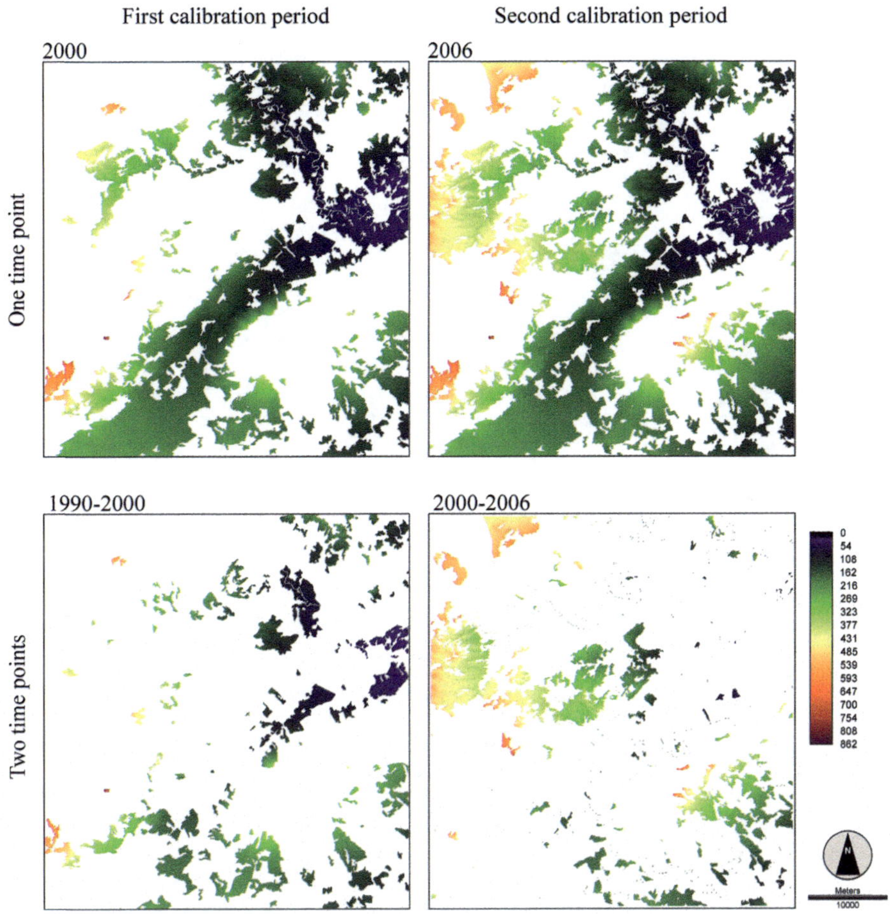

Fig. 3 Elevation variable (meters asl) in LUC state for irrigated crops in 2000 and 2006 (*above*) and in LUC transition to irrigated crops for 1990–2000 and 2000–2006 (*below*)

As commented earlier, in this study we discarded the transitions affecting small surface areas, and grouped together the transitions we selected according to their homogeneity, the normal procedure in transitions modeling. It is important to remember therefore that we are comparing LUC states with almost all, but not all, the LUC transitions.

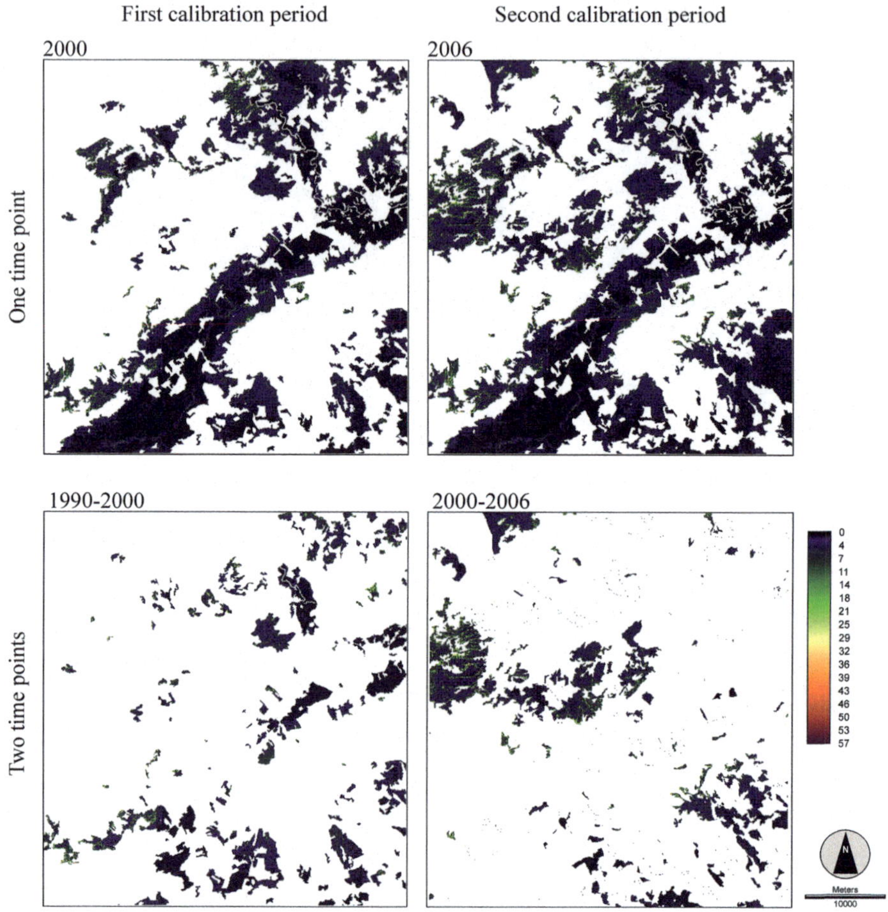

Fig. 4 Slope variable (degrees) in LUC state for irrigated crops in 2000 and 2006 (*above*) and in LUC transition to irrigated crops for 1990–2000 and 2000–2006 (*below*)

4 Results and Discussion

4.1 Collection of Factors

Four collections of factors were obtained for each LUC category: for *one time point* and for *two time points*, and for each of the two calibration periods. Figures 5 and 6 show the collection of factors derived from the elevation variable and from the slope variable in the reference maps for irrigated crops.

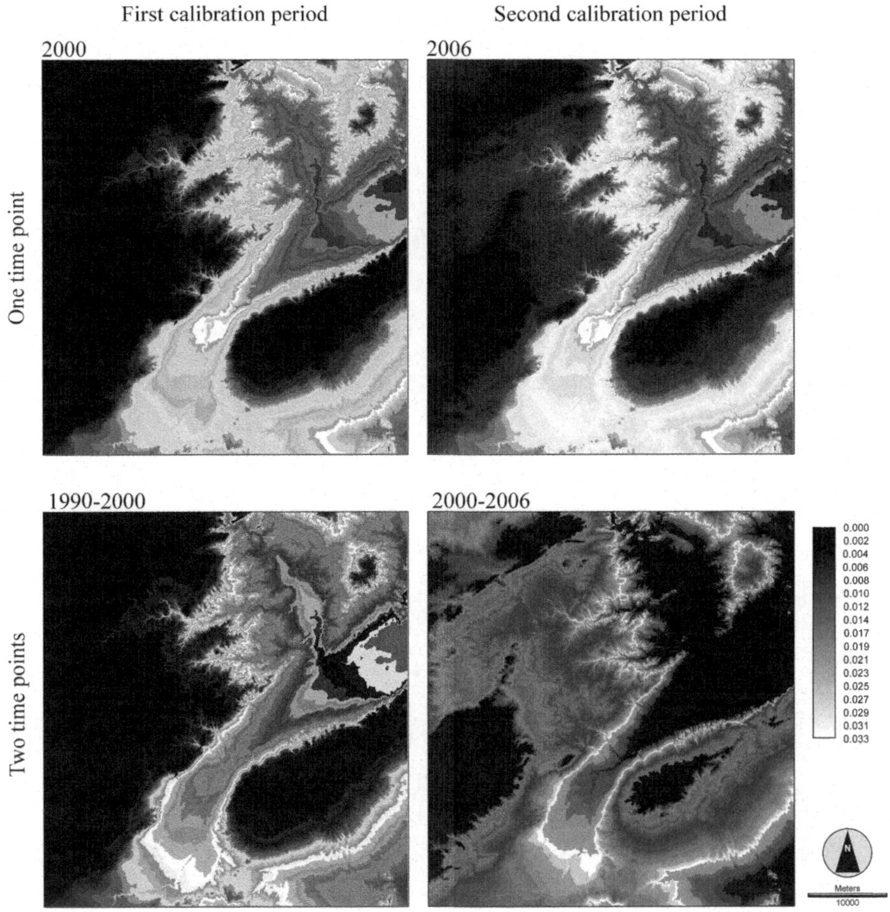

Fig. 5 Irrigated crops and elevation: Evidence likelihood of the LUC state for irrigated crops in 2000 and 2006 derived from the elevation variable (*above*) and of the LUC transition to irrigated crops over the periods 1990–2000 and 2000–2006 derived from the elevation variable (*below*). Values from 0.0 to 1.0

4.2 Comparison of the Collections of Factors Obtained in Two Different Calibration Periods

We calculated the Pearson correlation values for every pair of factors for the different LUC categories between the first calibration period t0–t1 and the second calibration period t1–t2. (Table 3).

Behavior can be analyzed by time point based models and LUC categories, and by factors.

All correlation values for factors obtained in *one time point* exceed 0.96 and the majority of them are over 0.99. The average values are over 0.98 for all LUC

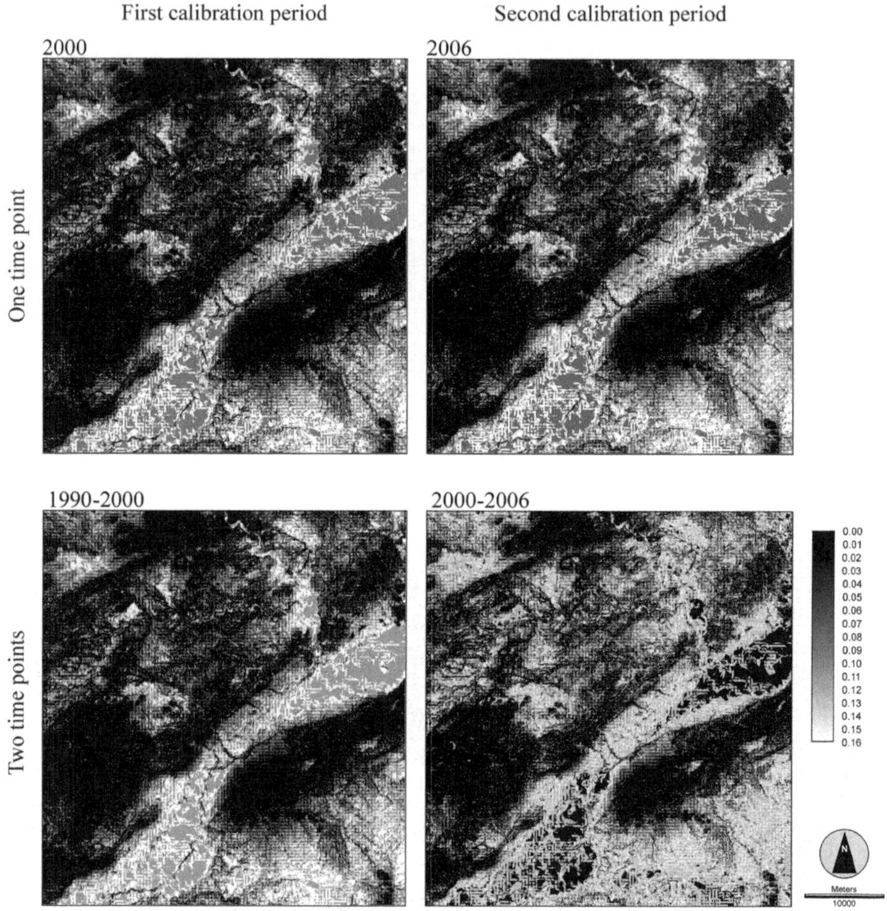

Fig. 6 Irrigated crops and slope: Evidence likelihood of the LUC state for irrigated crops in 2000 and 2006 derived from the slope variable (*above*) and of the LUC transition to irrigated crops over the periods 1990–2000 and 2000–2006 derived from the slope variable (*below*). Values from 0.0 to 1.0

categories. That means that factors obtained in the first calibration period are very similar to factors obtained in the second calibration period, which suggests a high degree of continuity in land use over time.

The correlation values for the factors obtained from *two time points* vary much more from negative (−0.13) and low positive values to high positive values (exceed 0.99). This means that some factors obtained in the first calibration period are different from those obtained in the second calibration period while others are similar. The LUC transitions in the first calibration period are not included in the LUC transitions in the second calibration period. If the transitions in the two calibration periods show similar patterns, factors are similar; if the transitions show

Table 3 Pearson correlation values for every pair of factors for the different LUC categories between the first calibration period t0–t1 and the second calibration period t1–t2

	Urban		Natural		Irrigated		Rainfed		
Factors	One time point	Two time points	One time point	Two time points	One time point	Two time points	One time point	Two time points	Average
Elevation	0.9861	0.7033	0.9947	0.3026	0.9813	0.4205	0.9697	0.3708	**0.7161**
Slope	0.9997	0.9937	0.9998	0.9739	0.9927	0.7972	0.9998	0.9089	**0.9582**
Aspect	0.9858	0.6716	0.9968	0.5528	0.9959	0.9015	0.9931	0.3635	**0.8076**
Accessibility to main road	0.9991	0.9176							**0.9584**
Accessibility to human settlements	0.9983	0.8464							**0.9224**
Distance to secondary road	0.9998	0.9939			0.9981	0.9517	0.9998	0.3820	**0.8876**
Distance to main irrigation channel	0.9970	0.8854			0.9894	0.1688	0.9607	-0.1341	**0.6445**
Distance to secondary irrigation channels					0.9846	0.2219			**0.6033**
Distance to network of rivers and streams					0.9991	0.9823			**0.9907**
Distance to network of ditches					0.9926	0.2394			**0.6160**
Distance to water catchments					0.9973	0.9606			**0.9790**
Average	**0.9951**	**0.8588**	**0.9971**	**0.6097**	**0.9923**	**0.6271**	**0.9846**	**0.3782**	

For the *one time point* based model, the calibration dates are 2000 and 2006; for the *two time points* based model, the calibration periods are 1990–2000 and 2000–2006. Averages for the different factors are in rows; averages for the different LUC/time point based models are in columns. Values below 0.6 are underlined

different patterns, factors are unrelated. If we look at the different LUC categories, urban scores the highest values, with an average of 0.8588, indicating that for this category the pattern in the first calibration period is very similar to that in the second calibration period. This similarity does not occur in the remaining LUC categories. The lowest average correlation values belong to rainfed crops with 0.3782, which means that the LUC transitions in the first calibration period show different patterns to those in the second calibration period (underlined values). Natural vegetation and irrigated crops have average correlation values of 0.6097 and 0.6271 respectively, showing a moderate degree of similarity between the two calibration periods. These average values mask a highly varied situation in which some factors show very high values and others very low ones (underlined values). This issue will be discussed in greater detail below in 4.4.

If we look at the different factors, even if the average values are all over 0.6, the specific behavior varies a great deal. Focusing only on common factors (used by at least two categories), the slope factor shows very high Pearson values for all categories and time based models (over 0.79); distance to a secondary road also scores high values except in rainfed crops in the *two time points* based model. Other factors, such as elevation and aspect, have widely varying scores by categories and specifically in the *two time points* based model.

4.3 Comparison of the Collections of Factors Obtained in One Time Point and Two Time Points Calibration

The Pearson correlation was calculated for every pair of factors for the different LUC categories between the *one time point* and *two time points* calibration methods. Table 4 shows the results for the first calibration period t0–t1 (Camacho Olmedo et al. 2013) and for the second calibration period t1–t2. For the *one time point* based model, the calibration dates are 2000 and 2006; for the *two time points* based model, the calibration periods are 1990–2000 and 2000–2006.

Table 4 Pearson correlation values for every pair of factors for the different LUC categories between the *one time point* and *two time points* calibration methods

	Urban		Natural		Irrigated		Rainfed		
Factors	First period	Second period	First period	Second period	First period	Second period	First period	Second period	Average
Elevation	0.9047	0.7907	0.4127	0.9472	0.9263	0.5613	0.1596	0.3764	**0.6349**
Slope	0.9988	0.9961	0.9746	0.9410	0.9959	0.8646	0.8637	0.9603	**0.9494**
Aspect	0.9131	0.8657	0.8028	0.2707	0.9235	0.9650	0.7378	0.6333	**0.7640**
Accessibility to main road	0.9620	0.9810							**0.9715**
Accessibility to human settlements	0.9703	0.9536							**0.9620**
Distance to secondary road	0.9851	0.9964			0.9992	0.9617	0.4663	0.9512	**0.8933**
Distance to main irrigation channel	0.9830	0.9392			0.9562	0.1431	0.2149	0.7211	**0.6596**
Distance to secondary irrigation channels					0.8174	0.7170			**0.7672**
Distance to network of rivers and streams					0.9850	0.9954			**0.9902**
Distance to network of ditches					0.9821	0.3936			**0.6879**
Distance to water catchments					0.9966	0.9705			**0.9836**
Average	**0.9596**	**0.9318**	**0.7300**	**0.7196**	**0.9536**	**0.7302**	**0.4885**	**0.7285**	

First calibration period is t0–t1 and second calibration period is t1–t2. For the *one time point* based model, the calibration dates are 2000 and 2006; for the *two time points* based model, the calibration periods are 1990–2000 and 2000–2006. Averages per factors are in rows; averages per LUC/time point based model are in columns. Values below 0.6 are underlined

These results can be analyzed by time point based models in two calibration periods and LUC categories, and by factors.

The average Pearson values for every pair of factors for the different LUC categories between the *one time point* and *two time points* calibration methods are higher in the first calibration period (1990–2000) than in the second calibration period (2000–2006) in all LUC categories except rainfed crops. This means that the transitions in the first calibration period are more similar to the states than in the second period. Urban category scores the highest values (over 0.93) in both calibration periods, which means that transition patterns are constant and close to the state patterns for this category. The natural category has average values of around 0.7, but with a greatly varying situation in which some factors show very high values and others very low ones (underlined values). With respect to irrigated crops, the average Pearson values show that transitions to this category in the first calibration period had the same pattern as the LUC state (over 0.95). In the second calibration period, however, the transitions showed different patterns to the state (underlined values). The lowest average Pearson value (under 0.5) in the first calibration period is for rainfed crops, i.e. transitions to this category occurred in different patterns from the state (underlined values). In the second calibration period, the Pearson value is medium-high.

If we look at the different factors, even if the average values are all over 0.6, the specific behavior varies widely, as commented also for Table 3. If we focus only on common factors (used by at least two categories), it is the slope factor that shows the highest scores and the most homogeneous behavior (average of 0.9494 and all Pearson values over 0.86). Distance to a secondary road obtained high values except in the first calibration period in rainfed crops. Aspect scored medium or high values except in the second period in natural category. Elevation and distance to irrigation channel show widely varying behavior in the different categories and in both calibration periods.

4.4 Comparison of Four Collections of Factors

Figures 7 and 8 show the Pearson correlation values for every pair of factors (each square corresponds to one comparison) also presented in Tables 3 and 4. Each cross tabulation matrix is composed of one column per variable grouped by LUC category (Fig. 7) or per LUC category grouped by variable (Fig. 8) and by four rows: *One time point* based model (first and second calibration period), *Two time points* based model (first and second calibration period), First calibration period (*one* and *two time points* based model), Second calibration period (*one* and *two time points* based model). In Fig. 8 only variables common to at least two LUC categories are shown.

Figure 7 confirmed the comments from Sects. 4.2 and 4.3, grouped by LUC categories. The collections of factors for the urban category are all very similar. This means that transitions patterns to this category are very close to the state

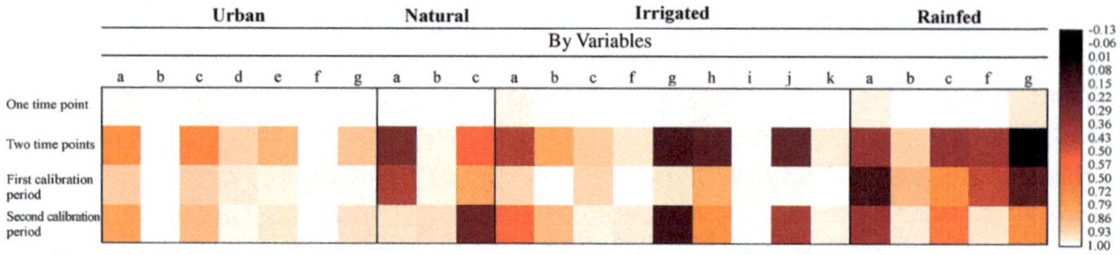

Fig. 7 Representation of Pearson correlation values for each pair of factors (each *square* corresponds to one comparison). Each cross tabulation matrix is composed of one column per variables grouped by LUC, and of four rows: One time point based model (first and second calibration period), Two time points based model (first and second calibration period), First calibration period (one and two time points based model), Second calibration period (one and two time points based model). Variables legend: elevation (**a**), slope (**b**), aspect (**c**), accessibility to main road (**d**), accessibility to human settlements (**e**), distance to secondary road (**f**), distance to main irrigation channel (**g**), distance to secondary irrigation channels (**h**), distance to network of rivers and streams (**i**), distance to network of ditches (**j**), distance to water catchments (**k**)

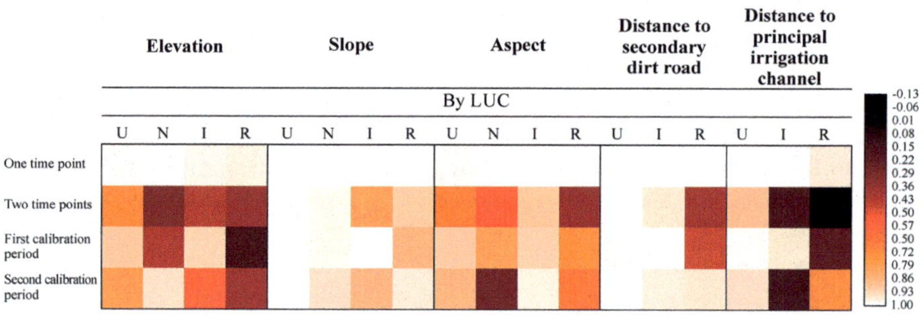

Fig. 8 Representation of Pearson correlation values for each pair of factors (each *square* corresponds to one comparison). Each cross tabulation matrix is composed by one column per LUC grouped by variables, and by four rows: One time point based model (first and second calibration period), Two time points based model (first and second calibration period), First calibration period (one and two time points based model), Second calibration period (one and two time points based model). LUC legend: urban (U), natural (N), irrigated (I), rainfed (R)

pattern for this category in both calibration periods. The only exceptions are the elevation and aspect factors. As an example, if we focus on the Pearson correlation values for elevation factors related to the urban category, we can see that for 2000 and 2006 the situations are almost identical (first row); the transitions between 1990–2000 and 2000–2006 are not so close (second row); the state in 2000 is very similar to the transitions over the period 1990–2000 (third row); and the state in 2006 is less similar to the transitions that took place over the period 2000–2006 (fourth row).

The factors for the natural category and the factors for irrigated crops vary more sharply: transition patterns in the first calibration period are not similar to those in the second. Transitions are not very close to the state pattern in either period. With respect to irrigated crops, in the second calibration period the transitions patterns are quite different from the state pattern. This is due to elevation, distance to a main irrigation channel and distance to a network of ditches. Finally, for the collection of factors for rainfed crops, a high dissimilarity is present in transition patterns for both calibration periods and with respect to the state pattern, particularly in the first calibration period. However, it is also important to emphasize that the state patterns are stable for all categories (first row in Fig. 7, *one time point* based model).

In brief, if we compare the two calibration methods, there is a medium to high linear relationship between LUC transitions and LUC states, which is higher in the first calibration period in all the categories except for one. Looking at each category, the urban patterns are very stable while at the opposite extreme, the patterns for rainfed crops show high variation. The situation also varies a great deal in the natural category and in irrigated crops: the transition patterns are not very stable and are not very similar to the state pattern.

In Fig. 8, the Pearson values are grouped by variables. A quick overview confirms that the state patterns are stable for all categories (first row, *one time point* based model). Aspect is the variable with the highest values in both calibration periods and both models, followed by distance to secondary road, except in the rainfed crops category. Elevation and aspect seem to be the most sensitive variables. They show widely varying behavior, with high, medium and low Pearson values, which means that transition patterns and state patterns are not regular with respect to these variables. As regards distance to main irrigation channel, the transition patterns for irrigated crops are not regular, although the most irregular are those for rainfed crops. In brief, when looking at the different factors, the homogeneity or heterogeneity of LUC locations can lead to widely varying behavior. Previous researchers observed a relationship between environmental and accessibility factors and the initial conditions in which LUC changes are carried out (Lambin et al. 2001; Yu et al. 2011; Osorio et al. 2015).

For a better understanding of these patterns, we focused on the collection of factors for irrigated crops. Figure 9 and 10 present the histograms (ha) for the LUC state for irrigated crops in 2000 and 2006 and for the LUC transition to irrigated crops over the periods 1990–2000 and 2000–2006, by elevation intervals and by slope intervals.

If we compare these two variables, we can conclude that irrigated crops behave in a more homogenous manner with respect to slope (only some slope intervals are affected) than to elevation, which explains the different Pearson values commented above. Figure 9 shows that irrigated crops were located at lower elevations in the first calibration period, 1990–2000, and that the new irrigated fields planted from 2000 to 2006, went up to higher elevations, in other words, transitions occurred at different altitudes. However, we do not know if this is a general dynamic or if it is due to the particular behavior of one of the LUC origin categories, in other words, natural or rainfed. We must remember that, in this study, we grouped some transitions (natural/irrigated/rainfed to urban; natural/rainfed to irrigated) together. Although this is a common procedure in modeling, it does not allow us to distinguish between the categories that have been grouped together.

Paegelow M, Camacho Olmedo MT, Mas JF, Houet T, Pontius RG Jr (2013) Land Change Modelling: moving beyond projections. Int J Geogr Inf Sci 27(9):1691–1695

Paegelow M, Camacho Olmedo MT, Mas JF, Houet T (2014) Benchmarking of LUCC modelling tools by various validation techniques and error analysis. Cybergeo, document 701, mis en ligne le 22 décembre 2014

Pérez-Vega A, Mas JF, Ligmann-Zielinska A (2012) Comparing two approaches to land use/cover change modeling and their implications for the assessment of biodiversity loss in a deciduous tropical forest. Environ Model Softw 29(1):11–23

Pontius RG Jr, Malanson J (2005) Comparison of the structure and accuracy of two land change models. Int J Geogr Inf Sci 19:243–265

Pontius RG Jr, Neeti N (2010) Uncertainty in the difference between maps of future land change scenarios. Sustain Sci 5:39–50

Pontius RG Jr, Petrova SH (2010) Assessing a predictive model on land change using uncertain data. Environ Model Softw 25:299–309

Pontius RG Jr et al (2008) Comparing the input, output, and validation maps for several models of land change. Ann Region Sci 42(1):11–27

Sangermano F, Eastman JR, Zhu H (2010) Similarity weighted instance based learning for the generation of transition potentials in land change modeling. Trans GIS 14(5):569–580

Soares-Filho B, Rodrigues H, Follador M (2013) A hybrid analytical-heuristic method for calibrating land-use change models. Environ Model Softw 43(2013):80–87

Turner ABL, Meyer WB, Skole DL (1994) Global land-use/land-cover change: towards an integrated study. Ambio 23(1):91–95

Villa N et al (2007) Various approaches for predicting land cover in Mediterranean mountains. Commun Stat 36(1):73–86

Wang J, Mountrakis G (2011) Developing a multi-network urbanization model: a case study of urban growth in Denver, Colorado. Int J Geogr Inf Sci 25(2):229–253

Yu J et al (2011) Cellular automata-based spatial multi-criteria land suitability simulation for irrigated agriculture. Int J Geogr Inf Sci 25(1):131–148

Chapter 7
Impact and Integration of Multiple Training Dates for Markov Based Land Change Modeling

M. Paegelow

Abstract Most geomatic land use/cover (LUC) simulation tools use two LUC maps as training dates, particularly prediction models based on Markov chains. In this paper we begin by listing the potential errors resulting from only considering two past dates. We then illustrate the consequences of this approach on quantitative model calibration using a dataset encompassing six LUC maps. This offers multiple Markovian combinations for input matrices generating a wide range of Markovian probability transitions. An even larger spectrum can be achieved by introducing limited confidence in data. The comparison of LUCC budgets and possible Markov chains offers a broad spectrum of results and randomness in the choice of only two dates. We propose two techniques for integrating the knowledge obtained from more than two training dates into forecasting scenarios. First we calculate an annual rate of change, which is weighted according to time distance from the present in order to fix expected total change in the simulation step and at the category level. We then produce alternatives to Markov chains at a transitional level. In this way we integrate all available LUCC-budgets and propose different methods for weighting observed transitions, so as to produce transition matrices that could act as alternatives to Markov chains based on just two dates.

Keywords Land change modeling · Training dates · Validation

1 Introduction

Land Use/Cover Change (LUCC) modeling simulates land use change in terms of quantity and category (Camacho et al. 2015). The quantitative aspect of simulation depends on the modeling objective. If the purpose is to design plausible scenarios, modelers simulate different hypotheses (e.g. a business-as-usual scenario, a massive

M. Paegelow (✉)
Département de Géographie, Aménagement, Environnement,
Université Toulouse Jean Jaurès, Toulouse, France
e-mail: paegelow@univ-tlse2.fr

M.T. Camacho Olmedo et al. (eds.), *Geomatic Approaches for Modeling Land Change Scenarios*, Lecture Notes in Geoinformation and Cartography,
https://doi.org/10.1007/978-3-319-60801-3_7

deforestation scenario, a sustainable development scenario) and then introduce various quantities of LUC in order to answer the question '*What will be the spatial impact if…?*' If the objective is prediction or forecasting, we can calculate the expected quantities of overall LUC or transition between categories. Quantity prediction is mostly based on probabilistic approaches such as Markov chains. Various geomatic LUCC modeling software programs such as CA-MARKOV, Land Change Modeler (both implemented in Terrset) and Dinamica EGO (Mas et al. 2014) can be used to calculate Markovian conditional transitions. They perform this extrapolation in time using only two training dates. This is a risky procedure because model training depends on only two time points in the past. These two dates have to match key points in the time series. What happens if at least one of the two training dates is not representative of the considered training period, if they represent atypical, unusual points in time (e.g. before/after fire, timber extraction)? Data error due to classification or photo interpretation is more serious when working with just two maps than with a long time series. Numerous studies have underlined the influence of temporal data resolution (Allen and Starr 1982; Kim 2013) and analyzed the impact of time intervals on the amount of change (Burnicki et al. 2007; Lee et al. 2009; Liu and Deng 2010).

After the presentation of test areas and data sets including time series for six LUC maps, we first illustrate randomness using just two training maps before considering alternative techniques to introduce multi-temporal knowledge into predictive models. This is done at global, category and transition level. In this way we present alternatives to Markov-chain-predicted transitions. Both proposed alternatives (coupling different training dates and confidence levels within a Markov chain process) involve more than two training dates, and may inform the modeler when looking for the best choices to anticipate a future LUC situation.

2 Test Areas and Dataset

Garrotxes is an 8750 ha catchment area in the western part of the Pyrénées Orientales, a *département* in southern France (Fig. 1). "The lowest area in this region is located in the SE and varies from 650 m at the confluence of the Têt River to 1000 m. A Mediterranean climate dominates this low area. By contrast, the upper region reaches 2400 m and is influenced by a mountain climate. The western area of this region is characterized by a ponderous geomorphological relief on granite. This area is composed of early terrace cultivation and coniferous forests (*Pinus uncinata* and *P. sylvestris*). The east bank forms a large, steep, south-facing area that overlies schist and is used as a pasture. The demographic maximum, which occurred during the 1820s, corresponds with intensive use of all natural resources in this area. According to the Napoleonic cadastre, a quarter of the Garrotxes catchment was terraced for crops in 1826. Today, the crops have entirely disappeared. The population fell from 1832 inhabitants in 1826 to 94 inhabitants in 2008. Crop terraces were transformed into pastures prior to becoming shrub or forest areas.

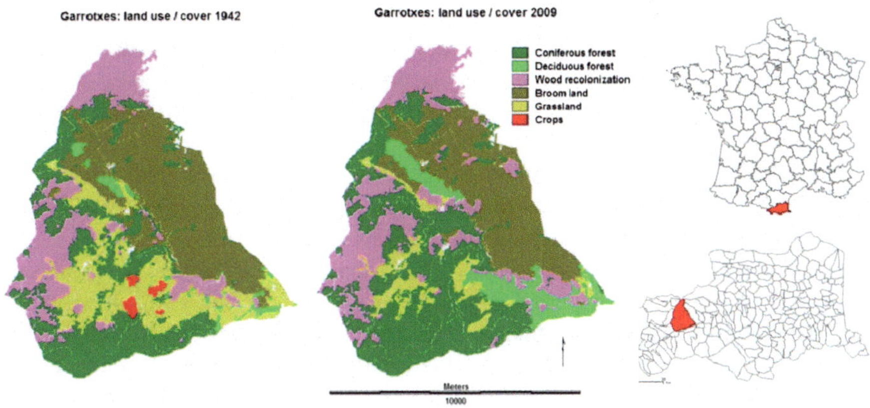

Fig. 1 The Garrotxes study area is in the *Département* of Pyrénées Orientales (map of France, *top right*) and is composed of five municipalities (map of municipalities *bottom right*). LUC in 1942 (*left*) and 2009 (*middle*)

Currently, the crops grown in this area are marginal. In addition, the near future likely depends on the intensity of pastoral activity and management, which will determine how far the forest spreads" (Paegelow et al. 2014).

The data set we used is a collection of six LUC maps for different years (1942, 1962, 1980, 1989, 2000 and 2009). LUC maps are produced by image segmentation and supervised classification of orthophotographs and visual post validation.

3 Methodology and Practical Application to the Data Sets

3.1 Use of/Impact of Multiple Calibration LUC

We used the LUCC budgets technique (Pontius et al. 2004b, 2008) to characterize land change and its components such as gain, loss, total change, net change and swap for the five training periods in the LUC maps time series. For comparison purposes, we converted coarse time-interval dependent indicators into mean annual rates.

Most quantity prediction in business-as-usual (BAU) simulation scenarios is performed using Markov chains (MC) based on only two training LUC maps (first order Markov chain), where t1 and t2 are training dates and T the simulation date. We noted that n-order Markov chains are frequently employed in a spatially non-explicit context. Generally, these n-order MCs are based on a rather eventful multi-temporal database (Hu et al. 2003). Nevertheless, n-order MCs are more complex to handle and are therefore not included in popular GIS software. In this study, we used a series of six LUC maps to test: (i) various combinations of two training dates to calculate MC transition matrices and (ii) the confidence level in these training data. We began by forming all possible pairs of training dates except the last one (2009). For each of these pairs we computed MC expected transitions for T (2009, model unknown). Various comparisons between observed and MC predicted change and persistence were analyzed. This was done with limited confidence in the data by using a confidence level of 90% (proportional error of 0.1), whilst bearing in mind that a confidence level of 100% (proportional error 0.0) is the default option in many software programs and sometimes the only available option for computing Markov chains (as with LCM).

In order to integrate more than two LUC maps into techniques for predicting expected LUC quantities or transition quantities, we considered two approaches. First we computed the annual change rates per period on two levels: total rate of change and category level. Second, on the transitional level, we integrated all the available LUC maps into transition matrices and weighted them to compute expected transition rates as alternatives to MC-predicted transitions based on only two training dates.

3.2 *Integration of Multiple Training Dates to Compute the Expected Annual Amount of Change*

First we calculate the mean amount of change (%) per period by dividing total change for the period by its number of years.

This enabled us to compute the overall and LUC specific annual change rates (%).

3.3 *Computing Transition Rates as an Alternative to Markov Chain Transitional Predictions*

Markov chains are the most common way to model the expected amount of LUC change. As mentioned earlier, the most popular software programs only integrate two training dates as a basis for simulating expected conditional transitions. Using a dataset encompassing six LUC maps, we propose a series of alternative techniques to simulate future LUC by computing transition matrices between 2009 (last known date) and 2020 (simulated LUC) using all known LUC maps. This means that our approach includes five training periods (six training dates). The only difference between these techniques is in the way they weight the impact of individual training periods. The starting point was observed for annual transition rates by period.

Weighting of multiple transition periods:

- Average: the sum of the transition rates divided by the number of periods considered. This was done for each cell in the transition matrix.
- Average weighted according to its closeness to the present: the impact of a period increases proportionally with its closeness to the present. For a series of n

time intervals, the weight of the oldest time interval = 1, the weight of the most recent time interval = n. Annual rates are multiplied by weights and summed together. The resulting sum is divided by the sum of weights. We are aware that this weighting scale could be enhanced either by considering equal time intervals or by varying individual weights by the corresponding length of interval.

where the sum of weights:

$$\sum_{n=1}^{n} = (1+n) * n/2$$

- Linear trend: here we use the best linear fit (linear regression)
- Exponential trend: weights are obtained by the geometric exponential trend.

Each weighting technique is applied to each transition, except persistence (diagonal cells). To compute cross-tabulation for expected changes, we:

- Fix a simulation date: 2020. 2020 means the situation in 2009 (last known LUC) with 11 times the expected annual rate of change. An exception will be made for the LUC category *crops* (5th row and column in the table), which disappeared completely between 1980 and 1989. We set the 5th column and the bottom row to zero by reporting proportionally missing pixels on the rest of the table.
- At this stage, persistence (diagonal cells) is not included in the transition matrix. We fill each diagonal cell by the number of cells in the relevant category in 2009 (starting date) *minus* the sum of transitions from this category to other categories.

To evaluate these alternatives to the MC transition matrix, we decided to test and compare them. The problem arises when, as in our case, you have a database with six dates, and you have to select just two of them for use as training dates. We tested two options: i) the most recent dates (i.e. 2000–2009 to simulate expected changes for 2020) and ii) a recent period with a change rate that is close to the average for all periods (i.e. 1989–2009).

4 Results

4.1 Use of/Impact of Multiple Calibration LUC T

4.1.1 LUCC Budgets

LUCC budgets were calculated for each period. In order to make comparable LUCC indicators for periods of different lengths, we divided the total change by the number of years for the period in question. As shown in Fig. 2, the mean annual

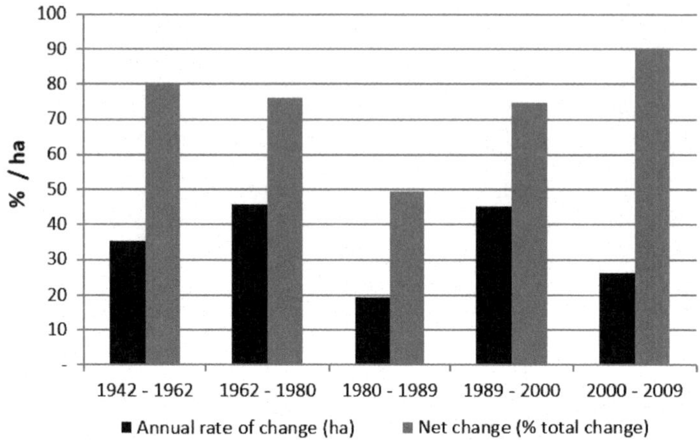

Fig. 2 LUCC budget indicators—mean annual rate of change (km^2) and percentage of net change—for each of the five periods in the data set

rate of total change varies from less than 20 ha/year during the 1980s to more than 40 ha/year during the periods 1962–1980 and 1989–2000. The amount of land change is not linear. The proportion of net change varies from less than 50% (1980s) up to 90% during the last period. In most cases, the proportion of net change was about 75% of total change.

4.1.2 Markov Chains and Variation in the Confidence Level

Table 1 shows all possible MC (Markov chain) combinations of two training dates, using the penultimate date (2000) as the last training date for extrapolation to the model unknown date (2009), so enabling comparison. For instance, if we take 1942 as the first training date, we have four possibilities (1962, 1980, 1989 or 2000) as the second training date for the Markov chain prediction, while if we use 1989 as the starting point the only possible second training date is 2000. The numbering of the ten possibilities for the dataset is purely arbitrary and is solely for indicating the total number of possibilities. There are 10 possible combinations. For each of the 10 Markov chains, Table 2 summarizes the corresponding lengths of the training period (TP) and simulation period (SP); both are summarized as a ratio SP/TP.

All the Markov chains were performed twice: first with a 100% confidence level and then with a 10% proportional error. We compared MC-predicted LUC with observed LUC for 2009 and for each category we summed up the absolute difference between predicted and observed LUC (both expressed as a percentage of area). Figure 3 shows the sums of these absolute differences. If we assume a 100% confidence level in the LUC maps, the quantitative prediction error may vary greatly from around 5% (choosing 1980 and 2000) to almost four times higher (1942 and 1962) depending on the pair of dates selected.

Table 1 Possible MC (Markov chain) combinations when excluding the last known date 2009, which is used as the (model unknown) simulation date (T)

	1942	1962	1980	1989	2000
1942		1	5	8	10
1962			2	6	9
1980				3	7
1989					4
2000					

Table 2 Duration of training periods, simulation periods and ratio of training/simulation period resulting from different MC (Markov chain) combinations for T (simulation date) 2009

		1st	2nd	TP	SP	ST/TP
t − t + 1	1	1942	1962	20	47	2.35
	2	1962	1980	18	29	1.61
	3	1980	1989	9	20	2.22
	4	1989	2000	11	9	0.82
t − t + 2	5	1942	1980	38	29	0.76
	6	1962	1989	27	20	0.74
	7	1980	2000	20	9	0.45
t − t + 3	8	1942	1989	47	20	0.43
	9	1962	2000	38	9	0.24
t − t + 4	10	1942	2000	58	9	0.16

1st—first training date; 2nd—second training date

TP—training period in years; SP—simulation period in years

TP/SP—ratio of duration simulation period/training period

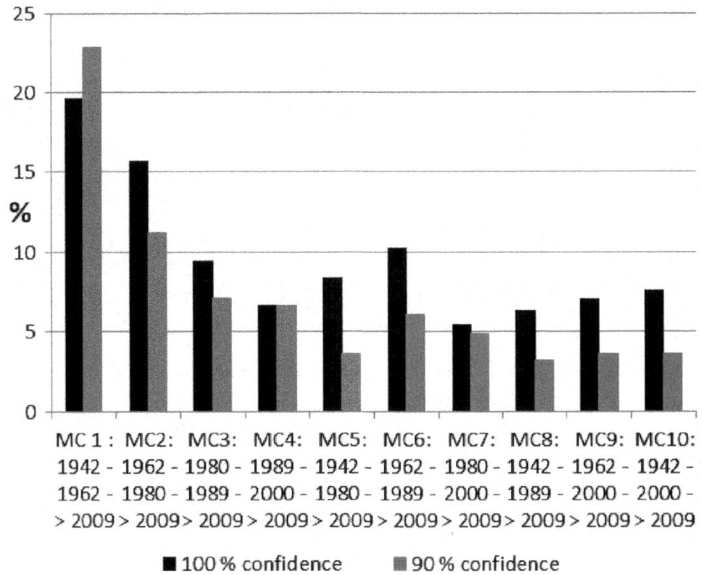

Fig. 3 Absolute differences between observed and MC predicted (proportional error 0.0 and 0.1) LUC for 2009

Centering on observed and MC predicted persistence assuming full confidence (100%) in data, Table 3 (left) shows that half of the Markov chains predict too much persistence (negative values) while the other half predict too much change (positive values). When assuming full confidence, the MC-predicted LUC results that were closest to the real levels we observed were obtained when the last known date (2000) was used as the second training date (right column, left table). MCs based on an earlier second training date generally underestimate expected change. Assuming a confidence level of 90% (Table 3, right-hand table), all MCs predict more change than observed. In contrast to 100% confidence, MCs based on the most recent training dates (last column 2000–2009) produce more change than earlier training dates.

4.2 Integration of Multiple Training Dates to Compute the Expected Annual Amount of Change

Figures 4 and 5 show average annual rates of change for each period. These rates vary, showing low rates of change over two of the last three periods, one period that was close to the average rate of change for all the periods (0.42%), and two periods with high rates of change.

If we split change rates by LUC categories, the graph obtained (Fig. 5) presents a more contrasting situation. First, the most dynamic periods (1962–1980; 1989–2000) for all the LUCC categories mixed together (Fig. 4) were only the most dynamic for coniferous forest. The other LUC categories show different trends: broom land becomes more dynamic with time while grassland becomes less so. Deciduous forest had two more dynamic periods: 1942–1962 and 1989–2000, while the crops graph is difficult to interpret because they disappeared as a LUC category between 1980 and 1989. Second, Fig. 5 shows rates of change for the different LUC categories. If we look at the average change rates, wood recolonization is the most dynamic category, while coniferous and deciduous forests are more stable.

4.3 Computing Transition Rates as an Alternative to Markov Chain Transitional Predictions

Table 4 shows the four alternative (average, time-distance weighted average, linear and exponential trend) simulated transition rates for 2020 and, below, Markov chain matrices for expected changes to 2020 based on the 1989–2009 training period (left) and the 2000–2009 training period (right).

Global persistence is uniformly high (varying from 95.06 to 96.71%).

In order to evaluate the alternative techniques for 2020 simulated transition rates, we compare them to MC-simulated transition matrices as shown in Fig. 6.

Table 3 Comparison of the differences between observed and MC predicted land use persistence depending on MC confidence level: 100% (left) and 90% (right). Negative values mean that the MCs predict less change than observed. Positive values mean that they predict too much change

	Confidence level: 100%				Confidence level: 90%			
	1962 → 2009	1980 → 2009	1989 → 2009	2000 → 2009	1962 → 2009	1980 → 2009	1989 → 2009	2000 → 2009
MC1: 1942–1962 → 2009	−6.67				2.14			
MC2: 1962–1980 → 2009		2.97				11.64		
MC3: 1980–1989 → 2009			−5.02				4.58	
MC4: 1989–2000 → 2009				1.66				11.20
MC5: 1942–1980 → 2009		−2.13				7.05		
MC6: 1962–1980 → 2009		−3.03				6.24		
MC7: 1980–2000 → 2009				0.16				9.85
MC8: 1942–1989 → 2009			−3.38				6.06	
MC9: 1962–2000 → 2009				1.08				10.68
MC10: 1942–2000 → 2009				0.90				10.52

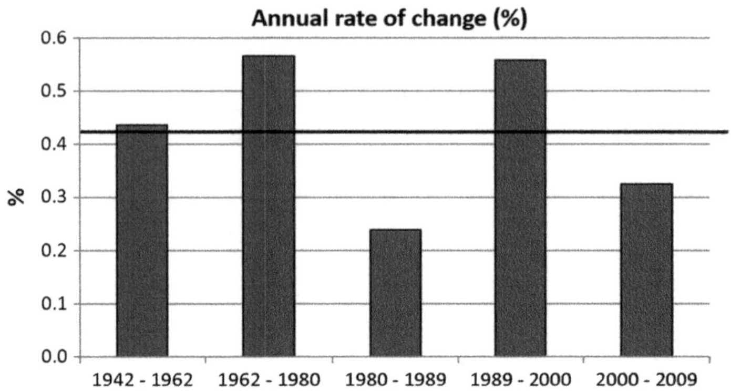

Fig. 4 Annual rates of change (%) by period. The *black line* is the average rate of change for all the periods

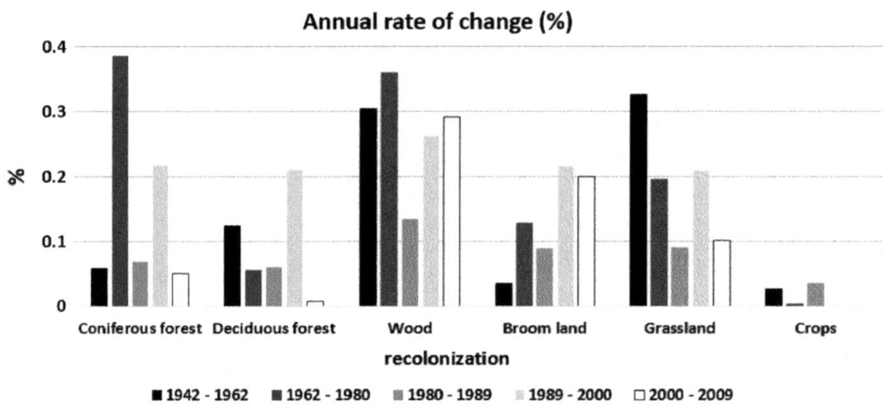

Fig. 5 Annual rates of change (%) for each LUC by period

The graphics in Fig. 7 show the difference between the MC-predicted transition rates (%) and those predicted in other ways. A positive number means that the Markov chain simulated a larger area of land use change than the alternative method. A negative number means that the Markov chain predicted a smaller area of change. Figure 7 shows that at the individual transition level:

- Differences do not exceed more than 2% of total area.
- Differences are greatest when using the 2000–2009 training period (right column).
- Of the four average-based integration techniques the greatest differences were obtained when using the trend formula (especially the linear trend).
- The LUC categories are affected in different ways: wood recolonization (third row) in particular is a 'gaining category'. This means that the Markov chain predicts a higher amount of change than alternative calculation methods. On the

Table 4 Simulated transition rates for 2020 computed by four alternative techniques (average, time-distance weighted average, linear and exponential trend) and, below, Markov chain matrices of expected changes to 2020 based on the 1989–2009 training period (L) and the 2000–2009 training period (R)

		From average									**From linear trend**						
		Conif. forest	Decid. forest	Wood recoloniz	Broom land	Grassland	Crops	∑ **2020**			Conif. forest	Decid. forest	Wood recoloniz	Broom land	Grassland	Crops	∑ **2020**
To 2020	Conif. forest	37.51	0	0.83	0.26	0.51	0	39.12	To 2020	Conif. forest	37.59	0	1.1	0.18	0.84	0	39.7
	Decid. forest	0	8.09	0.29	0.19	0.52	0	9.08		Decid. forest	0	8.09	0.51	0.07	0.59	0	9.26
	Wood recoloniz	0.07	0	20.34	0.89	0.8	0	22.11		Wood recoloniz	0	0	19.85	0.14	1.32	0	31.31
	Broom land	0.02	0	0.01	24.33	0.07	0	24.43		Broom land	0.01	0	0.01	25.32	0.14	0	25.48
	Grassland	0.02	0	0.01	0.03	5.2	0	5.26		Grassland	0.01	0	0.01	0.01	4.21	0	4.25
	Crops	0	0	0	0	0	0	0		Crops	0	0	0	0	0	0	0
∑ **2009**		37.61	8.1	21.49	25.71	7.1	0	100	∑ **2009**		37.61	8.1	21.49	25.71	7.1	0	100
		Persistance (%):					**95.6**				**Persistance (%):**					**95.06**	
		From time distance weighted									**From exponential trend**						
		Conif. forest	Decid. forest	Wood recoloniz	Broom land	Grassland	Crops	∑ **2020**			Conif. forest	Decid. forest	Wood recoloniz	Broom land	Grassland	Crops	∑ **2020**
To 2020	Conif. forest	37.48	0	0.74	0.29	0.4	0	38.92	To 2020	Conif. forest	37.55	0	0.04	0.1	1.02	0	38.71
	Decid. forest	0	8.09	0.21	0.23	0.49	0	9.02		Decid. forest	0	8.08	0.15	0.09	0.74	0	9.08
	Wood recoloniz	0.1	0	20.51	1.15	0.63	0	22.39		Wood recoloniz	0.02	0	21.28	0.2	0.8	0	22.31
	Broom land	0.02	0	0.01	24.01	0.04	0	24.07		Broom land	0.02	0	0.01	25.3	0.07	0	25.4
	Grassland	0.02	0	0.01	0.04	5.53	0	5.6		Grassland	0.02	0	0	0.01	4.47	0	4.51
	Crops	0	0	0	0	0	0	0		Crops	0	0	0	0	0	0	0
∑ **2009**		37.61	8.1	21.49	25.71	7.1	0	100	∑ **2009**		37.61	8.1	21.49	25.71	7.1	0	100

(continued)

Table 4 (continued)

		From average									**From linear trend**						
		Persistance (%):					**95.6**				**Persistance (%):**					**96.68**	
		Markov from 1889–2009									**Markov from 2000–2009**						
		Conif. forest	Decid. forest	Wood recoloniz	Broom land	Grassland	Crops	∑ **2020**			Conif. forest	Decid. forest	Wood recoloniz	Broom land	Grassland	Crops	∑ **2020**
To 2020	Conif. forest	37.45	0	0.77	0.27	0.31	0	38.81	To 2020	Conif. forest	37.35	0	0.04	0.23	0.01	0	37.62
	Decid. forest	0	8.09	0.24	0.38	0.52	0	9.23		Decid. forest	0	8.09	0	0.01	0.05	0	8.15
	Wood recoloniz	0.12	0	20.46	1.31	0.47	0	22.37		Wood recoloniz	0.23	0	21.44	1.77	0.9	0	24.34
	Broom land	0.02	0	0	23.68	0.01	0	23.72		Broom land	0.02	0	0	23.7	0.01	0	23.73
	Grassland	0.02	0	0.01	0.06	5.78	0	5.87		Grassland	0.01	0	0	0	6.14	0	6.16
	Crops	0	0	0	0	0	0	0		Crops	0	0	0	0	0	0	0
∑ **2009**		37.61	8.1	21.49	25.71	7.1	0	100	∑ **2009**		37.61	8.1	21.49	25.71	7.1	0	100
		Persistance (%):					**95.47**				**Persistance (%):**					**96.71**	

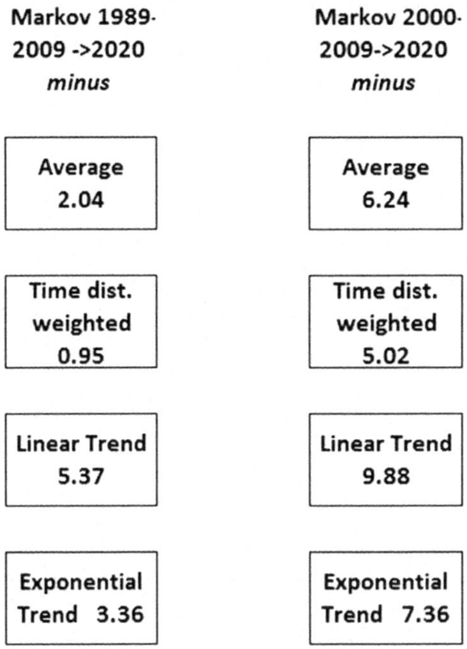

Fig. 6 Schema of Fig. 7 and overall sums of absolute differences between transition rates. The calculation is the MC-computed transition rate for 2020 minus the transition rate for 2020 integrating all LUC maps (alternatively computed by average, time-distance weighting, linear or exponential trend). Each cell expresses the transition rates obtained by MC minus the alternatively obtained transition rates: The *left column* shows the differences between the two transition rates based on a Markov chain from 1989–2009 to 2020 while the *right column* expresses differences based on the Markov chain for the 2000–2009 training period. The *number inside* the windows is the sum of the absolute differences

other hand, transitions from grassland to other LUC (fifth column) are generally negative (i.e. alternatively computed transition rates are higher than those calculated using the Markov chain) while persistence (bottom right cell) balances this.

5 Discussion

Many LUC modeling approaches analyze land change in the recent past as a means of simulating what might occur in the future. The wide range of techniques used to describe dynamics such as LUCC include budget (Pontius 2000; Pontius et al. 2004a, b) and intensity analysis (Aldwaik and Pontius 2012; Pontius et al. 2013). Other methods such as sensitivity analysis (Gómez Delgado and Tarantola 2006; Jokar Arsanjani 2012) test the robustness of model outputs by analyzing, among other aspects, the significance of the data used, including the training dates.

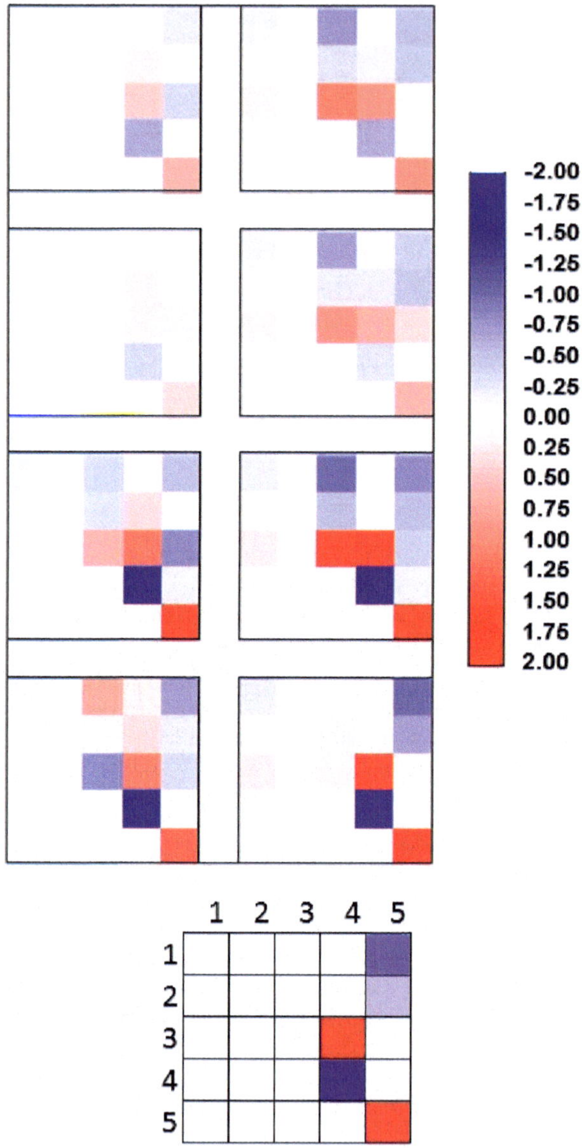

Fig. 7 Differences between the transition rates for 2020 predicted by the Markov chain technique and those calculated in other ways, laid out as explained in Fig. 6. Each *square* presents one comparison. Because crops were excluded, each cross tabulation matrix is composed of only five columns and five rows. From *left* to *right*/*top* to *bottom*: coniferous forest (*1*), deciduous forest (*2*), wood recolonization (3), broom land (*4*) and grassland (*5*), cf. schema on the right

5.1 Use of/Impact of Multiple Calibration LUC

The analysis of previous land change using the LUCC budgets technique reveals that land change is not a linear process and neither is its composition. It is important to notice that the LUCC budget indicators shown in Fig. 2 are quite average land change indicators. Individual LUC shows much more variety, as illustrated by the

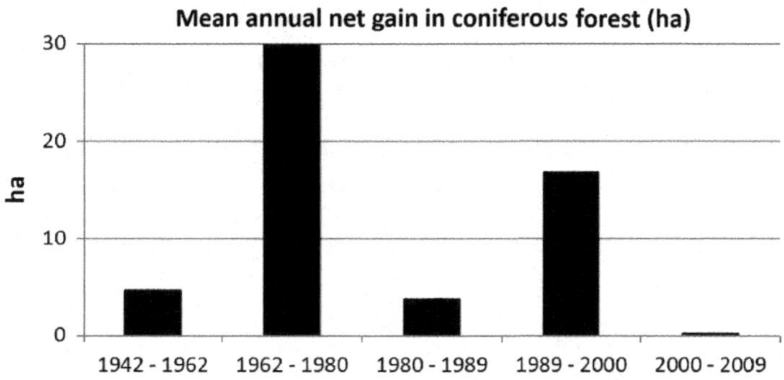

Fig. 8 Average annual net gain in coniferous forest per period (in hectares)

mean annual net gain in coniferous forest (expressed in hectares—Fig. 8). If the modeler chooses 2000 and 2009 as training dates for a BAU scenario, a smaller amount of land change will be simulated and specific net gain for coniferous forest will be near zero. However, over the period 2000–2009 land use change tended towards forest. The average net gain for wood recolonization was the highest of all the different categories for this period (cf. Fig. 5). At the opposite extreme, if the modeler chooses 1962 and 1980 as the training dates, the BAU scenario will be very dynamic, while wood recolonization registers an average net loss of about 9.6 ha/year.

This example shows that the most recent data are not representative and that changes in LUC categories cannot be understood separately.

Computed Markov chains (MC) and comparison with observed land change indicate that the most recent training dates are not per se the most realistic. At the opposite extreme, this data set shows that the use of the last available training date (2000) reduces the absolute difference between observed and MC-predicted LUC (MC 4, 7, 9 and 10) even though the training periods vary greatly in duration (from 11 to 58 years).

The choice of training dates for MC prediction affects the quantitative accuracy of BAU scenarios. Given that land change is not linear, using only two training dates may be pure lottery. In addition to Fig. 2 representing the average annual rate of change and showing differences ranging from single to double figures depending on the period being considered, Fig. 8 shows the average annual rate for one category, coniferous forest. The reader can easily see that variations are magnified at this level and that the choice of training dates is crucial for the accuracy of the model. The assumed level of data error is also an important factor. Using this data set, assuming a 10% error increases the amount of predicted change in comparison to 100% confidence in data.

5.2 Integration of Multiple Training Dates to Compute the Expected Annual Amount of Change

The comparison of average annual transition rates (%) at global level (all LUC categories mixed together) and category level (Figs. 4 and 5) illustrates the heterogeneity of change speed and tendencies. The choice of accurate training dates is more complex and selecting only two training dates may inordinately impoverish the real dynamics. To overcome this problem, we propose alternative methods involving the integration of multi-temporal data as a basis for quantitative simulation.

5.3 Computing Transition Rates as an Alternative to Markov Chain Transitional Predictions

The technique used to integrate multiple training dates demonstrates the possibility of overcoming the two-date restriction of commonly used Markov chains to predict quantities of land change. The results obtained depend on the weighting of the impact of individual dates, close to the MC generated transition matrices. This attempt to compare them underlines the methodological difficulty of relating a 2-dates-based approach to one with 2 + n dates. The choice of a pair of dates for the Markov chain unavoidably results in data reduction, whereas the proposed alternatives allow historical information to be taken into account in the simulation process, and are therefore, theoretically at least, an improvement. On the other hand, a process using all the available LUC maps must be supervised to avoid illogical transitions. Crops were used as a category in the original data set, although they disappeared completely during the third period. It seems unlikely that locals will begin planting them again, so unless there is a paradigm shift, their future presence may be excluded. This means that if we want to take all the dates into account, adjustments have to be made and the process must be supervised.

The weighting techniques we applied are still just a small sample amongst a wide range of possibilities, and lie somewhere between path-dependent land change prediction and forecasting scenarios that break with the observed trend.

6 Conclusion and Outlook

Generally, land change models undertake only two tasks: computing expected quantities and allocating them on the map. The first task is often accomplished using Markov chain simulated transitions based on only two training dates. This paper shows first the randomness of picking out two of a wider set of training dates, the uncertainty this produces and its consequences on Markov chain predicted land change. The complexity of LUCC is illustrated by computing annual transition rates on three levels: global, category and transitional. In an attempt to overcome the limitations of Markov chains that analyze the LUC dynamics in a single training period, we describe alternative methods that use all the available dates and weight them differently. There are two difficulties to this approach. First, modelers have to supervise and, if necessary adjust, the generation of transition matrices to avoid illogical transitions. Second, it seems difficult to evaluate the simulated land change transitions generated in this way by comparing them with those computed by Markov chains based on just two dates.

The range of results shows that both modelers and possible recipients (planners and other more general users) of this data must be cautious when interpreting a simulation that is never more than a plausible future. This means that we have a duty to act with the greatest possible transparency, indicating the available data, the data we used and the methodological choices we made.

Acknowledgements This work was supported by the BIA2013-43462-P Project funded by the Spanish Ministry of Economy and Competitiveness and by the FEDER European Regional Fund.

References

Aldwaik SZ, Pontius RG Jr (2012) Intensity analysis to unify measurements of size and stationaity of land changes by interval, category, and transition. Landsc Urban Plann 106:103–114

Allen THF, Starr B (1982) Hierarchy: Perspectives for ecological complexity. University of Chicago Press, Chicago 310 p

Burnicki AC, Brown DG, Goovaerts P (2007) Simulating error propagation in land-cover change analysis: The implications of temporal dependence. Comput Environ Urban Syst 31:282–302

Camacho Olmedo MT, Pontius RG Jr, Paegelow M, Mas JF (2015) Comparison of simulation models in terms of quantity and allocation of land change. Environ Model Softw 69:214–221

Gómez Delgado M, Tarantola S (2006) Global sensitivity analysis, GIS and multi-criteria evaluation for a sustainable planning of hazardous waste disposal site in Spain. Int J Geogr Inf Sci 20:449–466

Hu G, Hu J, Zhang C, Zhuang L, Song J (2003) Short-term traffic flow forecasting based on Markov chain model. In: Intelligent vehicles symposium 9–11 June 2003. Proceedings IEEE pp 208–212

Jokar Arsanjani J (2012) Dynamic land-use/cover change simulation: geosimulation and multi agent-based modelling. Springer Verlag, Springer Theses

Kim JH (2013) Spatiotemporal scale dependency and other sensitivities in dynamic land-use change simulations. Int J Geogr Inf Sci 27:1782–1803

Lee YJ, Lee JW, Chai DJ, Hwang BH, Ryu KH (2009) Mining temporal interval relational rules from temporal data. J Syst Softw 82:155–167

Liu J, Deng X (2010) Progress of the research methodologies on the temporal and spatial process of LUCC. Chin Sci Bull 55:1354–1362

Mas JF, Kolb M, Paegelow M, Camacho Olmedo MT, Houet T (2014) Inductive pattern-based land use/cover change models: a comparison of four software packages. Environ Model Softw 51:94–111

Paegelow M, Camacho Olmedo MT, Mas JF, Houet T (2014) Benchmarking of LUCC modelling tools by various validation techniques and error analysis. Cybergeo, document 701, mis en ligne le 22 décembre 2014. http://cybergeo.revues.org

Pontius RG Jr (2000) Quantification error versus location error in comparison of category maps. Photogramm Eng Remote Sens 66(8):1011–1016

Pontius RG Jr, Huffaker D, Denman K (2004a) Useful techniques of validation for spatially explicit land-change models. Ecol Model 179(4):445–461

Pontius RG Jr, Shusas E, McEachern M (2004b) Detecting important category land changes while accounting for persistence. Agric Ecosyst Environ 101(2–3):251–268

Pontius RG Jr, Boersma W, Castella JC, Clarke K, de Nijs T, Dietzel C, Duan Z, Fotsing E, Golstein N, Kok K, Koomen E, Lippitt CD, McConnell W, Sood AM, Pijankowski B, Pidhadia S, Sweeney S, Trung TN, Veldkamp AT, Verburg PH (2008) Comparing the input, output, and validation maps for several models of land change. Ann Region Sci 42(1):11–27

Pontius RG Jr, Gao Y, Giner NM, Kohyama T, Osaki M, Hirose K (2013) Design and interpretation of intensity analysis illustrated by land change in Central Kalimantan, Indonesia. Land 2(3):351–369. doi:10.3390/land2030351

Chapter 8
Land Use Change Modeling with SLEUTH: Improving Calibration with a Genetic Algorithm

K.C. Clarke

Abstract SLEUTH is a cellular automaton computer simulation model that uses historical land use and other data to project growth and land use change into the future. The model has seen over 100 applications worldwide, and has been among the leading cellular automaton (CA) models applied in simulating land use change at many different spatial scales. The model is highly dependent on the use of historical data to derive the behavioral parameters that best capture the structure and dynamics of the location-specific growth history. While several improvements have been made to the model to increase calibration speed, the current brute force calibration technique has proven popular, in spite of it requiring a multi-phase process and hundreds of CPU hours. This chapter reports on the use of a new alternative calibration method, in which the brute force method is replaced with a genetic algorithm (GA). A version of the model code that executes the GA calibration has been written and made public. The GA calibration process populates a "chromosome" with a set of parameter combinations (genes), of which five are required by the model, each with ranges from 0 to 100. These combinations are then used for model calibration runs, and the most successful (as measured by the Optimal SLEUTH metric) are selected for mutation (recombination of their values), while the least successful are replaced with new randomly selected values. Critical values that must be provided are the population size of the chromosome, the number of iterations or generations over which evolution will continue, the evolution mutation rate, and the number of offspring and replacements in each generation. To select suitable default values for these rates, two SLEUTH applications were used at the extremes of the model's calibration performance success. These were for San Diego, California where the model fit was very strong, and Andijan, Uzbekistan, where the model was most hard pressed to capture the complex growth process. In both cases, full model calibrations were completed using brute force calibration, followed by calibrations using the GA. It was found necessary to hold the GA parameters constant while repeatedly recalibrating the model using different values for the GA settings. In all cases, the GA model performed as well as the

K.C. Clarke (✉)
Department of Geography, University of California, Santa Barbara, USA
e-mail: kclarke@geog.ucsb.edu

M.T. Camacho Olmedo et al. (eds.), *Geomatic Approaches for Modeling Land Change Scenarios*, Lecture Notes in Geoinformation and Cartography,
https://doi.org/10.1007/978-3-319-60801-3_8

brute force method, but used vastly less computation time. There were also subtle but minor differences in the best SLEUTH forecasts that were explored by mapping the differences among results. The optimal values for GA calibration are given and set as the defaults for SLEUTH-GA, a new version of the SLEUTH model.

Keywords Land use change · Model · SLEUTH · Calibration · Genetic algorithm

1 The Purpose of Modeling

A model is an abstraction that embodies a simplification of reality, such that information can be gleaned about reality from the formalization inherent in the model. Models can simply represent, they can codify relations, they can simulate structure and form, or they can hypothesize the development of specific time-space relations. Models should be as simple as possible yet not so simple that they fail to capture the complexity of the system in question (Clarke 2004; Batty and Torrens 2001). A good model is accurate, accountable, explanatory, predictive, useful and simple (Benenson and Torrens 2004). Among the most complex of systems subjected to modeling is the geographical landscape of changing land uses over time. Land use change is the slow but persistent human modification of Earth's terrestrial surface. As humankind has gradually occupied much of the Earth's surface, the environmental consequences of such changes have become inescapable. Models of urbanization and land use change have been struggling to simplify the complexity of land use change using contrasting modeling methods and paradigms for decades (Verburg et al. 2004). Recently, calls have been made, and prototype systems designed, to integrate both time and space in a more dynamic modeling framework for land change science (An and Brown 2008).

The contemporary magnitude of land use change faced by the inhabitants of Planet Earth is unprecedented in history. Few places on Earth are untouched by human forces, and the number of places left where a photograph taken 100 years ago would be identical to one taken today is diminishing rapidly. Land use change is primarily driven by the conversion of natural lands to agriculture to feed the billions of humans, but also increasingly by the expansion of built-up land. Cities expand their extent of impervious surface outward and inward, and the evolution from village to town to city continues unabated. Land use and land cover change modeling asks what causes these changes, but far more importantly it also asks how they can be modified, diverted or prevented so as to ensure that future cities are more sustainable.

There are four purposes in modeling. First it can help us gain an understanding of a process, usually as revealed by its consequent spatial forms (Clarke 2014a). Secondly we can try to forecast the process by modeling, and so predict where and when changes will actually occur (NRC 2014). Thirdly, we can explore alternative futures by varying the forecasts to reflect anticipated changed circumstances (Xiang and Clarke 2003; Houet et al. 2016). Fourthly, we can use the model itself to help others understand the process, its outcomes and its consequences, using it as a vehicle to raise awareness and to educate. It is notable that all four of these purposes are highly dependent on the accuracy, reliability and effectiveness of the model.

A model whose behavior and properties have been mined from facts and data may be predictive, but fails to explain the how and why of modeling, just as its forecasts have little educational value. Good models make their assumptions about a process explicit, using facts and data as inputs, and then create accurate renditions of future system states. Such accurate models must use real data to fine-tune the constants and variables that determine model behavior. This happens in two ways: first the model design should incorporate knowledge or testing into the choices of constants and variables; and second, the model should use hindcasting, that is, be applied to historical data to effectively replicate the present. Accuracy can then be measured as the level of agreement between the forecasted present and the actual present (Pontius et al. 2007). It is this stage of modeling that assures the model's level of accuracy, reliability and effectiveness by expressing them as measurements, and attempting to maximize them. This process is called model calibration, and calibration remains the most critical phase of model design and application.

2 The Need for Calibration

Calibration of models covers the two steps above, but extends to include a vast array of tools and techniques to attempt the optimization of model performance. In all cases, calibration seeks to determine the impact of changes in a particular constant or variable in terms of the model outputs. Constants are the values that will remain internal to the model, and may be choices of particular values (such as the greatest topographic slope that can be built upon) or more structural elements of the model (such as the choice of Moore vs. Von Neumann neighborhoods). The determination of these constants is the first stage of calibration because it happens during model design. Methods used include visual inspection of the correspondence of outputs, simple match statistics and the computation of all outputs across a range of constant values. Critical in the latter case is determining the threshold values, that is, points at which a small change in the constant produces large differences in the output–what Houet et al. (2016) term "non path-dependent" and contrasted/breaking trends. Simple models generally seek to avoid these values, while complex systems models seek to exploit them. Crossing these thresholds is termed phase change in complexity theory, and often leads to emergence (Holland 1998).

The second type of calibration involves repeated application of the model, the measurement of model performance, degree of fit, or accuracy, and the adjustment of input variables and data until performance is maximized. This may involve accuracy of the model outputs as measured using historical data, or achievement of some other goal, such as tractability (Clarke 2003). Usually a model is started at some point in the past, and executed without further input until the last period of known data (the present), periodically matching its numerical and spatially distributed outputs with real data.

A third approach is to automate the calibration process. Given the matches described in the last paragraph, one or more measures can be compiled that represent the performance of a suite of parameters. Changing the parameters and repeating the process allows retention of the best performing settings. An easy way to optimize is to repeat the parameter changes for all possible combinations and permutations of their values, the so-called brute force method. Models now increasingly use machine-learning algorithms to achieve this maximization. For example, weights assigned in agent-based models can be selected using support vector machines, or cellular automata behavior rules selected using genetic algorithms (Clarke 2014b). The development of automated calibration methods is discussed in the following section. Needless to say, good calibrations derive the best set of input parameters that determine the model's performance, accuracy and behavior. Good models are almost always well calibrated.

The complete set of models of land use and land cover change is spread across a vast literature, including periodical reviews and surveys of the models and their applications (NRC 2014). All such models require calibration, but these calibrations depend on the model type and its intended purpose. A subset of land use change models is cellular automata (CA) models. These have been discussed at length (Torrens and O'Sullivan 2001) and even divided into types (Sante et al. 2010). A review of the calibration methods for all land use and land cover change models would be prohibitive, so in the remainder of this chapter we will focus on CA models only, then discuss a particular model and its improvement using a genetic algorithm to replace its current brute force calibration method. An advantage of this approach is that it removes human interaction entirely from the calibration process (Jafarnezhad et al. 2015).

3 Developments in CA Model Calibration

CA models are complex system models consisting of: (1) a set of mutually exclusive and non-overlapping states; (2) a framework of points, cells or a grid in which each element is in one and only one state; (3) a defined neighborhood, consisting of a set of cells usually surrounding or adjacent to a cell; (4) a set of rules that govern state changes as a function of the other states within the neighborhood; (5) a relation to discrete time, such that all cells are evaluated in each time step; and (6) an initial arrangement of the states within each of the cells. In land use change models, the states are the standard land use classes, such as forest, agriculture, urban and wetlands; the framework is a map, consisting of a grid of raster cells usually within a GIS; the neighborhood is the set of cells forming the Moore, Von Neumann or other neighborhood around each cell; the time steps are annual increments from a start time to a stop time, either in the past, the present or the future; and the initial arrangements reflect actual mapped distributions at some point in time. This leaves the rules to be determined during the model design stage. Rules can be created by following those of other models, by using some a priori assumption about system behavior, derived statistically using probabilities or from exogenous quotas, or derived from data mining past land use changes as functions of location, type and quantity.

The rule sets associated with land use and land cover change are often chosen to reflect the driving factors of land use change. For example, actual changes can be analyzed using logistic regression as being "caused" by environmental constraints such as the topographic slope, the distance from a city center, the distance from a road, or the zoning at that point. The factors that prove significant are then prioritized and assigned weights. Modeling then consists of taking an input model, combining the weighed input factors, deciding probabilistically whether a change from type A to type B could occur, then enacting the change at the most probable locations. Such models can be highly data dependent, and may not be transferable from one place to another, even though the methods can be reapplied if the rules are derived anew. Other models use rules determined by trial and error and by applying theoretical knowledge, such as the SLEUTH model (Clarke et al. 1998).

The use of two land use maps as inputs to derive a rule set for CA by data mining has led to numerous attempts to calibrate CA models with data reduction methods. These include multi-criterion evaluation (MCE) (Wu and Webster 1998; Wu 1998), multiobjective optimization (Cao et al. 2014), logistic regression (Wu 2002) and decision trees (Li and Yeh 2004). Most successful among these methods have been neural networks (Yang and Li 2007). A neural network uses a training subset of the data to compute the weights that map inputs to one or more model layers of hidden neurons, and then on to the outputs. The success of neural networks lies in the fact that no assumptions are made about the underlying distribution. Logistic regression, for example, assumes independence among the input variables. Some models use neural networks as the entire basis for land use change modeling (e.g. ANN-CA by Li and Gar-On Yeh 2002 and LTM by Pijanowski et al. 2002).

Other machine-learning algorithms have been used to help calibrate (and derive CA rules for) CA models on land use and land cover change. Long et al. (2009); Hu and Lo (2007) and Liu and Phinn (2003) used logistic regression to evaluate and select CA transition rules in the model design stage. Guan et al. (2005) used artificial neural networks for the same purpose. Another favored method is the support vector machine, a discriminative classifier defined by a hyperplane that divides the data in multidimensional space. The method uses supervised learning to output an optimal hyperplane which can be used to forecast future states (Yang et al. 2008). Others have used neural networks to optimize CA control parameters (Li and Yeh 2004). More recently, methods such as particle swarm optimization (Feng et al. 2011) and ensemble learning strategies (multiple methods combined in parallel) have also been introduced (Gong et al. 2012).

One of the most successful machine-learning methods to be used in both contexts (CA rules selection and parameterization) is genetic algorithms. A genetic algorithm (GA) is a method for solving optimization problems based on a process of natural selection that mimics evolution in plants and animals. The algorithm starts with an approximate initial set of solutions, and then repeatedly modifies the population of genes while assessing fitness. In each iteration, changes are made to create better solutions (evolution and mutation) and to allow the entry of random solutions that may outperform the current best "gene." Studies that have used GA to calibrate CA include Colonna et al. (1998), Goldstein (2004), Yang and Li (2007), Yang et al. (2008), Shan et al. (2008), Cao et al. (2011), Feng and Liu (2012), Clarke-Lauer and Clarke (2011), Garcia et al. (2013) and Jafarnezhad et al. (2015).

Genetic algorithms use the concept of fitness to determine evolution toward a best outcome. There are many possible measures of goodness of fit between a real map and a modeled map, and these include producer and user accuracy, the various Kappa measures, matching of landscape metrics, correlation, the Receiver Operating Characteristics curve and others. Many calibrations simply use the percent correct as a measure. As an example, the SLEUTH model produces 13 regression-based fit measures, which in the past were combined by multiplication, although many studies have used the Lee-Sallee metric alone (Silva and Clarke 2002). Current practice uses the Optimal SLEUTH Metric (Dietzel and Clarke 2007). This measure uses a subset of 7 of the 13 metrics, also combined by multiplication, selected to reduce interdependencies among the 13 metrics. The current study used the OSM as the fitness measure for calibrating the SLEUTH model.

Use of GA implies creation of the equivalent of a chromosome, with individual genes reflecting traits of an individual. In biology, typical traits include eye color, height, etc., but for this application the genome consisted of a set of feasible control parameters for a SLEUTH model run. SLEUTH has five control parameters, each of which varies from 0 to 100, with 0 meaning the absence of a behavior and 100 meaning unrestricted behavior. These parameters are termed diffusion, breed, spread, slope and road growth. Diffusion controls the amount of scatter displayed by new urban pixels. Zero means that no new growth can occur other than at urban edges, while 100 allows urban growth anywhere. Breed determines which newly urbanized pixel can immediately start spreading. A value of 0 leaves each new pixel isolated, while 100 turns every newly urbanized pixel into a growing cluster. Spread determines the rate of outward organic growth or infill. Zero permits no outward spread, while 100 spreads the edges of all urban pixels in each time period. Slope is a factor that controls the response of urbanization to topographic slope. At zero, there is no evaluation of the slope of a pixel before its urbanization; approaching 100, higher slopes are increasingly penalized up to a critical value at which development is impossible, termed the critical slope. Road growth emulates the attraction power of roads for new growth. At 0, roads have no impact on growth while at 100, all new development within a buffer distance is attracted to the roads.

A single run is then controlled by the five values {diffusion, breed, spread, slope, roads} within the integer range from {0,0,0,0,0} to {100,100,100,100,100}. The single set of five values forms a gene, and a population of P such sets is the chromosome. Each gene is evaluated, i.e. the model is run and the fitness (here the Optimal SLEUTH Metric or OSM) calculated. The genes are then sorted by OSM, so that those that performed best rise to the top. This is termed a generation. Between generations, new genes are created by combining the values of the best performing genes, after having pairs of genes "compete" to reproduce, and so share their gene. Combining parts of one gene with another is called crossover, and is an important part of biological evolution, determining the traits of the offspring produced by a particular mating pair. Next some of the genes in the chromosome are mutated, by altering their values. Mutation is applied at the mutation rate, in this case the proportion of the genome subjected to change. If 10 out of 100 genes in a chromosome are subjected to mutation, the rate would be 10/100 = 0.1. Mutation can be done by switching values, or by replacing one or more values with random numbers in the range 0–100. There are two levels of fitness associated with each generation: the total fitness of the chromosome and the specific fitness of an individual gene. In our case, we are interested in maximizing both total fitness to move the training process forward, and the fitness of the best performing gene, which is the best model fit at that generation. Evolution ends when a maximum number of generations is reached, or when successive generations have no better total fitness than their parents.

Blecic et al. (2010) performed a comparison of genetic algorithms to select the best for the automatic calibration of a constrained CA. Various strategies for gene selection were tested, such as generational genetics, elitist selection, steady state, and choice of the fitness metric. The chief variables in a GA include choosing the size of the population (number of genes in the chromosome), the maximum number of generations (or minimum improvement in fitness to continue evolution), the mutation rate, number of crossovers, the number of offspring, and the number of replacements. A second stopping criterion is the maximum number of evaluations of genes for possible inclusion as replacements. The GA populates the initial gene with chromosomes using random numbers within the individual chromosomes range, usually standardizing values between zero and one, or zero and one hundred. In one generation, each of the genes is used as model input, and the fitness criterion calculated. In the study by Blecic et al., the fitness values used were the Kappa coefficient and the Lee-Sallee metric (Silva and Clarke 2005), while others have used the Optimal SLEUTH Metric (Dietzel and Clarke 2007). This is repeated for all genes in the chromosome, and the results ranked. Elitism determines how many genes will survive to the next generation. Some proportion of the genes are crossed over, that is their values are switched by breaking a sequence and combining it with that from another gene in the chromosome. For example a set of SLEUTH input parameters may be {10, 20, 30, 40, 50}. After mutation, it may be {10, 20, 50, 40, 30} with 2 values switched and 3 remaining. Another form of mutation simply randomly or incrementally changes one or more gene values. Lastly, the lowest performing genes in terms of fitness are replaced, or "killed off" and replaced with new random values. Such a choice increases the number of evaluations, when a maximum number is reached or a maximum number of generations pass, the winning genome is selected.

This final replacement stage is important because there is always a possibility that the chromosome with the highest total fitness is not a global but only a local maximum. Both mutation and replacement ensure that a superior value either evolves or arrives by chance. The altered chromosome is then subjected to the next generation, and the process is repeated either until no further gain in fitness is achieved, or a maximum number of generations is exceeded. Blecic et al. (2010) compared two variations of a constrained cellular automaton growth model, one with 14 parameters and the other with 27. Note that the GA optimizes in two ways, firstly by seeking a chromosome with the highest total fitness, and secondly by seeking a gene with the highest individual fitness.

Veerbeek et al. (2015) noted that although GAs are successful as optimization algorithms for a large range of problems they generally demand "a large population of candidate solutions and a large number of iterations to reach a global optimum within a search space." They stated that GAs excel at quickly identifying good solutions but are often less ideal for finding the optimum. In their study, they used a memetic algorithm in which the standard GA is extended with a local search algorithm that slightly increases or decreases chromosome values toward locally better solutions, a modified form of mutation. Their test implementation for Beijing used a Dinamica-EGO-based LULC change model in a 2-stage modeling approach which separates the calculation of the urban-area growth from the extension into urban LULC classes. Their fitness measure was obtained by performing a cell-by-cell fuzzy set intersection, a method similar to using the Lee-Sallee value. One conclusion of the study is that "the choice and implementation of machine-learning algorithms for calibrating LULC models often seem arbitrary and are too often based on standard 'off-the-shelf' tools."

Another variation is the adaptive genetic algorithm (Srinivas and Patnaik 1994), in which crossovers and the mutation probability are automatically adjusted according to the individual and total gene fitness. The authors noted that this is an alternative way to ensure the solution does not remain a local maximum. Lastly, Li et al. (2013) focused on the calibration metric, using a pattern-calibrated method which is based on multiple landscape metrics. They used a pattern-calibrated GA–CA that incorporated the percentage of landscape, patch metric and landscape division into the fitness function of the GA. Many CA studies use landscape metrics after simulation to compute the similarity among outcomes, such as simulations using different future scenarios, but this method incorporates them into the model calibration itself.

While research continues on using GA as a means to calibrate CA models, relatively few studies have examined how the specifics of the GA affect the performance, accuracy and tractability of model calibrations. For example, what impact is there on the spatial and quantitative characteristics of forecasts when GAs are used for calibration? Obviously such a question can only be answered in the context of a single model. SLEUTH will be used for this purpose because it is one of the few instances where both brute force and GA calibration options are available in open source code.

4 Calibrating SLEUTH

SLEUTH is a land use and land cover change model based on two tightly coupled CA models: the Urban Growth Model that simulates how urban areas expand and change; and the Deltatron model that propagates urban changes into other land use types. The model was originally developed and applied to the San Francisco Bay area (Kirtland et al. 1994; Clarke et al. 1997) and then to the Washington-Baltimore area (Clarke et al. 1998). SLEUTH's initial calibration was by monolooping (trying all possible settings for a single parameter, holding the others constant), but this was replaced by brute force calibration (Clarke et al. 1996). The calibration methods were systematically improved over a long period (Clarke et al. 2007, 2008a, b; Chaudhuri and Clarke 2013). Successive improvements were made in the code, in the choices during sequencing of the brute force process, and in the choice of resolution paralleled improvements in the speed of calibration as CPUs became faster, and as the model moved into geocomputational and high performance computing solutions (Clarke 2003). Recently, research has examined the goodness of fit between SLEUTH simulations and actual data, usually using hindcasting and spatial metrics of various kinds (Wu et al. 2009; Rienow and Goetzke 2014; Sakieh 2013).

Noah Goldstein was the first to experiment with GAs to calibrate SLEUTH (Goldstein 2004). Others tried the same approach with more sophisticated tools (Clarke-Lauer and Clarke 2011; Jafarnezhad et al. 2015). Clarke-Lauer and Clarke used the Optimal SLEUTH Metric as the fitness criterion and replaced the brute force module in SLEUTH with a new code routine that employed a GA that was posted to SourceForge. Values that could be varied included choices on encoding, fitness evaluation, crossover, mutation and survival selection. Coding involved a random number between 0 and 4 to index the five SLEUTH control parameters (diffusion, breed, spread, slope and road growth) and to decide how many elements from the parent were to be reproduced in the offspring. Remaining elements were selected from the second parent, with the second offspring using the opposite genes used for the first. Parents were selected by tournament selection, with a random set selected and the parents chosen with the highest fitness. Each generation replaces the weakest genes in the old population with the strongest in the new. The SLEUTH-GA was tested using the demo city sample data set available on the SLEUTH website, with a population size of 25. The paper concluded that the GA produced a speed up by a factor of 5 over brute force calibration. This means that the model calibrated in 20% of the time taken by brute force.

Jafarnezhad et al. (2015) used the SLEUTH-GA code to apply SLEUTH to 3 cities in Golestan Province, Iran. They calibrated SLEUTH first using the standard brute force procedure, and then used GA with the fitness metric as the OSM. They coded their own GA procedures based on Goldstein's method (Goldstein 2004). Model outputs were then compared using the Receiving Operator Statistic (ROC), landscape metrics and two Kappa coefficients (K_{histo}) and $K_{location}$). They concluded that GA produced better results as evaluated by OSM, landscape metrics, and K_{histo}, while brute force produced slightly better values for $K_{location}$, with the methods equal for the ROC statistic. They used population sizes from 20 to 30, mutations rates of 0.1, and the GA required 71–133 generations. The highest OSM's attained were 0.8195 for Azadshahr, 0.5861 for Gonbadekavoos and 0.9213 for Gorgan. The equivalent best OSM values for the three cities from brute force calibration were 0.5346, 0.3664 and 0.7740. Speed up was 4–5 times, and the authors noted that the results could be improved by "testing different values for mutation rate and decreasing model tendency to elitism." This chapter responds to this challenge by seeking the best parameter settings for calibrating SLEUTH with a genetic algorithm.

5 Methods

To conduct such an experiment, data were extracted from the SLEUTH model archive on the SLEUTH website (www.ncgia.ucsb.edu/projects/gig/) for San Diego and Andijan, Uzbekistan. The San Diego dataset produced by Mark McGinnis was among those that produced the most successful brute force calibrations according to the OSM (Syphard et al. 2011). Conversely, the Andijan data set (compiled by Lola Gulyamova of Tashkent State University, Uzbekistan) produced the lowest OSM fits achieved by SLEUTH. In both cases these were the best model calibrations, but they varied substantially in predictive power. This is believed to be because of Andijan's extraordinary urban growth history. During WWII, Andijan received large numbers of Soviet citizens, including Jewish refugees from Poland. After a period of Soviet managed growth with rapid construction of housing, during the 1990s Andijan and the surrounding agricultural Fergana Valley became politically unstable, with border closures and Islamic fundamentalism depleting the economy, leading to widespread poverty. The result was an end to population growth, which until then had been high. The halting of growth and deurbanization of land are known problems for SLEUTH modeling, so the poor measures of fit are hardly surprising. Table 1 shows the SLEUTH modeling results for the two cities, while Fig. 1 shows two comparative versions of the most recent data sets, modeled and actual. Note that some results differ from the earlier published research due to the recalibration, and do not reflect the averaging conducted before forecasting.

Table 1 Brute Force Calibration Results. Values for constants are after calibration, with high and low coefficients in the top 8 solutions given, then after averaging to the last time period

City	Calibration period	Best OSM	Diffusion/derived	Breed/derived	Spread/derived	Slope/derived	Road gravity/derived	Calibration time (CPU Seconds)
San Diego	1960–1999	0.7414836	(100:98–100) 100	(97:97–99) 100	(25:24–25) 25	(15:15–18) 1	(53:45–53) 53	175589
Andijan	1934–2013	0.0773797	(63:60–63) 100	(100:85–100) 100	(1:1–2) 3	(80:75–79) 1	(25:15–25) 38	440715

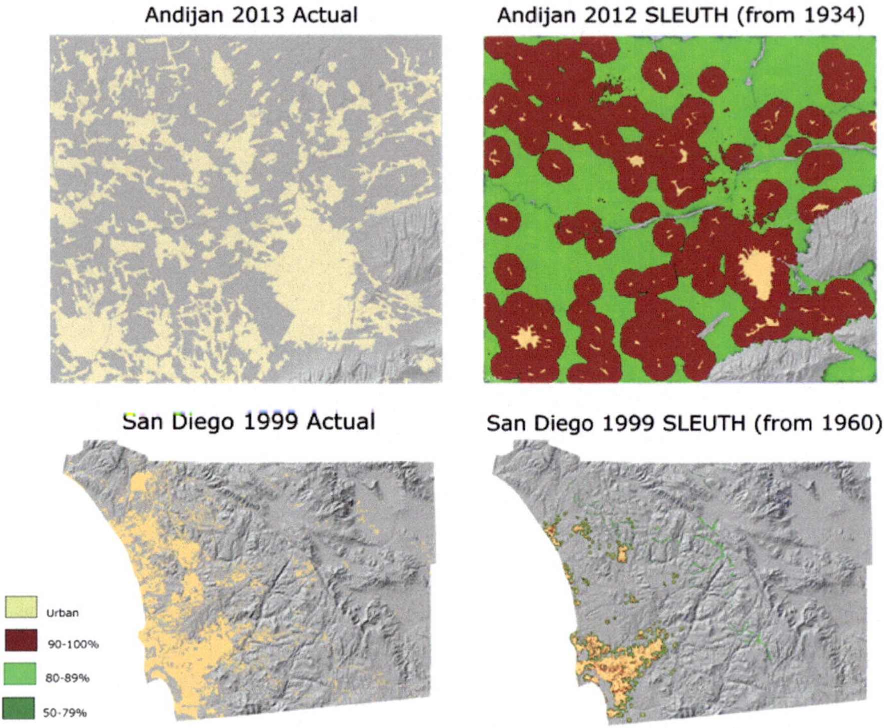

Fig. 1 Spatial extent of SLEUTH forecasts and actual urban growth during the calibration period

The two cities shown in Fig. 1 give an indication of the degree of model fit. In this figure, the cities are shown in the year prior to the last date of the calibration, i.e. SLEUTH has guided the growth starting from just the 1934 image for Andijan and 1960 for San Diego. In the Andijan modeled image, the modeled urban area is of low certainty (green, 50–90%) and almost entirely centered on the existing settlement in 1934 (yellow). The model overestimates uncertain growth, and underestimates the actual extent, which explains the poor model fit. In San Diego, the modeled certainties are higher (red > 90%) but the growth areas are almost all surrounding the existing urban class, and some roads. Growth in the flat parts of the interior valleys is overestimated, but generally the model fit is good.

6 Results

Both cities were then used with identical inputs in the SLEUTH-GA version of the model code. The SourceForge version was adjusted slightly to take six parameters from the shell to be passed to the code. These were the population size (genes in the

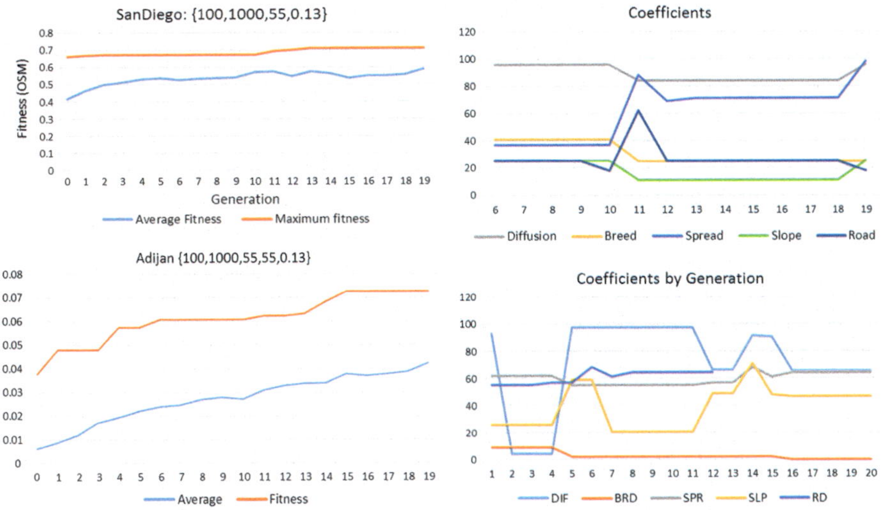

Fig. 2 Values during test GA calibration runs for San Diego and Andijan, showing coefficient evolution and fitness improvement

chromosome), the maximum number of generations, the mutation rate, the maximum number of evaluations per gene, the number of offspring, and the maximum replacement number. Population size, mutation rate, number of offspring, the replacement number and the maximum number of evaluations were varied, while the other values were held constant. The maximum number of generations was set to 100, but in fact the GA rarely used more than 20 generations in the calibration, contrary to the higher numbers determined by Jafarnezhad et al. (2015). The maximum number of evaluations for substitution per chromosome was found to give peak fitness at about 900, and this did not affect the calibration process, other than increasing the number of generations and CPU time. Figure 2 shows how the coefficients and fitness parameters adjusted during a single calibration. Specific chromosomes can be seen to persist, before eventually being dropped from the elite set as evolution continues.

The first experiments examined the best population size for the gene. Graphs in Fig. 3 show the results for the two cities. For Andijan the fitness was very low, with a slight peak at a population size of 70. For San Diego, the peak fitness occurred at a population size of 55, so this value was used for the next monoloops. Similarly for mutation rates, the peak fitness for both San Diego and Andijan was at a rate of 0.13, so this value was used for all further calibrations.

The information on calibration fine tuning for the GA was rather limited for the Andijan case, so testing of the ranges of the number of offspring and the

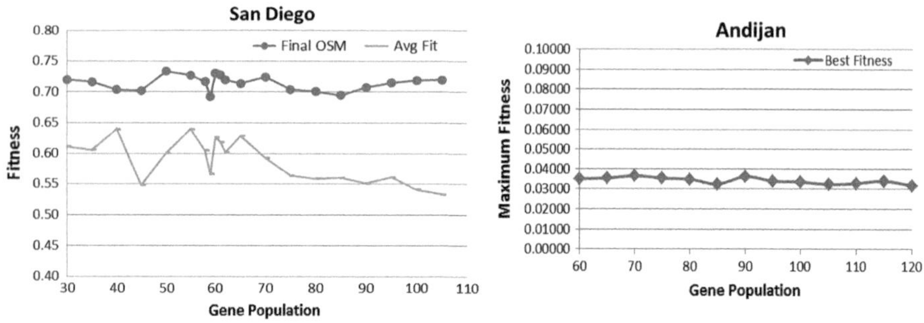

Fig. 3 Monolooping results for the two cities for the gene size, i.e. population of chromosomes

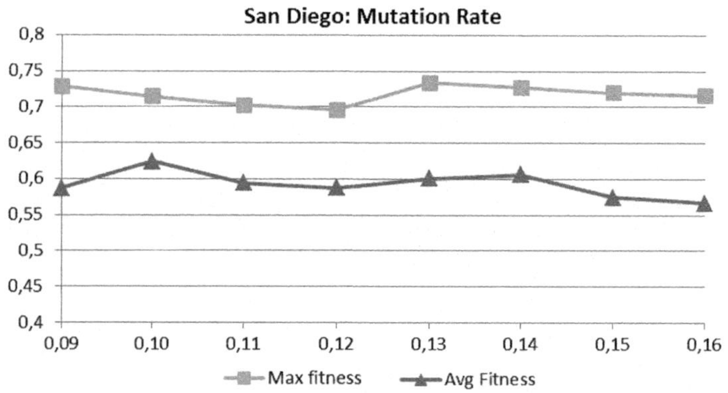

Fig. 4 Monolooping results for San Diego for the Mutation Rate

replacement number were restricted to the San Diego data. Their best fitness values were 55 and 50 respectively (Fig. 4).

Testing of the most suitable ranges for the GA parameters produced the final set of input parameters, shown in Table 2. The graphs of the values tested to select these values are shown in Figs. 5, 6 and 7. In particular, the maximum number of evaluations sets the computation cost for the run, and there appears to be a fine balance between too many generations versus achieving a good fit. A best value of 900 was selected, which creates about 10–12 generations of evolution.

It is hoped that this set of GA control parameters will enable universal application of SLEUTH-GA modeling. The values have been integrated into the SLEUTH-GA code as defaults, selected whenever the number of input parameters

Table 2 Genetic algorithm parameter monolooping calibration results

City	Calibration period	Best OSM	Maximum number of evaluations	Population (Gene size)	Mutation Rate	Number of offspring	Replacement number per generation	Calibration time (CPU Seconds)
San Diego	1960–1999	0.729724	900	55	0.13	55	50	55,588
Andijan	1934–2013	0.072920	900	55	0.13	55	50	19,866

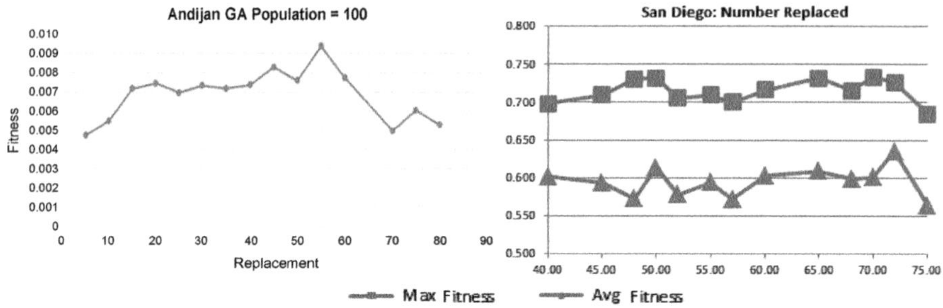

Fig. 5 Monolooping results for Andijan and San Diego for the gene replacement rate

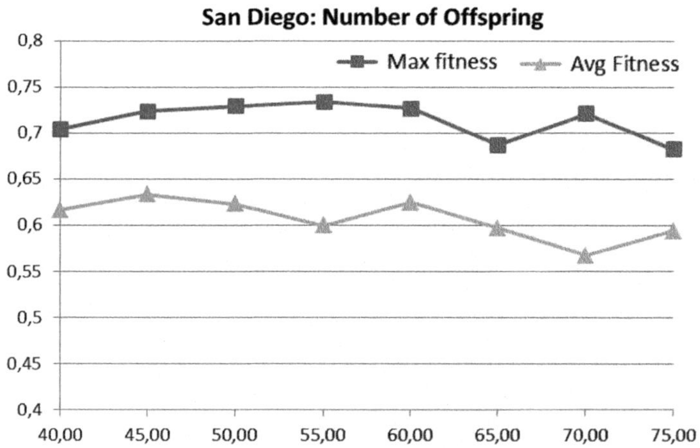

Fig. 6 Monolooping results for San Diego for the Number of Offspring

is not that expected to reset all of the values (i.e. the program execution mode is "evolve" and the parameter count is not 9). This goes a long way toward the fully automated and objective calibration of SLEUTH, without user intervention (Straatman et al. 2004). However, what range of parameters is there within the gene that might still be improved by brute force calibration over a smaller range, and what is the spatial impact of this difference on the actual forecasts? The ranges of parameters in the first subpopulation (highest performing individuals of the 8 most fit parents) for the best GA derived parameters are shown in Table 3.

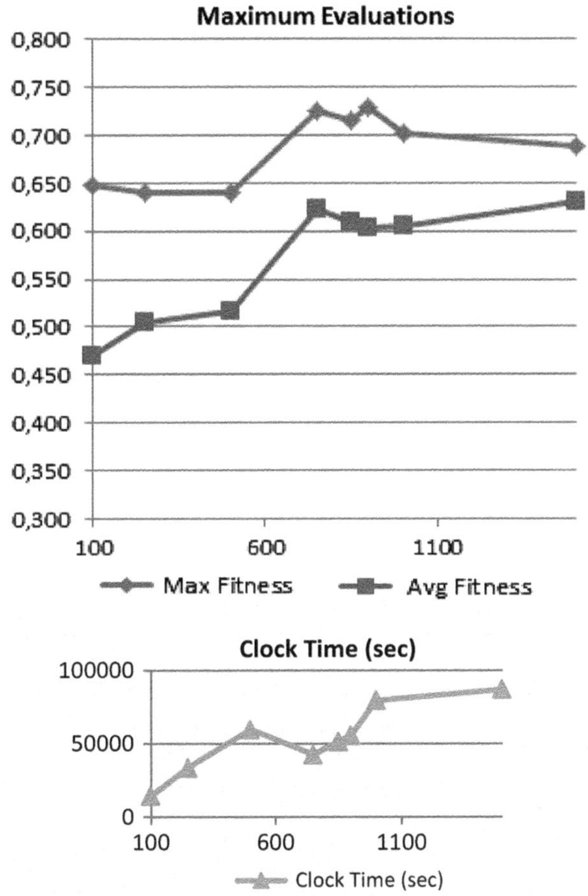

Fig. 7 Monolooping results for San Diego for the maximum number of evaluations and the corresponding clock based CPU time

Note that the maximum, average and total fitness of a genome tend to peak simultaneously, indicating that the best performing chromosome is led by the most fit gene.

To investigate the spatial impact of the differences in calibration mode, maps of forecast urbanization with a likelihood of over 50% were created for the two cities and shown for both methods of calibration (Fig. 8). It is evident that as in the calibrations, both cities are forecast with higher uncertainty and greater spread using brute force calibration, while the forecasts for both cities are more constrained but with greater certainty using GA. This appears to be the case for both high and low model fit, and may be a robust way of providing better forecasts.

Table 3 Genetic algorithm final calibration results

City	Calibration period	Best OSM	(Diffusion: Range) derived	(Breed: Range) derived	(Spread: Range) derived	(Slope: Range) derived	(Road gravity: Range) derived	Calibration time (CPU Seconds)	Speed Up
San Diego	1960–1999	0.729724	(90: 79–90) 100	(23:22–25) 26	(89:74–98) 100	(13:2–32) 1	(19:19–98) 30	55,588	3.16
Andijan	1934–2013	0.072920	(54:53–94) 82	(2:0–2) 3	88(62–94) 82	(70:9–70) 3	(47:14–47) 75	19,866	22.18

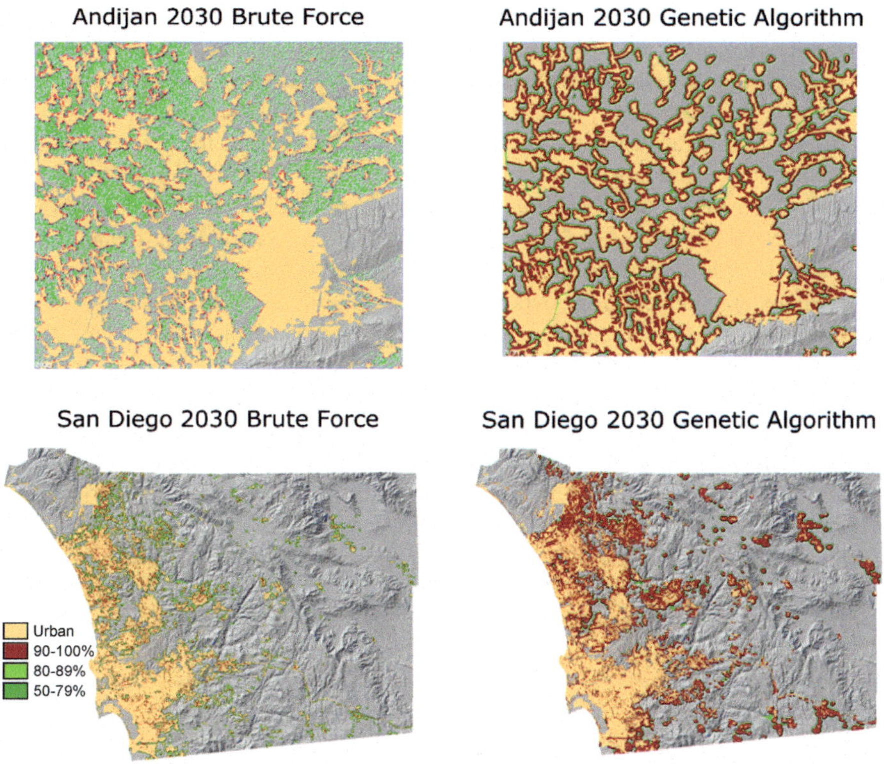

Fig. 8 Spatial Extent of SLEUTH forecasts to 2030 for the two cities using both calibration methods

7 Conclusion

Santé et al. (2010) pointed out the "need of making urban CA more flexible while keeping their simplicity by developing better calibration methods." This study has been in response to this challenge. Models of complex systems need to be complex, but not too complicated (Clarke 2004). An important move, as suggested by Jafarnezhad et al. (2015) is to eliminate human choices and judgements during the calibration process, replacing the subjective with the objective, or in Goldstein's terms, replacing brawn with brain (Goldstein 2004). On the surface, replacing the brute force calibration method for SLEUTH calibration just introduces a new set of calibration problems, i.e. dealing with the characteristics of the gene and determining how the evolutionary process yields the best results. Confirming prior research cited above, this study shows that GA leads to at least as good and often better calibration results. The results here indicate lower modeling uncertainty. The differences in the calibration parameter sets are small, and the differences among model forecasts are also small. The two advantages are the objectivity and the obvious benefits of speed-up, and therefore tractability. At the very least, GA can provide a convergent set of chromosomes that can be further optimized by brute force over a much more limited parameter set, such as the range over the top 8 genes listed in Table 2.

This study set out to review the importance of calibration for CA modeling in land use change models. Calibration performs important functions for models because it ensures their accuracy, integrity, reliability and trustworthiness. Well-calibrated models are defensible and objective, and use real world data instead of assumptions in their properties, constants, variables and behavior types. When these values are derived from data and by machine-learning methods, there is an obligation to perform sensitivity analyses and to run controls. Only when objectivity and accuracy can be assumed to coexist within a model can the model's forecasts, experiments and predictions be believable. By moving SLEUTH calibration from brute force to GA, the level of objectivity is further improved. As a bonus, the amount of CPU time that must be devoted to calibration was reduced by about a factor of 3 for San Diego and 22 for Andijan. This goes a long way toward solving the issue of the amount of raw CPU time required for SLEUTH application, which in turn should enable new applications and new cities to be modeled and simulated. The final version of the SLEUTHGA software is posted on the SLEUTH website at: www.ncgia.ucsb.edu/projects/gig/Dnload/download.htm and is available as open source code for modelers.

References

An L, Brown D (2008) Survival analysis in land change science: Integrating with GIScience to address temporal complexities. Ann Assoc Am Geogr 98(2):323–344

Batty M, Torrens PM (2001) Modeling complexity: the limits to prediction.12th European colloquium on quantitative and theoretical geography. St-Valery-en-Caux, France, 7–11 Sept 2001

Benenson I, Torrens P (2004) Geosimulation—object-based modeling of urban systems. Comput Environ Urban Syst 28(1/2):1–8

Blecic I, Cecchini A, Trunfio GA (2010) A comparison of evolutionary algorithms for automatic calibration of constrained cellular Automata. In: Taniar D et al (eds) ICCSA 2010, Part I. LNCS, vol 6016. Springer, Berlin, pp 166–181

Cao K, Huang B, Li M, Li W (2014) Calibrating a cellular automata model for understanding rural–urban land conversion: a Pareto front-based multiobjective optimization approach. Int J Geogr Inf Sci 28(5):1028–1046. doi:10.1080/13658816.2013.851793

Cao K, Wang S, Li X, Chen R (2011) Modeling conversion of rural-urban land use based on cellular automata and genetic algorithm geoinformatics. In: 19th IEEE international conference on, Shanghai 1–5. doi:10.1109/GeoInformatics.2011.5981029

Clarke KC, Hoppen S, Gaydos L (1996) Methods and techniques for rigorous calibration of a cellular automaton model of urban growth. proceedings. In: Third international conference/workshop on integrating geographic information systems and environmental modeling, 21–25th Jan 1996, Santa Fe, NM

Clarke KC, Hoppen S, Gaydos L (1997) A self-modifying cellular automaton model of historical urbanization in the San Francisco Bay area. Environ Plan 24:247–261

Clarke KC, Gaydos L (1998) Loose coupling a cellular automaton model and GIS: long-term growth prediction for San Francisco and Washington/Baltimore. Int J Geogr Inf Sci 12(7): 699–714

Clarke KC (2003) Geocomputation's future at the extremes: high performance computing and nanoclients. Parallel Comput 29(10):12811295

Clarke KC (2004) The limits of simplicity: toward geocomputational honesty in urban modeling. In: Atkinson P, Foody G, Darby S, Wu F (eds) GeoDynamics. CRC Press, Florida

Clarke KC, Gazulis N, Dietzel CK, Goldstein NC (2007) A decade of SLEUTHing: Lessons learned from applications of a cellular automaton land use change model. In: Fisher P (ed) Classics from IJGIS. twenty years of the International Journal of Geographical Information Systems and Science. Taylor and Francis, CRC, Boca Raton, FL, pp 413–425

Clarke KC (2008a) Mapping and Modelling Land Use Change: an Application of the SLEUTH Model. In: Pettit C, Cartwright W, Bishop I, Lowell K, Pullar D, Duncan D (eds) Landscape analysis and visualisation: spatial models for natural resource management and planning. Springer, Berlin, pp 353–366

Clarke KC (2008b) A Decade of Cellular Urban Modeling with SLEUTH: Unresolved Issues and Problems. In: Brail RK, Lincoln RK (eds) Planning support systems for cities and regions rail. Lincoln Institute of Land Policy, Cambridge MA, pp 47–60

Clarke KC (2014a) Why simulate cities? GeoJournal 79:129–136. doi:10.1007/s10708-013-9499-5

Clarke KC (2014b) Cellular automata and agent-based models. In: Fischer MM, Nijkamp P (eds) Handbook of regional science. Springer, Berlin Heidelberg

Clarke-Lauer MD, Clarke KC (2011) Evolving simulation modeling: Calibrating SLEUTH using a genetic algorithm. In: Proceedings of the 11th international conference on geo computation. Univ. College London, London

Chaudhuri G, Clarke KC (2013) The SLEUTH land use change model: a review. Int J Environ Resour Res 1(1):88–104

Colonna A, Di Stefano V, Lombardo S, Papini L, Rabino GA (1998) Learning urban cellular automata in a real world: the case study of Rome metropolitan area. In: ACRI'98 third conference on cellular automata for research and industry. Trieste, 7–9 Oct 1998. Springer, London, pp 165–218

Dietzel C, Clarke KC (2007) Toward optimal calibration of the SLEUTH land use change model. Trans GIS 11(1):29–45

Feng Y, Liu Y, Tong X, Liu M, Deng S (2011) Modeling dynamic urban growth using cellular automata and particle swarm optimization rules. Landsc Urban Plann 102(3):188–196. doi:10.1016/j.landurbplan.2011.04.004

Feng Y, Liu Y (2012) Anoptimised cellular automata model based on adaptive genetic algorithm for urban growth simulation. In: Shi W, Yeh A, Leung Y, Zhou C (eds) Advances in spatial data handling and GIS: 14th international symposium on spatial data handling. Springer, Heidelberg, Germany, pp 27–38

García A, Santé MI, Boullón M, Crecente R (2013) Calibration of an urban cellular automaton model by using statistical techniques and a genetic algorithm. Application to a small urban settlement of NW Spain. Int J Geogr Inf Sci 27(8):1593–1611. doi:10.1080/13658816.2012.762454

Goldstein NC (2004) Brains vs. Brawn: Comparative strategies for the calibration of a cellular automata-based urban growth model. In: Atkinson P, Foody G, Darby A, Wu F (eds) GeoDynamics. CRC Press, Boca Raton, FL

Gong Z, Tang W, Thill JC (2012) Parallelization of ensemble neural networks for spatial land-use modeling. In: Proceedings of the 5th ACM SIGSPATIAL international workshop on location-based social networks. ACM, pp 48–54

Guan Q, Wang L, Clarke KC (2005) An Artificial-Neural-Network-based, Constrained CA model for simulating urban growth. Cartogr Geogr Inf Sci 32(4):369–380

Guan Z, Wang Y, Zou Y, Yu F, Jia Z, Li Z (2010) Study on urban spatial structure changes of Jiaozuo City based on SLEUTH Model. In: Third international symposium of computer science and computational technology (ISCSCT 2010) 14–15 Aug 2010. Acad Publisher Finland

Houet T, Aguejdad R, Doukari O, Battaia G, Clarke K (2016) Description and validation of a 'non path-dependent' model for projecting contrasting urban growth futures. Cybergeo: Eur J Geogr. Systèmes, Modélisation, Géostatistiques, document 759 http://cybergeo.revues.org/27397. doi:10.4000/cybergeo.27397

Holland JH (1998) Emergence: from chaos to order. Addison-Wesley, Redwood City, CA

Hu Z, Lo C (2007) Modeling urban growth in Atlanta using logistic regression. Comput Environ Urban Syst 31(6):667–688

Jafarnezhad J, Salmanmahiny A, Sakieh Y (2015) Subjectivity versus objectivity: Comparative study between Brute Force method and Genetic Algorithm for calibrating the SLEUTH urban growth model. Urban Plann Dev. doi:10.1061/(ASCE)UP.1943-5444.0000307

Kirtland D, Gaydos L, Clarke KC, DeCola L, Acevedo W, Bell C (1994) An analysis of human-induced land transformations in the San Francisco Bay/Sacramento area. World Resour Rev 6(2):206–217

Li X, Lin J, Chen Y, Liu X, Ai B (2013) Calibrating cellular automata based on landscape metrics by using genetic algorithms. Int J Geogr Inf Sci 27(3):594–613. doi:10.1080/13658816.2012.698391

Li X, Yeh AGO (2002) Neural-network-based cellular automata for simulating multiple land use changes using GIS. Int J Geogr Inf Sci 16(4):323–343

Li X, Yeh AGO (2004) Data mining of cellular automata's transition rules. Int J Geogr Inf Sci 18:723–744

Liu Y, Phinn SR (2003) Modelling urban development with cellular automata incorporating fuzzy-set approaches. Comput Environ Urban Syst 27:637–658

Long Y, Mao Q, Dang A (2009) Beijing urban development model: Urban growth analysis and simulation. Tsinghua Science and Technology 14(6):782–794

National Research Council (2014) Advancing land change modeling: opportunities and research requirements. Geographical Sciences Committee. National Academy Press, Washington DC

Pijanowski BC, Shellito B, Pithadia S (2002) Using artificial neural networks, geographic information systems and remote sensing to model urban sprawl in coastal watersheds along eastern Lake Michigan. Lakes Reservoirs 7:271–285

Pontius RG Jr, Boersma W, Castella J-C, Clarke K, de Nijs T, Dietzel C, Zengqiang D, Fotsing E, Goldstein N, Kok K, Koomen E, Lippitt CD, McConnell W, Sood AM, Pijanowski B, Pithadia S, Sweeney S, Trung TN, Veldkamp AT, Verburg PH (2007) Comparing the input, output, and validation maps for several models of land change. Ann Region Sci 42(1):11–37

Rienow A, Goetzke R (2014) Supporting SLEUTH—Enhancing a cellular with support vector machines for urban growth modeling. Comput Environ Urban Syst 49:66–81

Sakieh Y (2013) Urban sustainability analysis through the SLEUTH urban growth model and multi-criterion evaluation. A case study of Karaj City. PhD Dissertation, University of Tehran, Iran

Santé I, García AM, Miranda D, Maseda RC (2010) Cellular automata models for the simulation of real-world urban processes: a review and analysis. Landsc Urban Plann 96(2):108–122

Shan J, Alkheder S, Wang J (2008) Genetic algorithms for the calibration of cellular automata urban growth modeling. Photogramm Eng Remote Sens 74(10):1267–1277

Silva EA, Clarke KC (2002) Calibration of the SLEUTH urban growth model for Lisbon and Porto, Portugal. Comput Environ Urban Syst 26(6):525–552. doi:10.1016/S0198-9715(01)00014-X

Silva EA, Clarke KC (2005) Complexity, emergence and cellular urban models: Lessons learned from applying SLEUTH to two Portuguese metropolitan areas. Eur Plan Stud 13(1):93–115

Syphard AD, Clarke KC, Franklin J, Regan HM, Mcginnis M (2011) Forecasts of habitat loss and fragmentation due to urban growth are sensitive to source of input data. J Environ Manage 92(7):1882–1893

Srinivas M, Patnaik LM (1994) Adaptive probabilities of crossover and mutation in genetic algorithms. IEEE Trans Syst Man Cybern 24(4):656–667

Straatman B, White R, Engelen G (2004) Towards an automatic calibration procedure for constrained cellular automata. Comput Environ Urban Syst 28:149–170

Torrens PT, O'Sullivan D (2001) Cellular automata and urban simulation: where do we go from here? Environ Plan 28(2):163–168

Veerbeek W, Pathirana A, Ashley R, Zevenbergen C (2015) Enhancing the calibration of an urban growth model using a memetic algorithm. Comput Environ Urban Syst 50:53–65

Verburg PH, Schot PP, Dijst MJ, Veldkamp A (2004) Land use change modelling: current practice and research priorities. GeoJournal 61:309–324

Wu F (1998) SimLand: a prototype to simulate land conversion through the integrated GIS and CA with AHP-derived transition rules. Int J Geogr Inf Sci 12(1):63–82

Wu F, Webster CJ (1998) Simulation of land development through the integration of cellular automata and multi-criteria evaluation. Environ Plan 25(1):103–126

Wu F (2002) Calibration of stochastic cellular automata: the application to rural-urban land conversions. Int J Geogr Inf Sci 16(8):795–818

Wu X, Hu Y, He H, Bu R, Onsted J, Xi F (2009) Performance evaluation of the SLEUTH model in the Shenyang Metropolitan Area of Northeastern China. Environ Model Assess 14(2):221–230

Xiang WN, Clarke KC (2003) The use of scenarios in land use planning. Environ Plan 30:885–909

Yang QS, Li X (2007) Calibrating urban cellular automata using genetic algorithms. Geogr Res 26 (2):229–237

Yang Q, Li X, Shi X (2008) Cellular automata for simulating land use changes based on support vector machines. Comput Geosci 34(6):592–560

Chapter 9
The Importance of Scale in Land Use Models: Experiments in Data Conversion, Data Resampling, Resolution and Neighborhood Extent

J. Díaz-Pacheco, H. van Delden and R. Hewitt

Abstract The investigation and modeling of land use dynamics can be conducted at different scales based on the objective of the study. However, few studies have looked at comparing various scale aspects, such as spatial resolution and the related neighborhood effect, for practical case study applications. In this chapter, we contribute to this under-explored area with a detailed study of how changes in the data preparation procedures and the scale decisions made in setting up a land use model can affect its performance. For these purposes we used a Cellular Automata (CA) based land use model, which we applied to the Madrid region in Spain. In order to discover the most appropriate method for preparing input data, different vector-to-raster conversion and resampling strategies were tested with reference to 4 statistics. For vector-to-raster conversion, the cell center method was found to give the best results across all of the statistics. Furthermore, direct conversion from the original vector map to raster format at the desired cell size was found to give better results than resampling to the desired cell size from a different cell size. We also tested the effect of changing spatial resolution and cell neighborhood distance on a model's goodness-of-fit to real data using a range of location and pattern metrics. Although differences were noted in the simulations, all the applications fitted the

J. Díaz-Pacheco (✉)
Universidad Complutense de Madrid, Madrid, Spain
e-mail: jdiazpac@ull.es

J. Díaz-Pacheco
Cátedra RRD. Ciudades Resilientes, Universidad de La Laguna, San Cristóbal de La Laguna, Spain

H. van Delden
Research Institute for Knowledge Systems, Maastricht, The Netherlands
e-mail: hvdelden@riks.nl

H. van Delden
The University of Adelaide, Adelaide, Australia

R. Hewitt
Observatorio para una Cultura del Territorio, Madrid, Spain
e-mail: richardjhewitt@hushmail.com

M.T. Camacho Olmedo et al. (eds.), *Geomatic Approaches for Modeling Land Change Scenarios*, Lecture Notes in Geoinformation and Cartography,
https://doi.org/10.1007/978-3-319-60801-3_9

data satisfactorily. Nevertheless, the 50 × 50 m cell resolution applications were visually much more realistic, perhaps because this resolution was used in the initial calibration of the model. The results indicate that data conversion issues have a major effect on the quality of the input data. Additionally, models of this type appear to be much less sensitive to scale changes, either through cell resolution changes, neighborhood changes, or both, than is usually suggested by the literature.

Keywords Land use models · Land use change · Scale · Cellular automata · Data conversion

1 Introduction

Researchers in land use/land cover change (hereafter LUCC) modeling approach their research with different objectives for different regions, and as a result work at different cartographic spatial scales. The observation and modeling of spatial phenomena should be carried out at a scale appropriate to the phenomena in question (Woodcock and Strahler 1987). However, the same phenomenon can be modelled at various scales depending on the spatial context of the analysis. When investigating the urbanization processes in a specific city, a high level of spatial detail might be required, because urban land use units (facilities, parks, residential areas, etc.) are small in size and specific local information about socio-economic and physical characteristics and planning regulations can play a key role in understanding dynamics. However, for modeling the urban growth of Europe, the analysis should not be too detailed or the generality of the process at the continental scale could be missed. The geographical scale of models depends therefore on both the phenomenon to be modelled and the spatial scale (local, regional, national, global…) of the analysis. From a geographical point of view, variability of scale can be regarded as both a strength and a weakness of the discipline (Lam and Quattrochi 1992). LUCC models, as geographical tools to understand phenomena (Longley and Batty 2003 p. 5) and provide policy support (Van Delden et al. 2010), are clearly subject to the same considerations. In fact, the results of dynamic spatial models are strongly influenced by scale, and results derived from a model developed for one spatial scale may not be applicable at another. When moving from one scale to another, land use patterns may disappear or emerge (de Koning et al. 1998), significant processes may lose their significance, and rates of change may vary (Kok and Veldkamp 2001). The Modifiable Areal Unit Problem (MAUP), a concept which describes '*the variation in results that can often be obtained when data for one set of areal units are progressively aggregated into fewer and larger units for analysis*' (Openshaw and Taylor 1979; Openshaw 1983) helps to understand some of these key issues. On the other hand, variability of scale could be considered an advantage since data and information can be adapted to suit the context in which the analyzed process are occurring (i.e. operational scale), or to take into account the way humans may perceive different spaces at differing levels of detail in a hierarchical fashion according to their proximity (Van Vliet et al. 2009).

Although the importance of scale in LUCC models is recognized by many researchers (e.g. Jenerette and Wu 2001; Theobald and Hobbs 1998; Ménard and Marceau 2005; van Delden et al. 2011), experimental studies in which the implications of different scale options are directly compared, are generally lacking (Jantz and Goetz 2005). In this paper, we contribute to this under-explored topic. We assess the accuracy and quality of the data when they are converted from vector to raster or resampled by different methods, to be used on spatially-explicit land use models. In particular, we explore the data conversion and rasterization options, and the impact of spatial resolution on the calibration of a land use model for Madrid.

The paper is organized as follows. First, we present the background and the method applied for each of the two components of this study, data preparation and model resolution, followed by details of the application including the GIS and the land use model used in this study. Finally we present and discuss the results, draw the relevant conclusions and make recommendations for further research.

2 Finding the Right Scale

2.1 Background

Although several authors have proposed ways to find the best scale for setting up raster-based LUCC models (e.g. Tobler 1988; Lam and Quattrochi 1992), no single widely-agreed method has emerged, and the final decision is usually taken on the basis of the researcher's own specialist knowledge. Such a decision may not always be the result of a rigorous procedure, but is not usually arbitrary. For example, in policy-relevant LUCC models, the decision about which scale to use is often a trade-off between the scale required by intended users, the scale at which processes are best represented, and practical considerations like data availability or computational resources (van Delden et al. 2011).

The scale decisions a LUCC modeler needs to make include the spatial and temporal extent, the spatial level(s) and the hierarchy by which they are ordered, and the amount of detail incorporated. Levels refer to locations along a scale (Gibson et al. 2000) and detail relates to the spatial, temporal and thematic resolution(s) and the complexity by which processes are represented (van Delden et al. 2011). When focusing on CA based land use models, other important factors related to spatial resolution, such as neighborhood size and type, must also be taken into account (Ménard and Marceau 2005).

Previous work on the effect of changing scale in a LUCC model for Central America by Kok and Veldkamp (2001) found that coarsening the resolution from 15 $\times$ 15 km to 75 $\times$ 75 km led to improved model explanatory power (r^2), but did not significantly affect the explanatory variables (i.e. land change drivers identified were broadly the same at both resolutions tested). However, changing the extent of the model produced a strong variation in performance (poorer fit for all Central America, better fit for individual countries). Though these authors do not say so explicitly, this is an excellent example of the scale problem (Openshaw 1983) in a land use model, since the land change dynamics modelled by these authors relate to national, not supra-national drivers, and are not generalizable across borders. Though it is common practice in land use models to work at larger spatial extents, the findings of these authors are a clear warning of the perils that this may entail. However, various modeling approaches overcome this issue by dividing the modelled area into smaller subdivisions, as for example in the case of the regional model incorporated into the Metronamica Modeling framework (RIKS 2014).

Jantz and Goetz (2005) investigated the behavior of different types of urban growth rules at different cell sizes in the popular SLEUTH model, concluding that cell resolution was a major determinant in model performance and that some types of urban growth rule produced much more growth at coarser resolutions than at finer ones. Though their findings are quite specific to the SLEUTH model, the implication is that neighborhood effects for urban land, which are fundamental in all CA models, may vary non-linearly across scales.

Ménard and Marceau (2005) observed how changing the size of the neighborhood radius and the resolution produced a non-linear relationship between the spatial scale and the simulation results. Their work was based on a dataset derived from remote-sensing images for two time periods and focused on land cover change. The study area was dominated by forest and agriculture, so the phenomenon of urban expansion was not considered (Ménard and Marceau 2005). Samat (2006) undertook sensitivity analysis of a CA-based urban model with the aim of finding the appropriate scale for the modelled region (Seberang Perai, Malaysia). The study found that the model performed well at 30, 90, and 270 m cell resolution, but at coarser resolutions (810, 2430 m), accuracy declined rapidly. These findings appear to contradict the findings of Kok and Veldkamp. However, these studies are difficult to compare for a number of reasons. Firstly, Kok and Veldkamp compared only two resolutions, while Samat investigated five. Secondly, the studies do not compare the same cell resolutions and address different spatial extents. Thirdly, the statistical comparison methods used were quite different (Kok and Veldcamp used the coefficient of determination (R^2) of a regression model, whereas Samat used cell-by-cell map comparison techniques). Finally, Samat employed standard Kappa for comparing real and simulated maps, an approach which has since been found to be inadequate (Pontius and Millones 2011; Van Vliet et al. 2011).

As an aid to determining the appropriate spatial scale for the general case, Samat's work (2006) has some limitations. On the one hand, the analysis comprised only two land use classes (urban and non-urban), so the type of urban land use was not a determining factor for selecting the scales for the tests. Moreover, the land use dataset employed was drawn from different sources for each of the two time periods (1990 and 1998). In addition, the cartographic scale chosen for the smallest cell resolution tested (30 × 30 m) does not seem to respect, at least for 1990, the general rules for transformation of a scaled vector map (1:75,000) to a raster map (see Tobler 1988).

The various studies show that the choice of scale, and, in particular, of the spatial resolution, is key in setting up a land use model, as these can have a large impact on the model results. With limited work being carried out in urban environments, this paper aims to contribute to an enhanced understanding in this area by exploring the effects of spatial resolution and neighborhood extents on a land use model's capacity to simulate land use change.

3 Approach

3.1 Scale in Geography and Remote Sensing

Three techniques of land use data conversion from vector to raster, and two techniques of aggregation by resampling from a high cell resolution to a lower one are tested. The data conversion and resampling techniques used are those implemented inside the popular ArcGIS 10.0 software. ArcGIS was chosen because it is widely used and provides a detailed description of the procedures in the user manual. Testing was undertaken by developing a series of land use maps as input data for a LUCC model generated by each technique and then comparing the results using statistical map comparison algorithms. In the following section we describe the data conversion and cell aggregation methods used to obtain the most appropriate data for use in the LUCC applications at different resolutions, together with the metrics used to evaluate the maps generated by the various techniques.

3.1.1 Vector-to-Raster Conversion

In the vector-to-raster conversion, some loss of accuracy is unavoidable, due to classification errors where the irregular polygon boundary coincides with a regular grid (Carver and Brunsdon 1994). Three techniques implemented in ArcGIS 10.0 for direct conversion from a vector polygon coverage to a regular grid were analyzed, namely Cell Center, Maximum Area and Maximum Combined Area (the names used in the software) (Fig. 1).

Using the cell center (Cc) algorithm the final categorical value of every cell in the grid is the attribute value which coincides with the center of the cell. In the case of the maximum area algorithm (Ma) the final value of the cell is established by assigning the value of the largest polygon coincident with the cell. The maximum combined area algorithm (Mca) works in a similar way to the Cc algorithm, except that the value of the cell is taken from the total area of different polygons with the same attributes coincident with the cell.

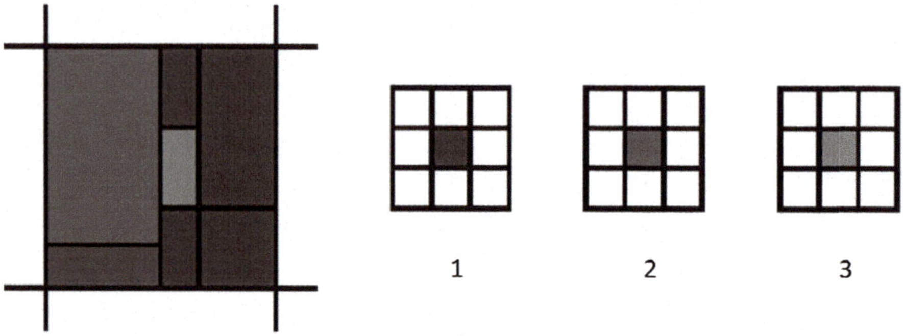

Fig. 1 Vector polygon to raster. (*1*) Mca. Maximum Combined Area; (*2*) Ma. Maximum Area; (*3*) Cc. Cell center. *Source* Adapted from ESRI (2010)

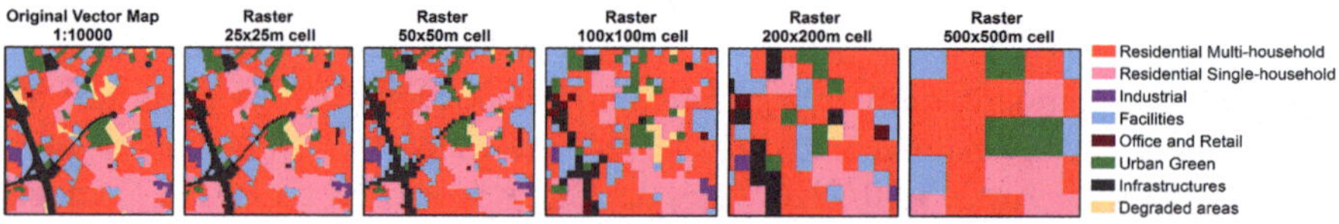

Absolute variation on urban land patches

Land Use	**Vector**	**raster 25**	**raster 50**	**raster 100**	**raster 200**	**raster 500**
Facilities	5887	4546	4188	2991	1360	346
Industrial	1931	1747	1778	1463	848	251
Office and Retail	540	555	547	381	198	55
Residential Multi-household	5512	1563	1552	1228	663	217
Residential Single-household	4538	3380	3620	2897	1695	575
Urban Green	2038	2214	2429	1624	678	195
Total land patches	20446	14005	14114	10584	5442	1639
% of missed patches		31.50	30.97	48.23	73.38	91.98

Relative variation on urban land patches (%)

Land Use	**raster 25**	**raster 50**	**raster 100**	**raster 200**	**raster 500**
Facilities	22.78	28.86	49.19	76.90	94.12
Industrial	9.53	7.92	24.24	56.08	87.00
Office and Retail	-2.78	-1.30	29.44	63.33	89.81
Residential Multi-household	71.64	71.84	77.72	87.97	96.06
Residential Single-household	25.52	20.23	36.16	62.65	87.33
Urban Green	-8.64	-19.19	20.31	66.73	90.43

Fig. 2 Variation on urban land patches after conversion vector information and resample up to a resolution of 500 × 500 m

Vector-to-raster conversions were performed from an original land use vector dataset to a grid of 25 × 25 m, 50 × 50 m, 100 × 100 m resolution successively. We considered that at lower resolution the general land use structure is missed (Fig. 2).

The most detailed resolution (25 × 25 m) was selected following recommendations given by Switzer (1975) in which 50% of the area of the cell should be larger than the smallest mapped polygon. In the MLU geodatabase the smallest mapped polygon is 30.4 m^2 and 50% of a 25 × 25 m cell is 312.5 m^2, which complies with the requirements of this rule.

3.1.2 Resampling

In the same way as for the vector-to-raster conversion analysis, for the resampling of grid maps, two techniques, Nearest Neighborhood Assignment and Majority, implemented in the software ArcGIS 10.0 were applied consecutively. The former assigns the categorical value to the new cell according to the value of the cell closest to the center of the new cell and the latter assigns the most popular values of the cells in the input map that fall inside the new cell in the output map. A simple example of both algorithms is shown in Fig. 3.

For the nearest neighborhood assignment method, the maximum spatial error must be one-half of the cell size, while for the majority method the results of the resampling tend to create higher compactness (ESRI 2010).

The techniques were applied to the 25 × 25 m raster map obtained from the vector polygon land use map using the vector to raster method that provided the best results. Aggregations were carried out into grids with 50 × 50 m and 100 × 100 m cell sizes, each one from the 25 × 25 m raster map.

3.1.3 Assessment Procedure for Vector to Raster Conversion and Resampling

Comparisons of the maps resulting from application of the various shape-to-raster and aggregation techniques were carried out at 25 m, 50 m and 100 m resolution. In order to compare 50 m and 100 m resolution maps, all the resulting maps were disaggregated to a 25 m resolution.

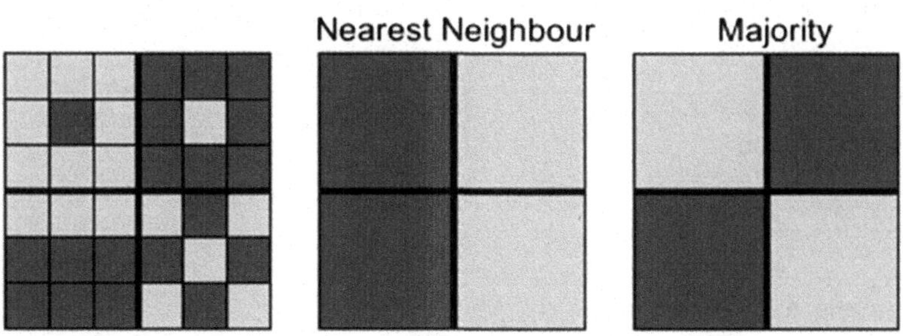

Fig. 3 Different techniques for resampling. Nearest Neighbour and Majority. *Source* Adapted from ESRI (2010)

The similarity of the land use maps resulting from the different conversion and resampling methods was analyzed using metrics to assess the similarity in location and the similarity in the resulting landscape pattern.

3.2 Assessing the Impact of Spatial Resolution and Neighborhood Extent on the LUCC Model

In order to evaluate the effects of the spatial resolution and the size of the neighborhood, a set of applications was developed, in which the cell size and neighborhood were varied while the extent remained constant. To keep the work manageable, it was decided to apply the model at three different spatial resolutions. Resolutions of 25 × 25 m, 50 × 50 m, and 100 × 100 m were selected based on the urban context and the authors' interest in investigating whether higher spatial resolutions, possible due to the availability of detailed land use data sets, would also result in improved model calibration and validation results.

We began by developing an application at 50 m, with a neighborhood of 8 cells (400 m). This application was calibrated over a first historic time period and validated over a second. Once this application was considered suitable for reproducing the (historic) land use dynamics, some of its scale characteristics were modified in order to evaluate their effects on the model results. To this end, two applications were developed with a modified resolution of the cells (25 m and 100 m), using the most appropriate methods found for data preparation, while maintaining the cell radius for the neighborhood effect at 8 cells. Next, two additional applications at 25 m and 100 m cell-resolution applications were created with respectively larger (16 cells) and smaller (4 cells) radii, so as to maintain the equivalent cell neighborhood distance as in the original 50 m application. All the applications were run using the same parameter settings employed in the original 50 m application (Table 1).

As with the assessment of the different conversion and resampling methods, the results of the calibration and validation have been analyzed using metrics for assessing similarity in location and in the resulting landscape pattern.

4 Applications

4.1 Study Area

The area selected for analysis is the Madrid region (Fig. 4), an area of around 6 million inhabitants. This region was chosen because of the large increase in urban development that it has experienced over recent decades (until the beginning of the

Table 1 Scale changes on neighborhood for each application

Resolution	100 × 100 Doubled resolution	100 × 100 Doubled resolution	50 × 50 Original	25 × 25 Halved resolution	25 × 25 Halved resolution
Feature of changes	Doubled resolution Equal radius in cells Unequal radius in meters	Doubled resolution Unequal radius in cells Equal radius in meters		Halved resolution Unequal radius in cells Equal radius in meters	Halved resolution Equal radius in cells Unequal radius in meters
Cell radius	8	4	8	16	8
Meter radius	800	400	400	400	200
Number of cells	196	48	196	796	196
Area in m^2	1,960,000	480,000	490,000	497,500	122,500

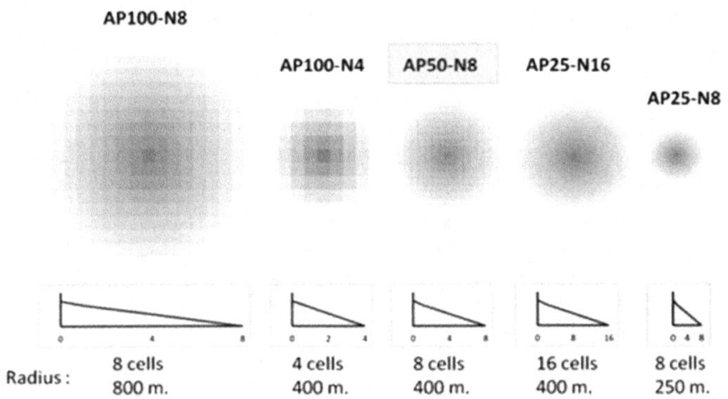

current economic crisis around 2008) and because a highly detailed land use database documenting this change has recently become available (Díaz-Pacheco and García Palomares 2014).

The expansion of urban land use in the Madrid metropolitan area during the 1990s was extraordinary, at least by European standards. According to CORINE land cover (EEA 2014), artificial land cover increased by more than 30,000 ha, an annual growth rate of 4.77%, while over the same period, the population of around 6 million grew by only 0.8% a year. Furthermore, over this decade the area under construction (mines, dumps, and construction sites) grew by 200% (Rocha et al. 2009; Hewitt and Escobar 2011). This growth in urban land, in a situation of demographic stability, produced a notable increase in the amount of artificial land per person, which in only 5 years (1996–2001) shot up from 153 to 179 m^2 per inhabitant (de Lucio 2011).

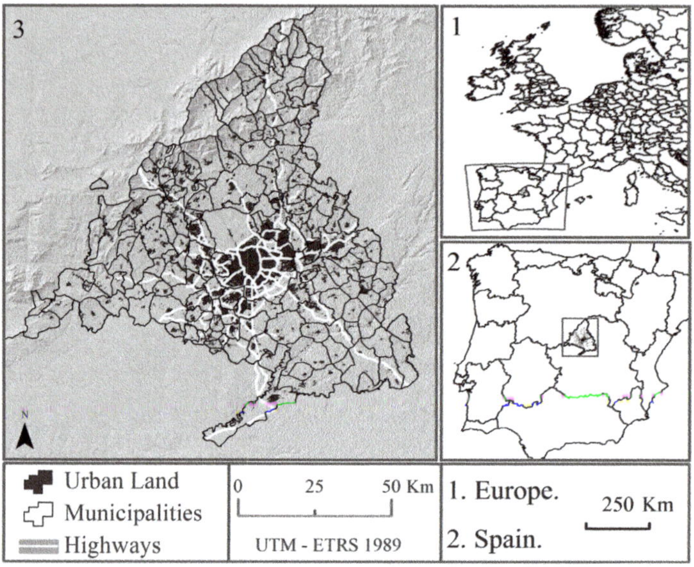

Fig. 4 Location of Madrid Region. *Source* Díaz-Pacheco and Gutiérrez (2013)

4.2 Land Use Data Set

Madrid Land Use (MLU) is a cartographic database with land use and land cover information for the Madrid Region, covering the time periods 2000, 2006 and 2009. The MLU dataset comprises 22 land use classes of which 7 are urban. Mapping was undertaken at a highly detailed basic reference scale of 1:10,000. The technical process did not include automatic or computer-assisted classification tasks, and the mapping work was undertaken entirely by photo-interpretation of high resolution (0.5 m) aerial orthophotographs, supported by large scale cartographic and cadastral information (1:5,000 and 1:1,000, respectively). Identical criteria were used for the digitization and thematic classification for each of the land use dataset periods. MLU clearly represents an excellent cartographic dataset for assessing urban land use in Madrid and outperforms CORINE land cover in this area in a number of respects (see Díaz-Pacheco and Gutiérrez 2013).

4.3 Land Use Model

The LUCC model applications were built using the well-known "Metronamica" framework, developed by RIKS (e.g. White and Engelen 1993, 2000; Van Delden and Hurkens 2011) and widely used around the world for simulating urban land transformation (Barredo et al. 2004; van Delden et al. 2005; Lajoie and Hagen-Zanker 2007).

In this model, the distribution of land use in a given area is represented as a raster map in which each cell has a value that represents a land use. Not all land uses are modelled in the same way and individual land use classes must be assigned to one of three land use states. They may be either active (dynamic, changing as a result of external demands), passive (dynamic, without an external demand), or static (inert throughout the model runtime).

Metronamica calculates land use changes over time according to a set of transition rules computed by simple equations in which the geographic effect of a cell over its neighbors (attraction or repulsion between land use cells, representing economic and political power to obtain locations of interest, inertia and ease of conversion) is the main driving force of change in the system. Three additional factors are included to reflect the heterogeneity of the area: accessibility and suitability drivers are introduced to align the model with the characteristics of the study region and zoning is included to incorporate the influence of policies or planning. The model includes a stochastic component to reflect uncertainty in the allocation process. Cells are allocated at each step of the model on the basis of the transition potential until cell demand (determined exogenously) is exhausted or all suitable and available cell space is used up (see the Metronamica documentation (RIKS 2014) for more information).

For the application to Madrid, we combined some of the MLU land use classes to create a set of 12 land use classes of which 7 are urban and 6 are actively simulated (Fig. 5). This permits the observation of the effects of the change of scale

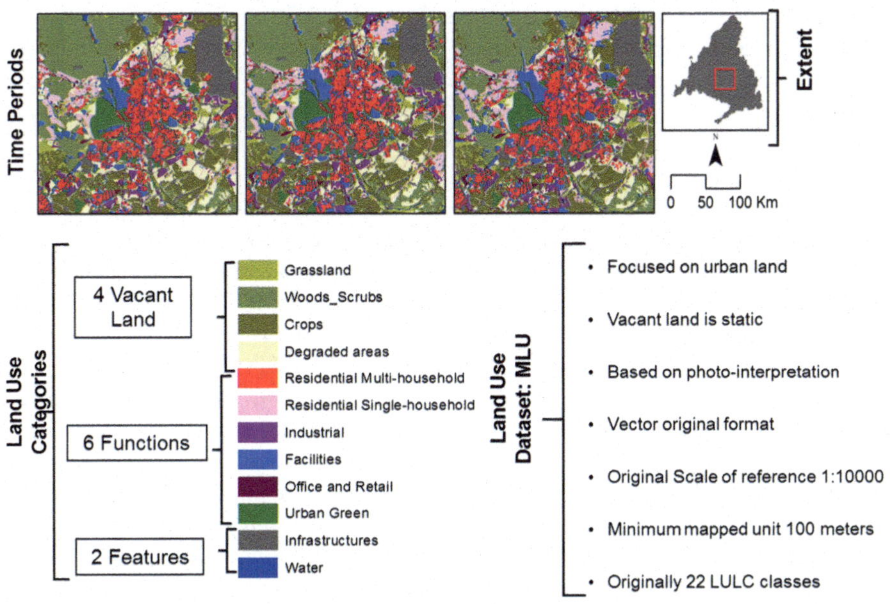

Fig. 5 Characteristics of the Madrid Model

on the model (cell size and neighborhood) for different urban land uses with dissimilar spatial behavior and dissimilar clustering.

The application for the Madrid region used in this research does not incorporate zoning so as to give the system much greater freedom. The only suitability factor included is the slope of the terrain, as this was found to be the only physical factor affecting urban land change in this region. Infrastructure networks and nodes (highways, roads, train stations and metro stations) are included and accessibility is empirically calibrated for each simulated land use through a distance decay function. The amount of randomness was set by trial-and-error during calibration.

A manual calibration was also performed on the 50 m resolution application using the common Metronamica calibration procedure (Van Delden et al. 2010, 2012). The transition rules were determined by trial-and-error, informed by previous analysis of land change processes, and by comparing the resulting simulations with historical data, until they achieved an acceptable goodness of fit (according to plausible parameters and map statistics).

The accessibility values were introduced in a similar way for each application, but in this case the values between the nearest and the furthest distance considered to the network (roads, rail, highway, metro stations…) were automatically computed by the software through a distance decay function. The only change made in this case was doubling or halving the distances in order to adapt the function for each application, e.g. if in the 50 × 50 m application a value for the road influence at 200 m to the residential land cells was considered, in the 100 × 100 m application this value was doubled to 400 m to respect the proportionality demanded by the size of the cell.

Following common practice, the transition rules thus obtained were tested for validation purposes by running the model over a different historical period than that over which the calibration was performed.

4.3.1 Metrics

Calibration and validation results were assessed through visual inspection of result maps and temporal dynamics, assessment of the plausibility of the parameters (structural validation) and a number of objective metrics to assess similarity between result maps and historic data (t_f data and t_f simulated).

The map comparison methods and techniques used during the calibration and validation processes are currently implemented in the software Map Comparison Kit (MCK), initially created by RIKS for the Netherland Environment Assessment Agency (Visser and De Nijs 2006). Three statistical tests were used to determine model accuracy, namely Kappa simulation (Ksim), clumpiness, and mass fractal dimension. The first of these, Ksim, is useful for determining the number of cells that have been correctly simulated, while the remaining two measures are used for determining the degree of spatial similarity between elements in the simulated map and the real map (White 2006). The extent (in cells) occupied by every land use on each map is also measured. In addition, a previous qualitative visual assessment based on the research criteria is generally included in the examination.

The Kappa coefficient of agreement (Cohen 1960) is a widely used index to calculate the rate of agreement between two images or two maps (categorical datasets). The Kappa simulation (Ksim) is a modification of the traditional Kappa coefficient, which is useful for evaluating simulations over short time periods (van Vliet et al. 2011). Most land use models usually simulate changes over years or decades during which time many locations do not undergo any land use change. Unfortunately, under standard Kappa, locations which do not change are also included in the calculation, which means that very high Kappa scores can be obtained regardless of the degree of accuracy of the simulation (Pontius and Millones 2011). Standard Kappa is therefore not a useful measure of the goodness of fit of simulations produced by land use models. Ksim takes values from −1, meaning total disagreement, to 1, for total agreement. The value 0 represents a special situation where the agreement is as good as can be expected by chance given a random distribution of the given class transitions (see van Vliet et al. 2011).

Clumpiness and mass fractal dimension are often employed in landscape ecology to analyze landscape structure. In this research, these metrics allow the pattern similarity of the simulated map and reference map to be assessed. Clumpiness is a measure of the degree of dispersion/aggregation of the patches in an image according to their type (McGarigal 1994). Mass fractal dimension measures the degree of "linearity" of elements in the map in which plane filling objects like circles or squares will have a value of 2.0 and a line will have a value of 1.0 (Gardner et al. 1987).

5 Results and Discussions

5.1 Results of Resample/Conversion Comparison

To examine the results (Table 2), five land use classes selected from the land use map for the year 2000 were analyzed. These classes were chosen in order to provide the greatest possible diversity of patch size for the experiment. The crops category has a very large mean patch size (107.70 ha) compared to the facilities category (3.07 ha). Residential multi-household (10.89 ha), industrial (7.56 ha) and urban green (5.57 ha) were selected to provide intermediate patch sizes between the two extremes.

Results of the vector to raster conversion and resampling tests are given in Table 2. It can be rapidly appreciated that the *CELL CENTER METHOD* gives the best results for direct conversion and the *NEAREST NEIGHBORHOOD METHOD* gives the best results for resampling. However, for the Crops and Urban Green category, the *MAXIMUM AREA* and *MAXIMUM COMBINED AREA* direct conversion methods give acceptable results, at least on the basis of the fractal

Table 2 Comparison of results from the conversion/resample operations

Land use class mean patch size	Index	Direct conversion to 50 m. Central cell	Direct conversion to 50 m. Maximum area	Direct conversion to 50 m. Maximum combined area	Resample to 50 m. Nearest neighbourhood	Resample to 50 m. Majority
Crops	Clumpiness difference	0.0060	0.0089	0.0092	0.0060	0.0165
107.70	Fractal dimension difference	0.0090	0.0212	0.0193	0.0236	0.0271
	Area difference ha	36.7394	4.2394	4.2394	19.2394	870.0106
	Kappa index	0.9535	0.9555	0.9556	0.9463	0.9500
R. Multihousehold	Clumpiness difference	0.0117	0.0179	0.0193	0.0113	0.0345
10.89	Fractal dimension difference	0.0004	0.0118	0.0137	0.0022	0.0485
	Area difference ha	20.7797	176.0297	177.2797	7.5297	325.7797
	Kappa index	0.9136	0.9168	0.9177	0.9003	0.9082
Industrial	Clumpiness difference	0.0063	0.0102	0.0100	0.0064	0.0179

(continued)

Table 2 (continued)

Land use class mean patch size	Index	Direct conversion to 50 m. Central cell	Direct conversion to 50 m. Maximum area	Direct conversion to 50 m. Maximum combined area	Resample to 50 m. Nearest neighbourhood	Resample to 50 m. Majority
7.56	Fractal dimension difference	0.0038	0.0107	0.0103	0.0005	0.0271
	Area difference ha	5.5083	108.5083	108.5083	5.2417	447.9917
	Kappa index	0.9217	0.9236	0.9077	0.9088	0.9146
Urban green	Clumpiness difference	0.0121	0.0158	0.0170	0.0128	0.0307
5.57	Fractal dimension difference	0.0266	0.0375	0.0427	0.0247	0.0642
	Area difference ha	17.3125	9.1875	9.1875	3.8125	212.0625
	Kappa index	0.9048	0.9074	0.9077	0.8907	0.8986
Facilities	Clumpiness difference	0.0120	0.0168	0.0177	0.0122	0.0313
3.07	Fractal dimension difference	0.0116	0.0178	0.0203	0.0120	0.0341

(continued)

Table 2 (continued)

Land use class mean patch size	Index	Direct conversion to 50 m. Central cell	Direct conversion to 50 m. Maximum area	Direct conversion to 50 m. Maximum combined area	Resample to 50 m. Nearest neighbourhood	Resample to 50 m. Majority
	Area difference ha	4.3011	69.9489	69.9489	6.6989	494.6989
	Kappa index	0.8930	0.8961	0.8966	0.8773	0.8852

dimension index, and the area difference in hectares. This could be related to the larger clusters found in this category (the mean patch size is 107.7 ha), but this argument does not apply to Urban Green land. The could be because Madrid contains unusually large, non-parcelled green areas such as the 'Casa de Campo' or 'Dehesa de la Villa', whose geometry is more like agricultural or natural areas than classical urban green land (parks, squares, gardens…).

For the resampling operations, the *MAJORITY METHOD* produces the lowest degree of similarity with the original data. As this method is widely used by researchers, this is a key finding.

For both clumpiness and mass fractal dimension, the calibrated application achieved similar values to the validated application and both outperformed a random land use map used as a benchmark (Table 3).

5.2 Results of Calibration and Validation of the Initial 50 × 50 m Application

Calibration was considered to be complete once values of 0.144 had been obtained for Ksim. The values considerably outperform a null model. The model was considered to have been acceptably validated at 0.113 (Table 3). These values are comparable with published values considered acceptable in other applications of the model (e.g. Hewitt et al. 2014).

5.3 Results of Testing the Changes on the Scale of the Applications

The results of the comparison of data for 2006 with simulations for the same year are shown in Table 4. According to the map comparison indices in use, the simulation results from all the different applications (apps) for 2006 (2000–2006) could be considered acceptable. Both the 25 m app with the 8 cell neighborhood radius and the 100 m app with the 4 cell neighborhood radius actually improve on the original 50 m, 8 cell neighborhood radius app (Table 4). If we look at the values for clumpiness, the difference between the clumpiness of the data and the clumpiness of the simulations is comparable across all the applications, and better than the random simulation used as a benchmark. The same is true for the fractal dimension index. In some cases, the scale-modified apps achieve slightly better values than the initial 50 m app (e.g. AP100-N4 clumpiness for multi-household and facilities classes). However, better performance of some categories tends to be compensated by poorer performance of others. Taken overall, the differences between the scale-modified apps and the original app are not large enough to be able to claim

Table 3 Values of the used indices for calibration and validation of the 50 m.application

Index	AP50 00-06	RAMD50 00-06	AP50 05-09	RAMD50 06-09
Kappa simulation	0.144	–	0.113	–
R. Multi-household Clumpiness difference	−0.0226	−0.0811	−0.002[illegible]	−0.0509
R. Single-household Clumpiness difference	−0.0018	−0.0754	0.0071	−0.0281
Industrial Clumpiness difference	0.0235	0.1196	0.0029	−0.0382
Facilities Clumpiness difference	−0.0183	−0.0856	−0.009[illegible]	−0.0331
Office and Reatil Clumpiness difference	0.0112	−0.2593	0.0081	−01271
Urban Green Clumpiness difference	−0.0361	−1115	−0.008[illegible]	−0.0352
Fractal Dimension difference	0.0070	−0.0268	0.0013	0.0040
	CALIBRATIONData06-Sim06	BENCHMARK	VALIDATIONData09-Sim09	BENCHMARK

AP = application; 50 = 50 m.; 00–06 = 2000–2006; 06–09 = 2006–2009; RAMD = random simulated map that any of the modified applications are significantly worse or better than the original 50 m app.

If we visually compare the land use data maps for 2006 and the simulations produced by the different applications, there are obvious differences that can be quickly detected, even though it is not really possible to specify the precise degree of similarity between the maps using this method. Figure 6 shows simulated (right) and real (left) land use for an enlargement of a highly urbanized (mainly residential) area in 2006. It is clear that the visual appearance of the maps reinforces the results of the statistical comparisons, i.e., none of the scale-modified apps looks significantly better or worse than any other for the land use changes simulated for the year 2006, despite the modifications in the scale. The middle column of Fig. 6 shows the apps with different resolutions (25, 50 and 100 m) and the same neighborhood radius in cells (N8). In these cases, despite the difference in resolution, all three simulations are quite alike, something that can be confirmed by consulting the results of the statistical indices (Table 4). Some classes, e.g. Residential Multi-household are not simulated very successfully in any application. This is probably because all locations in this residential area were equally favorable, being close to existing urban areas, on suitable land and close to transport networks. In such cases, identification of the "real" location tends to be difficult and is effectively made at random. Further discussion of this interesting topic is, however, beyond the scope of this paper.

The relationship between cell-size and the size of the land parcels is also clearly shown. Nonetheless, as the statistics do not show remarkable differences between the results of the apps at different resolution, a visual analysis of the 50 m resolution simulation seems to provide more realistic-looking results than the 25 m and 100 m resolution simulations, probably because the cell size is a closer match to the actual size of the land parcels, although the fact that the original calibration focused on this resolution could also be a factor. This emphasizes the importance of visual inspection when choosing the right resolution for a given application. It also suggests that pattern-based map comparison measures like clumpiness and fractal dimension have their limitations, as do all statistical measures.

This is a rather surprising result. Since the scale modifications were only applied to the maps themselves, and not to the neighborhood rules, neighborhood influence is different in all three applications. The maximum cell neighborhood of 8 cells corresponds to a distance of 400 m (8 × 50) away from the central cell in the original 50 m app, 200 m away from the central cell in the 25 m app, and 800 m away from the central cell in the 100 m app. Three possible explanations can be provided for this; (1) the cell neighborhood is not the key change driver (contrary to most known studies of urban change); (2) the neighborhood influence declines very steeply and all important interactions take place at close distances, or (3) the distance in cells is more important than the actual distance (in meters) in the calculation of the neighborhood effect. Further experimental work (see, e.g. Hewitt and Díaz-Pacheco 2017) would be needed to confirm or reject these hypotheses.

Table 4 4. Map comparison results for applications. Abbreviations: AP50-25-100: applications and resolutions; N8-4: neighborhood and radius in cells; RAMD50-100-25: random simulations and resolutions

	Applications					Benchmarks		
Index	AP50-N8	AP25-N8	AP100-N8	AP25-N16	AP100-N4	RAMD50	RAMD100	RAMD25
Kappa simulation	0.144	0.149	0.116	0.146	0.158	–	–	–
R. Multi-household Clumpiness difference	0.0226	0.0430	0.0050	0.0400	0.0007	0.0811	0.0602	0.1017
R. Single-household Clumpiness difference	0.0018	0.0079	0.0111	0.0192	0.0250	0.0754	0.0643	0.0867
Industrial Clumpiness difference	0.0235	0.0020	0.0409	0.0111	0.0409	0.1196	0.0991	0.1352
Facilities Clumpines Difference	0.0183	0.0290	0.0065	0.0567	0.0044	0.0856	0.0609	0.1054
Office and Reatil Clumpiness difference	0.0112	0.0156	0.0731	0.0284	0.0835	0.2593	0.1981	0.3007
Urban Green Clumpiness difference	0.0361	0.0642	0.0108	0.0849	0.0099	0.1115	0.0752	0.1550
Fractal Dimension difference	0.0070	0.0108	0.0053	0.0157	0.0003	0.0268	0.0182	0.0299

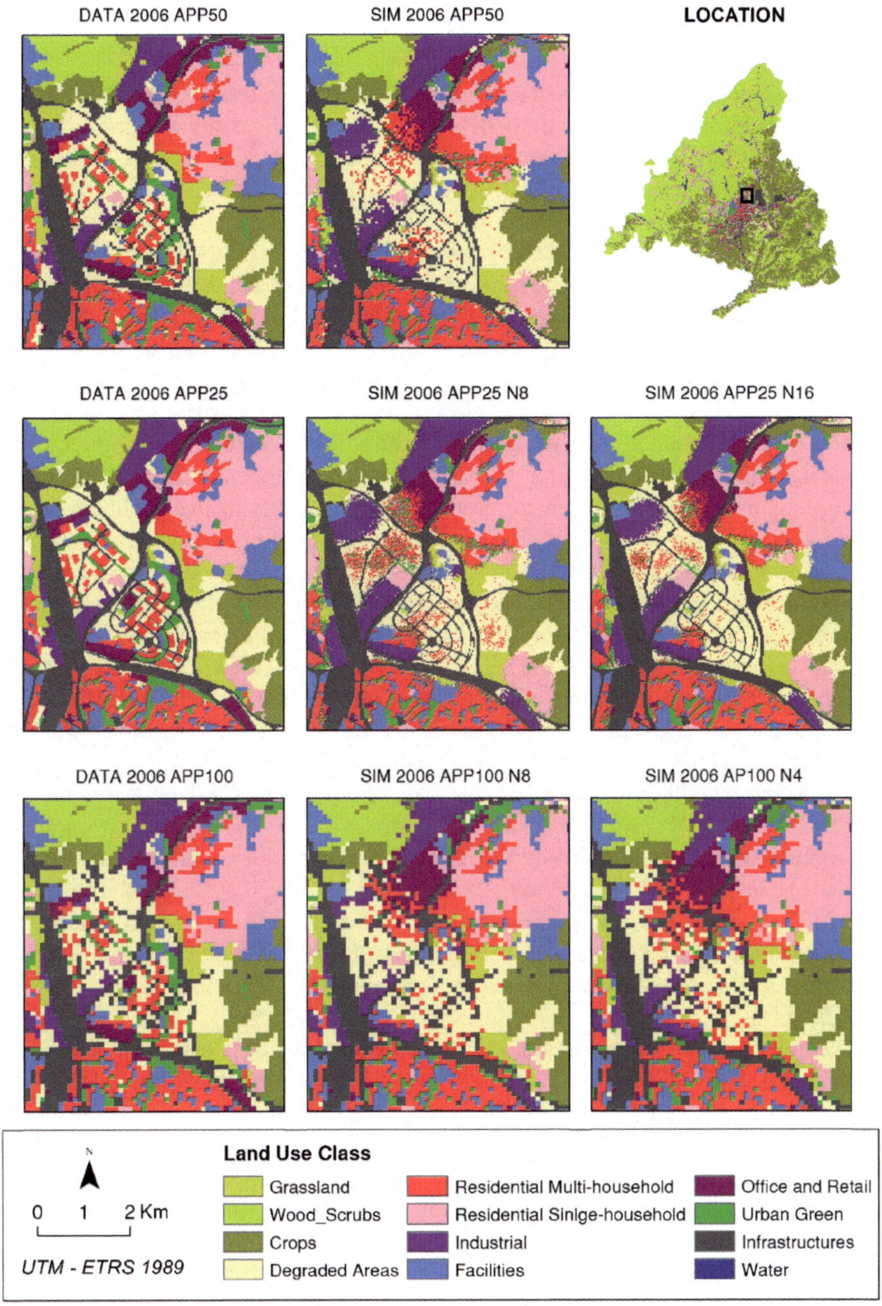

Fig. 6 Comparison of data 2006 and simulations for 2006 from the different apps. Abbreviations: SIM: simulation; APP25-50-100: application and resolution; N4-8: neighborhood and radius

6 Conclusions and Outlook

Standard GIS operations like vector-to-raster conversion and raster resampling have considerable influence on scale in models of land use and land cover change, and the MAUP (Openshaw 1983) would seem to be relevant. In CA models, which are highly dependent on the cell neighborhood for simulating land use conversions, both operations have a significant effect on the initial land use map and hence the cell neighborhood. The work presented in this paper has examined the influence of these operations, the first referred to the resolution transformation of the input data and the second to the land use model's capacity to simulate land use change for a case study application based on a large and detailed land use database for Madrid, Spain. Some important conclusions can be drawn that are likely to be extremely useful for researchers working with cell-based land use models. It is clear from this work that the use of one particular data preparation method over another can produce quite different results, both for vector to raster conversion operations and for raster resampling from one resolution to another. In the underlying research, for urban patch types (smaller mean patch sizes), better results (a closer match to the original land use dataset) are obtained by converting directly from the original vector coverage to a raster with the desired resolution than by converting to a scale equivalent to the original vector coverage and subsequently resampling up or down to obtain the desired resolution. Amongst the resampling methods themselves, the nearest neighbor technique gives improved agreement with regard to the original land use dataset than most other procedures. Future research could try to discover whether similar results would be found if the same methods were applied to different datasets.

Regarding the effects of changing the scale of a dynamic CA land use model, as reflected by the cell resolution and neighborhood radius, no significant variation was obtained in the accuracy of the final simulations measured by the metrics applied, at least in the urban context considered and for the range of resolutions tested (25, 50, and 100 m). A calibrated and validated land use model based on a 50 m resolution raster gave very similar results to applications with identical transition parameter settings but mapped at higher (25 m) and lower (100 m) resolutions.

The goodness-of-fit evaluation techniques (cell statistics, pattern comparison, visual inspection) showed that all of the applications acceptably reproduced the relevant land use change patterns. Despite this result, the 50 m resolution model looked more realistic than 25 m or 100 m resolution applications. This is likely to be because the 50 m cell size is a better fit to the size of the real land parcels, although the fact that the original calibration focused on this resolution may also be a factor.

The most surprising discovery is that doubling or halving the neighborhood distance radius did not produce any significant variation in the model's performance over the validation period. This indicates that for the applications we investigated the transition rules are rather insensitive to neighborhood distance effects. For future research it would be useful to investigate whether similar results are obtained for applications to different regions and datasets and if so, whether the following possible hypotheses could be confirmed or denied: the cell neighborhood is not the key driver for change, neighborhood effects all occur at close distances, the distance in cells is more important than the actual distance (in meters) in the calculation of the neighborhood effect.

Acknowledgements This work has been supported in part by project SIGEOMOD-II, BIA2013-43462-P (Spanish Ministry of Economy and Competitivity and European Regional Development Fund FEDER). The authors also gratefully acknowledge funding received under remit of the EU FP7 project COMPLEX (project no. 308601). The MLU geodatabase was built by researchers from the Department of Human Geography, Universidad Complutense de Madrid with financial support from the Ministerio de Ciencia e Innovación (project TRA2008-06682). Finally, we thank the anonymous reviewers for their helpful comments.

References

Barredo JI, Demicheli L, Lavalle C, Kasanko M, McCormick N (2004) Modelling future urban scenarios in developing countries: an application case study in Lagos, Nigeria. Environ Plann B 31(1):65–84

Carver S, Brunsdon C (1994) Vector to raster conversion error and feature complexity: an empir-ical study using simulated data. Int J Geogr Inf Syst 8(3):261–270

De Lucio RL (2011) Transformaciones territoriales recientes en la región urbana de Madrid. Urban 8:124–161

De Koning GHJ, Veldkamp A, Fresco LO (1998) Land use in Ecuador: a statistical analysis at different aggregation levels. Agric Ecosyst Environ 70:231–247

Díaz-Pacheco J, Gutiérrez J (2013) Exploring the limitations of CORINE land cover for monitoring urban land use dynamics in metropolitan areas. J Land Use Sci 9(3):243–259. doi:10.1080/1747423X.2012.761736

EEA (2014) Report of Content of CORINE Land Cover.European Environmental Agency. http://www.eea.europa.eu/publications/COR0-landcover

ESRI (2010) Manual ArcGIS [software GIS]. Versión 10.0. Redlands, CA: Environmental Sys-tems Research Institute, Inc

Gardner RH, Milne BT, Turnei MG, O'Neill RV (1987) Neutral models for the analysis of broad-scale landscape pattern. Landsc Ecol 1(1):19–28

Hewitt R, Escobar F (2011) The territorial dynamics of fast-growing regions: unsustainable land use change and future policy challenges in Madrid, Spain. Appl Geogr 31(2):650–667

Hewitt R, Van Delden H, Escobar F (2014) Participatory land use modelling, pathways to an integrated approach. Environ Model Softw 52:149–165

Hewitt R, Díaz-Pacheco J (2017) Stable models for metastable systems? Lessons from sensitivity analysis of a Cellular Automata urban land use model. Comput Environ Urban Syst 62:113–124.

Jantz C, Goetz SJ (2005) Analysis of scale dependencies in an urban land-use-change model. Int J Geogr Inf Sci 19(2):217–241

Jenerette GD, Wu J (2001) Analysis and simulation of land-use change in the central Arizona-Phoenix region, USA. Landsc Ecol 16(7):611–626

Kok K, Veldkamp A (2001) Evaluating impact of spatial scales on land use pattern analysis in Central America. Agr Ecosyst Environ 85(1):205–221

Lajoie G, Hagen-Zanker AH (2007) Modeling urban sprawl on the island La Réunion: contribu-tion of the Metronamica® cellular automata for territorial prospect. Cybergeo 405

Lam SN, Quattrochi DA (1992) On the issues of scale, resolution, and fractal analysis in the mapping sciences. The Professional Geographer 44(1):88–98

Longley P, Batty M (2003) Advanced spatial analysis: the CASA book of GIS. ESRI, Inc

McGarigal K, Marks BF (1994) Spatial pattern analysis program for quantifying landscape structure reference manual. Forest Science Department, Oregon State University, Corvallis Oregon

Ménard A, Marceau DJ (2005) Exploration of spatial scale sensitivity in geographic cellular automata. Environ Plan 32(5):693–714

Openshaw S, Taylor P (1979) A million or so correlation coefficients: three experiments on the modifiable areal unit problem. In: Wrigley N (ed) Statistical applications in the spatial sciences. Pion, London, pp 127–144

Openshaw S (1983) The modifiable areal unit problem, vol 38. In: Geo books, Norwich

Plata Rocha W, Gómez Delgado M, Bosque Sendra J (2009) Cambios de usos del suelo y expansión urbana en la Comunidad de Madrid (1990–2000). Scripta Nova: revistaelectrónica de geografía y cienciassociales 13

Pontius JRG, Millones M (2011) Death to Kappa: birth of quantity disagreement and allocation disagreement for accuracy assessment. Int J Remote Sens 32(15):4407–4429

RIKS (2014) Metronamica Documentation.Research Institute for Knowledge Systems, The Netherlands. http://www.riks.nl/

Samat N (2006) Characterizing the scale sensitivity of the cellular automata simulated urban growth: A case study of the SeberangPerai Region, Penang State, Malaysia. Computers, environment and urban systems 30(6):905–920

Switzer P (1975) Estimation of the accuracy of qualitative maps. Wiley, London

Theobald DM, Hobbs NT (1998) Forecasting rural land-use change: a comparison of regression-and spatial transition-based models. Geogr Environ Modell 2:65–82

Tobler W (1988) Resolution, resampling, and all that. Building databases for global science, vol 12, pp 9–137

Van Delden H, Escudero JC, Uljee I, Engelen G (2005) METRONAMICA: A dynamic spatial land use model applied to Vitoria-Gasteiz. In: Virtual Seminar of the MILES Project. Centro de Estudios Ambientales, Vitoria-Gasteiz

Van Delden H, Stuczynski T, Ciaian P, Paracchini ML, Hurkens J, Lopatka A, Shi Y, Gomez Prieto O, Calvo S, Van Vliet J, Vanhout R (2010) Integrated assessment of agricultural poli-cies with dynamic land use change modelling. Ecol Model 221(18):2153–2166

Van Delden H, Hurkens J (2011) A generic integrated spatial decision support system for urban and regional planning. Keynote presented at MODSIM11 international congress on modelling and simulation, Perth, Australia

Van Delden H, Van Vliet J, Rutledge DT, Kirkby MJ (2011) Comparison of scale and scaling issues in integrated land-use models for policy support. Agric Ecosyst Environ 142(1):18–28

Van Delden H, Díaz-Pacheco J, Shi Y, Van Vliet J (2012) Calibration of cellular automata based land use models: lessons learnt from practical experience, CAMUSS (Cellular Automata Modeling for Urban and Spatial Systems) conference, November 8-10. Oporto, Portugal

Van Vliet J, White R, Dragicevic S (2009) Modeling urban growth using a variable grid cellular automaton. Comput Environ Urban Syst 33(1):35–43

Van Vliet J, Bregt AK, Hagen-Zanker A (2011) Revisiting Kappa to account for change in the accuracy assessment of land-use change models. Ecol Modell 222(8):1367–1375

Visser H, De Nijs T (2006) The map comparison kit. Environ Model Softw 21(3):346–358

White R (2006) Pattern based map comparisons. J Geogr Syst 8(2):145–164

White R, Engelen G (1993) Cellular automata and fractal urban form: A cellular modelling approach to the evolution of urban land-use patterns. Environ Plann A 25:1175

White R, Engelen G (2000) High-resolution integrated modelling of the spatial dynamics of urban and regional systems. Comput Environ Urban Syst 24(5):383–400

Woodcock CE, Strahler AH (1987) The factor of scale in remote sensing. Remote Sens Environ 21(3):311–332

Chapter 10
The Influence of Scale in LULC Modeling. A Comparison Between Two Different LULC Maps (SIOSE and CORINE)

D. García-Álvarez

Abstract Scale is one of the most interesting issues in land change science. Although much research has been done on this topic, our understanding of its effects on data and models is still sketchy. We therefore decided to investigate how cartographic scale and minimum mapping unit (MMU) influence modeling results, for which purpose we chose a heterogeneous, dynamic study area in central Asturias (Spain). As opposed to most of the literature on this subject, which focuses on the grain component of scale comparing the same map resampled at different spatial resolutions, we used two different land use and land cover (LULC) maps (SIOSE and CORINE) at different resolutions (12.5 and 50 m) and with minimum mapping units of 0.5–2 and 25 ha respectively. We compared the input and simulated maps using spatial metrics and the matrix proposed by Pontius and Millones to find out the quantity and allocation disagreement. The results can provide a better understanding of the implications of the choice of input maps in LULC modeling.

Keywords Scale · LULC modeling · CORINE · SIOSE · Minimum mapping unit

1 Introduction

Scale has been described as a priority topic of research in relation to spatial information analysis and representation (Turner et al. 1989; Quattrochi and Goodchild 1997b; Castilla et al. 2009) and in modeling issues (Ménard and Marceau 2005; Lesschen et al. 2005; Houet et al. 2010; van Delden et al. 2011; Committee on Needs and Research Requirements for Land Change Modeling et al. 2014). Mike Goodchild (2001) even went so far as to say that "scale is perhaps the most important topic of geographical information science", and to view scale as a

D. García-Álvarez (✉)
Departamento de Análisis Geográfico Regional y Geografía Física,
Universidad de Granada, Campus de Cartuja s/n, 18071 Granada, Spain
e-mail: dagaral@ugr.es

M.T. Camacho Olmedo et al. (eds.), *Geomatic Approaches for Modeling Land Change Scenarios*, Lecture Notes in Geoinformation and Cartography,
https://doi.org/10.1007/978-3-319-60801-3_10

science in itself (Quattrochi and Goodchild 1997a; Wu and Qi 2000; Quattrochi et al. 2001).

Scale can be understood in a wide variety of ways (Lam and Quattrochi 1992; Wu 2004; Ménard and Marceau 2005), such as cartographic scale (ratio), observational or geographical scale (map size or study area size) or operational scale (scale at which certain processes operate in the environment). When we talk about scale we may also be referring to the level of detail represented in a map (Agarwal et al. 2002; Verburg et al. 2004). There is also the relation between scale and spatial, temporal and thematic resolution. Some researchers refer to this as spatial scale (spatial resolution) or temporal scale (temporal resolution) (O'Sullivan and Perry 2013).

When assessing the relation between scale and spatial resolution, a key concept is the minimum mapping unit (MMU), i.e. the minimum area of the smallest unit in a map. This will be smaller at fine scales and bigger at coarse scales. Castilla et al. (2009) considered the MMU to be an important parameter, which should be included in the concept of scale together with thematic resolution, while Saura (2002) stressed its importance in the variance of spatial data representation.

The main objective of this chapter is to analyse how the scale of the selected LULC (Land Use Land Cover) maps can affect modeling. This involves analysing how scale can cause maps to show different information and studying the ways in which model behaviour varies depending on the data detail (MMU). In order to achieve these objectives, we compared two LULC maps (SIOSE and CORINE) with different cartographic scales (1:25,000 vs. 1.100,000) and different MMU (0.2–2 ha vs. 25 ha).

The chapter is divided into five main sections. After an initial introduction, the second section describes the study area and data sets. Section 3 explains how we adjusted the LULC maps to obtain two comparable sources and then goes on to describe the model we used and how it was calibrated. It also has a short introduction to the methods we used in the analysis and assessment of data. In section four we present the results and in section five we discuss the main findings of the research.

2 Test Area and Data Sets

2.1 Test Area

Our chosen study area was the Asturias Central Area, which as its name suggests, is in the centre of the Asturias region, in northern Spain (Fig. 1). Although from a geographical point of view it is not a coherent homogeneous space that is clearly differentiated from its surroundings, there is a functional link between its different component parts (Fernández García et al. 2007; Carrero de Roa 2012) and as a

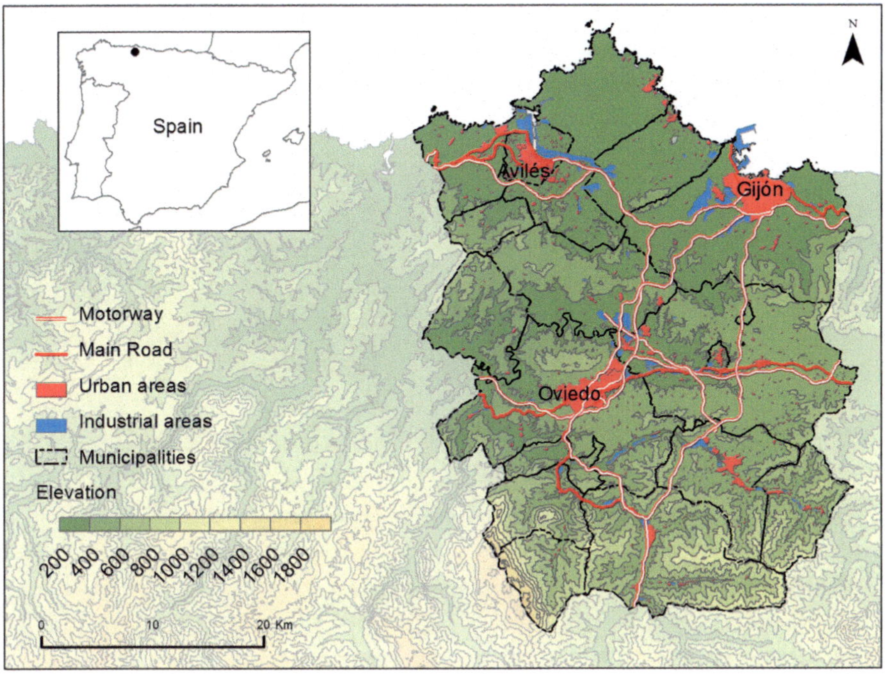

Fig. 1 Map showing the location of the Asturias Central Area (2015). *Source* National Topographic Map 1:200,000, DEM 25 m (National Geographic Institute)

result it has been declared a special planning area within the Asturias Territorial Plan (*Directrices Regionales de Ordenación del Territorio*).

Most of the LULC (Land Use Land Cover) changes taking place in Asturias are represented in this Central Area. These changes are essentially rural-urban and rural-industrial, and take place in the main centres of economic activity. Within the study area there are several sub-areas with a specific economic profile (Rodríguez Gutiérrez et al. 2009), which are therefore affected by specific LULC dynamics. For the whole study area, transport infrastructures, and above all roads, are the main drivers of change.

2.2 *Data Sets*

We selected CORINE and SIOSE as our two data sets (Fig. 2). CORINE (Coordination of Information on the Environment) is a well-known LULC resource that has often been used in modeling studies (Verburg and Overmars 2009; Camacho Olmedo et al. 2013; Renwick et al. 2013). It is updated every six years and covers most European Union countries. This standardized approach allows

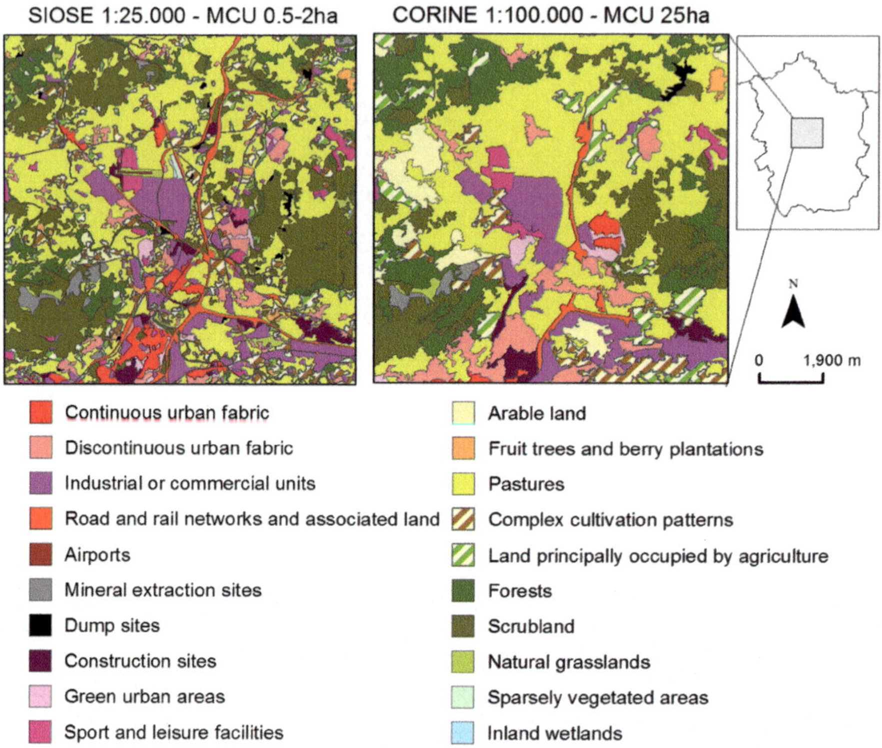

Fig. 2 LULC for an example area (Lugones-Llanera) as represented by SIOSE (*left*) and CORINE (*right*). This area is one of the most dynamic spaces in the study area

users to make comparison studies across the EU. Nevertheless, some authors have noted that each member state has its own CORINE team applying slightly different criteria, each of which has produced a CORINE map with a different validity rate (Waser and Schwarz 2006).

The CORINE scale of reference is 1:100,000. It has a minimum mapping unit (MMU) of 25 ha and a minimum polygon width (MPW) (minimum distance between two polygon sides) of 100 m. As a general rule, only changes affecting areas of over 5 ha are drawn. The final product is a map made by photointerpretation in which each polygon is assigned to a single category (*classification system*) (Hernández 2016).

CORINE is very useful for studies at regional and national levels. However, when studying urban land covers with complex patterns and a heterogeneous nature, this scale may be insufficient (Antrop 2004; Herold et al. 2005). This is also true for studies which require more detailed analysis (Chas-amil and Touza 2015), especially those focusing on local or sub-regional areas.

SIOSE (*Sistema de Información sobre Ocupación del Suelo de España*—Information System about Land Cover in Spain) is an LULC map made at a scale of

1:25,000. This allows for more detail, with a MMU of between 0.5 and 2 ha depending on the particular land cover (e.g. 0.5 ha for wetlands, 1 ha for urban areas and 2 ha for forests). The MPW in SIOSE is 15 m. Nevertheless, in certain cases SIOSE accepts even lower widths and areas below the specified minimum. This is very important because it makes a big difference in the spatial representation of the terrain, allowing infrastructures such as roads to be represented on the map. These infrastructures create barriers that produce additional divisions of the covers that cannot be seen on maps at smaller scales. When updating SIOSE maps, small changes affecting areas of less than 0.4 ha are not included.

The SIOSE data set has two different parts. Firstly, a polygon map which marks out homogeneous areas with similar characteristics (either single categories or mosaics of various different categories) and secondly, a database which compiles all the land cover information for each polygon without scale restrictions, i.e. each polygon has a log in the database where the different land covers that make up the polygon and their proportions are registered. This database model is called Application Schema (ISO 19101) (ISO/IEC 2014) and this way of gathering information is known as a description system (Hernández 2016).

One of the main advantages of the SIOSE database model is the enormous detail it provides about the earth's surface. Nevertheless, this wealth of detailed information must be generalized in order to produce a map in which each polygon is allocated to a single class or category, i.e. to move from a *description system to a classification system.* The resulting map will vary according to the thresholds we use in the generalization process. This introduces uncertainty in the analysis that must be addressed in further research.

Since 2012 the Spanish CORINE has been obtained from a generalization of SIOSE. The new CORINE production method has caused important changes in the CORINE map, which has many striking differences from the previous version (2006). Every time CORINE is updated, the previous map is reviewed. Hence with the updating of CORINE in 2012, the National Geographic Institute of Spain produced a new version of CORINE 2006 that is coherent with CORINE 2012, so enabling comparisons to be made. However, as SIOSE only started in 2005, this retroactive adjustment of CORINE cannot be performed for earlier versions, making it impossible to analyse change over longer periods (from 1990 to 2012).

Both data sets are vector data, which must be rasterized to be used as input in the chosen model (Dinamica EGO). Vector data allows for more precise feature allocation, but modeling is more difficult in this structure because each spatial entity has a different shape and form; e.g. computing neighbourhood interactions is much more complex than with rasters, characterized by a simple, regular shape (Burrough et al. 2015).

In our case study, we have selected the two dates for which both maps were available: 2005–2006 and 2011–2012. As CORINE has been generalized from SIOSE for both these dates, the baseline information is the same for both maps. This enables us to compare the two maps and analyse the changes in Land Use Cover over this period.

Summarizing, the uncertainty arising from the use of diverse data sets is dependent on the following variables: cartographic scale, minimum mapping unit, generalization process and dates.

3 Methodology

We began by adjusting the two data sets (SIOSE and CORINE) to produce two comparable LULC maps with the same legend (see Fig. 3). These maps were then compared to identify any differences between them. Once we had obtained the model inputs, one model was calibrated for each input map using criteria based on expert knowledge. Finally, both models were run to obtain a simulation for the year 2020. Differences between the two simulations were analysed and compared with real changes over the calibration period (2005–2011).

3.1 Data Set Processing

SIOSE and CORINE were adjusted to make them comparable (same legend). The legend we chose is a slight modification of the Level 3 CORINE legend, which we simplified in order to focus on the most important types of cover in the study area.

As SIOSE was designed to serve as a basis for the production of CORINE by generalization, there are no significant problems in the equivalence between the two legends; each SIOSE category fits well with the meaning of broader categories in CORINE. The associations between the categories on the two maps were made in accordance with similar category meanings. The definitions of the categories were obtained from the technical guide.

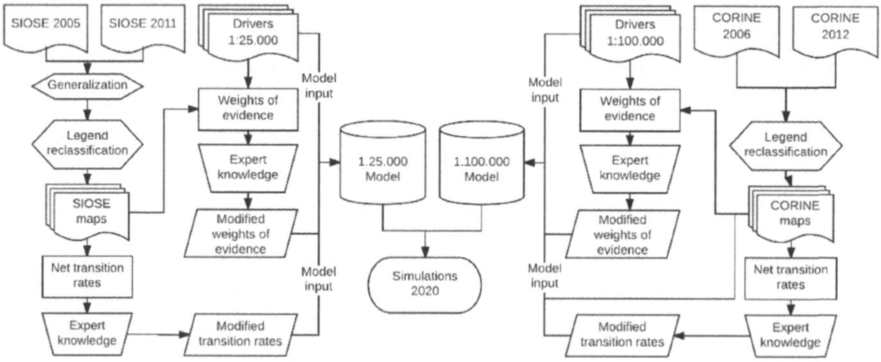

Fig. 3 Flowchart of the methodological procedure we followed to produce two calibrated models, one for each input map (SIOSE and CORINE)

The most crucial step in the processing of the data sets was the generation of a SIOSE map from the numerical information provided in the SIOSE database, that is, the respective proportions of each of the categories that make up each polygon. We have established a set of rules which, for each land cover, establish the thresholds above which a specific proportion of land cover or an association of several land covers with different proportions can be given an individual label (e.g. when three quarters of a polygon is occupied by a particular LULC category, the entire polygon will be allocated to this category).

The final step in the data set treatment was the rasterization of the maps, given that modeling software requires raster data. We chose the resolution following the method established by Tomislav Hengl (2006). Bearing in mind the cartographic scales of the reference maps (1:25,000 and 1:100,000) we opted for a resolution of 12.5 m for SIOSE and 50 m for CORINE.

All this data set processing inevitably introduces uncertainty into the analysis. Even the choice of legend can affect the results of the model (Dietzel and Clarke 2006; Conway 2009). However, this question would be best addressed in additional future research. We have tried to maintain the maximum thematic detail by avoiding introducing external factors (such as thematic resolution) into the analysis.

The vector to raster conversion has also introduced new uncertainty. Comparison of the final raster maps with the vector data has shown a maximum difference per category of 7.6 ha for SIOSE and 22 ha for CORINE, with a mean difference of 3.9 and 1.3 ha respectively. This uncertainty is minimal, although some studies have also noted the important influence of rasterization in the analysed pattern (Dendoncker et al. 2008). Finally, the way we processed the SIOSE maps (generalization) also introduced additional uncertainties.

It is therefore necessary to compare the two input maps in order to distinguish their initial differences from the simulations, differences caused by the fact that the model behaves differently depending on the specific data set being modelled. This analysis was carried out for maps for 2011–12 according to methods explained at the end of this section (Data analysis and assessment).

3.2 *Modeling*

3.2.1 Modeling Framework

We used the DINAMICA EGO software, which has been widely tried and tested in recent modeling research (Maeda et al. 2011; Ahmed and Bramley 2015). Furthermore, several model comparison studies praised DINAMICA's architecture and flexibility (Mas et al. 2011; Pérez-Vega et al. 2012; Mas et al. 2014).

It is a stochastic cellular automata model that models transitions through two different functions: expander and patcher. The expander function models new pixels as an expansion of previous patches, whereas the patcher function models one or several new pixels as a new patch for the category. In addition, the model uses the weights of evidence method to produce a probability map of occurrence for each transition. More information about the components of the model and its characteristics can be found in Soares et al. (2002). It is compared with other models in Follador et al. (2008) and in the technical notes at the end of this book.

3.2.2 Model Calibration and Simulation

We set up two models (Fig. 3), one for each of the LULC maps we compared (SIOSE and CORINE). These maps were calibrated against real data for the period 2005-06 – 2011-12, the only available dates. The model then ran a simulation up to the year 2020, which fits well with the short calibration period (six years).

The land covers modelled were the most dynamic artificial (soil-sealing) covers in the study area. We selected the most significant transitions (i.e. those affecting the largest areas of land) during the calibration period (2005-06 – 2011-2012). Usually, a threshold of 10 ha was applied. Different transitions were selected for each model because of the different input maps (SIOSE and CORINE). These measure a different amount of change between categories.

Transition rates were computed by Dinamica Ego using a Markov matrix, which was obtained by comparing LULC maps for the selected dates (2005-06 – 2011-12). Since each pair of LULC maps measure a different amount of change, the transition rates are also dissimilar.

For each transition, we chose a set of driving forces according to information collected in interviews with experts in this field together with information provided by academic studies about the study area (Fernández García et al. 2007; Rodríguez Gutiérrez et al. 2009; Alonso Ibáñez and Pérez Fernández 2012). The selection of explanatory variables was based on expert knowledge, methodological orientation (Pontius Jr. et al. 2008) and the availability of data. Although the drivers were the same for both models, when the source offered the same data at different scales we created separate variables according to the input data scale (Table 1). Therefore, there are slight differences between the two models because of the different scale used in these separate variables.

The models were calibrated using the *Weights of Evidence* method according to the knowledge provided by the experts we interviewed and information from previous research. When strange or incorrect behaviour was detected, we corrected it manually. Variables with a correlation (Cramer's Coefficient) greater than 0.5 were discarded according to thresholds established by similar studies (Quiroz Ortuño 2009; Maeda et al. 2011).

The flexibility of the Dinamica modeling framework, which allows the user to manually modify the calculated weights of evidence, enabled us to adjust the model parameters in order to obtain the maximum similarity between the two simulations. Thus, both models were run with the same weights of evidence, according to the expert criteria.

Table 1 List of drivers used in the two models

Drivers	Source		Year
	SIOSE	CORINE	
Euclidean distance to regional roads	NTN 1:25,000	NTM 1:100,000	2011
Euclidean distance to highways	NTM 1:25,000	NTM 1:100,000	2011
Euclidean distance to train stations	NTM 1:25,000	NTM 1:100,000	2011
Euclidean distance to residential buildings	NTM 1:25,000	NTM 1:100,000	2011
Euclidean distance to industrial buildings	NTM 1:25,000	NTM 1:100,000	2011
Euclidean distance to coastline	NTM 1:25,000	NTM 1:25,000	2015
Euclidean distance to leisure facilities	SIOSE	CORINE	2011
Population density	Basic geographical name index of Spain		2011
Slopes	DEM 5 m	DEM 25 m	2015
Future road development	NTM 1:100,000	NTM 1:100,000	2015
Rural settlements limits	Asturias planning maps		2015
Urban settlements limits	Asturias planning maps		2015
Building land	Asturias planning maps		2015
Protected areas	Asturias planning maps		2015
Mining area restructuring plans	Asturias planning maps		2015
Substratum	Geology Map 1:50,000		2015
Industrial ports	NTM 1:100,000		2011

*NTM: National Topographic Map; DEM: Digital Elevation Model

The patcher and expander function parameters (proportion of changes simulated as expansion, patch isometry and patch mean and variance) were established according to the pattern of real changes in the calibration period.

Transition rates for the simulation year (2020) were modified (Table 2) in accordance with the different trends of change forecast for the coming years (economic crisis) in the study area by the experts we interviewed.

Summing up, the two models differ because of the transitions selected, the transition rates (quantity of change), the input data (SIOSE and CORINE maps and driving forces maps) and the expander and patcher parametrization (proportion of changes simulated as expansion, patch isometry and patch mean and variance). However, the land use change logic is the same for both models in that they follow the same methodological procedure with the same drivers of change.

3.3 *Data Analysis and Assessment*

There is a great deal of academic research about scale and its influence on data. The most frequently used methods include fractal analysis, local variance method, variograms, wavelets or texture analysis methods (Cao and Siu-Ngan Lam 1997;

Table 2 List of transitions for the two models, showing the categories involved in each transition and the corresponding transition rates

SIOSE				CORINE			
Modeled cover	Transition from	% of original rate	Transition rate	Transition rate	% of original rate	Transition from	Modeled cover
Continuous urban fabric	Construction sites	50	0.015410	0.002338	50	Construction sites	Continuous urban fabric
	Pastures	33	0.000119	0.000021	33	Pastures	
Discontinuous urban fabric	Construction sites	100	0.016205	0.033069	100	Construction sites	Discontinuous urban fabric
	Pastures	33	0.000060	0.000183	33	Pastures	
Industrial or commercial units	Construction sites	33	0.014230	0.007034	33	Construction sites	Industrial or commercial units
				0.000037	25	Arable lands	
	Pastures	25	0.000042	0.000087	25	Pastures	
	Complex cultivation patterns	25	0.000038				
				0.000198	25	Land principally occupied by agriculture	
	Forests	25	0.000019	0.000009	25	Forests	
				0.000150	25	Natural grasslands	
	Shrubland	25	0.000070				
Road and rail networks and associated land	Construction sites	150	0.011951	0.009200	150	Construction sites	Road and rail networks and associated land
Mineral extraction sites	Forests	33	0.000099	0.000094	33	Arable lands	Mineral extraction sites
	Shrubland	33	0.000103				
Dump sites	Pastures	33	0.000059				Dump sites
	Forests	33	0.000104	0.000036	33	Forests	

(continued)

Table 2 (continued)

SIOSE				CORINE			
Modeled cover	Transition from	% of original rate	Transition rate	Transition rate	% of original rate	Transition from	Modeled cover
Construction sites	Dump sites	15	0.001848				Construction sites
	Arable lands	15	0.000213				
	Pastures	15	0.000186	0.000193	15	Pastures	
	Complex cultivation patterns	15	0.000113	0.000035	15	Complex cultivation patterns	
	Land principally occupied by agriculture	15	0.000123	0.000115	15	Land principally occupied by agriculture	
	Forests	15	0.000054	0.000027	15	Forests	
	Natural grasslands	15	0.000070				
	Scrubland	15	0.000077	0.000045	15	Scrubland	

The Percent of original rate column refers to the selected proportion of the original transition rates. These were obtained by cross tabulation of calibration dates

Zong-Guo and Clarke 1997; Oliver 2001). Several studies have also used spatial metrics (Saura 2002; Wu 2004; Uuemaa et al. 2005), which allow us to analyse the effects of the scale of the input maps on the patterns being modelled. Numerous validation techniques have also been created in modeling research (Paegelow et al. 2014). They usually compare simulated maps with real maps in order to assess the fitness of the model. However, they can also be used to compare two maps and highlight their differences. The method proposed by Pontius and Millones (2011) to evaluate quantity and allocation disagreement is a good validation tool to analyse the differences between two maps. It shows two types of error: quantity and allocation (omission and commission) errors.

We compared the input maps (SIOSE and CORINE) and real (2005-6 – 2011-12 cross tabulation) and simulated changes (2011-12 – 2020-21 cross tabulation). When comparing the input maps, we only analysed the land covers that took part in the modelled transitions. When comparing the changes, we only analysed the land covers that are actively modelled (i.e. the categories allocated by the model).

Spatial metrics were calculated using FRAGSTATS 4.2 for the evaluation of the pattern difference between input maps (SIOSE and CORINE maps) as well as for the analysis of the pattern difference between real (2005-6 – 2011-12) and simulated changes (2011-12 – 2020-21). The metrics were selected according to the information that they provided, that is, according to the variability of their results. These metrics are: Total Area, Number of Patches, Largest Patch Index, Weighted Mean Patch Area, Area Coefficient of Variation, Mean Fractal Dimension, Proportion of Like Adjacencies and Patch Cohesion Index. A detailed description of each metric can be found in the FRAGSTATS help guide (McGarigal et al. 2015) and in Leitão et al. (2012).

The matrix proposed by Pontius and Millones was applied for the comparison of input maps and simulated changes. The results highlight the differences between simulations in terms of quantity and allocation disagreement compared with the same disagreements between the SIOSE and CORINE maps.

4 Results

4.1 Quantity Disagreement

4.1.1 Input Maps

There is an important quantity disagreement between the input maps. This disagreement varies from one category to the next, as shown in Fig. 4, where confusion bars represent the confusion of one input map with regard to the other for each category. Each confusion bar is divided into various sections, one for each category on the map. Each section of the bar represents the proportion of pixels that are allocated to a different category on the other map. When the section for any particular category

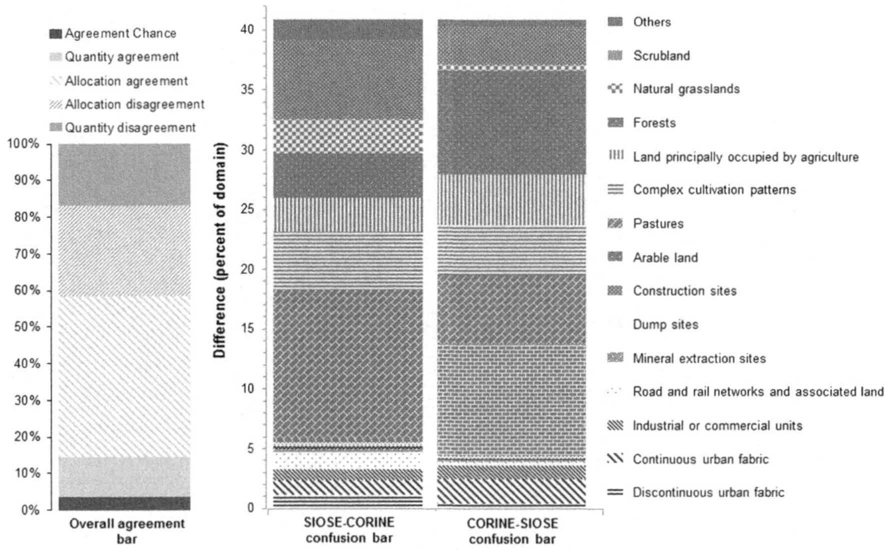

Fig. 4 Agreement bars for input maps. The *first bar* depicts the overall components of agreement between the two maps. The *second bar* depicts the SIOSE disagreements with regard to CORINE, and the *third bar* depicts the CORINE disagreements with regard to SIOSE

(e.g. arable land) is larger on one confusion bar than on the other, this means that there is a quantity disagreement, which is equal to the difference between the two sections for that category. The rest of the disagreement is allocation disagreement.

These differences in quantities result from the fact that different minimum mapping units (MMU) were used in the two input maps. When cartographic scale is reduced and the MMU is bigger, the polygons that do not meet the mapping criteria (MMU) must be absorbed by others with similar definitions (generalization process).

Categories with flexible definitions incorporate polygons from other categories and increase their area (e.g. discontinuous urban fabric, forests) whereas categories with rigid definitions lose polygons and their total area diminishes (e.g. areas with sparse vegetation, dump sites). João (2001) called this process "competition for map space". It results in a disproportionately high representation of categories with flexible definitions and a low representation of categories with rigid definitions in maps at smaller scales with bigger MMU.

The categories that absorb other categories during the generalization process are different for the two input maps (e.g. arable land for CORINE and pastures for SIOSE). This results in large quantity disagreements and allocation disagreements between the maps. This disagreement can be attributed to the use of different criteria in the generalization process for SIOSE (conducted by the author, as explained in Sect. 3.1) and for CORINE (conducted by the National Geographic Institute of Spain, as explained in Sect. 2.2).

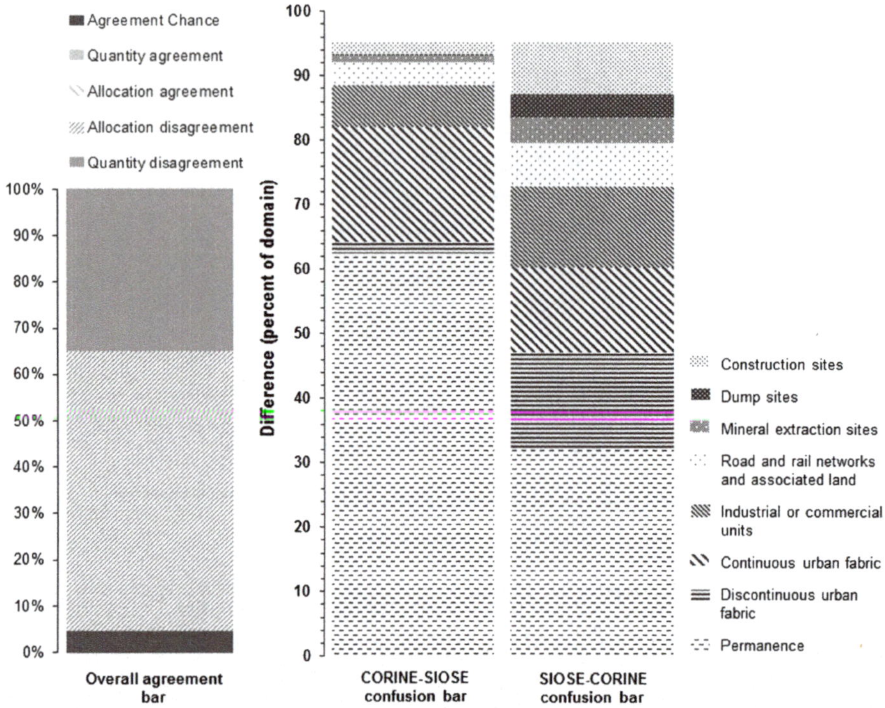

Fig. 5 Agreement bars for simulated changes. The *first bar* depicts the overall components of agreement between the two maps. The *second bar* depicts the CORINE simulation disagreements with regard to the SIOSE simulation, and the *third bar* depicts the SIOSE simulation disagreements with regard to the CORINE simulation

4.1.2 Simulated Changes

The simulated changes for the period 2011-12 – 2020-21 show a high degree of quantity confusion as illustrated in the bar chart below (Fig. 5). The simulated changes are greater in SIOSE than in CORINE because when the two calibration maps (2005–2006 and 2011–2012) were compared, there was a larger amount of change in SIOSE than in CORINE. The MMU for SIOSE allows us to detect small changes, in addition to the changes it also detects for CORINE. In consequence, the transition rates (Markov matrix obtained through comparison of the input maps) are greater for SIOSE than for CORINE.

4.2 Allocation Disagreement

4.2.1 Input Maps

There is also a significant level of allocation disagreement between the input maps that varies from one category to the next. Mixed categories (complex cultivation patterns and land principally occupied by agriculture) show a high degree of error (Fig. 6). Because of the vague definition of these categories, real covers labelled as such are very different in the two input maps.

Allocation disagreement analysis also provides information about which categories are confused with the quantity disagreement of other categories. We observed for example a relation in the error between pastures and arable lands and between continuous and discontinuous urban fabric. Visual comparison of the two LULC maps have proved this relation: some locations labelled as pastures in one map are classified as arable lands in the other. The same relation can be observed for continuous and discontinuous urban fabric. Therefore, the same covers are assigned to different categories on each map.

The model was set up on the basis of the definitions of the categories and expert knowledge. Same category definitions were considered independent of the model and the same criteria were applied to them. Since the same categories do not represent the same covers in both models, different results were expected in the simulations. e.g. we used the same drivers for the transition from pastures to continuous urban fabric. However, the land allocated to these two categories is different in the two input maps (allocation disagreement in Fig. 6).

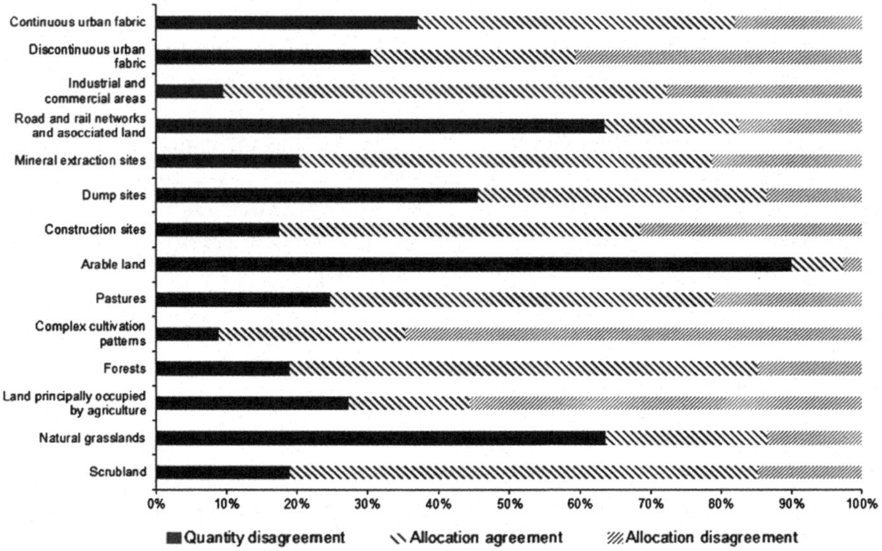

Fig. 6 Agreement, quantity and allocation disagreement for categories in input maps (SIOSE and CORINE)

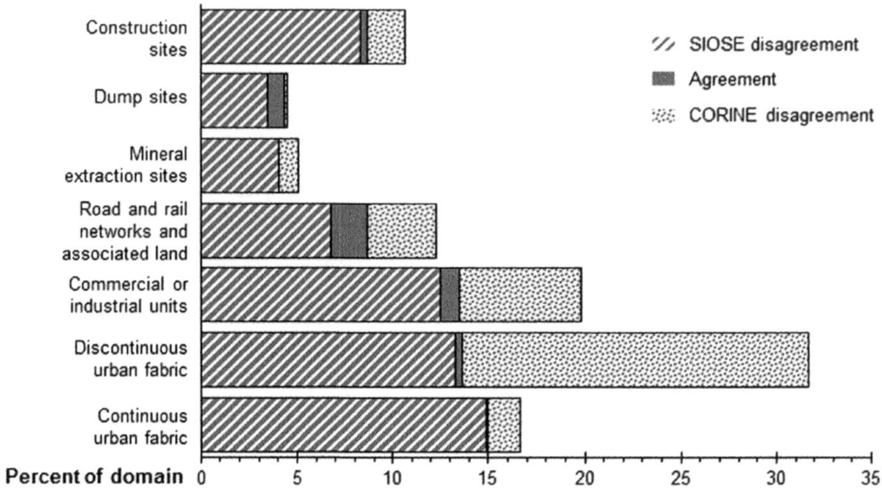

Fig. 7 Agreement bars per category for changes in SIOSE and CORINE simulations. SIOSE and CORINE disagreements refer to changes simulated by each model that do not correspond to the same change in the other model

4.2.2 Simulated Changes

The agreement between changes in the two simulations is minimum (Fig. 7) and can be attributed purely to chance (Fig. 5). Most of the pixels allocated by the two calibrated models are in different positions.

The disagreement is mostly (94% of total disagreement) related to persistence (i.e. areas with no change) on the other map: 62% of the confusion in the SIOSE simulation is with areas that do not change in the CORINE simulation, whereas 32% of the confusion in the CORINE simulation is with areas that do not change in the SIOSE simulation (Fig. 5). This is because of the allocation disagreement between input maps due to different category definitions, as pointed out in the previous section. The candidate areas (i.e. those in which transition is possible) are different in both models, whereas the drivers of change and therefore the candidate areas in which change is most likely, are very similar.

4.3 Pattern Disagreement

4.3.1 Input Maps

As expected, fragmentation (the number of patches or polygons) is greater in SIOSE than in CORINE (Table 3). The smaller the MMU, the higher the number of

Table 3 Class metrics from SIOSE and CORINE maps

	Number of patches (NP)		Largest patch index (LPI)		Mean Patch Area-Weighted (MPAW)		Area coefficient of variation (ACV)	
	SIOSE	CORINE	SIOSE	CORINE	SIOSE	CORINE	SIOSE	CORINE
Continuous urban fabric	100	22	0.468	0.2109	521.2025	269.9869	379.8538	135.419
Discontinuous urban fabric	815	94	0.0418	0.2762	15.9712	181.065	171.8313	160.946
Industrial or commercial units	657	59	0.129	0.4861	66.8406	427.5172	295.8965	201.6491
Road and rail networks and associated land	91	26	0.9313	0.0696	1645.6102	84.8598	705.1368	106.82
Mineral extraction sites	83	15	0.0428	0.0497	34.2393	68.5387	147.5139	68.1162
Dump sites	108	5	0.0464	0.0404	30.3907	57.7592	243.8299	56.7203
Construction sites	95	17	0.07	0.0694	52.9917	88.9033	187.5387	81.9385
Arable land	186	117	0.0773	2.1861	45.2817	2418.446	207.8323	401.5608
Pastures	1651	256	1.8602	6.0199	746.3756	6122.6768	516.1228	676.0464
Complex cultivation patterns	1221	145	0.4077	1.3211	140.598	1098.1782	387.2766	391.0417
Land principally occupied by agriculture	1371	174	0.0246	0.1352	10.4921	92.3475	119.5047	97.9302
Forests	1378	318	0.7864	1.263	359.9393	1051.2964	370.1635	264.7445
Scrubland	583	46	0.2976	0.1016	146.7024	104.3517	366.4032	114.0313
Natural grasslands	1046	199	0.498	0.6947	289.356	509.3823	350.1007	221.4825

patches. And the higher the number of patches, the smaller the weighted mean area (mean area corrected according to polygon size).

Therefore, SIOSE maps have more polygons and smaller polygons than CORINE maps. In consequence, SIOSE has more potential pixels to be modelled as an expansion of previous patches, whereas CORINE is more sensitive to the patcher function. Any new patch modelled in CORINE should meet the MMU criteria. If not, the simulation would result in a more fragmented landscape.

In short, in the case of input maps, more fragmented and complex patterns are expected at larger scales. At these scales the land cover information resembles the real situation more closely. This means that the SIOSE maps are more realistic, but also more complex and there is no relation between complexity and the good performance of the model (Clarke 2004; Wainwright and Mulligan 2013).

4.3.2 Simulated and Real Changes

Real changes (2005–6—2011–12) and simulated changes (2011–12—2020–21) show a different pattern. While the effect of MMU is evident in real changes (fewer and larger changing patches for CORINE), it is practically non-existent in the simulations. The number of CORINE patches increases with the simulation whereas the number of SIOSE patches falls (Table 4).

The proportion of like adjacencies and patch cohesion index, which measure the grouping of patches that belong to the same class (McGarigal et al. 2015), also show different results. Simulated changes are more disaggregated than real changes, especially in the CORINE simulation.

These different patterns between simulated and real changes are a consequence of the logic of the model. The pixel is the essential unit of work of any raster LULC model, such that the location of the first change simulated will be the most suitable pixel for each specific transition. The pixel area is much smaller than the MMU (156 m^2 vs. 0.2–0.5 ha in SIOSE and 0.25 ha vs. 25 ha in CORINE). The simulated changes will therefore be characterized by smaller (area-weighted mean patch area), more poorly connected patches (proportion of like adjacencies and patch cohesion index), when compared with the real changes, which are affected by the MMU rules.

The bigger the contrast between the MMU and the spatial resolution (pixel size), the more evident the fragmentation in the simulation. This means that compact input maps are more sensitive to model allocation change than fragmented input maps.

As regards the total area, in both cases, the proportion of modelled changes compared to real changes (Table 4) is higher than the proportion of selected transition rates with regard to original rates (Table 2). This is due to the transition rates function, which estimates the real changes through a Markov Matrix. This introduces a new source of uncertainty into the model. The estimated changes are different depending on the model and method used (Mas et al. 2011).

Table 4 Class metrics from SIOSE (S) and CORINE (C) simulations

Simulated changes 2011–2020	Total area		Number of patches		Mean patch area-weighted		Mean fractal dimension		Proportion of like adjacencies		Patch cohesion index	
	S	C	S	C	S	C	S	C	S	C	S	C
Continuous urban fabric	191.5	23.0	44	13	10.7	3.4	1.1091	1.0618	89.7944	45.6522	95.0723	71.0631
Discontinuous urban fabric	174.9	235.0	79	88	11.9	16.6	1.1092	1.046	88.2103	65.4787	93.7738	80.4932
Industrial or commercial units	172.5	93.8	81	42	15.0	8.1	1.1241	1.0401	83.3288	58.8	95.7527	79.9843
Road and rail networks and associated land	111.3	71.0	15	4	14.5	32.4	1.1997	1.1959	84.6364	61.7958	96.4041	92.4082
Mineral extraction sites	51.4	13.5	35	5	1.9	3.5	1.1194	1.0565	81.6109	61.1111	90.3935	72.052
Dump sites	55.2	13.5	11	4	7.7	5.2	1.111	1.0502	89.8641	64.8148	94.9689	77.3581
Construction sites	111.3	30.8	118	26	3.3	2.9	1.1351	1.0488	76.8815	43.4959	90.2173	63.8639
Total	868.1	480.5	383	182								
Real changes 2005–2011	Total Area		Number of patches		Mean Patch Area-Weighted		Mean Fractal Dimension		Proportion of like adjacencies		Patch cohesion index	
	S	C	S	C	S	C	S	C	S	C	S	C
Continuous urban fabric	316.5	40.3	72	4	19.5	12.5	1.0915	1.0616	91.7189	77.9503	96.4369	86.0809
Discontinuous urban fabric	164.1	304.8	130	19	11.6	43.5	1.071	1.0651	87.813	82.5267	93.5797	92.4059
Industrial or commercial units	410.3	265.5	130	23	26.3	34.0	1.0821	1.0906	91.8907	76.2241	95.7127	89.8172
Road and rail networks and associated land	56.6	35.8	8	1	13.3	35.8	1.205	1.1439	81.9299	79.021	96.6727	93.0449
Mineral extraction sites	102.4	28.0	64	2	5.3	18.9	1.0926	1.1069	86.3387	75.8929	92.89	88.8021
Dump sites	75.7	27.0	34	8	5.6	7.3	1.1018	1.0448	87.4329	64.3519	93.4537	80.6492
Construction sites	628.3	391.5	95	10	37.7	83.3	1.1118	1.0904	91.2315	84.7063	97.3597	95.4769
Total	1754.0	1092.8	533	67								

5 Discussion

5.1 Data Sets Uncertainty

The study area can be displayed in very different ways depending on the cartographic source used, even though they are apparently similar. Previous studies have proved this by analysing several data sets for the same area (Waser and Schwarz 2006; Schmit et al. 2006; Chasamil and Touza 2015).

In our study case, SIOSE and CORINE showed big dissimilarities despite the fact that one map was obtained from the other by generalization. The only differences between the input maps are the cartographic scale, minimum mapping unit (MMU), minimum polygon width (MPW) and generalization criteria. It is these factors therefore that cause the maps to provide dissimilar information.

More transparency is required with regard to the generalization process used in CORINE. Although we have based our SIOSE generalization on CORINE class definitions, as described in its technical specifications, important disagreements appear due to the different generalization criteria. More information about how CORINE was obtained from SIOSE would result in more correspondence between the two maps.

Uncertainty analysis of the input data should be a critical step in modeling research, as shown by Verburg et al. (2011) and Pai and Saraswat (2013). Nevertheless, when addressing this problem of uncertainty, researchers usually present accuracy rates obtained by error analysis based on data gathered in the field. These general rates can vary widely across local areas and categories. This means that the extent and the thematic resolution of the analysis must be chosen carefully.

In our study case, most of the errors come from confusion between the following classes: scrub and grasslands; pastures and agricultural areas; and continuous and discontinuous urban fabric. A simplified legend which grouped these categories together into larger, more broadly-defined classes would remove these errors. This confirms the ideas of Verstegen et al. (2012), who pointed out that uncertainty is usually lower at coarser scales since local changes are omitted. Given that some level of generalization is always needed because of the impossibility of representing the real situation exactly on a scale map, it is sometimes better for input maps to ensure greater accuracy even if this means less detail. Hence, smaller scales and bigger MMU are sometimes preferable.

5.2 Model Parameters

Maps at finer scales (smaller MMU) provide more detailed information and, ergo, show more changes. Consequently, different rates of change and potential transitions are obtained. Similar results have been noted for the analysis of other components of scale such as extent (Bhatti et al. 2015) or thematic resolution (Pontius Jr and Malizia 2004; Conway 2009).

Fine scale maps can be useful for local studies, because they provide essential detailed information about local dynamics (Wang and Marceau 2013; Zhao 2013). However, the probability of error as noise in maps at finer scales is higher. The probability of introducing this noise into the model is also greater (Blanchard et al. 2015).

At finer scales, transitions rarely occur alone and different processes or transitions happen simultaneously (Wang and Marceau 2013). The patterns of change are also more complex. This makes calibration of the model more challenging and errors more likely. Sometimes simpler models work better (Clarke 2004; Wainwright and Mulligan 2015). In fact, if the drivers of change are simple and we cannot explain the local changes, as in SIOSE, input maps at coarser scales and with bigger MMU are advisable.

Applying the same criteria to the two calibrated models results in different simulated changes because the models do not show the same dynamics. More detailed knowledge is required to enable the model to be properly calibrated with SIOSE data since the experts only considered the main dynamics in the study area and, therefore, ignored most of the small changes. However, CORINE mapping rules do not fit well with the size of most of the changes in the area. Neither system offers a perfect data set and consequently the modeller must try to strike a balance between generality, precision and realism (O'Sullivan and Perry 2013).

Explanatory drivers of change vary with the scale (Verburg et al. 2003; Moreira et al. 2009; Bhatti et al. 2015). However, in this case the two scales of analysis (both local-regional) are too similar to be affected by contrasting driving forces. Nonetheless, additional driving forces could be included in the model with SIOSE as input maps because of the additional local processes involved in this model.

Explanatory factors were the same for both models and the maps were very similar: same source but different scale. The variations resulting from these different scales are essentially a question of the degree of precision in the location of attributes. Consequently, the areas with the greatest transition potential for each land cover are similar in both models despite the substantial differences in land cover information. As a result, an area with high transition potential could be located under a particular land cover in one model and under a different one in the other. This results in significant incoherence between LULC maps and driving forces. Making driver maps from LULC maps is an alternative way to achieve coherence in the datasets that define the model. However, this would limit the variety of drivers.

5.3 Modelled Pattern

The changes in the LULC pattern are similar in both models, regardless of the MMU. The simulated pattern is always more fragmented than the initial pattern and this is more obvious when the pattern of the initial map is more compact.

As explained in Sect. 4.3.2, model allocation function allocates changes as pixels whereas input maps only measure changing polygons that meet the MMU criteria. Input map resolutions do not fit the MMU rules. Consequently, the simulated landscape is more fragmented than the real one and this is clearer for the CORINE simulation because of its larger MMU.

The patcher and expander functions of DINAMICA can be parametrized (mean area and variance of simulated polygons) to achieve the simulation pattern we want. Although some studies have proved effective (Soares-Filho et al. 2003), this did not work in our study area. The transitions modelled are only possible when there is a suitable area (obtained from the drivers) inside the polygons of the destination category for the transition. If this happens, the model then considers the user's parameters. Suitable areas for specific transitions are going to be smaller in models at large scales (smaller MMU) than in models at coarse scales (bigger MMU) because of the respective size of the polygons in each model. It is therefore more difficult to vary the modelled patterns in models with fine scales.

Finer resolution models show problems in the allocation process because real changes tend to take place in a group of pixels. Models at coarser resolutions are more suited to deal with these problems (Kocabas and Dragicevic 2006; Blanchard et al. 2015). Thus, there is a relation between the behaviour we observed in our models and their spatial resolution, which is related to the cartographic scale and MMU of the input data. Although the rasterization method we followed (Hengl 2006) links the chosen spatial resolution with the cartographic scale and the MMU, there is still an inconsistency between the two in the model. Some authors have proposed patch-based models to solve this problem (Wang and Marceau 2013). There is also a wide body of literature about spatial resolution influence in LULC modeling (mainly CA-based models), which reaches similar conclusions (Marceau et al. 2005; Pan et al. 2010; Blanchard et al. 2015).

5.4 Allocation Differences

Although one might imagine that the higher quantity of change detected by input maps would result in a simulation that was closer to the true situation, this is not always the case. The transition rates could result from changes caused by different processes. However, the driving forces defining the most likely transition areas were the same for both models. As a result, the model at finer scale (SIOSE) could extrapolate changes from one process to changes from other processes, e.g. both simulations allocated changes in 'Dump sites' in the area around the central dump in Asturias, forecasting expansion of this dump. Over the calibration period CORINE only detected changes in this area, while SIOSE also detected changes in other parts of the study area, resulting from other processes such as road building works.

Regarding the spatial resolution, models at finer spatial resolutions simulate more pixels than models at coarser spatial resolutions. This means that the possibility of error in the two models is different. The larger the quantity of pixels to allocate, the more likely the model is to make a mistake. Validation is therefore dependent on the spatial resolution of the model. We must not compare errors in models with different spatial resolutions.

Finally, the difference between pixel-based modeling and MMU rules must also be borne in mind in the validation step. Many models are validated by LULC maps that follow MMU rules (Paegelow and Olmedo 2005; Pontius Jr. and Malanson 2005; Pérez-Vega et al. 2012). Changes that do not obey these rules do not appear on the map. However, changes simulated by the model will be located in the most suitable area, regardless of the minimum area of these changes. This means that real changes could be considered as errors in validation analysis.

6 Conclusion and Outlook

Model results can vary greatly depending on the cartographic scale and minimum mapping unit of the input data.

Most of the uncertainty comes from the differences between the input maps. The choice of a suitable cartographic source is therefore crucial. In-depth research must be done on this issue, comparing different cartographic sources and their influence on land cover representation. Great care must also be taken with the generalization of LULC maps, since most of their dissimilarities result from different generalization criteria.

Minimum mapping unit affects the quantities of change obtained and selected transitions since maps with smaller MMU measure more changes and more varied ones. This makes the resulting model more complex. If the user cannot manage this complexity properly, it will produce more uncertainty, because most of the analysed change is not interpreted in the model. Thus, the modeller must strike a balance between model complexity and its explanatory power.

Modelled patterns are also dependent on spatial resolution. This is linked with the minimum mapping unit: small MMU imply finer spatial resolution than larger units. Big differences between pixel size and MMU result in more fragmented scenarios. Patch-based models can be a solution to this problem.

Other components of scale, such as extent or thematic resolution, also influence modeling results and changes in these components could help resolve some of the problems we encountered, such as LULC input map disagreement. Therefore, in-depth research must focus on the influence of scale in modeling and, especially, on the relation between the different meanings of scale and their general influence on modeling.

Acknowledgements This work has been supported in part by project SIGEOMOD_2020, BIA2013-43462-P (Spanish Ministry of Economy and Competitiveness and the Feder European Regional Development Fund). The author is also grateful to the Spanish Ministry of Economy and Competitiveness and the European Social Fund for the funding of his research activity (Ayudas para contratos predoctorales para la formación de doctores 2014).

References

Agarwal C, Green GM, Grove JM, et al (2002) A review and assessment of land-use change models: dynamics of space, time, and human choice. Newtown

Ahmed S, Bramley G (2015) How will Dhaka grow spatially in future? Modelling its urban growth with a near future planning scenario perspective. Int J Sustain Built Environ 4:359–377. doi:10.1016/j.ijsbe.2015.07.003

Alonso Ibáñez M rosario, Pérez Fernández JM (2012) Espacio metropolitano y difusión urbana: su incidencia en el medio rural. Consejo Económico y Social del Principado de Asturias, Oviedo

Antrop M (2004) Landscape change and the urbanization process in Europe. Landsc Urban Plan 67:9–26. doi:10.1016/S0169-2046(03)00026-4

Bhatti SS, Tripathi NK, Nitivattananon V et al (2015) A multi-scale modeling approach for simulating urbanization in a metropolitan region. Habitat Int 50:354–365. doi:10.1016/j.habitatint.2015.09.005

Blanchard SD, Pontius RG Jr, Urban KM (2015) Implications of Using 2 m versus 30 m spatial resolution data for suburban residential land change modeling. J Environ Inform 25:1–13. doi:10.3808/jei.201400284

Burrough PA, McDonnell RA, Lloyd CD (2015) Principles of geographical information systems. OUP, Oxford

Camacho Olmedo MT, Paegelow M, Mas JF (2013) Interest in intermediate soft-classified maps in land change model validation: suitability versus transition potential. Int J Geogr Inf Sci 27:2343–2361. doi:10.1080/13658816.2013.831867

Cao C, Siu-Ngan Lam N (1997) Understanding the scale and resolution effects in remote sensing and GIS. In: Quattrochi DA, Goodchild MF (eds) Scale in remote sensing and GIS. CRC Press, pp 57–72

Carrero de Roa M (2012) Transformaciones en los espacios rururbanos del área metropolitana de asturias. Espacio metropolitano y difusión urbana: su incidencia en el medio rural. Consejo Económico y Social del Principado de ASturias, Oviedo, pp 135–154

Castilla G, Larkin K, Linke J, Hay GJ (2009) The impact of thematic resolution on the patch-mosaic model of natural landscapes. Landsc Ecol 24:15–23. doi:10.1007/s10980-008-9310-z

Chas-amil ML, Touza J (2015) Assessment of the Spanish land cover information to estimate forest area in Galicia. Boletín la Asoc Geógrafos Españoles 69:333–350

Clarke KC (2004) The limits of simplicity: toward geocomputational honesty in urban modeling. In: Atkinson P, Foody G, Darby S, Wu F (eds) Geodynamics. CRC Press, Boca Raton, pp 215–232

Committee on Needs and Research Requirements for Land Change Modeling, Geographical Sciences Committee, Board on Earth Sciences and Resources et al (2014) Advancing land change modeling. National Academies Press, Washington, DC

Conway TM (2009) The impact of class resolution in land use change models. Comput Environ Urban Syst 33:269–277. doi:10.1016/j.compenvurbsys.2009.02.001

Dendoncker N, Schmit C, Rounsevell M (2008) Exploring spatial data uncertainties in land-use change scenarios. Int J Geogr Inf Sci 22:1013–1030. doi:10.1080/13658810701812836

Dietzel C, Clarke K (2006) The effect of disaggregating land use categories in cellular automata during model calibration and forecasting. Comput Environ Urban Syst 30:78–101. doi:10.1016/j.compenvurbsys.2005.04.001

Fernández García A, Ortega Montequín M, Sevilla Álvarez J et al (2007) Población, administración y territorio en Asturias. Consejo Económico y Social del Principado de Asturias, Oviedo

Follador M, Villa N, Paegelow M (2008) Tropical deforestation modelling: comparative analysis of different predictive approaches. The case study of Peten, Guatemala. In: Paegelow M, Camacho Olmedo MT (eds) Modelling environmental dynamics. Springer, Berlin, Heidelberg, pp 77–107

Hengl T (2006) Finding the right pixel size. Comput Geosci 32:1283–1298. doi:10.1016/j.cageo.2005.11.008
Hernández JD (2016) Methodology of classification extraction from descriptive systems of land cover and land use
Herold M, Couclelis H, Clarke KC (2005) The role of spatial metrics in the analysis and modeling of urban land use change. Comput Environ Urban Syst 29:369–399. doi:10.1016/j.compenvurbsys.2003.12.001
Houet T, Verburg PH, Loveland TR (2010) Monitoring and modelling landscape dynamics. Landsc Ecol 25:163–167. doi:10.1007/s10980-009-9417-x ISO/IEC (2014) ISO 19101-1. 48
João E (2001) Measuring scale effects caused by map generalization and the importance of displacement. In: Tate NJ, Atkinson PM (eds) Modelling scale in geographical information science. Wiley, Chichester, New York, Weinheim, Brisbane, Singapore, Toronto, pp 161–179
Kocabas V, Dragicevic S (2006) Assessing cellular automata model behaviour using a sensitivity analysis approach. Comput Environ Urban Syst 30:921–953. doi:10.1016/j.compenvurbsys.2006.01.001
Lam NS-N, Quattrochi D (1992) On the issues of scale, resolution, and fractal analysis in the mapping sciences. Prof Geogr 44:88–98. doi:10.1111/j.0033-0124.1992.00088.x
Leitão AB, Miller J, Ahern J, McGarigal K (2012) Measuring landscapes: a planner's handbook. Island press, Washington, Covelo, London
Lesschen JP, Verburg PH, Staal SJ (2005) Statistical methods for analysing the spatial dimension of changes in land use and farming systems. Citeseer
Maeda EE, de Almeida CM, de Carvalho Ximenes A et al (2011) Dynamic modeling of forest conversion: Simulation of past and future scenarios of rural activities expansion in the fringes of the Xingu National Park, Brazilian Amazon. Int J Appl Earth Obs Geoinf 13:435–446. doi:10.1016/j.jag.2010.09.008
Marceau DJ, others, Ménard A, Marceau DJ (2005) Exploration of spatial scale sensitivity in geographic cellular automata. Environ Plan B Plan Des 32:693–714. doi:10.1068/b31163
Mas J-F, Kolb M, Houet T et al (2011) Éclairer le choix des outils de simulation des changements des modes d'occupation et d'usages des sols. Une approche comparative. Rev Int Géomatique 21:405–430. doi:10.3166/RIG.15.297-322
Mas J-F, Kolb M, Paegelow M et al (2014) Inductive pattern-based land use/cover change models: A comparison of four software packages. Environ Model Softw 51:94–111. doi:10.1016/j.envsoft.2013.09.010
McGarigal K, Cushman SA, Neel MC, Ene E (2015) FRAGSTATS: spatial pattern analysis program for categorical and continuous maps
Ménard A, Marceau DJ (2005) Exploration of spatial scale sensitivity in geographic cellular automata. Environ Plan B Plan Des 32:693–714. doi:10.1068/b31163
Moreira E, Costa S, Aguiar AP et al (2009) Dynamical coupling of multiscale land change models. Landsc Ecol 24:1183–1194. doi:10.1007/s10980-009-9397-x
O'Sullivan D, Perry GLW (2013) Spatial simulation: exploring pattern and process. Wiley, Chichester
Oliver MA (2001) Determining the spatial scale of variation in environmental properties using the variogram. In: Modelling scale in geographical information science, pp 193–219
Paegelow M, Camacho Olmedo MT, Mas J-F, Houet T (2014) Benchmarking of LUCC modelling tools by various validation techniques and error analysis. Cybergeo. doi:10.4000/cybergeo.26610
Paegelow M, Olmedo MTC (2005) Possibilities and limits of prospective GIS land cover modelling—a compared case study: Garrotxes (France) and Alta Alpujarra Granadina (Spain). Int J Geogr Inf Sci 19:697–722. doi:10.1080/13658810500076443
Pai N, Saraswat D (2013) Impact of land use and land cover categorical uncertainty on SWAT hydrologic modeling. Trans ASABE 56:1387–1397. doi:10.13031/trans.56.10062
Pan Y, Roth A, Yu Z, Doluschitz Reiner R (2010) The impact of variation in scale on the behavior of a cellular automata used for land use change modeling. Comput Environ Urban Syst 34:400–408. doi:10.1016/j.compenvurbsys.2010.03.003

Pérez-Vega A, Mas J-F, Ligmann-Zielinska A (2012) Comparing two approaches to land use/cover change modeling and their implications for the assessment of biodiversity loss in a deciduous tropical forest. Environ Model Softw 29:11–23. doi:10.1016/j.envsoft.2011.09.011
Pontius RG Jr, Boersma W, Castella J-C et al (2008) Comparing the input, output, and validation maps for several models of land change. Ann Reg Sci 42:11–37. doi:10.1007/s00168-007-0138-2
Pontius RG Jr, Malanson J (2005) Comparison of the structure and accuracy of two land change models. Int J Geogr Inf Sci 19:243–265. doi:10.1080/13658810410001713434
Pontius Jr. RG, Malizia NR (2004) Effect of category aggregation on map comparison. In: Geographic information science. Springer, pp 251–268
Pontius RG Jr, Millones M (2011) Death to Kappa: birth of quantity disagreement and allocation disagreement for accuracy assessment. Int J Remote Sens 32:4407–4429. doi:10.1080/01431161.2011.552923
Quattrochi DA, Emerson CW, Siu-Ngan Lam N, Qui H (2001) Fractal characterization of multitemporal remote sensing data. In: Tate NJ, Atkinson PM (eds) Modelling scale in geographical information science. Wiley, Chichester, New York, Weinheim, Brisbane, Singapore, Toronto, pp 13–34
Quattrochi DA, Goodchild MF (1997a) Scale, multiscaling, remote sensing, and GIS. In: Quattrochi DA, Goodchild MF (eds) Scale in remote sensing and GIS. CRC Press, pp 1–11
Quattrochi DA, Goodchild MF (1997b) Scale in remote sensing and GIS. CRC Press
Quiroz Ortuño Y (2009) Modelo dinámico de cambio de cobertura y uso del suelo en una zona de transición urbano-rural, entre la ciudad de Morelia y el ejido Jesús del Monte. Universidad Autónoma de México
Renwick A, Jansson T, Verburg PH et al (2013) Policy reform and agricultural land abandonment in the EU. Land Use Policy 30:446–457. doi:10.1016/j.landusepol.2012.04.005
Rodríguez Gutiérrez F, Menéndez Fernández R, Blanco Fernández J (2009) El área metropolitana de Asturias. Ciudad Astur: el nacimiento de una estrella urbana en Europa. Trea, Oviedo
Saura S (2002) Effects of minimum mapping unit on land cover data spatial configuration and composition. Int J Remote Sens 23:4853–4880. doi:10.1080/01431160110114493
Schmit C, Rounsevell MD a., La Jeunesse I (2006) The limitations of spatial land use data in environmental analysis. Environ Sci Policy 9:174–188. doi:10.1016/j.envsci.2005.11.006
Soares-Filho BS, Cerqueira GC, Pennachin CL (2002) DINAMICA—a stochastic cellular automata model designed to simulate the landscape dynamics in an Amazonian colonization frontier. Ecol Modell 154:217–235. doi:10.1016/S0304-3800(02)00059-5
Soares-Filho BS, Corradi Filho L, Cerqueira GC, Leite Araujo W (2003) Simulating the spatial patterns of change through the use of the dinamica model. In: XI Simpósio Brasileiro de Sensoriamento Remoto - SBSR. Belo Horizonte, pp 721–728
Turner MG, O'Neill RV, Gardner RH, Milne BT (1989) Effects of changing spatial scale on the analysis of landscape pattern. Landsc Ecol 3:153–162. doi:10.1007/BF00131534
Uuemaa E, Roosaare J, Mander Ü (2005) Scale dependence of landscape metrics and their indicatory value for nutrient and organic matter losses from catchments. Ecol Indic 5:350–369. doi:10.1016/j.ecolind.2005.03.009
van Delden H, van Vliet J, Rutledge DT, Kirkby MJ (2011) Comparison of scale and scaling issues in integrated land-use models for policy support. Agric Ecosyst Environ 142:18–28. doi:10.1016/j.agee.2011.03.005
Verburg P, Schot P, Dijst M, Veldkamp A (2004) Land use change modelling: current practice and research priorities. GeoJournal 61:309–324. doi:10.1007/s10708-004-4946-y
Verburg PH, de Groot WT, Veldkamp AJ (2003) Methodology for multi-scale land-use change modelling: concepts and challenges. In: Dolman AJ, Verhagen A, Rovers CA (eds) Global environmental change and land use. Springer, Dordrecht, pp 17–51
Verburg PH, Neumann K, Nol L (2011) Challenges in using land use and land cover data for global change studies. Glob Chang Biol 17:974–989. doi:10.1111/j.1365-2486.2010.02307.x

Verburg PH, Overmars KP (2009) Combining top-down and bottom-up dynamics in land use modeling: exploring the future of abandoned farmlands in Europe with the Dyna-CLUE model. Landsc Ecol 24:1167–1181. doi:10.1007/s10980-009-9355-7

Verstegen JA, Karssenberg D, van der Hilst F, Faaij A (2012) Spatio-temporal uncertainty in spatial decision support systems: a case study of changing land availability for bioenergy crops in Mozambique. Comput Environ Urban Syst 36:30–42. doi:10.1016/j.compenvurbsys.2011.08.003

Wainwright J, Mulligan M (2013) Environmental modelling: finding simplicity in complexity, 2nd edn. Wiley, New York

Wang F, Marceau DJ (2013) A patch-based cellular automaton for simulating land-use changes at fine spatial resolution. Trans GIS 17:828–846. doi:10.1111/tgis.12009

Waser LT, Schwarz M (2006) Comparison of large-area land cover products with national forest inventories and CORINE land cover in the European Alps. Int J Appl Earth Obs Geoinf 8:196–207. doi:10.1016/j.jag.2005.10.001

Wu J (2004) Effects of changing scale on landscape pattern analysis: scaling relations. Landsc Ecol 19:125–138. doi:10.1023/B:LAND.0000021711.40074.ae

Wu J, Qi Y (2000) Dealing with scale in landscape analysis: an overview. Ann GIS 6:1–5. doi:10.1080/10824000009480528

Zhao G (2013) Effects of spatial scale in cellular automata model for land use change. In: Gaol FL (ed) Recent progress in data engineering and internet technology. Springer, Berlin, Heidelberg, pp 101–106

Zong-Guo X, Clarke KC (1997) Approaches to scaling of geo-spatial data. In: Quattrochi DA, Goodchild MF (eds) Scale in remote sensing and GIS. CRC Press, pp 309–360

Chapter 11
Who Knows Best? The Role of Stakeholder Knowledge in Land Use Models—An Example from Doñana, SW Spain

R.J. Hewitt, V. Hernández Jiménez, L. Román Bermejo and F. Escobar

Abstract Participatory processes are increasingly used for understanding human-environment interaction problems and for developing common strategies for land resource management. These approaches are particularly important in areas where resources are shared by many stakeholders and yet there is no general agreement about how these resources should be managed. In many of these cases, detailed quantitative information about human-environment interaction problems (e.g. land degradation, erosion, water contamination etc.) is available to scientific institutions and land managers, but not easily accessible to other stakeholders. Conversely, key information, such as historical evolution of the landscape in the locality or the probable drivers of historic land change is often embedded informally in stakeholder communities but may not be accessible via conventional knowledge sharing pathways (scientific literature, reports, directives, policy briefs etc.). Land use models, in which qualitative and quantitative data can be combined at multiple levels and scales, provide an ideal bridge between highly detailed quantitative knowledge available from scientific stakeholders, and informal or unstructured knowledge about dynamics, evolution and change held by other parts

---

R.J. Hewitt (✉) · V. Hernández Jiménez · L. Román Bermejo
Observatorio para una Cultura del Territorio, Madrid, Spain
e-mail: richard.hewitt@observatorioculturayterritorio.org

V. Hernández Jiménez
e-mail: vero.hj@observatorioculturayterritorio.org

L. Román Bermejo
e-mail: lara.rb@observatorioculturayterritorio.org

R.J. Hewitt
Information and Computational Sciences Group,
James Hutton Institute, Aberdeen, Scotland, UK

F. Escobar
Department of Geology, Geography and Environmental Sciences,
University of Alcalá, Alcalá de Henares, Spain
e-mail: francisco.escobar@uah.es

M.T. Camacho Olmedo et al. (eds.), *Geomatic Approaches for Modeling Land Change Scenarios*, Lecture Notes in Geoinformation and Cartography,
https://doi.org/10.1007/978-3-319-60801-3_11

of the stakeholder community. Many essential land use modeling activities, traditionally carried out by a single scientist in front of a computer, such as map comparison and subdivision or aggregation of land use categories, may in fact be better accomplished by working in groups with key stakeholders. Involvement of stakeholders in basic model decisions not only makes for a better model, it may also increase stakeholder confidence in the model and makes it more likely that the results of the model will be applied. We argue, with reference to the recent participatory modeling work undertaken in Doñana, south-west Spain, that stakeholder information can be incorporated into land use models by engaging stakeholders as model co-developers, and structuring activities, where possible, so as to include their knowledge directly as parameters and variables. A participatory land use model is thus conceived as a cycle of alternating analytical and discursive activities from which useful results may be obtained, but which does not presuppose an optimum or "right answer", or prioritize scientists' knowledge above other kinds of knowledge available to the community.

Keywords Participatory modeling · Land use change · Stakeholder · Knowledge co-generation · Doñana

1 Introduction

The Doñana natural area is an internationally renowned coastal dune and marshland ecosystem of outstanding importance for biodiversity at the mouth of the Guadalquivir River in South West Spain (Fig. 1). Intensive agriculture and tourism have transformed the economy of the area over the last 60 years but have led to severe degradation of ecosystems and habitats, including the loss of large areas of wetland. In response to these serious threats, Doñana has been the subject of major conservation efforts since the 1960s, and today enjoys diverse statutes of protection (National Park, Natural Park, UNESCO world natural heritage site, amongst others). These conservation efforts have undoubtedly been highly successful in preventing, for example, further wetland habitat loss, expansion of invasive species such as eucalyptus and urban development along the coast. The natural area remains, nonetheless, highly threatened, and continues to decay. This is principally because the highly intensive land uses in the watershed of the Guadiamar River, which supplies the Doñana marshes, are incompatible with the maintenance of pristine natural habitat downstream. However, since the wider watershed is not included in the protected area, land use is not subject to strict controls. Yet while many stakeholders might be unwilling to suspend business-as-usual in the watershed, few would regard the loss of Doñana as an acceptable price to pay. For this reason, there is a strong case for dialogue about land use in the watershed involving all the relevant actors (policy makers, farmers, conservationists, tourism representatives etc.), with a view to securing a more sustainable future.

Fig. 1 Location of study area

This chapter presents the results of a participatory scenario modeling cycle in which a range of land use futures were explored for Doñana for the year 2035. The aim of the modeling process was threefold:

(1) To bring together key stakeholders from different sectors with different and (sometimes opposing) perspectives and engage them in a shared discussion about Doñana's future.
(2) To enable knowledge co-generation and social learning between stakeholders about land use changes to facilitate transition to a more sustainable model of development for the area.
(3) To push the boundaries of participatory modeling by increasing model transparency and inclusiveness and reducing the model's dependence on researchers' decisions.

To achieve these aims, detailed input from key stakeholders obtained through a series of workshops was used to parameterize, calibrate and critically evaluate a Cellular Automata (CA) model of land use change in the Guadiamar river watershed. Despite the strong emphasis on social learning and knowledge co-generation, as opposed to enhancing the accuracy of model simulations, the model meets all commonly accepted metrics for assessment of goodness-of-fit. Nonetheless, we argue that integration of qualitative and quantitative data through a participatory process leads to a better model, and that acceptance by stakeholders should be considered a valid criterion for determining a "successful model".

2 Test Areas and Data Sets

The study area corresponds approximately to the watershed of the Guadiamar River (Fig. 1). The principal data sets used in the work were the 1:25,000 scale vegetation cover and land use map series developed by the Government of Andalusia and freely available for download from the website of the regional government environmental information service, REDIAM. Other information necessary for model calibration included topographic relief, municipal boundaries, rivers and watercourses, and natural protected areas, all downloaded from the REDIAM website. For detailed information regarding these datasets, see:

http://www.juntadeandalucia.es/medioambiente/site/rediam/.

3 Methodology and Practical Application to the Data Sets

3.1 Cellular Automata Models of Land Use Change

The model employed in this research is a Cellular Automata (CA) based land use model. CA models integrate mathematical theories of self-reproduction in automata

(Von Neumann and Birks 1966) and stochasticity (Ulam 1952) with the 2 dimensional cellular-grid or raster cartographic space familiar to present-day users of Geographical Information Systems (GIS). The concept of a dynamic geographical cellular automata was proposed by Tobler (1979) and developed during the 1990s by researchers interested in modeling urban growth and change (e.g. White and Engelen 1993; Batty and Xie 1994; Clarke et al. 1997; Phipps and Langlois 1997). Well-known examples of CA modeling frameworks include SLEUTH (Clarke et al. 1997) and members of the Metronamica family such as SimLucia (White et al. 2000) and Xplorah (Van Delden et al. 2008). CA modeling systems aim to simulate the aggregate behaviour of multiple change agents by developing land use transition rules and testing these rules against data. For this application, we used the well-known modeling software Metronamica, developed by White and collaborators (e.g. White and Engelen 1993; White et al. 2000). Detailed methodological description of the functioning of the model is published in the user guide (RIKS 2012) and will not be repeated here.

When using the model, the first step is to introduce a GIS raster map representing land use at a given moment in time (t_n). Cells can change from one land use to another over the course of a time sequence (t_1, t_2, t_3… t_n) on the basis of the relationship between their own land use and the land use of the cells that are immediately adjacent or in proximity, known as the cell neighbourhood (N). However, the potential of each cell to transition is not determined exclusively by its neighbourhood, and Accessibility (A) (i.e. the influence of lines of communication such as transport, irrigation, and electricity network) must also be taken into account. A cell's transition potential also takes into account Suitability (S) or the biophysical eligibility (e.g. rainfall, slope) of each land area for a particular use, and Zoning (Z), the current legislative and planning restrictions (for example, protected areas, urban spatial plans). Finally, since human activity in the landscape is not purely deterministic, a stochastic parameter is added (v). This type of model is referred to as an N,A,S,Z Cellular Automata model or NASZCA (Hewitt et al. 2014).

To calibrate the model, parameter values for the N, A, S, and Z blocks are set and the model is run from an initial map t_1 (1956 in this case) to a second date *n* time steps (i.e. years) forward for which a map is available for comparison (1999 in this case).This second map is referred to as t_2. The number of cells which are to be allocated for each land use at each time step t_n is known as the *demand*. Once the total number of cells corresponding to land use demand has been allocated to all suitable locations at model time step t_n, the next step ($t_n + 1$) is computed from t_n and so on until time t_2 is reached. The time period between t_1 and t_2 is known as the calibration period. If data are available, the use of a third period t_3 known as the validation period, subsequent to t_2, is recommended. In this case, the most recent available land use map, the map for 2007, was used for validation.

To validate the land use model, the rules developed to simulate land change evolution between t_1 and t_2 (the calibration period) were applied to t_3. Successful replication of land-change tendencies at both t_2 and t_3 against accepted benchmarks reinforces the main assumption of the model, i.e. that the past change processes being modelled will hold true in the future. Once calibration and validation have

Table 1 Calibration milestones and their relationship to individual simulations evaluated by stakeholders (Workshop 2, Activity 1) in bold

Milestone number:	Simulation run:	Calibrationsub-step
1	1	Simple neighbourhood rules only (benchmark model)
2	5	Calibrate neighbourhood rules
3	**11**	**Add Accessibility**
4	**15**	**Adjust Accessibility Parameters**
5	21	Add Suitability
6	**23**	**Add Zoning**
7	**34**	**Adjust Suitability Parameters**
8	**35**	**Adjust Neighbourhood parameters**

been carried out successfully, the model is considered to be ready to generate simulation for future dates. Calibration and assessment of the quality of the calibration was a continuous and iterative process managed around a series of milestones relating to the determination of parameters for the key model drivers, N,A,S, Z. Attaining a calibration milestone required experimentation with different parameter settings, major adjustments were recorded with a unique simulation number (Table 1).

Assessment of the model's goodness-of-fit was undertaken using three types of assessment: (1) visual inspection, in which stakeholders played a key part (see results of Workshop 2); (2) cell-by-cell comparison measures using the kappa simulation statistic K_{sim} (van Vliet et al. 2011); and (3) map pattern and structure evaluation through the clumpiness statistic, derived from the Fragstats package (McGarigal et al. 2002) and implemented in the Map Comparison Kit (MCK) software, developed by Alex Hagen-Zanker for the Netherlands Environment Assessment Agency (Visser and De Nijs 2006).

3.2 The Participatory Modeling Process

A geographical land use model is a useful tool for integrating qualitative and quantitative data (Fig. 2).

Land use models are sometimes thought of as purely quantitative tools, in that they are based on numerical inputs and outputs. However, they also contain abundant qualitative information (e.g. land use category decisions, study area decisions, transition rules etc.) that is essential to their successful operation. Often these qualitative inputs are decided in a subjective or arbitrary way by the researcher, who may not always have detailed knowledge of the study area at hand, and, in many cases, it may be more appropriate to consult with, and directly elicit information from, the appropriate stakeholder, especially if the model is intended for use outside the scientific community. However stakeholder contributions to

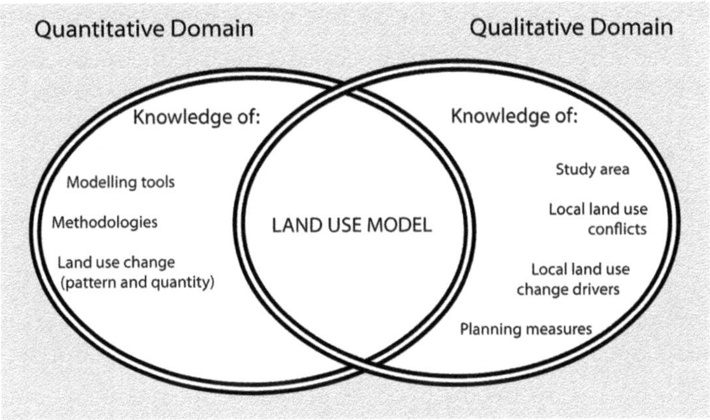

Fig. 2 A land use model as the intersection of quantitative and qualitative knowledge domains

models need not be limited to qualitative data, but can also include approximate or estimated *quantitative* data. For example, in the modeling case study presented here, stakeholders estimated suitability of terrain for each land use through a scoring procedure linked to a numerical value. While such approaches are clearly subjective, they can be surprisingly accurate, especially if the task is carefully matched to areas of stakeholder knowledge. Farmers, for example, are likely to be adept at accurately valuing terrain suitability or estimating areal quantities for crop types.

On the one hand it is likely that the successful incorporation of local stakeholders' knowledge will improve the model. For example, in this case, analysis of land use dynamics by stakeholders allowed researchers to differentiate between land use change due to the degradation of natural areas and land use change as a result of deliberate policy. One example was the elimination of eucalyptus plantations, the cause of which was not evident prior to the participatory process (Hewitt et al. 2014). By including stakeholders in the modeling procedure as co-developers, the applicability, utility and validity of the model is likely to be enhanced by promoting a sense of ownership and developing trust in the modeling procedure and model outputs among the stakeholder community as a whole (Voinov et al. 2016). In the model presented here, stakeholders were involved in model development from the very beginning of the process through three one-day workshops held in the study area between February 2012 and September 2013, during the key phases of model development; workshop 1; parameterisation, workshop 2: calibration; workshop 3; evaluation. The stakeholders who participated in the workshops are listed in Table 2. With respect to the process of stakeholder identification, we were very fortunate in that a team of researchers from the socio-ecosystems laboratory of the Madrid Autonomous University were already active in the area, and kindly shared their extensive list of contacts with us.

Table 2 Workshop participants by sector. In addition 3–4 researchers attended each workshop in the capacity of workshop coordinators

Stakeholder, by sector	WS1	WS2	WS3
Science			
Researcher, Autonomous University of Madrid	Yes	Yes	Yes
Researcher and University Lecturer, University of Seville	Yes	Yes	Yes
Researcher, Doñana Biological Station (National Scientific Institute)	Yes	Yes	Yes
Agriculture			
Director, federation of rice farmers, Seville	Yes	Yes	Yes
Representative, young farmers agricultural association (ASAJA)	Yes	Yes	Yes
Representative, Andalusian Farmers and Livestock keepers union, Huelva division	No	Yes	Yes
Tourism			
Tourism representative, Doñana natural area	No	Yes	Yes
Local policy makers			
Moguer Municipal council, Environment technician	Yes	No	Yes
Representative Doñana 21 Foundation	Yes	Yes	Yes
Almonte Municipal council, Environment technician	No	No	Yes
Regional policy makers			
Regional administration, environmental research division	No	Yes	Yes
Regional administration, environmental research division	No	Yes	Yes
Natural area managers			
Autonomous Body for National parks, head of project monitoring	Yes	Yes	No
Director, Doñana Natural Area	Yes	Yes	yes
Sub-director, Doñana Natural Area	Yes	No	No
Director of Conservation, Doñana Natural Area	Yes	No	No
Director of Public Use, Doñana Natural Area	Yes	Yes	Yes
Guide, Doñana Natural Area	No	No	Yes
Monitoring division, Doñana Natural Area	Yes	Yes	Yes
Environmentalists			
Ex- Ecologistas en acción (Environmental group)	Yes	No	Yes
World Wildlife Fund	No	No	Yes

Through these three workshops, the stakeholder community undertook a wide range of model co-development tasks at each major stage of the modeling process (Table 3). They helped to define the study area, the database and land use categories to use in the model, and they reclassified the land use data set to determine the most important land use dynamics for modeling their region. They provided information about the effect of the different biophysical factors on each land use category, evaluated the goodness-of-fit of simulations during the model calibration phase and estimated land use demand for the model scenarios. Finally, they explored the model results, developed a prioritized list of indicators for environmental monitoring on the basis of the model outputs and evaluated the applicability and utility of the model as well as the overall success of the participatory process. Note that, while “stakeholders” have been treated as separate to “researchers” in the preceding text in order to emphasize their participation in modeling decisions usually carried out by researchers alone, the researcher is of course a stakeholder. Although it is evident that the researchers, as organisers of the process, have greater power over the information than other stakeholders, we attempted to balance this by presenting, in detail, prior to each workshop, the way in which the information we obtained had been used. This is important in order to establish a relationship of trust and to allow the researcher to become properly embedded in the stakeholder community. In addition, while researchers led the process, the physical results of workshop activities (ideas written on post-it-notes and wall charts, completed activity sheets) always came directly from the participants themselves. At the writing up stage, the extensive video material recorded during the workshops helped to reduce bias caused by researchers’ unintentional interpretation of the primary data—where possible we actually transcribed verbatim the dialogue recorded on video before interpretation.

The participatory modeling process was developed through a chain of iterating participatory/non-participatory activities in which 9 separate steps were identified (Table 3).

4 Results

Results of the participatory modeling process developed through the three stakeholder workshops are summarized as follows. For more detailed description of all the workshop activities and the results obtained see Hewitt et al. (2012, 2014), Escobar et al. (2015) and Hewitt et al. (2016). Full descriptions of the Workshops can also be found on the project webpage.[1]

Table 3 details the key modeling stages and the input received at each stage from the participatory process and the corresponding non-participatory (analytical-technical) activity subsequently undertaken by the research team.

4.1 Workshop 1—22nd February 2012

The aim of the first workshop, at which 14 stakeholders were present, was to inform participants about the DUSPANAC project and the proposed modeling activities, and to collect information from them directly in order to parameterize the model. In the first activity (discussion and reclassification of land use categories in groups) the most relevant land use categories for explaining the dynamics of change in Doñana

[1]http://www.geogra.uah.es/duspanac/taller.html.

Table 3 Step-by-step model procedure, together with the relevant participatory and analytical-technical tasks

Modeling step#	Modeling step	Sub-step	Participatory method	Analytical-technical method
1	**Decisions on setting up an application**	Delineation of modelled region	Workshop 1: stakeholder assessment of most suitable study area to reflect dynamics	Researchers' decision based on dynamics observed and own understanding
		Selection of land use classes for modeling	Workshop 1: stakeholders select and reclassify land use categories based on their understanding of land use in the natural area	Selection of land use classes according to land change dynamics observed in cross tab analysis, process understanding and expected model use
		Assign land use classes to behaviour types: dynamic vs. static	Workshop 1: stakeholder evaluation of dynamics (drivers of LUC) stakeholder responses help to understand which classes are most important for dynamic modeling	Assignation of land use classes to types according to land change dynamics observed in cross-tab analysis
		Choose spatial resolution	No consultation	Chosen by researchers on the basis of own knowledge and available datasets
2	**Analysis of dynamics of land use change in the territory to be modelled**	Workshop 1: stakeholder evaluation of dynamics (drivers of LUC, category losses and gains, assessment of map quality)	Cross-tabulation analysis of LUC, neighbourhood analysis and landscape pattern analysis	
3	**Data preparation and setting up the model for the calibration period**	Input land use maps Prepare accessibility, suitability and zoning layers	No consultation until parameters need to be defined (stage 4, below)	Data preparation and incorporation of above defined parameters into modeling environment

(continued)

Table 3 (continued)

Modeling step#	Modeling step	Sub-step	Participatory method	Analytical-technical method
4	**Calibration**	Set neighbourhood rules Set random parameter Set accessibility parameters, Set suitability parameters	Parameters defined by stakeholders from information gathered in Workshop 1	Model manipulation and data handling, statistical testing (kappa sim, clumpiness, visual inspection)
5	**Analytical testing/evaluation of calibration**	Workshop 2: participatory visual inspection of cell-by-cell accuracy & spatial patterning	Statistical testing of model goodness of fit (cell-by-cell accuracy & spatial patterning)	
6	**Fine-tune calibration**	Adjust parameter set in step 4	Apply results of participatory model evaluation to reconfigure model	Re-configure model with new datasets or parameters
7	**Scenario development**	Workshop 2: participatory estimation of demand for land uses	Input stakeholder demand and run scenarios	
8	**Evaluation**	Workshop 3: Explore model in workshop session, evaluate model and modeling process		
9	**Indicator development**	Workshop 3: Brainstorming of indicators	Development of prioritised indicators in GIS	

were defined. Prior to the workshop, the research team had undertaken a preliminary reclassification of the Andalusian Regional Government land use and vegetation cover database (Moreira 2007), reducing 107 land use categories to 48. This was still too many for a workable land use model, but further aggregation clearly required local knowledge. To undertake this task, participants were divided into 3

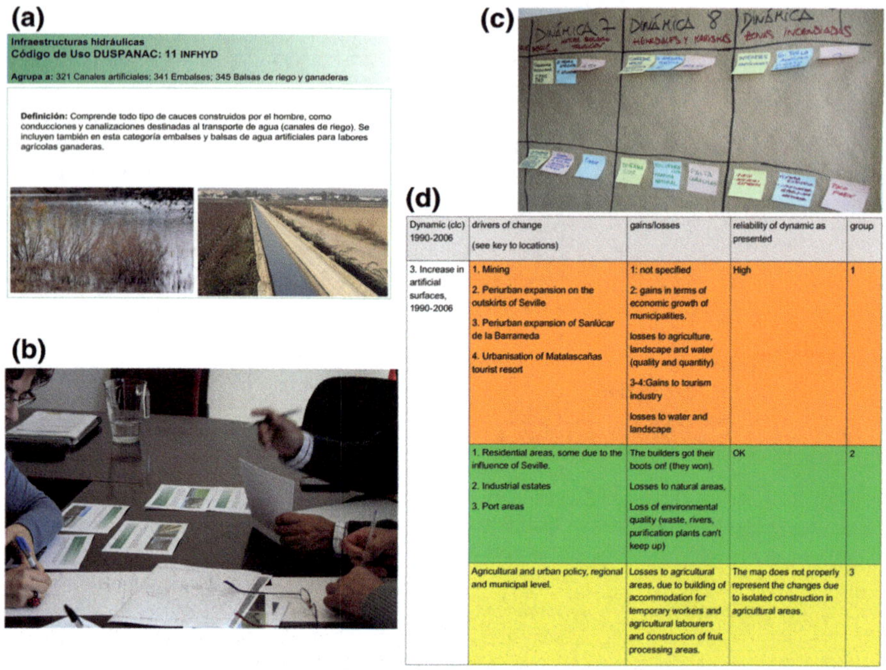

Dynamic (clc) 1990-2006	drivers of change (see key to locations)	gains/losses	reliability of dynamic as presented	group
3. Increase in artificial surfaces, 1990-2006	1. Mining 2. Periurban expansion on the outskirts of Seville 3. Periurban expansion of Sanlúcar de la Barrameda 4. Urbanisation of Matalascañas tourist resort	1: not specified 2: gains in terms of economic growth of municipalities. losses to agriculture, landscape and water (quality and quantity) 3-4:Gains to tourism industry losses to water and landscape	High	1
	1. Residential areas, some due to the influence of Seville. 2. Industrial estates 3. Port areas	The builders got their boots on! (they won). Losses to natural areas. Loss of environmental quality (waste, rivers, purification plants can't keep up)	OK	2
	Agricultural and urban policy, regional and municipal level.	Losses to agricultural areas, due to building of accommodation for temporary workers and agricultural labourers and construction of fruit processing areas.	The map does not properly represent the changes due to isolated construction in agricultural areas.	3

Fig. 3 Workshop 1 activities; **a** example of land use cards used in activity 1; **b** stakeholders debating the land use classification in groups; **c** Stakeholder analysis of land use dynamics on post -it notes; **d** Results of stakeholder analysis of land use dynamics in digital form

groups of 4/5 and each group was tasked with building a new, simpler land use categorisation to represent synthetically and as realistically as possible key land use dynamics from the 48 initial categories. A set of cards illustrating and describing the 48 initial categories was handed out to each group to support the group discussion. Each group chose a spokesperson who presented their categorisation to all participants. Finally, a common and representative "consensus" classification was obtained through general discussion with all participants (Fig. 3). The tangible result of this first activity was the definition of the final 23 land use categories to be introduced into the model.

In the next activity, stakeholders were asked to analyse land use changes that have been observed in Doñana on the basis of previous studies. As in the previous activity, participants worked in groups and later shared their conclusions with the whole group. This activity gave the research team a much better understanding of the land use dynamics of the study area, and enabled us to set the model transition rules (neighbourhood, zoning and suitability) (Table 1), allowing model calibration to begin. In the final activity, participants evaluated the influence of particular suitability factors (rainfall, slope, temperature etc.), on each land use class in the study area through a simple qualitative scoring system—strong (mucho), weak (poco), or no influence at all (nada). On the basis of this information, an agreement or confidence index (C) was calculated by allocating a value of 0 where all three groups disagreed, a value of 1 where two groups disagreed with the third group, and 2 where all groups agreed. These values were then added together to create a total agreement index for each suitability factor. The categorical responses (strong, weak and no influence) given by the stakeholders for each land use regarding the influence of a given factor were translated into a simple scoring system of 2 (strong), 1 (little) and 0 (no influence) to create what we called the influence index (I). Finally, the confidence index (C) was multiplied by the influence index (I) to give a total overall score by land use for each suitability factor. Thus, for example, in assessing the PLASTIC (intensive crops grown under plastic) land use, all three groups felt slope to be important and responded strong, a score of 2 for each group, giving $(2 + 2 + 2) = 6$. Since all groups agreed about the importance of slope for this land use, the highest confidence score (2) was allocated. Thus the total score for the slope factor for the PLASTIC land use category was 12 (6×2), indicating that the stakeholders felt, with a high degree of confidence, that slope was influential in determining the location of crops under plastic, for which flat and slightly-sloping land was preferred. Suitability parameter settings inside the model were estimated on the basis of this information. For example, in the case of the Industrial (IND), Rice (RICE), Intensive crops under plastic (PLASTIC) and Intensive woody crops (INTWOOD) categories, high suitability parameter values were given to areas with slopes of less than 5%. These values were subsequently modified using an iterative trial-and-error approach which involved experimenting with various different suitability values for different slope categories in relation to these land use classes until some improvement could be seen in the location and spatial pattern according to the analytical assessment methods used.

4.2 *Workshop 2—11th December 2012*

The second participatory workshop was held on December 11, 2012. Its main objective was to assist in model calibration through three main participatory activities; (1) visual assessment of the quality of model simulations at different phases of calibration; (2) estimation of land use demand for 4 model scenarios adapted from the Doñana Eco-Futures (Palomo et al. 2011), and; (3) estimation of new land use location through a participatory cartography exercise. For the first activity, after a brief explanation and introduction, participants were tasked with evaluating model goodness-of-fit (Fig. 4a, b). To do this, participants were divided into 3 groups and given four simulations representing different stages of the calibration process (Table 1), though only 2 simulations (11 and 35) were successfully evaluated by all participants in the time available. Simulations were provided both on screen and as paper printouts. Participants explored the simulations and then debated the merits of each one in groups (Fig. 4a). Subsequently they rated the similarity of location and degree of clustering of simulated land use categories compared to real land use categories on a pro-forma worksheet. This activity aimed,

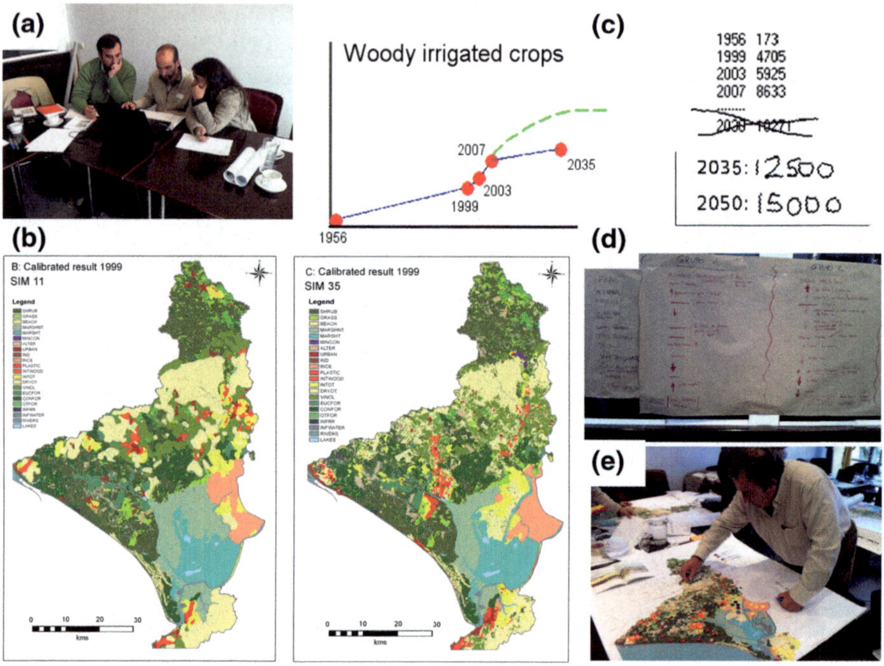

Fig. 4 Workshop 2 activities; **a** Stakeholders evaluating simulation goodness-of-fit on the computer; **b** Simulations 11 and 35 which were evaluated by stakeholders; **c** Land use demand with example of modified tendency curve and demand estimated by stakeholders; **d** Paper wall chart with land change tendencies from open discussion of land use demand; **e** Stakeholder using counters to locate land use demand on a paper map

first of all, to better acquaint the stakeholders with the detailed process of creating land use simulations and to remove a little of the mystery surrounding the operation of the model. It also sought to enhance the validity of the process of visual inspection of calibrations, a job researchers normally do themselves. Clearly, while evaluation of goodness-of-fit with the human eye is highly subjective, subjectivity decreases when simulations are evaluated by many people. Taken together, stakeholder goodness-of-fit evaluation broadly agrees with the results of the statistical tests (Fig. 5).

The second activity of the workshop was intended to link the CA land use model with the Doñana Eco-Future scenarios. The Doñana Eco-Futures were developed prior to our research team's work by stakeholders working with a group of researchers in socio-ecosystems. Full details about the scenario development process can be found in Palomo et al. (2011). Though the research team was very fortunate in being able to work with existing scenarios, these were not suitable, in their existing form, for direct inclusion in the model, since the land use categories selected by stakeholders in Workshop 1, and later used to build the model, were not explicitly present in the scenario narratives. To address this, stakeholders first

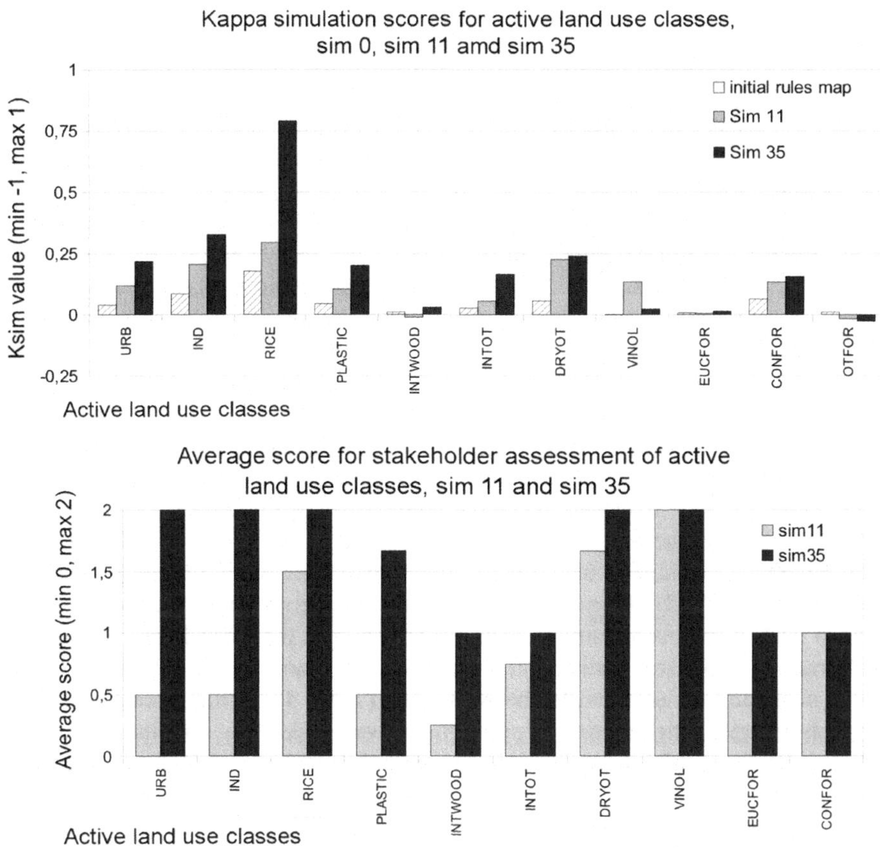

Fig. 5 Results of statistical tests (*top*) and stakeholder evaluation (*bottom*) for two calibration phase simulations, 11 and 35

analysed the scenario narratives looking for references to land use change which they then related to a category. On the basis of this qualitative information for each of the 4 Eco-Future scenarios (1. Doñana globalized knowledge; 2. Trademark Doñana; 3. Arid Doñana; 4. Adaptive Doñana: wet and wild), stakeholders were able to estimate land use demand for the model for each category under each scenario (Fig. 4c, d). In the final activity of the workshop, participatory mapping, stakeholders located the land use demand they had previously estimated using coloured buttons representing different quantities of land (e.g. large button, 50 ha, small button 10 ha) on a A0 paper plot of the 2007 land use map (Fig. 4e). Results were recorded photographically.

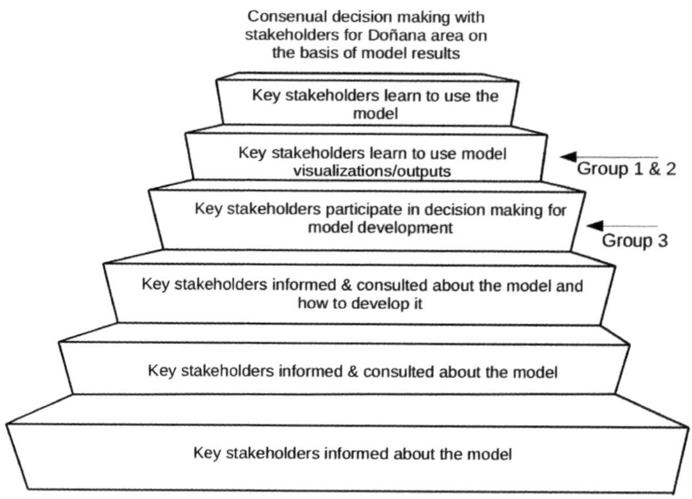

Fig. 7 Results of the "participation stairway" exercise

informative workshops as "participatory processes". Our approach is different for the following reasons:

Real information elicitation takes place. Specific information is provided to researchers by other stakeholders for a specific purpose. Researchers return this information to stakeholders in its processed form, and invite discussion and criticism, which is then used to improve the model. Knowledge is thus co-generated by all stakeholders.

The participatory activities have given rise to genuine changes in the way the work was carried out. Had researchers simply developed the model unaided and presented the results to stakeholders, the model would have been completely different, with different land use categories, different drivers, different scenarios and different indicators.

The participatory process and its most visible output, the land use model, have been subjected to a rigorous process of evaluation using a range of methods. All stakeholders have been given the opportunity both to contribute information and to criticise the process and its outcomes. Not all of this information is presented here, for reasons of space, but it is provided in full on the website (http://www.geogra.uah.es/duspanac/taller.html).

The participatory modeling approach presented here has a number of important benefits with respect to non-participatory modeling approaches, particularly where, as in this case, the aim is to influence decisions about the management of natural resources. Firstly, the participatory process is likely to lead to a better model. Researchers obtained a large quantity of useful information that was used directly for model development, for example, advice about the most important land use changes to represent in the model, group assessment of model calibration goodness-of-fit, and the most useful indicators that the model should try to produce.

Secondly, workshop participants clearly felt the process to be an enriching and worthwhile experience, as can be seen from their responses to the evaluation activities. The land use categorisation and land use dynamics activities were highly rated by participants, who said that they had learned a great deal, especially about the concept of drivers of land use change.

The participatory evaluation of model simulations was found to be very useful as a goodness-of-fit testing procedure. Though visual inspection is clearly a subjective process, incorporating multiple opinions increases the degree of confidence in the results, and in any case, the model also performs adequately with respect to comparable studies (e.g. Van Vliet 2013) for two widely accepted metrics, K_{sim} and *clumpiness*, for both calibration and validation periods.

The participatory indicator development exercise, discussed in detail in a separate paper (Hewitt et al. 2016), was highly successful since it focussed stakeholders' attention on model outputs and possible future utility of the model. We recommend that there should be at least one output-focussed model activity in order to establish a link between the development process and the model's potential use.

There were two activities (Ws 2, Activity 2, and Ws 3, Activity 1) in which participants worked directly with model simulations on the computer. Though these activities were perceived to be the most difficult, they were very useful to communicate the way the model actually worked, much more useful in fact than lecturing stakeholders on model design and operation. In general, the very visual nature of the model makes it attractive for hands-on learning and information sharing.

One of the most interesting reflections emerged during the scenario-based activities in Workshop 2. These activities generated some revealing discussions, predominantly centred around the level of realism that should be represented in the scenarios. Two clear positions emerged, one, which we might call "fantasising about paradise is not the solution", was highly critical of the strongly environmental scenario 4—adaptive Doñana: wet and wild—which this group saw as unhelpfully idealistic. The opposing position, which we might call "dare to dream", strongly emphasised the need for "outside the box" thinking that transcended currently accepted possibilities in order to look for new, sustainable, alternatives. It is possible that the strongest proponents of these positions, in the first case, a regional government employee, and in the second case, a researcher, may be broadly representative of the attitude of their peer groups in society. However, this is not simply a clash between idealism and realism, since actively searching for new, creative options is probably a more realistic approach to finding solutions to the alarming degradation of the natural area than continuing with business as usual. It may be that the regional government representative was not prepared to consider that the extensive environmental measures taken by their organisation could be insufficient or unsuccessful. This is understandable, since these measures, compared with many other areas, are exemplary. Unfortunately, they are unlikely to be successful in the long term, unless an effective alternative to the reigning ideology of "growth without limits" can be found. The incoherence of this position is starkly illustrated by the recent decision of the regional government to push for the reopening of the Aznalcóllar mine, upstream from the Doñana wetlands, in which a disastrous industrial accident occurred in 1998.[2] As a consequence of this decision, during the preparation of this chapter, Doñana was added to the World Wildlife Fund's list of UNESCO world heritage properties threatened by industrial development (El País 2016). Unless we blindly accept that "all development is good development", this decision is hard to understand. In the words of one workshop participant (Table 4, new reflections) "The trends in most cases indicate a difficult future for the natural area".

5.2 *The Role of Participation in Land Use Models*

Even a participatory process like the one described above, with various in-depth workshops, does not really allow time for the detailed workings of the model to be completely assimilated. It is therefore important to find out to what extent the stakeholders had come to understand the model by the end of the participatory process. Unfortunately, this is difficult to measure without a specific test, and developing such a test was not felt to be a priority in this research given the pressures of time and the fact that quite extensive participatory evaluation procedures had been carried out anyway. However, despite the numerous activities aimed at improving participants understanding, it seems highly likely that most stakeholders probably had an incomplete picture of the model and its capabilities. This is important, because some sources caution that poorly understood models are useless and can even be dangerous, in the sense that they may lead to significant misunderstandings about important issues (see, for example, http://www.fund-model.org/). However, the extent to which this is a significant problem depends very much on the individual case, and the experiences that have been shared in this chapter lead us to think otherwise, for the following reasons:

- The model does not actually make decisions itself, and can instead be used to support policies over the long term. In the short and medium term, the modeling and embedded participatory process is used to inform discussion and share knowledge about a particular issue. Clearly, while a poorly understood decision-making robot might be dangerous, a poorly understood discussion support tool is simply less useful than a well understood one. For our purposes, an incomplete understanding of the model is a significant improvement on no knowledge at all.

---

[2]On April 25, 1998, the collapse of part of a tailings dam flooded the Agrio and Guadiamar Rivers with high pyrite content mine tailings and acid water containing dissolved heavy metals. The spill affected a branch of the Guadiamar river basin measuring 62 km long with a width of between 500 and 1000 m between the village of Aznalcóllar and the border of the Doñana National Park, with catastrophic effects on flora and fauna (Hernández et al. 2004).

- Though misunderstandings of the model are frequent, they rarely have serious consequences. The commonest misconception among stakeholders was a tendency to attribute to the model capabilities it does not have. This is, in itself, a constructive process, since it can lead participants to come up with interesting and useful hypotheses. For example, in the Business as Usual (BAU) scenario one group of participants noted the disappearance of areas of crops under plastic in the municipality of Lucena, even though the BAU scenario shows an overall increase of this crop type. This was interpreted as an attempt to change the strong economic dependence of the municipalities in this area on these crops, by searching for substitute crops with less environmental impact or refocussing the economy towards other services. However, this explanation does not explain the behaviour of the model, which did not include these factors (in fact, other model areas are simply more attractive, so this crop type moves location). Nonetheless, the interaction of the participants with the model simulation has allowed an interesting real world dynamic to emerge that could be used to inform future policy in this area. This is a nice example of what Cartledge et al. (2009) call "constructive ambiguity", very useful for developing new policies or expanding the "option spaces" in which policy actions can take place (Oxley et al. 2002). In practical terms, this is also useful for improving the modeling tools and approaches themselves, e.g. the software, since stakeholders will typically have preconceptions about what a model can or cannot do which will differ substantively from the modeller's or software developer's vision of the way the model should work.
- Involving stakeholders in building and using the model may indeed lead to misconceptions, but it also reduces them, as well as informing more widely about what models are and what they do. We argue that the "danger" lies in the uncritical acceptance of scientific data, whether from models or from other sources, and that our aim as researchers is to reduce this danger by encouraging participation in, and critical reflection on, the scientific process.
- Human societies are already in serious danger from an economic model that makes the planet's life support systems subservient to the accumulation of monetary capital and consumer goods. Given the urgent need to secure sustainable future for threatened ecosystems like Doñana, risks that may arise from misunderstanding a land use model are outweighed by the benefits that the participatory model development process brings, e.g. uniting stakeholders with different or opposing views to discuss the future of shared territory, transparent dissemination of the inner workings of a scientific process, building confidence in cross sector collaboration on environmental issues etc.
- Ideas about models being "dangerous" in the "wrong hands" carry with them some questionable assumptions that risk reinforcing the persistent myth of scientists as objectively separate from society as a whole. In fact, scientists are an inseparable part of the society in which they operate and bring to their work many conscious and unconscious biases that mean they are unlikely to be more

objective than any other stakeholder (see, for example, Marshall 2015). In addition, while scientists can provide many tools and approaches, they are likely to lack specific local knowledge. The assumption that scientists are somehow omniscient is rarely intentional or conscious, but it is important to challenge it, nonetheless. Much information, especially relating to land and natural resources, is informal and unwritten, and is not easily accessible to outsiders without formal knowledge sharing procedures of the kind discussed in this chapter.

6 Conclusions and Outlook

6.1 Limitations of the Study and Future Work

As with any such study, some limitations can be identified, leading to questions for future research.

Although the scenarios were constructed by stakeholders as part of a prior participatory process (Palomo et al. 2011), not all of the participants from this earlier process attended the participatory modeling workshops. In future, it would be interesting to try to seamlessly integrate the whole scenario modeling chain, from narrative construction to scenario building through to land allocation for the scenarios as they appeared in the model, ending with participatory scenario evaluation.

Stakeholders (Table 2) represented a wide variety of sectors and professional skills, but did not specifically include local people, park visitors or religious tourists (an important group due to Doñana's importance as the site of a famous local pilgrimage). There is no doubt that the inclusion of these other stakeholders would have brought to light new and different perspectives that did not emerge here. Additionally, although the organizers aimed to strike a balance in terms of the stakeholders contacted, when it came to the stakeholders that actually attended, policy makers were over-represented. This was almost certainly because of the choice of venue for the workshops (national park offices). In future, choosing a more neutral venue might help to even out the balance.

Finally, it is clear that the participatory process could be enriched to include a role-playing game, with stakeholders encouraged to play roles that have nothing to do with their professional responsibilities (e.g. farmer as policy maker, scientist as religious tourist etc.).

6.2 Who Knows Best?

This paper presents a land use model for the Doñana natural area which was co-developed through a series of participatory workshops held with key stakeholders from a variety of sectors. Fundamental decisions about model set up and calibration were made consensually by the whole group. The participatory modeling process was extremely useful to the research team, but also clearly beneficial to all stakeholders, as evidenced from the results of the stakeholders' own evaluation.

The answer to the question "Who knows best?" posed in the title of this chapter, is probably "it depends", i.e. it depends on the context of the modeling exercise to be undertaken. Clearly there are many situations in which a researcher with the relevant training and experience is the best person to undertake a scientific task. But for the insights obtained from a scientific approach, such as a land use change modeling process, to be properly integrated into decision-making, it is insufficient for scientists to carry out the task alone, and then baldly state that "the results are likely to be of interest to land use planners and resources managers". Even stakeholders with similar perspectives, such as environmental researchers and protected area managers, are often unable to communicate effectively through standard information sharing channels. When, as is normally the case, the stakeholder community is much more diverse than this, these problems are compounded. For example, while the agricultural sector has the potential to be well-informed about relevant scientific developments e.g. through trade periodicals, we are not aware of any standard procedures that ensure that knowledge flows the other way—e.g. that farmers' own knowledge is routinely available to scientists. This is no doubt because of the implicit presumption that farmers are not generators of "useful" knowledge. If these preconceptions are not challenged, it is likely that misunderstandings between stakeholder communities will persist and environmental problems will continue to resist long-term solutions.

A participatory modeling process is a very useful way to bridge the significant differences between these different stakeholder communities and spark a genuine process of social learning, in which the importance of stakeholders' knowledge depends on its relevance to the question, rather than on conventional social structures or traditional knowledge hierarchies.

Acknowledgements This work has been supported in part by project Simulaciones geomáticas para modelizar dinámicas ambientales II.Horizonte 2020, Spanish Ministry of the Economy and Competitiveness and European Regional Development Fund FEDER (ref. BIA2013-43462-P). The authors also wish to acknowledge funding received from the Organismo Autónomo de Parques Nacionales (OAPN) a dependency of the Spanish Ministry for the Environment, for the DUSPANAC project (ref. 118/2010), and from the EU FP7 project COMPLEX (ref. 308601). Some of the ideas presented in this chapter were presented in a lecture at the IALE 9th World Congress "Crossing Scales, Crossing Borders: Global Approaches to Complex Challenges" Portland, Oregon, US, 5–10th July 2015. We are grateful to Calum Brown, Jasper Van Vliet and Nynke Schulp, organisers of the session entitled "Modeling Drivers and Consequences of Land Change: Data-based vs. Process-based Approaches" for inviting us to participate and for the interesting and stimulating discussions that ensued

References

Batty M, Xie Y (1994) From cells to cities. Environ Plan 21(Supplement):31–48

Cartledge K, Dürrwächter C, Jiménez VH, Winder NP (2009) Making sure you solve the right problem. Ecol Soc 14(3)

Clarke K, Hoppen S, Gaydos L (1997) A self-modifying cellular automaton model of historical urbanization in the San Francisco Bay area. Environ Plan 24(2):247–261

Escobar F, Hewitt, R, Hernández-Jiménez V (2015) Usos Del Suelo En Los Parques Nacionales Españoles. Evolución y Modelado Participativo. Proyectos de Investigación en Parques Naciones 2010–2013. OAPN/MAGRAMA Madrid

El País (2016) La industria amenaza a la mitad de los parques Patrimonio Mundial. http://elpais.com/elpais/2016/04/05/ciencia/1459868131_008351.html. Accessed: 6 April 2016

Heras López M (2015) Towards new forms of learning. Exploring the potential of participatory theatre in sustainability science. University of Barcelona. PhD thesis

Hernández E, Carmona J, Schmidt G (2004) Report on the situation of the Aznalcóllar Mine and the Guadiamar Green Corridor. WWF/Adena Report http://assets.wwf.es/downloads/report_on_the_aznalcollar_mine_2004_def.pdf

Hewitt R, Hernández-Jiménez V, Encinas M, Escobar F (2012) Land use modelling and the role of stakeholders in natural protected areas: the case of Doñana, Spain. In: Proceedings of the 2012 international congress on environmental modelling and software: managing resources of a limited planet, Leipzig, Germany

Hewitt R, Van Delden H, Escobar F (2014) Participatory land use modelling, pathways to an integrated approach. Environ Model Softw 52:149–165

Hewitt R, Hernández Jiménez V, Lasierra Navarro, M, de la Cruz Lecanda A, Escobar, F (2016) ¿Qué futuro queremos? Generación de indicadores medioambientales para escenarios de futuro a través de modelado espacial y procesos participativos. XVII Congreso Nacional de Tecnologías de Información Geográfica. Communication. 28, 29th June–1st July 2016. Malaga

Marshall D (2015) Unconscious bias: an introduction to how our biases affect our lives and our work. Ebook. http://marshallelearning.com/wp-content/uploads/2015/11/Unconscious-Bias-Whitepaper-Marshall-Elearning.pdf. Accessed: March 2016

McGarigal K, Cushman SA, Neel MC, Ene E (2002) FRAGSTATS: spatial analysis program for categorical maps. Computer software program produced by the authors at the University of Massachusets, Amherst. www.umass.edu/landeco/research/fragstats/fragstats.html

Moreira J (2007) 1:25000 scale cartographic series, Junta de Andalucía, REDIAM, Retrieved from: www.juntadeandalucia.es/medioambiente/rediam, September 2011

O'Brien N, Moules T (2007) So round the spiral again: A reflective participatory research project with children and young people. Educ Action Res 15(3):385–402

Oxley T, Jeffrey P, Lemon M (2002) Policy relevant modelling: relationships between water, land use, and farmer decision processes. Integrated Assessment 3(1):30–49

Palomo I, Martín-López B, López-Santiago C, Montes C (2011) Participatory Scenario Planning for Natural Protected Areas management under ecosystem services framework: the Doñana social-ecological system, SW, Spain. Ecol Soc 16:23

Phipps M, Langlois A (1997) Spatial dynamics, cellular automata, and parallel processing computers. Environ Plan 24(2):193–204

Riks BV (2012) Metronamica documentation. http://www.riks.nl/resources/documentation/Metronamica%20documentation.pdf

Tobler W (1979) Cellular Geography. In: Gale S, Olsson G (eds) Philosophy in geography. Reidel, Dortrecht, pp 379–386

Ulam SM (1952) Random Processes and Transformations. In: Proceedings of the international congress of mathematicians, vol 2, pp 264–275

VanDelden H, Gutiérrez ER, van Vliet J, Hurkens J (2008) Xplorah, A multi-scale integrated land use model. In: Sànchez-Marrè M, Béjar J, Comas J, Rizzoli A, Guariso G (eds) Proceedings of IEMS 2008 (iEMSs) http://www.iemss.org/iemss2008/index.php?n=Main

Van Vliet J, Bregt A, Hagen-Zanker A (2011) Revisiting Kappa to account for change in the accuracy assessment of land-use change models. Ecol Model 222(8):1367–1375
Van Vliet J, Naus N, van Lammeren Bregt AK, Hurkens J, van Delden H (2013) Measuring the neighbourhood effect to calibrate land use models. Comput Environ Urban Syst 41:55–64
von Neumann J, Birks AW (1966) On the theory of self-reproducing automata. Univ. of Illinois Press, Urbana IL
Visser H, De Nijs T (2006) The map comparison kit. Environ Model Softw 21(3):346–358
Voinov A, Kolagani N, McCall MK, Glynn PD, Kragt ME, Ostermann FO, Ramu P (2016) Modelling with stakeholders–Next generation. Environ Model Softw 77:196–220
White R, Engelen G (1993) Cellular automata and fractal urban form: a cellular modelling approach to the evolution of urban land-use patterns. Environ Plann A 25(8):1175–1199
White R, Engelen G, Uljee I (2000) Modeling land use change with linked cellular automata and socio-economic models: a tool for exploring the impact of climate change on the island of St. Lucia. In: Hill M, Aspinall R. Spatial information for land use management. Gordon and Breach
World Agroforestry Center (WAC) (2003) Evaluation Dartboard. http://www.cglrc.cgiar.org/icraf/toolkit/Evaluation_dartboard.htm. Accessed March 2016

Chapter 12
Land Use and Cover Change Modeling as an Integration Framework: A Mixed Methods Approach for the Southern Coast of Jalisco (Western Mexico)

M. Kolb, P.R.W. Gerritsen, G. Garduño, E. Lazos Chavero, S. Quijas, P. Balvanera, N. Álvarez and J. Solís

Abstract The rapid loss of forests with negative consequences for biodiversity and ecosystem services has drawn the attention of scientists and decision makers to deforestation and land use change. Over the last two decades, a broad range of models of land use and cover change (LUCC) have been developed to assist in land management and to better understand, evaluate and project the future role of LUCC. Pattern-based LUCC models are empirical approaches based on the observation of past LUCC, including the spatial dimension of change patterns from which the

M. Kolb (✉)
Instituto de Geografía, Universidad Nacional Autónoma de México (UNAM), Mexico City, Mexico
e-mail: melanesien@gmail.com

P.R.W. Gerritsen · N. Álvarez
Departamento de Ecología y Recursos Naturales, Centro Universitario de la Costa Sur Universidad de Guadalajara (UdG), Autlán, Mexico
e-mail: petergerritsen@cucsur.udg.mx

G. Garduño
Comisión Nacional para el Conocimiento y Uso de la Biodiversidad (CONABIO), Tlalpan, Mexico
e-mail: ggarduno@conabio.gob.mx

E. Lazos Chavero · J. Solís
Instituto de Investigaciones Sociales, Universidad Nacional Autónoma de México (UNAM), Mexico City, Mexico
e-mail: lazos@unam.mx

S. Quijas
Centro Universitario de la Costa, Universidad de Guadalajara (UdG), Mexico City, Mexico
e-mail: squijas@gmail.com

P. Balvanera
Instituto de Investigaciones en Ecosistemas y Sustentabilidad (IIEES), Universidad Nacional Autónoma de México (UNAM), Mexico City, Mexico
e-mail: pbalvanera@cieco.unam.mx

M.T. Camacho Olmedo et al. (eds.), *Geomatic Approaches for Modeling Land Change Scenarios*, Lecture Notes in Geoinformation and Cartography,
https://doi.org/10.1007/978-3-319-60801-3_12

underlying behavior can be inferred, through the statistical relationships of model parameters. Even though these models present a number of drawbacks such as data intensity and limited capacity to connect to other driver scales, they offer a framework to integrate data from multiple disciplines. In this chapter, we present a case study that shows land use and cover change modeling as an integrative framework for cross-referencing among different data sources. Spatial information on LUCC, econometric models and stakeholder perceptions were generated in an interdisciplinary working group in order to obtain insights into LUCC at the regional level. Land use and cover (LUC) maps were the starting point for the spatial analyses of historic changes, which together with ancillary data were used to establish change probabilities for the main change processes. Econometric models showed historic tendencies of agricultural production and a panel analysis clarified the relation between variables. Local stakeholder perception gave the historic background and participatory fuzzy cognitive maps shed light on the underlying drivers of change. By cross-referencing the different data sources, we show that for this particular region the official LUC maps do capture the main change processes. Both local stakeholder perceptions and econometric models confirm deforestation and agricultural expansion, especially livestock farming, as the main processes. The econometric models confirm the difference in magnitude between the large growth in areas for livestock farming and much more restricted growth of agricultural areas and show that beef production and pasture for cattle ranching is displacing the production of maize and beans. As regards the drivers of change, the different data sources complement each other quite well as they cover different scales: the stakeholder elicitations revealed a set of indirect drivers related to the direct drivers identified in the spatial analysis of historic change. The indirect drivers included the political, social, cultural and economic forces behind agricultural expansion, especially cattle ranching. The analysis of the spatial factors related to change showed that a large array of variables play a role in LUCC. The mixed method approach is helpful in unravelling the different levels of connection between drivers.

Keywords Mixed methods approach · LUCC · Drivers · Data-driven · Perceptions · Participatory models · Mexico

1 Introduction

Deforestation and the resulting land use change have negative consequences for biodiversity and ecosystem services. Scientists and decision makers have therefore become increasingly concerned about the rapid loss of forests in recent decades (Vitousek et al. 1997; Lepers et al. 2005; UNEP 2012). In the Neotropics, there were high rates of deforestation and environmental degradation throughout the second half of the twentieth century (Lambin et al. 2003), leading to important habitat loss and depletion of natural resources in tropical and temperate forests

(CBD 2006). Deforestation in Mexico is ranked ninth in the world (Bradshaw et al. 2010), and in Latin America is exceeded only by Brazil, with large regional disparities in LUCC and its drivers.

Over the last two decades, a broad range of land use and cover change (LUCC) models have been developed to assist in land management and to better understand, evaluate and project the future role of LUCC (Mas et al. 2014). In order to model human-induced land-use change using pattern-based modeling frameworks, it is necessary to have reliable historic data on land use and cover change (LUCC), plus a large array of ancillary variables that can be used to fit the model. One of the problems of parameterizing simulation models for prospective analysis has been the lack of sufficient, locally specific data that could inform modelers about the particular settings of the study area that influence the different situations in the proposed qualitative scenario information.

Pattern-based LUCC models are empirical approaches based on the observation of past LUCC, including the spatial dimension of change patterns from which the underlying behavior can be inferred through the statistical relationships of model parameters. Their functioning has been strongly influenced by development in geographic information science and while they can use a wide array of data and information, from remotely sensed data sets to socio-economic variables and field surveys, the focus is usually on biophysical and landscape structure aspects (Cheong et al. 2012). Usually an analysis of historic changes helps determine the types and magnitude of LUCC, as part of model calibration. The fact that these models are based on historic spatial patterns makes them scale dependent, both temporally and spatially. This means that the LUCC dynamics to be modeled are determined by the spatial resolution, the extent of the study area and the time period being considered. It cannot be assumed that the same patterns can be explained by the same processes in different spatial and temporal settings.

The strong dependency of LUCC dynamics on the historic land use and cover (LUC) information used in model calibration can also be seen as a drawback, since LUCC is normally detected by comparing two LUC maps. There has been substantial discussion about how map accuracy affects the detection of change processes, since the base maps often come with thematic and positional errors (Pontius 2000; Mas 2005; Pontius and Petrova 2010). These map errors can lead to spurious or incorrect evaluation of LUCC (Pontius and Lippitt 2006). Since the official land use and cover maps for Mexico have not been officially validated, there is a question mark regarding their quality and usefulness for change detection, despite their widespread use. In this context, a combination of different data sources (such as stakeholder perceptions and econometric models) can be used to validate LUCC information based on map comparison.

Another limitation of pattern-based models has been the evaluation of drivers. These models rely on ancillary spatial data to establish rules for change probabilities or suitability. However, these data represent biogeophysical constraints or direct drivers of LUCC, making it very hard to identify underlying drivers. By generating complementary information and analysis, such as stakeholder elicitations, we can shed light on the connections between underlying and direct drivers.

Participatory research in the social sciences has traditionally been developed for specific local contexts and thus represents phenomena that often cannot be visualized on a map. The result is that only very limited spatial modeling of LUCC has been carried out within the social sciences (Goeghegan et al. 2004). Nevertheless, their focus on human-environment interactions offers an excellent opportunity to advance the understanding of human effects on ecosystems and the implication of changes. For their part, econometric models have been extensively used in LUCC modeling, since economic factors are directly related to land use dynamics and some data, e.g. agricultural production figures, can be used directly to infer LUCC tendencies. While these models have been extended to include spatial information (spatial econometrics), the level of spatial information is usually limited because of the inherent aggregation of input data usually reported for administrative units (e.g. census, Brady and Irwin 2011). Interdisciplinary studies can help fill this gap to generate relevant spatial information that can be used for local and regional policy-making (Gerritsen 2012).

Even though pattern-based models present a number of drawbacks, such as data intensity and a limited capacity to connect to other driver scales, they offer an effective framework for integrating data from multiple disciplines. Despite the widely recognized potential of integration in a land use change model, successful integration is difficult to achieve (Cheong et al. 2012). In this chapter, a case study shows land use and cover change modeling as an integrative framework for cross-referencing among different data sources. We explore ways of using a mixed methods approach at the interface between land science, social science and participatory stakeholder work to remedy the lack of information about map accuracy. We started out with historic land use and cover maps to analyse change dynamics in a GIS, with which a conceptual LUCC model is established to generate probability maps of change. Model calibration data were then cross-referenced with econometric models (LUCC tendencies and direct drivers). The last ingredient that makes this study truly multi-disciplinary is the analysis of stakeholder perceptions and a participatory process model that aims to validate historic change processes and dynamics and to provide complementary information, especially on underlying drivers. With this mixed methods approach we intend to enhance the understanding of LUCC processes in the region.

2 Test Area

The Southern Coast of Jalisco comprises a set of municipalities located along a 300 km strip of the Pacific coastline of Western central Mexico, with elevations of up to 2800 m (Maass et al. 2005). The study area includes the important Long-Term Ecological Research (LTER) site of Río Armería, Río Cutzmala, Río San Nicolás (also known as Chamela). This LTER site was established because of the importance of these tropical dry forests in terms of their biological, topographic and cultural diversity. The LTER research studies are being carried out in the basins

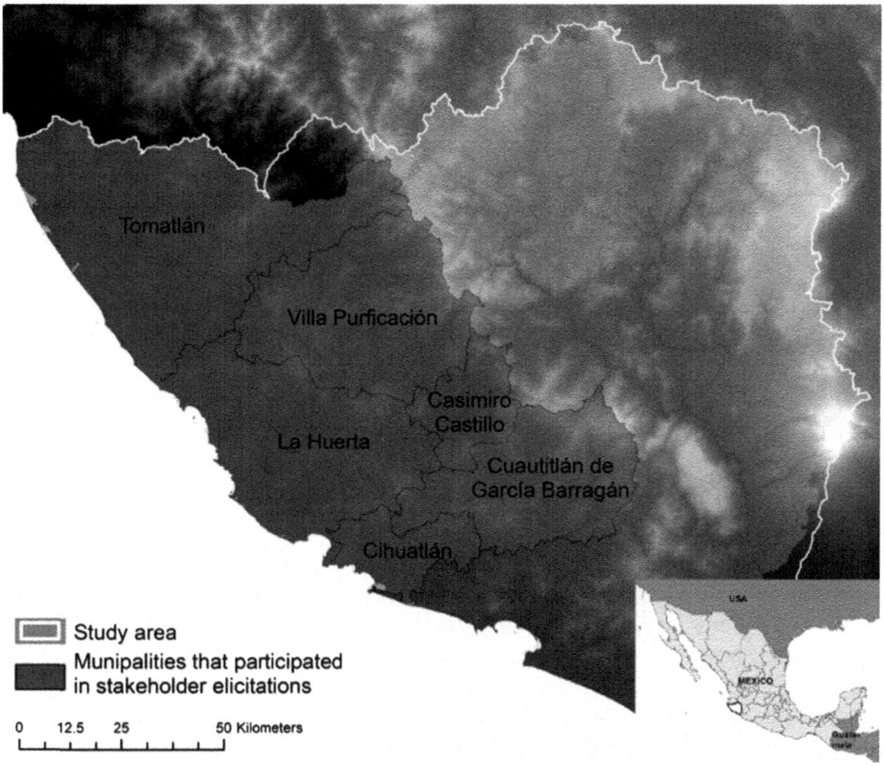

Fig. 1 Study area with municipalities that participated in stakeholder elicitations

of three rivers: Río Armería, Río Cutzmala and Río San Nicolás. In this case, the study area was defined by selecting a set of watersheds that connect the three river basins in order to fill a coherent geographic space (Río Marabasco, Río Purificación, Arroyo Maderos, Arroyo El Pedregal).

To make sure that the study area is meaningful for all types of analyses in the ROBIN project, it was extended to include all of the municipalities located inside the 26 watersheds. In this way we established a study area of 21659.11 km^2, which was divided into two blocks of municipalities: the ones where the stakeholder research was carried out (Villa Purificación, La Huerta, Cihuatlán, Casimiro Castillo) and the rest, to which the models were extrapolated (Fig. 1). LUCC analysis and modeling was done for the whole study area.

The main vegetation types are distributed following a topographic and climatic gradient ranging from coastal ecosystems at sea level up to temperate forests (TF, oak and pine forests) in the upper part of the watersheds (above 1000 m). Tropical dry forest (TDF) is dominant in the region between 0 and 300 m on hilly terrain (max. height 700 m). Tropical semi-deciduous forest (TSF) is found on the alluvial terraces along the channels of ephemeral and permanent streams, and in more elevated, more humid areas above the TDF. These TDF and TSF are the most diverse dry forest in the Neotropics, and 40% of their plant and 10% of their bird and mammal species are endemic to Mexico (Lott 1993; Ceballos and García 1995; Gentry 1995; Arizmendi et al. 2002).

Precipitation is strongly seasonal, which restricts the length of the productive season. The climate in this region ranges from warm (coast) to temperate (higher elevations), and annual precipitation from 800 to 2100 mm, most of which falls between June and October. Rainfall patterns vary greatly due to tropical cyclones and the occurrence of El Niño and La Niña events but there are usually three months of severe drought (February−April, Maass et al. 2005).

Main land uses include pasture for extensive cattle ranching on hill slopes and agriculture on the alluvial plains. Most of the TSF in the alluvial areas have been converted into agricultural lands. In some cases, the size of fields allows for intensive agriculture using heavy machinery, irrigation, and fertilization. Rain-fed agriculture and pasture face a high probability of failure because of frequent droughts. The deep fertile soils, which are close to the water table and suitable for agriculture (e.g., phaeozems and fluvisols), are only found at lower altitudes along the few streams and rivers. Most of the area is dominated by young, shallow, rocky, nutrient-poor soils (e.g., regosols) found on predominantly sloping land (moderately to extremely steep) and is not suitable for nutrient –and water– demanding agro-pastoral activities (Maass et al. 2005).

Human presence in the region remained low until the end of the 19th century and the first transformations of ecosystems occurred when "Haciendas" were consolidated during the 1850s. In 1950, a federal government initiative to colonize the coastal areas of Mexico began. This policy, together with the land reforms that distributed land to landless peasants led to the immigration of landless peasants between 1950 and 1970 into this, for them, unfamiliar environment with low agricultural potential, no public services and no job opportunities. In the subsequent years, road and communication infrastructure was built to improve accessibility. Nowadays, population is distributed in mainly small settlements (<1000 inhab.) all over the study area. General income levels are low, unemployment is high but disguised by informal activities, marginalization is medium-high, living conditions are poor and migration is common.

Land ownership varies a great deal in the different municipalities with *Ejidos* (communal lands) and private estates. The Ejidos are generally used for basic grain and cattle.

As in other parts of Mexico, rural areas in the study region face a prolonged crisis that has persisted for almost 50 years resulting in widespread poverty and degradation of natural resources (Bray et al. 2006; Morales 2011). The rural population is forced to supplement their income with other activities, and many migrate to the United States (Magaña 2003). Conflicts over land and resources are very common between the *ejido* members, and are often related to external pressures on resources, such as mines, beaches and fresh water (Bray et al. 2007; Tetrault et al. 2012; Gerritsen et al. 2015).

3 Methodology and Practical Application to the Data Sets

Here we use a suite of data, i.e. all the available data sets for the region, in order to explore the possibility of advancing towards a mixed method approach. Each data source gives an independent account of LUCC at regional level from a different point of view and focus; together, they were used to cross-reference and validate spatial LUCC information and to complement aspects that were impossible to infer from pure spatial data. An LUCC analysis based on several time steps provided spatially detailed information, which was used to establish historic change processes and their magnitude. The official LUC maps for Mexico have been extensively used for different purposes, despite the fact that their accuracy has not been assessed (Couturier et al. 2012) and questions about map quality have led to severe criticism of many LUCC studies. In this case, economic data was used to cross-reference spatial results, while stakeholder perceptions were used to validate general tendencies. Direct spatial drivers were analyzed in a spatial LUCC model and probability maps of change were generated, while underlying drivers were evaluated in a participatory process model. As such, our study was based upon three methodological approaches to explore LUCC processes in the region, before mixing the methods to reach integration in the discussion section.

3.1 LUCC Dynamics and Direct Drivers—LUCC Analysis and Modeling

A spatial analysis of land use and cover change was carried out in order to characterize the main processes in land use change in the study area and to derive basic parameters for the land use change model. After a preliminary analysis of change by comparing official land use and cover maps produced by National Institute of Statistics and Geography (INEGI), LUC classes for the analyses were defined by grouping the original classes into meaningful categories. Since there were differences in the dynamics regarding forest degradation for temperate and tropical dry forests, the final categories maintain this division between temperate and tropical forests and their conservation state (Table 1). The data sets did not include any forest plantations and urbanization was excluded from the analysis. The base maps form a 1:250 000 series from 1993, 2002, 2008 and 2011 INEGI (2001, 2005, 2008, 2013, see Kolb and Galicia 2012 for details). These maps were converted into rasters with a cell size of 100 m for geo-processing. Spurious changes were partially corrected by re-assigning the categories (e.g. all natural vegetation categories other than the analyzed one were assumed to be false changes: a change from coniferous to broad leaf forest was considered as a false change and the amount of broad leaf forest was added to the amount of coniferous forest).

Based on the quantity of changes for each LUC category, several statistical measures and dominant change processes were established. Deforestation rates and

Table 1 Land use and cover change modeling legend and the original types of vegetation in each class

Original land use and cover classes (INEGI)	Modelled land use and cover classes	Code
Oak forest	Temperate forest	1
Fir forest		
Pine-oak forest		
Mountain cloud forest		
Low tropical dry forest and low tropical spiny forest	Tropical dry forest	2
Medium tropical semi-dry forest		
Rain-fed agriculture	Agriculture	5
Permanent crops		
Irrigation agriculture		
Pasture	Pasture	6
Secondary oak forest	Secondary temperate forest	3
Secondary fir forest		
Secondary pine-oak forest		
Secondary low tropical dry forest	Secondary tropical dry forest	4
Secondary medium tropical semi-dry forest		
Urban areas	Other	7
Mangroves		
Palms		
Coastal dune vegetation		
Without vegetation		
Water bodies		

change rates were calculated using the following formula which expresses the proportion of change with respect to the initial area for each year (FAO 1996):

$$R = (1 - (A1 - A2)/A)^{1/t}) - 1)) * 100$$

where R is the annual change rate in percentage, A1 is the area at t_1, A2 the area at t_2 and t the number of years in the period. For deforestation rates, primary and secondary forest classes were aggregated and the results were multiplied by –1 to obtain positive numbers for negative change rates.

In addition, change matrices (change probabilities) were generated using DINAMICA-EGO to establish the main change processes and change was mapped to capture the patterns of change and permanence.

In order to generate maps of the areas with the highest probability of change, an LUCC model was established. Spatial model calibration was done using DINAMICA-EGO by comparing land cover maps for 4 time steps over 18 years so as to obtain information on the temporal behavior of change dynamics and to establish potential non-stationary scenario tendencies. The aggregated LUC classes from the LUCC analysis were maintained, so enabling the model to take into account different physiological constraints to LUCC (temperate broad leaf and tropical dry forests types), distinguishing between primary and secondary forest types, in order to consider forest degradation, as well as deforestation. As the analysis of LUCC showed, agriculture and pasture for livestock were the main agents of deforestation and were therefore also included in the modeling legend. We then determined the quantity of change and the change probabilities for each transition by Markov chain matrices.

In order to generate change probability maps, a set of explanatory variables (predictors) related to biophysical, topographic, demographic and socio-economic characteristics was compiled. The selection of explanatory variables and the derivation of testable hypotheses of land use change are also based on social science theories, such as accessibility (von Thuenen) and land quality (Ricardian model). Findings reported in the scientific literature and in governmental studies (SEMADET 2015) were considered together with the results of stakeholder research. After compiling all 40 data sets, measures for each variable were established (Table 2).

Using spatial statistics we identified those variables that determine the highest probability of change. We began by conducting an exploratory correlation analysis in ArcGIS 10.2 (ESRI), based on which water bodies, total population 1995 and 2010 and very high marginalization 1995 were excluded from further analyses. Between two highly correlated variables describing the risk of deforestation on a national scale, the original version (INECC 2015) was excluded in favor of a modified version (CONABIO 2012) that showed higher correlation to observed changes. Finally, in DINAMICA-EGO the Weight of Evidence (WoE) approach was used to establish the relationships between variables and observed changes, which included the editing of WoE categories with irregular behavior. In a last step, the correlation between maps was reviewed using the Cramer test before probability maps were generated using DINAMICA-EGO.

3.2 LUCC Tendencies—Econometric Model

Econometric models were generated to investigate trends in agricultural and livestock production, to see if stakeholder perceptions and trends derived by map comparison can be validated with official statistics. In particular, data on the production of beef in carcasses, planted area for beans, planted area for maize, planted area for grazing, and the total planted area were used to run the econometric models. Datasets were taken from the INEGI (2016) and comprise the period from 2002 to 2011. Models were run in Stata (StataCorp 2013). The following regression model was applied to estimate the coefficients for the trends in agricultural production or growth rates, once the variables were defined for each case:

Table 12.2 Spatial factors analyzed for their relationship with observed changes

Environmental and biophysical variables		Unit/measure	Original scale/resolution	Source
Topography	Altitude	m	90 m	INEGI (1998)
	Slope	°	90 m	INEGI (1998)
	Hydro1 k topo wetness index (cti)		1000 m	USGS (1999)
	Altimetric zone/basin	Categorical	1:250 000	
Climate	Evapotranspiration	mm	1:4 000 000	Maderey (1990)
	Bio 5, 7, 9	°C/mm	1000 m	http://www.worldclim.org/
Hydrology	Rivers	Distance	1:250 000	CONABIO (2008)
	Permanent rivers	Distance	1:250 000	CONABIO (2008)
	Intermittent rivers			
	Water bodies	Distance	1:250 000	CONABIO (2009)
Geomorphology	Geomorphology	Categorical	1:250 000	INE, SEMARNAT, IG-UNAM (2003)
Edaphology	Soil texture	Categorical	1:250 000	INIFAP, CONABIO (1995)
	Soil erosion	Categorical	1:250 000	INIFAP, CONABIO (1995)
Ecosystem productivity	NDVI (HANTS)		250 m	Kooistra et al. (2015)
	Biomass map for woody vegetation (above ground)	t•C/ha		Alianza MREDD+ (2013)
Human impact variables				
Human impact	MEXBIO 1995	0–1	1:1 000 000	Kolb (2009)
	MEXBIO 2011	0–1	1:1 000 000	Kolb (2009)
	Fragmentation: entropy 1993	Entropy	30/100 m	Kolb and Morales Luque (2016a)
	Fragmentation: entropy 2010	Entropy	30/100 m	Kolb and Morales Luque (2016b)
	Fragmentation: MSPA 1993	Categorical	30/100 m	Kolb and Morales Luque (2016c)

(continued)

Table 12.2 (continued)

Environmental and biophysical variables		Unit/measure	Original scale/resolution	Source
Topography	Altitude	m	90 m	INEGI (1998)
	Fragmentation: MSPA 2010	Categorical	30/100 m	Kolb and Morales Luque (2016d)
Contagion	Temperate forest	Distance	1:250 000	INEGI (2001, 2013)
	Tropical forest	Distance	1:250 000	INEGI (2001, 2013)
	Secondary temperate forest	Distance	1:250 000	INEGI (2001, 2013)
	Secondary tropical forest	Distance	1:250 000	INEGI (2001, 2013)
	Agriculture	Distance	1:250 000	INEGI (2001, 2013)
	Pasture	Distance	1:250 000	INEGI (2001, 2013)
Deforestation risk	IRDef	Categorical	300 m	INECC (2015)
	IRDef modified	Categorical	300 m	CONABIO (2012)
Infrastructure	All roads	Distance	1:1	IMT (2007)
	Paved roads	Distance	1:2	IMT (2007)
	Unpaved roads	Distance	1:3	IMT (2001)
Socio-economic variables				
Political factors	Tenure	Distance		
Agricultural production	Cattle	t		INEGI 2016
Marginalization	Marginalization 1995	Categorical		
	Marginalization 2010	Categorical		
Demography	Population growth 1995–2015	no.	census data/ interpolated	INEGI (1995, 2010)
Ethnicity	Population density of indigenous language speakers (1995, 2010)	no.	census data/ interpolated	INEGI (1995, 2010)
Protected areas	Federal PA	Distance		CONANP (2010)
	Management units for wild life conservation	Distance		SEMARNAT (2010)

$$\ln y_t = b_0 + b_1 t + u_t.$$

where (ln y) is the natural logarithm of the selected variables, t is the time (the years from 2002–2011), u_t is the error term, b^1t is the regression coefficient and (b_1t) times 100 gives the growth rate in y.

Once the model coefficients were estimated for all the selected variables, a growth forecast was done for the period 2011 to 2030 for each study case. The study area was stratified into three sections: all the municipalities (case 1); only municipalities that were part of the workshops (case 2); and all the other municipalities (case 3). This stratification was used to explore and compare the tendencies in the coastal area with the municipalities where stakeholder work had been carried out and the rest. This was done to make sure that no sub-regional disparities existed and that the stakeholder elicitations could be extrapolated to the rest of the study area. We therefore defined the value for the selected variables as the average value over the total number of municipalities for each case. For instance, to obtain the value for the total planted area for year t for case 1, all the values of this variable at year t were summed over the municipalities that belong to case 1. This value was then divided by the total number of municipalities to obtain the final value.

In order to test whether the behavior of the area used for grazing could explain the trends in the other variables, a panel data model was used for the three cases stated above:

$$Y_{it} = b_{1i} + b_2 X_{2it} + b_3 X_{3it} + b_4 X_{4it} + b_5 X_{5it} + u_{it}.$$

where i represents the municipalities, t the time (years from 2002 to 2011), y_{it} the planted area for grazing; X_{it} represents the other variables—e.g., the total planted area and planted area for beans, maize and grazing—and u_{it} is the error term.

Projections of trends were used to provide quantitative estimations of future demands for the main land uses and these were later used to set the parameters of the LUCC model.

3.3 LUCC Tendencies and Indirect Drivers—Local Stakeholder Perceptions and Participatory Process Model

Data on stakeholder perceptions on land-use and cover change and the elaboration of participatory scenarios were obtained in a series of six workshops that were held from October, 2012 to February, 2015. The regional actors that took part in these workshops were selected on the basis of previous socioecological research performed by the authors and social capital existing in the region.

The methodology of the workshops consisted of group discussions, using the Metaplan technique, combined with participatory mapping (Schnelle 1979; Gerritsen 2016).

The following general questions were asked:

- What changes have you observed over the last 50 years?
- What were the drivers related to those changes?
- Who are the main actors in these changes and what roles do they play?
- Who benefitted from the changes?
- What would be needed to build sustainability?

The workshops not only served to reconstruct land use trans-formations and its drivers, but also to assess the relative importance of the different factors involved. This was done by drawing up Fuzzy Cognitive Maps, which consist of collectively constructed flow diagrams, diagrams, indicating the relative importance of direct and indirect drivers.

4 Results

4.1 LUCC Dynamics and Direct Drivers—LUCC Analysis and Modeling

LUCC in the Southern Coast of Jalisco is dominated by agricultural expansion. The main change processes were deforestation (defined as the transition of any forest category to any agricultural category), forest degradation (defined as the transition of primary to secondary forests) and natural regeneration (defined as the transition of any agricultural category to any forest type). Deforestation can be detected for all forest types, but is biggest for tropical dry forests, especially in their primary state. In general, a loss of primary vegetation and an increase in secondary vegetation can be observed (Fig. 2).

LUCC dynamics generally decreased over time: The nineties were clearly the period with most LUCC resulting in an important loss of primary vegetation from 42% to 28.5% of the study area, while the lowest LUCC was observed from 2007 to 2011. Since 2002, the area of primary vegetation has been stable, but there has been a decrease in secondary vegetation. Overall, the increase in pastures has been stronger than the increase in agriculture. The most widespread change during the earliest period was forest degradation (1993–2002: 1.76, 2002–2007: 1.98, 2007–2011: 0.86 times the deforested area), with deforestation in second place. The decrease in forest degradation was less pronounced for temperate forests, while deforestation has decreased most in temperate primary forests.

Among the different LUCC transitions, deforestation of tropical forests (especially primary tropical semi-dry, dry and oak forests) for pasture is the main deforestation process in all the different periods. Expansion of agriculture was also

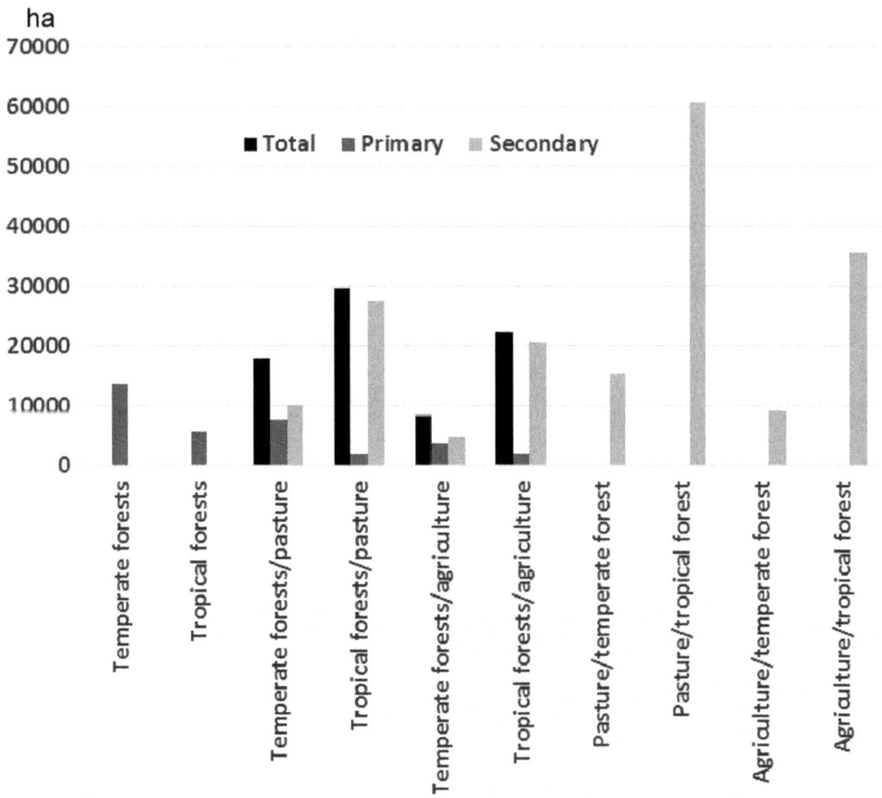

Fig. 2 Area (ha) of transitions for the period 1993–2011

important, especially from 2002 to 2007. Forest degradation is strongest for oak and pine-oak forests, in the case of the latter an ongoing process. The strongest decrease in the magnitude of change can be observed for regeneration, especially from pasture to secondary tropical forests. Since the first period is the one with the highest LUCC rates, it dominates the general picture of decreasing primary vegetation and increasing secondary vegetation. In the later periods deforestation was much more intense in secondary vegetation than in primary (most pronounced in secondary tropical semi-dry forests). In the last period deforestation rates are generally low. Mangroves, cloud forests and hydrophilic vegetation are more or less stable over time.

The transition potential maps generated with weights of evidence represent the probabilities of change for each set of main transitions, namely deforestation, forest degradation and regeneration (considering temperate and tropical forests, Fig. 3). The transition potentials express the statistical relationship of a set of spatial factors and observed transitions. These relationships are defined by the direction (positive or negative), are not necessarily linear and show the importance of the factors for

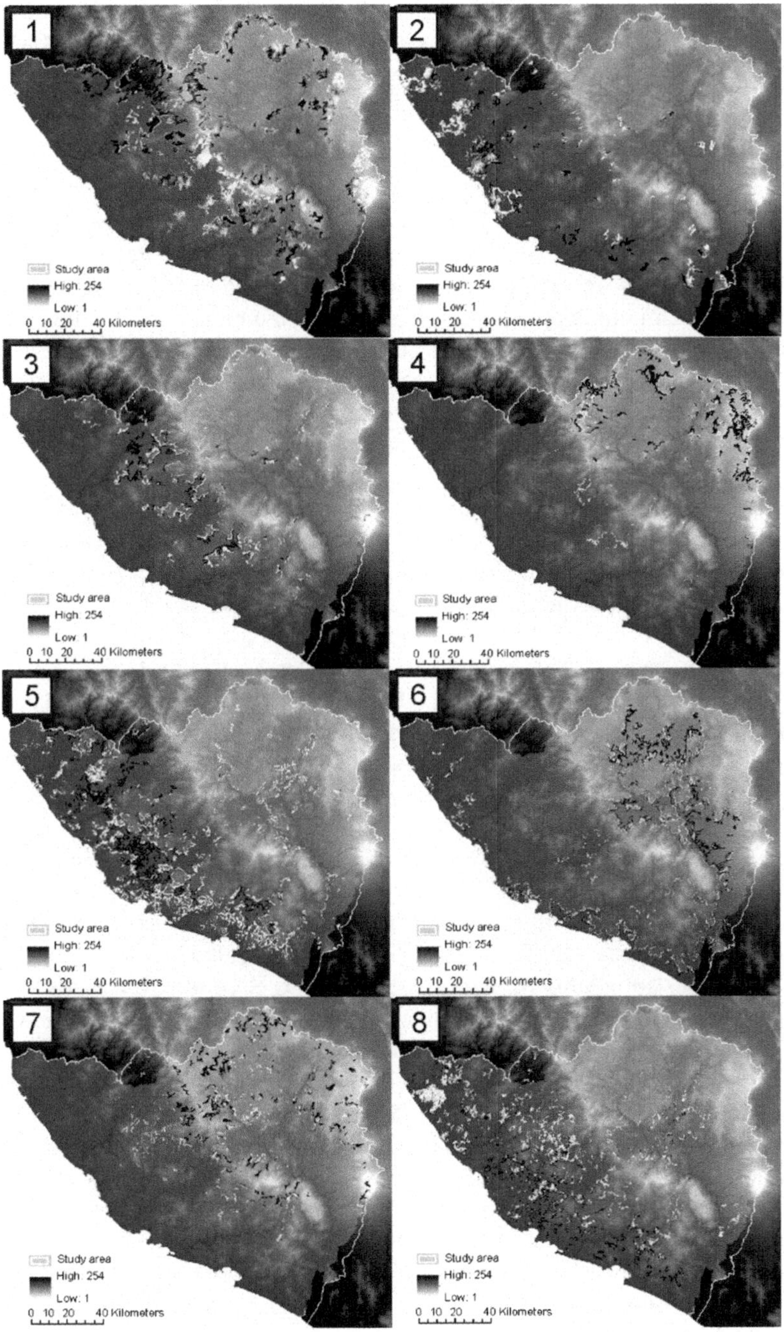
1
Study area
High: 254
Low: 1
0 10 20 40 Kilometers
2
Study area
High: 254
Low: 1
0 10 20 40 Kilometers
3
Study area
High: 254
Low: 1
0 10 20 40 Kilometers
4
Study area
High: 254
Low: 1
0 10 20 40 Kilometers
5
Study area
High: 254
Low: 1
0 10 20 40 Kilometers
6
Study area
High: 254
Low: 1
0 10 20 40 Kilometers
7
Study area
High: 254
Low: 1
0 10 20 40 Kilometers
8
Study area
High: 254
Low: 1
0 10 20 40 Kilometers

◀**Fig. 3** The figure shows transition probabilities for selected transitions*: (*1*) temperate forest degradation, (*2*) tropical forest degradation, (*3*) deforestation of temperate forest for pasture, (*4*) deforestation of temperate forest for agriculture, (*5*) deforestation of secondary tropical forest for pasture, (*6*) deforestation of secondary tropical forest for agriculture, (*7*) regeneration of secondary tropical forests from pasture and (*8*) regeneration of secondary tropical forests from agriculture. *Transitions have been selected based on their extent

each transition (Table 3). Bioclimatic variables prove to be important; annual temperature range positively correlates with temperate forest degradation, deforestation of secondary forests for pasture and regeneration of temperate forests from agriculture. Maximum temperature of the warmest month and mean temperature of the driest quarter are related to deforestation, degradation and regeneration of tropical forests. Topographic factors show the expected relation of less deforestation and forest degradation in higher and steeper areas, while regeneration shows the opposite tendency. Areas closer to permanent rivers suffer more deforestation and forest degradation. Areas with volcanic relief are prone to deforestation of secondary temperate forests for agriculture and regeneration of secondary forests from pasture; areas within lacustrine and alluvial plains are prone to deforestation, especially for agriculture and show little regeneration in general. Fine soil texture favors deforestation for agriculture. Biomass is an important factor with mainly negative correlations for most deforestation and forest degradation transitions. In tropical forests, regeneration is generally favored in areas with high biomass. The amplitude of phenology follows this pattern in many cases (both deforestation and regeneration of secondary forests occur in areas with a high amplitude). Fragmentation shows the classic tendency in which more fragmented areas (entropy) with patches of vegetation (MSPA) are at greater risk of deforestation and forest degradation, while regeneration is more likely in less fragmented areas. Distance to roads in general shows that areas closer to roads are prone to deforestation and forest degradation. The opposite is true for regeneration, except for regeneration of tropical forests from pasture. Distance from paved roads is important for deforestation of secondary forests for agriculture (in temperate forests, there is a higher probability of deforestation in areas close to paved roads, while in tropical forests deforestation is less likely in these areas) and for regeneration of secondary forests from agriculture (in both cases positive correlation). The level of cattle production is especially important for tropical deforestation and forest degradation, but is also related to the deforestation of temperate forest for agriculture and the deforestation of secondary forests for pasture. Medium to high levels of marginalization relate to the deforestation of tropical forests for agriculture and temperate forest degradation, while lower levels of marginalization are negatively related to deforestation for agriculture, but positively related to deforestation of secondary tropical forests for pasture and the regeneration of secondary temperate forest from pasture. Population growth is negatively related to deforestation and forest degradation of temperate forests (except for secondary temperate forest to pasture), but positively related to the deforestation of tropical forests (except for secondary tropical forest to pasture). The probability of deforestation and forest

Table 3 Importance of spatial factors for the modelled LUCC transitions. Signs indicate direction of correlation (+: positively correlated, –: negatively correlated), background signs details indicate importance levels (white: weak relations, light grey: moderately strong relations, dark grey: strong relations). See Table 1 for the codes used to describe the transitions. * indicates that the variable has been analyzed as a distance surface

	1 to 3	1 to 5	1 to 6	2 to 4	2 to 5	2 to 6	3 to 5	3 to 6	4 to 5	4 to 6	5 to 3	5 to 4	6 to 3	6 to 4
Altitude	−+	−+	−	−	−	−	−+−	−	−+−	−	+	−	−+	−
Slope	−	−	−	−	−	−	−	−	−	−	+	+	+	+−
Hydro1k topo wetness index (cti)	−	−+		−		−	−+	−	−+				+−	−
Altimetric zone of the basin (high)		−	−	−	−	−	+	+	−	+	+	−	+	−
Altimetric zone of the basin (intermediate)	+	+	+	+	−	−	−	−	−	−	−	+	−	+
Altimetric zone of the basin (low)	+	+	−	−	+	+			+	+	−	−		−
Bio 5 (max temperature of warmest month)		+−	+	+	−+	−+	+	+−				+		+
Bio 7 (temperature annual range)	+	−+	−+−		−	+	−+	−+−	−	+−	+	−	+−	−
Bio 9 (mean temperature of driest quarter)		+−+	+	−+		+	+−					+		+
Permanent rivers*	−	−	−	−	−	−	−	−	+				−	−
Geomorphology (aclinal/table structure)		+	+	+	+	+	+	−	+	−	−	−		−
Geomorphology (coast)				−	+	−			+	−		−		−
Geomorphology (volcanic relief)	−	−	−				+	−	+	−	−	−	+	+
Geomorphology (low elevations)				+	+	+	−	−	+	+	−	+	−	+
Geomorphology (lava flow)	−	+	−				+	+	+	−	+	−	+	−
Geomorphology (modelled slope)	−	+	−	−	−	−	+	+	−	−	+	+	+	−
Geomorphology (lacustrine plain)	−	+	+	+	+	+	+	−	+	−	−	−	−	−
Geomorphology (mountain range)	+	−	+	−	−	−	−	−	−	+	+	+	−	+
Geomorphology (foothills)	−	+	−	+	+	+	−	−	+	+	−	−	−	+
Geomorphology (karst relief)	+	−	−	−	−	−	−	−	−	−	+	+	+	+
Geomorphology (alluvial plain/valley)	+	+	+	+	+	+	+	+	+	+	−	−	−	−
Soil texture (fine)	−	+	+	+	+		+	−	+	−	−	−	+	−
Soil texture (medium)	−		−	−	+		−	−	−	−	+	+	+	−
Soil texture (coarse)	+	−	+		−		−		−	+	−	+	−	+
Biomass	−	−	−	−+	−	−	−	−	−+	−+	+	+−+	+	+−+
NDVI (HANTS)	−	−	−	−+	−	−	+−	+−	+−	−+−	+		+	+
Fragmentation 1993 (entropy)	+	+	+	+	+	+	−+	+	+	+	−	−	−	−
Fragmentation 1993 (branch)	+	+	+	+	+	+	+	+	+	+	+	−	−	−
Fragmentation 1993 (edge)	+	+	+	+	+	+	+	+	+	+	−	+	−	−
Fragmentation 1993 (perforation)	+	+	+	+	+	+	+	+	+	+	+	+	−	−
Fragmentation 1993 (islet)	+	+	+	+	+	+	+	+	+	+	+	−	−	−
Fragmentation 1993 (core)	−		−	−	−	−	−	−	−	−	+	+	+	+
Fragmentation 1993 (bridge)	+	+	+	+	+	+	+	+	+	+	−	−	−	−
Fragmentation 1993 (loop)	+	+	+	+	+	+	+	+	+	+	−	+	−	−
Temperate forest 1993*				−+	+				+	−+				
Tropical forest 1993*		+−	−				+−	+−	+	+		−	+	
Temperate secondary forest 1993*	−	−		+−	+−	+−	+				−		−+	+
Tropical secondary forest 1993*		+	−	−	−+	−	+−				+	−	+	−
Agriculture 1993*	+	−	+	−	−	−	−+	−+	−	+			+	
Pasture 1993*	−	−	−	−+	+	−	−	−		−	−	−		
Roads*	−	−	−	−	−	−	−	−	−	−	+	+	+	−
Paved roads*	−+	−+	−−	−	−	−	+−+	−+−	−+	+	+	+	+	
Not paved roads*	−	−	−+	−+	−+	−+	−	−+	−	−	−		+	−
Ejido*	+	+	+		−	−	+−	+		−			+−+	
Comunidad*	−	+		+	+	+−		+	+−				−	
Private property*	−+	−			−	−+	−	−	−+	+−+			−+	+
Cattle production	+	+	+	+−+	+−	+	+	+	+	−+		−	−	
Marginalization 1995 (very low)	−	−	−	+	−	+−	−	−	−	+−			+	+−
Marginalization 1995 (low)	−	−	−	+	−+	−	−	+	−	+		+		
Marginalization 1995 (medium)	+	−	−	+	−+	−+−	−	−	−	−+			+	+
Marginalization 1995 (high)	−+	−	−		−+	−	−	−	−	−+				+
Population growth 1995-2005	−	−	−	−	+	+	−	+−+		−			−	
Federal PA*		+	+	+−	+−	+−	+−		+−	+	−	−	−	+
Environmental friendly managed unit*	+		+		+		+−	+	+	+−	−	−	−	−

degradation increases with the distance from protected areas and environmental friendly managed areas (Table 3). Together, these spatial factors lead to the general pattern, both for temperate and tropical forests that the probability for deforestation for pasture is high in the western part of the study area on the Pacific slope, while the probability of deforestation for agriculture is elevated on the eastern or inland side of the study area (Fig. 3). The highest probabilities for temperate forest degradation are in the northern half of the study area (Fig. 3).

4.2 LUCC Tendencies—Econometric Model

The econometric models highlight a negative growth rate for the planted area for beans and maize and a positive growth rate for the production of beef, the planted area for grazing and the total planted area (Table 4). Stratification of the study area into blocks of municipalities to check for sub-regional differences confirmed that these tendencies for traditional crops and beef production are consistent all over the study area.

The area of cultivated pastures for cattle ranching is positively related to the production of beef in carcasses (not significant for the random effects model) and the total cultivated area, but negatively related with the planted area for maize (Table 5), as shown by the panel data analysis. The increase in grazing area was thus associated with the expansion of the total cultivated area and the reduction of the cultivated area for maize, indicating that pastures expanded at the expense of beans and maize.

The random effects model indicates that the planted area for grazing is statistically significant but negatively related to the production of beef in carcasses which is opposite to the results for the other models.

Table 4 Growth rates of beef in carcasses, area for maize, area for beans, area for grazing and the total cultivated area for 2002 to 2012

Cultivated area maize			Cultivated area beans			Total cultivated		
a	b	c	a	b	c	a	b	c
−0.0188*	−0.022	−0.0180*	−0.127**	−0.134**	−0.116**	0.00443+	0.0104+	−0.004
(−3.03)	(−1.09)	(−3.06)	(−9.78)	(−6.44)	(−9.85)	−1.86	−2.15	(−0.90)
Cultivated area pasture			Beef production					
0.0168**	0.0116*	0.0383**	0.0271**	0.0409**	0.0114**			
−5.88	−2.52	−5.63	−7.05	−5.58	−4.28			

(a) For all municipalities, (b) all the municipalities that participated in the workshops and (c) all the other. T statistics in parentheses, +p < 0.10, * < 0.05, **p < 0.01, number of observations in all cases was 10

Table 5 Results of the panel data model

	Fixed effects			Random effects		
	a	b	c	a	b	c
Beef production	0.166+	−0.0323	0.897**	0.106	−0.335*	0.758**
	−1.72	(−0.34)	−3.27	−1.11	(−2.45)	−2.93
Beans	0.573	1.419	−6.444**	0.638	1.831	−6.848**
	−0.69	−1.62	(−4.34)	−0.76	−1.25	(−4.69)
Maize	−0.934**	−0.842**	−0.556**	−0.949**	−0.936**	−0.540**
	(−17.53)	(−8.05)	(−9.69)	(−18.13)	(−5.60)	(−9.80)
Total cultivated area	0.852**	0.884**	0.490**	0.867**	0.962**	0.475**
	−64.49	−66.52	−12.58	−70.04	−51.39	−13
N° of observations	290	60	230	290	60	230

(a) For all municipalities, (b) all the municipalities that participated in the workshops and (c) all the other municipalities. T statistics in parentheses, +p<0.10, *<0.05, **p<0.01

4.3 *Local Stakeholder Perceptions and Participatory Scenarios*

The workshops we organized revealed that those taking part had detailed knowledge of the socioenvironmental changes in their region. This knowledge is of an observational nature and has been nurtured through oral history. When asked about the changes they had observed over the last 50 years, workshop participants said that forest had decreased in both area and quality and there had also been a reduction in total rainfall, water availability and water quality. The number of wild fauna was also reported to have fallen.

Farmers described how they had stopped growing maize for subsistence and had begun to grow commercial crops. Yet, they considered that the use of agrochemicals led to reduced yields and increases in pest frequency and abundance. The most frequently mentioned commercial crops included rice and sugar cane. Rice was said to have high yields because it is grown in areas with rich soils and high water quality.

Deforestation occurred mainly because of the establishment of agricultural fields, and especially because of pasture for livestock: a third, less important factor was mining projects. Other factors mentioned were wood extraction and intentional forest fires. Over time, the traditional life style of the *rancherías* (small rural communities) has been lost and nowadays few people cultivate maize because of the high cost of external inputs.

Several political and social drivers were mentioned. Governmental programs fostered the colonization of the coastal areas (1950s) and led to the initial LUCC. The construction of a sugar processing facility resulted in a widespread monoculture of sugar cane (1960s). Road construction in the 1970s onset major LUCC in the region. Large landowners, external actors and governmental incentives promoted deforestation for livestock farming.

indicates that the land is being allocated to the activity that creates most profit at the margin, in this case pasture for livestock. Another point mentioned by the local stakeholders is the displacement of traditional agriculture and subsistence crops by industrialized agricultural practices and commercial crops. This tendency cannot be clearly observed solely by analyzing LUC maps, but an increase in irrigation agriculture (as a proxy for industrialized agricultural practices) can be noted. Forest degradation is one of the main change processes and has been clearly identified by local stakeholders. Since no statistics are available regarding the extraction of wood and other forest products in this region, in the case of forest degradation only LUC maps and stakeholder perceptions can be used to compare tendencies, and although stakeholders mentioned a decrease in forest quality, the huge extent of this process discovered in the spatial analysis of LUC maps is not represented in local perceptions.

As regards the drivers of change, the different data sources complement each other quite well as they account for different scales, but also show some interesting cross-referencing aspects. The stakeholder elicitations revealed a set of indirect drivers underlying the direct drivers identified in the spatial analysis of historic change. The combination of data sources reveals indirect drivers which are very hard to consider in a spatial-only setting, since spatial information on distant drivers is often not available and the relationships between these distant drivers and change are difficult to detect. The lack of mentioned direct drivers could be attributed to the separation in the local perceptions of the changes from the agents of change, i.e. the local stakeholders themselves. This apparent disjunction arises because the actions that lead to change are part of their daily lives. Demography (population increase) and technology (industrialized agriculture) were not mentioned for similar reasons. The indirect drivers included political, social, cultural and economic aspects, which promoted agricultural expansion, especially cattle ranching. The spatial analysis of the direct drivers of change showed that a large array of variables play a role in LUCC. These include biophysical (bioclimate, topography, geomorphology, hydrology, edaphology, ecosystem productivity), socio-economic (marginalization, population growth, protected areas) and human impact (fragmentation, deforestation risk, roads) factors. The LUCC probability maps make use of these spatial direct drivers that influence spatial patterns of land use, in order to account for spatial variation in the biophysical and socio-economic environment. The differences between the locations with the highest probabilities of change for agriculture and pasture could be further explained by the differences in remoteness and market incorporation. The areas belonging to the Pacific slope are not completely integrated into the market and the mountain chain that separates this region from the more populated and developed inland regions is a major obstacle to cross. The econometric models showed that pasture for cattle and beef production is a significant driver of change, since the traditional crops are being replaced with pasture for cattle. When these models are combined, a detailed picture of regional LUCC drivers emerges in its historical context.

In general, different data sources are quite consistent, which could be attributed to the strong and steady tendencies of the main change processes over the last decades. This meant that even the temporal differences between the data sources did not prevent cross-referencing. Stakeholder perceptions of what happened in the earlier period for which there is no LUCC data provides a historic context for the changes observed in the last two decades in the other data sets. Unfortunately, agricultural production data in Mexico only started being collected in 2002 and thus only shares one decade with the other data sets. Nevertheless, this decade is sufficient to capture LUCC tendencies based on econometric data. LUCC is a long-term broad-scale disturbance related to public politics, which determines regional landscape dynamics that were validated by cross referencing the different data sources.

Data sources also provide different views of change. One of the most basic differences was the temporal dynamics of change. While only mentioned very generally in the local elicitations, the spatial analysis is very explicit about changes in the magnitude of change processes over time. Elicitations about these temporal changes could be included in future work with stakeholders to discuss the direct and indirect drivers behind the observed decrease in LUCC. A possibility for data based cross-referencing could be the temporal analysis of various measures implemented for environmental conservation (protected areas and payment for ecosystem services) and compare those with subsidies from the agricultural sector on a regional level.

Studies at a national scale are necessary for certain objectives and are informative in general terms, but those at a sub-national regional scale offer interesting insights into specific change processes, their magnitude and relative importance. For this region, several local scale LUCC studies have been conducted, but this is the first time that regional scale analysis has been carried out. Most studies focus on a small watershed (Cuitzmala) and evaluate the effects (such as functional disruption and soil erosion) of LUCC on ecosystems and their biotic and abiotic components (Maass et al. 2005; Cotler y Ortega-Larrocea 2006). Burgos and Maass (2004) analyzed LUCC by establishing transitions validated by local perceptions and reported forest-agriculture transitions in flatlands, forest-pasture on slopes and wood extraction on hilltops as the main pathways. Natural regeneration is reported to start one to three years after abandonment and to remain in a successional phase dominated by *Acacia* and *Mimosa* spp for at least 20 years. Farfán et al. (2016) analyzed LUCC in the Manantlán Biosphere reserve and found that temperate forests are more stable than tropical forests, which are mainly converted into pastures.

Even though overall deforestation rates are not so high as to qualify the region as a deforestation hot spot (defined as more than 2% annual deforestation of total forest area), primary tropical dry forests, oak forests and pine-oak forests suffered very high deforestation rates at the beginning of the study period (1993–2002). The forest type with most relative losses was the tropical semi-dry forest. Deforestation rates plummeted in subsequent years and only oak and tropical dry forests continue to undergo considerable deforestation. If besides deforestation the successional state and forest degradation are taken into account, a much more severe picture of LUCC emerges: Primary vegetation cover decreased from 42% in 1993 to 28% in 2011.

Even though not all of this loss is due to deforestation, the increase of secondary vegetation from 26 to 38% of the same period indicates an important loss of ecological integrity. In the study area, the coastal zone where the stakeholder workshops took place has been the main focus of attention from the academic sector, and there is also a wealth of information on tropical dry forests. However little research has been done on the temperate forests covering a large section of the study area. It would be a good idea to widen the research approach to include the higher sections of the study area, because of their ecological connectivity with the low-lying tropical regions (e.g. watershed management). One of the options to encourage the maintenance of these temperate forests could be a sustainable forestry plan, allowing the forests to be partially used for timber and non-timber products, as opposed to tropical dry forests, which are not suitable for commercial forestry because of their high species diversity and patchy distribution (Maass et al. 2005). This could also be combined with poverty alleviation programs, since the most marginalized settlements are located in the higher mountainous part of the study area. Nowadays, the majority of the population living in these settlements are engaged in agricultural activities, which puts a lot of pressure on natural vegetation (Gerritsen 2012).

6 Conclusion and Outlook

In this case study LUCC modeling has been used as an integration tool for local and landscape scale data sources, spatial and non-spatial data, qualitative socio-economic and quantitative biogeophysical data. The subdivision of the study area in the econometric models showed that the main processes are consistent all over the region and that the information for the municipalities participating in the stakeholder elicitations can be used to indicate regional scale tendencies. This congruence of sources implies that LUCC probability maps are based on validated LUCC dynamics, so diminishing doubts about map accuracy and the resulting effects for LUCC modeling. This also means that the outcomes of this study can be used by the research team and the local stakeholders in subsequent steps towards spatial participatory scenarios for decision-making. Besides the cross-referencing and validation effect, a consequence of the interdisciplinary working group and the integration of otherwise disconnected data is that the information is enriched via complementarities among data sources.

Acknowledgements This study was partly funded by the European Union's Seventh Framework Programme ([FP7/2007–13]) under grant agreement no 283093; Role of Biodiversity in climate change mitigation (ROBIN). PB acknowledges support from PAPIIT IN211114. This project (BIA2013-43462-P) is financed by the Ministry of the Economy and Competitiveness and by the FEDER European Regional Fund. We would also like to thank the members of the Mexican National Commission for Knowledge and Use of Biodiversity (CONABIO) for their opinions, ideas and comments.

References

Alianza MREDD+ (2013) Mapa y base de datos sobre la distribución de la biomasa aérea de la vegetación leñosa en México. Versión 1.0. Woods Hole Research Center, USAID, CONAFOR, CONABIO, Proyecto México Noruega, México

Arizmendi MC, Márquez-Valdelamar L, Ornela JF (2002) Avifauna de la región de Chamela, Jalisco. In: Noguera FA, Vera Rivera JH, García Aldrete AN, Quesada Avendaño M (eds) His-toria Natural de Chamela. Universidad Nacional Autónoma de México, México, DF, Instituto de Biología, pp 297–328

Aspinall R (2004) Modelling land use change with generalized linear models-a multi-model analysis of change between 1860 and 2000 in Gallatin Valley, Montana. J Environ Manag 72(1–2):91–103

Bradshaw CJA, Giam X, Sodhi NS (2010) Evaluating the relative environmental impact of countries. PLoS ONE 5(5):1–16

Brady M, Irwin E (2011) Accounting for spatial effects in economic models of land use: Recent developments and challenges ahead. Environ Resour Econ 48:487–509

Bray D, Antinori C, Torres-Rojo JM (2006) Mexican model of community forest management: The role of agrarian policy, forest policy and entrepreneurial organization. Forest Policy Econ 8:470–484

Bray D (2007) El manejo comunitario de los bosques en México: veinte lecciones aprendidas y cuatro senderos para el futuro. In: Bray D, Merino L (coords) (2007) Los bosques comunitarios de México: manejo sustentable de paisajes forestales. Instituto Nacional de Ecología. Mexico City, Mexico pp 417–435

Burgos A, Maass JM (2004) Vegetation change associated with land-use in tropical dry forest areas of Western Mexico. Agric Ecosyst Environ 104(3):475–481

Ceballos G, García A (1995) Conserving neotropical biodiversity: the role of dry forest in western Mexico. Conserv Biol 9:1349–1356

Collins SL et al (2011) An integrated conceptual framework for long-term social-ecological research. Front Ecol Environ 9(6):351–357

CONABIO (2008) Red hidrográfica Nacional modificada por CONABIO, escala 1:250000. Comisión Nacional para el Conocimiento y Uso de la Biodiversidad, Mexico

CONABIO (2009) Cuerpos de agua para el análisis de vacíos y omisiones en conservación de la biodiversidad epicontinental. Comisión Nacional para el Conocimiento y Uso de la Bio-diversidad, Mexico

CONABIO (2012) Desarrollo territorial sustentable: Programa especial de gestión en zonas de alta biodiversidad. Comisión Nacional para el Conocimiento y Uso de la Biodiversidad, Mexico

CONANP (2010) Mapa de áreas naturales protegidas federales. Comisión Nacional de Áreas Natural Protegidas, Mexico City, Mexico

Couturier S, Núñez JM, Kolb M (2012) Measuring tropical deforestation with error margins: a method for REDD monitoring in southeastern Mexico. In: Sudarshana P (ed) Tropical Forests. EEUU, INTECH, Nueva York, NY, pp 269–296

Cotler H, Ortega-Larrocea MP (2006) Effects of land use on soil erosion in a tropical dry forest ecosystem, Chamela watershed. Mexico. Catena 65(2):107–117

Farfán Gutiérrez M, Rodríguez-Tapia G, Mas JF (2016) Análisis jerárquico de la intensidad de cambio de cobertura/uso de suelo y deforestación (2000-2008) en la Reserva de la Biosfera Sierra de Manantlán, México. Investigaciones Geográficas, Boletín del Instituto de Geografía, UNAM 90:89–104

Gentry AH (1995) Diversity and floristic composition of neotropical dry forests. In: Bullock SH, Mooney HA, Medina E (eds) Seasonally dry tropical forests. Cambridge University Press, Cambridge UK, pp 146–194

Geoghegan J, Schneider L, Vance C (2004) Temporal dynamics and spatial scales: Modeling deforestation in the southern Yucatán peninsular region. GeoJournal 61(4):353–363

Gerritsen PRW (2010) Perspectivas campesinas sobre el manejo de los recursos naturales. Mundi-Prensa, Mexico

Gerritsen PRW (2012) Investigación sociológica, manejo de recursos naturales y vinculación social. Notas introductorias. In: Gerritsen PRW (ed) Sociología rural y medio ambiente. Tópicos selectos de Jalisco. Plaza y Valdés/Universidad de Guadalajara, México pp 23–49

Gerritsen PRW, Alvarez GN (2015) Paisajes rurales en disputa: co-producción y conflictos ambientales en el sur de Jalisco. In: Guzmán M, Garibay C, Tyrtiana L (eds) Conocimiento, poder y naturaleza. RESMA, Mexico

Gerritsen PRW (2016) Tarjetas Participativas. Plaza y Valdés/Universidad de Guadalajara, Mexico, Un método participativo para la discusión grupal

Gerritsen PRW, Lazos E, Álvarez N, Solis J (in press) Voces de la Tierra Pródiga: Testimonios Campesinos sobre las Transformaciones Socio-ambientales en la Costa de Jalisco, Occidente de México. In: Macias Macias A (ed) Voces de América Latina. Plaza y Valdés, Mexico

IMT (2007) Red nacional de carreteras, IMT (INIT-2007), escala 1:1. Unidad de Sistemas de Información Geoespacial, Instituto Mexicano del Transporte (IMT), Primera edición. México DF

IMT (2001) Caminos rurales, escala 1:1. Unidad de Sistemas de Información Geoespacial, Instituto Mexicano del Transporte (IMT), México DF

INE, SEMARNAT, IG-UNAM (2003) Sistema Clasificatorio del Relieve de México, escala 1:250000. Instituto Nacional de Ecología, Secretaria de Medio Ambiente y Recursos Naturales, Instituto de Geografía, Universidad Nacional Autónoma de México, Mexico

INECC (2015) Índice de Presión Económica (Riesgo) de Deforestación. In: Instituto Nacional de Ecología y de Cambio Climático. http://www.inecc.gob.mx/irdef. Accessed 15 Dec 2015

INEGI (1995) Conteo de población y vivienda. Resultados definitivos, Comisión Nacional para el Conocimiento y Uso de la Biodiversidad, Mexico

INEGI (1998) Modelo Digital del Terreno, escala 1:250 000. Comisión Nacional para el Conocimiento y Uso de la Biodiversidad, Mexico

INEGI (2000) Anuario Estadístico del Estado de Jalisco. Instituto Nacional de Estadística, Geografía e Informática, Mexico

INEGI (2001) Conjunto de datos vectoriales de la carta de uso de suelo y vegetación, escala 1:250,000. Serie II continuo nacional, Instituto Nacional de Estadística, Geografía e Informática, Aguascalientes, Mexico

INEGI (2005) Conjunto de datos vectoriales de uso del suelo y vegetación, escala 1:250,000, serie 3 continuo nacional. Instituto Nacional de Estadística, Geografía e Informática, Aguas-calientes, Mexico

INEGI (2008) Conjunto de datos vectoriales de uso del suelo y vegetación, escala 1:250,000, serie 4 continuo nacional. Instituto Nacional de Estadística, Geografía e Informática, Aguas-calientes, México

INEGI (2010) Conteo de población y vivienda. Resultados definitivos, Comisión Nacional para el Conocimiento y Uso de la Biodiversidad, Mexico

INEGI (2013) Conjunto de datos vectoriales de uso del suelo y vegetación, escala 1:250,000, Serie 5 continuo nacional. Instituto Nacional de Estadística, Geografía e Informática, Aguas-calientes, Mexico

INEGI (2016) Banco de información INEGI. http://www.inegi.org.mx/biinegi. Accessed 13 March 2016

INEGI (2016) Censo Agrícola, Ganadero y Forestal 2007. Instituto Nacional de Estadística, Geografía e Informática, Aguascalientes, Mexico. http://www.inegi.org.mx/est/contenidos/proyectos/Agro/ca2007/Resultados_Agricola/. Accessed 15 Dec 2015

INIFAP, CONABIO (1995) Mapa edafológico, escalas 1:250000 y 1:1000000. Instituto Nacional de Investigaciones Forestales y Agropecuarias/Comisión Nacional para el Conocimiento y Uso de la Biodiversidad, Mexico

Kolb M (2009) Model prototype to estimate human impact on Mexican biodiversity. Technical report, National Commission for the Knowledge and Use of Biodiversity, Mexico City, Mexico

Kolb M, Galicia L (2012) Regional patterns and processes of deforestation and regeneration in the Neotropic: recent tendencies in Southern Mexico. The Geographical Journal 178(2):147–161
Kolb M, Morales Luque G (2016a) Measures of fragmentation of natural vegetation: level of patch dispersion based on MAD-Mex 4.1 data for 1993, scale: 1:100000. Comisión Nacional para el Conocimento y Uso de la Biodiversidad, Mexico City, Mexico
Kolb M, Morales Luque G (2016b) Measures of fragmentation of natural vegetation: level of patch dispersion based on MAD-Mex 4.1 data for 2010, scale: 1:100000. Comisión Nacional para el Conocimento y Uso de la Biodiversidad, Mexico City, Mexico
Kolb M, Morales Luque G (2016c) Measures of fragmentation of natural vegetation: patch types based on MAD-Mex 4.1 data for 1993" scale: 1:100000. Comisión Nacional para el Conocimento y Uso de la Biodiversidad, Mexico City, Mexico
Kolb M, Morales Luque G (2016d) Measures of fragmentation of natural vegetation: patch types based on MAD-Mex 4.1 data for 2010"scale: 1:100000. Comisión Nacional para el Conocimento y Uso de la Biodiversidad, Mexico City, Mexico
Kolb M, Morales Luque G, Neuwirth B (2016) Influence of roads on surrounding ecosys-tems"scale: 1:100000. Comisión Nacional para el Conocimento y Uso de la Biodiversidad, Mexico City, Mexico
Kooistra L, Dutrieux, L, Equihua J, Kolb M, Roerink G, Gerard F, Peña Claros M (2015) Current contributions of biodiversity and ecosystems to climate change mitigation-an analysis using remote sensing datasets. Deliverable 1.1.3. Remotely sensed datasets, ROBIN project
Lazos E, Gerritsen PRW (in press) Entre promesas y esperanzas: actores regionales de la Costa Sur de Jalisco y la construcción de su futuro socioambiental. In: IIS (ed) El Futuro de México en 2035. Una visión prospectiva. Universidad Nacional Autónoma de México, Mexico
Lepers E, Lambin EF, Janetos AC, DeFries R, Achard F, Ramankutty N, Scholes RJ (2005) A synthesis of information on rapid land-cover change for the period 1981–2000. Bioscience 55 (2):115–124
Lott E (1993) Annotated checklist of the vascular flora of the Chamela Bay Regoin, Jalisco, Mexico. Occasional Papers of the California Academy of Sciences 148:1–60
Maderey LE (1990) Evapotranspiración real. In: Hidrogeografía IV.6.6. Atlas Nacional de México. Vol. II Escala 1 4000000. Instituto de Geografía UNAM, Mexico
Maass M et al (2005) Ecosystem services of tropical dry forests: insights from long-term ecological and social research on the Pacific Coast of Mexico. Ecology and Society 10(1):17
Magaña MA (2003) Actitudes y percepciones de productores rurales y sus familias hacia la conservación de la selva y el área natural protegida: Reserva de la Biosfera Chamela-Cuixmala, Jalisco, México. Dissertation, Universidad Michoacana de San Nicolás de Hidalgo
Mas JF, Kolb M, Paegelow M, Camacho Olmedo MT, Houet T (2014) Inductive pattern-based land use/cover change models: a comparison of four software packages. Environ Model Softw 51:94–111
Mas JF (2005) Change estimates by map comparison: A method to reduce erroneous changes due to positional error. Trans GIS 9(4):619–629
Morales J (2011) La crisis global y sus impactos en la vida rural. In: Morales J (ed) La agroecología en la construcción de alternativas hacia la sustentabilidad rural. ITESI/Siglo XXI, México, pp 17–50
Pontius RG, Lippitt CD (2006) Can error explain map differences over time? Cartogr Geogr Inf Sci 33(2):159–171
Pontius RG, Petrova SH (2010) Assessing a predictive model of land change using uncertain data. Environ Model Softw 25(3):299–309
Pontius RG (2000) Quantification error versus location error in comparison of categorical maps. Photogramm Eng Remote Sens 66(8):1011–1016
SEMADET (2015) Modelo de ordenamiento ecológico territorial de Jalisco. In: Secretaría de Medio Ambiente y Desarrollo Territorial del Estado de Jalisco
http://siga.jalisco.gob.mx/moet/ Accessed 18 Dec 2015

SEMARNAT 2010 Unidades de manejo para la conservación de la vida silvestre. Secretaría de Medio Ambiente y Recursos Naturales (SEMARNAT), Dirección General de vida silvestre. Mexico

Schnelle E, Thiersch M (1979) The Metaplan-method: communication tools for planning and learning. Metaplan-GmbH

StataCorp (2013) Stata Statistical Software: Release 13. StataCorp LP, College Station TX

Tetreault D, Ochoa H, Hernández E (eds) (2012) Conflictos socioambientales y alternativas de la sociedad civil. Colección Alternativas de Desarrollo, ITESO, Guadalajara, Mexico

UNEP (2012) Global Environmental Outlook 5 (GEO 5). Environment for the future we want, United Nations Environment Programme

USGS EROS Center (1999) HYDRO1 k Elevation Derivative Database. In: North America data set. US Geological Survey Center for Earth Resources Observation and Science

Vitousek PM, Mooney HA, Lubchenco J, Melillo JM (1997) Human domination of earth's ecosystems. Science 277:494–499

Part III
Methodological Developments and Case Studies: Case Studies

Chapter 13
Urban Land Use Change Analysis and Modeling: A Case Study of the Gaza Strip

B. Abuelaish

Abstract Analysis of land use and land cover change is of prime importance for understanding the ecological dynamics resulting from natural and human activities, and for the assessment and prediction of environmental change. The population of the Gaza Strip will have grown to more than 2.4 million by 2023 all of whom are forced to live within an area of some 365 km^2. This growth in population will lead to an increase in land demand, and will far exceed the sustainable land use capacity. The Gaza Strip is a relatively small area in which land use planning has not kept up with land development. Continued urban expansion and population growth in the future will place additional stress on land cover, unless appropriate integrated planning and management decisions are taken immediately. Decision-makers need further statistics and estimation tools to achieve their vision for the future of the Gaza Strip based on sound, accurate information. This study combines the use of satellite remote sensing with geographic information systems (GISs). The spatial database was developed by using six Landsat images taken in 1972, 1982, 1990, 2002, 2013 and 2014, together with different geodatabases for those years. Five past trend scenarios were selected for simulation to be completed by the year 2023 using the Land Change Modeler in the Idrisi Terrset software. These different scenarios, one of which takes into account the damage incurred during the 2014 War, try to cover the possible variations in areas and spatial distribution resulting from changes in land use. As an average over the five scenarios, by 2023 the projected urban area will have increased to 206.24 km^2 or 57.13% of the Gaza Strip.

Keywords Land use and land cover change · Scenario · Urban · Land change modeler

B. Abuelaish (✉)
Department of GIS, Environment Quality Authority, Gaza, Palestine
e-mail: abuelaish@correo.ugr.es

B. Abuelaish
Departamento de Análisis Geográfico Regional y Geografía Física,
Universidad de Granada, Granada, Spain

M.T. Camacho Olmedo et al. (eds.), *Geomatic Approaches for Modeling Land Change Scenarios*, Lecture Notes in Geoinformation and Cartography,
https://doi.org/10.1007/978-3-319-60801-3_13

1 Introduction

Understanding, predicting and analyzing land use and cover change is enormously important for future planning. One of the major factors affecting land use in the Gaza Strip is rapid population growth, one of the most significant issues in Palestinian society today. According to the Palestinian Central Bureau of Statistics (PCBS), with the recent growth rates of 3.44% in mid-2013, and 3.41% in mid 2014 (PCBS 2014) the population of the Gaza Strip will have grown to over 2.4 million by 2023. This area already has one of the highest population densities in the world with an estimated 3,956 persons/km^2 in 2006. This figure is even higher in the Gaza governorate (around 6,834 persons/km^2) where most of the population is concentrated. Another serious problem in Gaza is urban sprawl. The number of housing units in the Gaza Strip increased from 116,445 in 1997 to 147,437 in 2007 (PCBS 2012). Many human and natural factors have increased pressure on land use in this region, resulting in deteriorating quality and quantity of land (Abuelaish and Camacho 2016). Urbanization leads to increasing pressure on natural ecosystems (Taubenbock et al. 2012; Haas and Ban 2014) and brings with it soil, water and air pollution (Duh et al. 2006; Ren et al. 2003).

The Gaza Strip has been a theatre of conflict for decades. Each of these conflicts has left its mark, and a significant environmental footprint has developed in the Gaza Strip over time (UNEP 2009). The population growth rate and the urban expansion it drives affect the whole region. In general people prefer to live close to the urban facilities and infrastructures, usually found in the center of the residential areas, and to avoid the dangerous areas. The Gaza Strip has been directly involved in many wars, most recently in 2008, 2012 and 2014. The 2014 war was the most destructive in terms of buildings and infrastructure. The Israeli offensive against the Gaza Strip was launched on 8th July and continued until 26th August 2014. It left devastation all across this region, ranging from damage to complete destruction of thousands of homes. Post-war reconstruction is likely to exacerbate the normal urban growth rate, so adding a greater burden on this already congested country.

Several monitoring techniques, such as Remote Sensing, are very useful for gathering the data required for land use change assessment, urban planning, urban sprawl and other environmental issues. Land use changes must be monitored at suitable intervals in order to update the knowledge required to support decision making. Monitoring of land use and land cover requires the support of two parameters: spatial resolution and temporal frequencies (Curran 1985; Janssen 1993; Hualou et al. 2007). Modeling can be defined within the context of geographic information systems (GISs) as occurs whenever GIS operations attempt to emulate processing in the real world, at one point in time or over an extended period (Goodchild 2005; Paegelow et al. 2013). GIS models go beyond simply evaluating the future and are used to assess different scenarios, on the basis of the historical data retrieved from multiple resources. Scenarios have emerged as useful tools to explore uncertain futures in ecological and anthropogenic systems (Sleeter et al. 2012). Scenarios typically lack quantified probabilities (Nakicenovic and Swart 2000;

Swart et al. 2004), functioning instead as alternative narratives or storylines that capture important elements about the future (Nakicenovic and Swart 2000; Peterson et al. 2003; Swart et al. 2004). Alcamo et al. (2008) define scenarios as "descriptions of how the future may unfold based on 'if-then' propositions." Scenarios provide a structured framework for the exploration of alternative future pathways, and are used to assist in the understanding of possible future developments in complex systems that typically have high levels of scientific uncertainty (Nakicenovic and Swart 2000; Raskin et al. 1998). Plausible scenarios generally require knowledge of how drivers of change have acted to influence historical and current conditions (Sleeter et al. 2012).

This study aimed to analyze urban growth and monitor the spatial and temporal changes from 1972 to 2014 within five past trend scenarios using a model based on GIS techniques and remote sensing data. One of these scenarios takes into account the damage caused by the 2014 war. Scenarios are proposed to 2023, which can be helpful for planning decisions to be taken in the Gaza Strip within this timeframe. These decisions will be have an enormous impact on the future of environmental issues and urban development.

This paper begins by presenting the study area of the Gaza Strip. We then explain the methodology, including data processing, land change analysis, simulation and modeling of the study area before and after the 2014 war in the Gaza Strip. Finally, we present the results, discussion and conclusions.

2 Study Area and Dataset

2.1 Study Area

The Gaza Strip is a narrow area on the Mediterranean coastal plain. It is approximately 41 km long, and from 6 to 12 km wide, with a total area of 365 km^2. It shares a 12 km border with Egypt to the southwest and is surrounded by Israel to the east and north (the rest of the Strip—51 km of borders), as shown in Fig. 1. The Gaza Strip has a temperate climate, with mild winters (about 13 °C) and hot summers with frequent droughts (high 20 s °C). Average rainfall is about 300 mm a year (MOAg 2013). The terrain is flat or rolling, with dunes near the coast. In terms of topography the Gaza Strip slopes gradually downwards from east to west with the land surface elevation varying between 10 m above sea level in the west to 110 m above sea level in the east.

In 1948, the Gaza Strip had a population of less than 100,000 people (Ennab 1994), however by 2007, it had risen sharply to around 1.4 million (Census 2007). The total population in 2014 was estimated to be in excess of 1.79 million and, at the end of 2015, about 1.82 million inhabitants, distributed across five Governorates (PCBS 2015), of whom almost 1.3 million were UN-registered refugees. Gaza City, which is the biggest governorate, has some 625,824 inhabitants. The other two

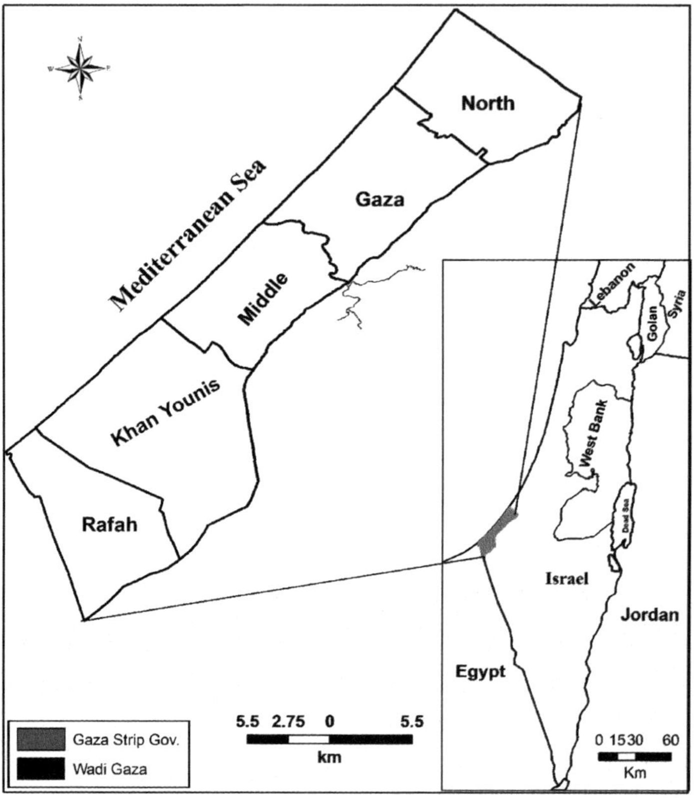

Fig. 1 Location on the Gaza Strip

main governorates are Khan Younis and Rafah in the southern part of the Gaza Strip, which have 341,393 and 225,538 inhabitants, respectively. There is also the Northern Governorate, with a population of about 362,772, and the Middle Governorate, which has 264,455 inhabitants. The smallest governorate in terms of area is the Middle Governorate, with 55.19 km^2. This is followed by Rafah (60.19 km^2), the Northern Governorate (60.66 km^2), and Gaza (72.44 km^2). The largest governorate is Khan Younis with an area of 111.61 km^2, as shown in Fig. 1.

Agriculture is the economic mainstay of the employed population, and nearly three quarters of the land area is under cultivation. On the Gaza coastal plain the original Saharo-Sindian flora has been almost completely replaced by farmland and buildings. Gaza has six main vegetation zones: the littoral zone along the coast, the stabilized dunes and blown-out dune valleys, the Kurkar, alluvial and grumosolic soils in the northern part, the loessial plains in the east, and three wadi (valley) areas (UNEP 2006).

2.2 Dataset

The spatial database has been produced using the historical Landsat images from 1972, 1982, 1990, 2002, 2013 and 2014, as shown in Table 1. The images were rectified from the aerial photos for 2007 using Erdas Imagine 2013. Generalized digitalization was used to build the urban GIS database using ArcGIS 10.2; interpretation was mainly visual, and both supervised and unsupervised classifications were used for more control and interpretation. The cell size of the entire dataset was converted to 15 × 15 meters. The database was validated before starting the analysis by tracking data with high resolution aerial photographs taken before and after a particular year. As no aerial photographs were available for the last year, we validated some points that were in doubt using the UNITAR database.

For the purposes of this research, we considered the whole Gaza Strip area as suitable for agriculture and classified the land into two classes: urban and agricultural areas (non-urban areas). Some other land uses and land covers in the study area were also considered as agricultural in this study.

On 20th November 2014, the UNRWA, UNDP and the Ministry of Public Works and Housing (MOPWH) announced the conclusion of their assessment of the damage caused to housing during the 2014 War, which they had conducted jointly over a two month period. 6,761 residential buildings were totally destroyed (including more than 11,000 housing units), 3,565 were severely damaged and 4,938 units were moderately damaged, as shown in Table 2. The UNITAR/UNOSAT

Table 1 Landsat data used in this study

Sensor	Row	Data type
Landsat MSS	188/38	22/10/1972
Landsat 3 TM	188/38	13/08/1982
Landsat 5 TM	174/38	11/06/1990
Landsat 7 ETM+	175/38	05/07/2002
Landsat 8	175/38	25/06/2013
Landsat 7 ETM+	175/38	24/09/2014
GIS dataset of UNOSAT/UNITAR	Geodatabase	Field Survey, October/2014

Table 2 Buildings damaged during the 2014 War

	Destroyed	Severely Damaged	Moderately Damaged	Total Structures Affected	Crater Impact
North	1,253	761	1,000	3,014	1,702
Gaza	1,963	1,127	1,378	4,468	1,765
Middle	809	406	683	1,898	553
KhanYounis	1,749	898	1,379	4,026	1,549
Rafah	987	373	498	1,858	1,904
TOTAL	6,761	3,565	4,938	15,264	7,473

Source UNITAR/UNOSAT (2014)

created geodatabases based on field work and high resolution satellite images. All images used to analyze the conflict were taken by the Pleiades satellites operated by Airbus Defense and Space, which provide 50 cm resolution images (UNITAR/UNOSAT 2014). The UNITAR/UNOSAT Geodatabase contains a total of 22,745 sites with crater impacts or some form of damage to housing after attacks during the war.

3 Methodology and Practical Application to the Datasets

The flow chart in Fig. 2 shows the methods used in the research reported in this paper, including the definition and creation of a database using remote sensing and GIS, land changes analysis, proposal and testing of explanatory variables, modeling and scenarios development of scenarios in the Gaza Strip.

3.1 Land Use Model and Data

3.1.1 Land Change Analysis

The chronological series of LUC maps was analyzed to detect changes. A quantitative assessment of category-wise land use changes in terms of net changes, swap, gains, losses and total changes (Eastman 2012) was extracted from several pairs of data, and the results are shown in maps and statistics. The change analysis was performed specifically between two images from 1972, 1982, 1990, 2002, 2013 and 2014 to understand the transitions in land-use classes over the years. A multiple regression line was created to predict the future urban area, and statistical values for the changes occurred in the area were represented on a scatter diagram.

3.1.2 Proposal and Testing of Explanatory Variables

Five static drivers were selected to simulate and predict the future urban area in 2023. The first driver is the distance from the main and regional roads in 2013, given that the population prefers to buy and live in houses overlooking the roads, which are also considered good investments. The second driver is elevation, because people prefer high locations which are considered to be safe from floods during rainfalls and have a more temperate climate in summer. The third driver is the distance from the urban area in 2013, since people prefer to live close to well-established urban areas with better infrastructure and services, which are safer during Israeli military attacks. The fourth driver is the 1 km wide buffer zone along

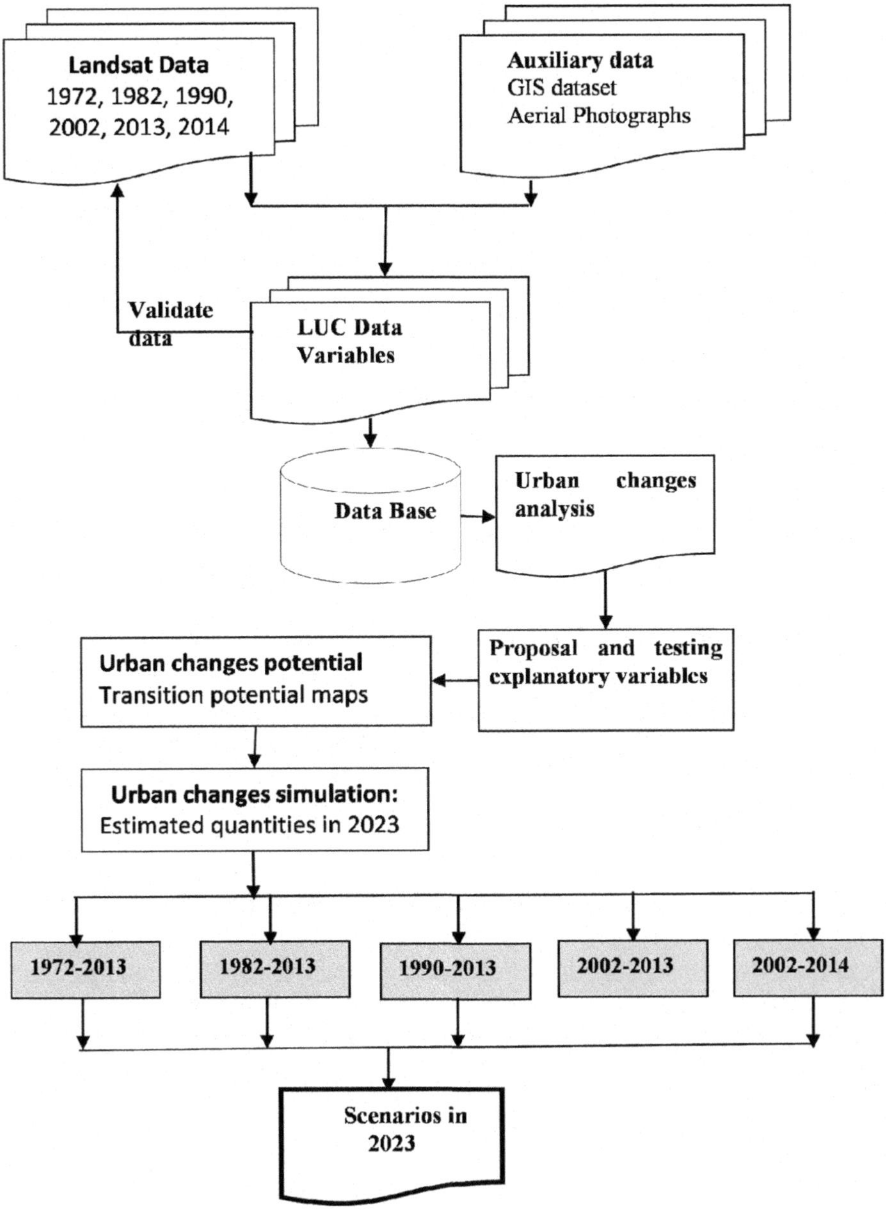

Fig. 2 Methodology flow chart of land-change analysis, potential and simulation

the border between the Gaza Strip and Israel. This is a restricted area which people are forbidden to enter. The fifth driver is the buildings destroyed during the 2014 War. It is only used in the 2002–2014 scenario.

The quantitative measure of the influence of the variables can be obtained using Cramer's V. A high Cramer's V value indicates that the variable has good explanatory potential, but does not guarantee a strong performance since it cannot take into account the mathematical requirements of the modeling approach or the complexity of the relationship. However, a very low Cramer's V value is a good indication that a variable can be discarded (Eastman 2012). The Cramer's V values for these drivers for all the calibrated periods were Elevation 0.142, Roads 0.169, Distance to built-up areas 0.707 and Border buffer zone 0.230.

We noticed that the Cramer's V values were similar for all the calibrated periods using the same latest land cover map (2013). Cramer's V values for the calibrated period (2002–2014) were also very similar, and its fifth key driver of "buildings destroyed during the war" obtained a value of 0.0294.

Even though the "buildings destroyed during the war" variable could be discarded because it has a very low Cramer's V value of 0.0294, we decided to include it because it is a key driver for reconstruction.

The results from the categories revealed that the distance to built-up areas and away from the border buffer zone were the main drivers for all predictions.

3.2 *Methodology for Modeling and Scenario Development*

In order to project the urban area in 2023, we selected a GIS model in Idrisi Terrset software called the Land Change Model (LCM). This model is used to analyze land cover change, empirically modeling its relationship to explanatory variables and projecting future changes (Eastman 2012).

3.2.1 Land Change Potential: Transition Potential Maps

To predict the change, each land use transition must be modeled empirically on maps called transition potential maps. These maps are used together with driver maps. A collection of factors are obtained from these drivers by the Natural Log Transformation. The Natural Log Transformation is effective in linearizing distance decay variables (e.g., proximity to roads) (Eastman 2012).

The transition potential maps are in essence potential maps for each transition in LCM. A collection of transition potential maps is organized within an empirically evaluated transition sub-model that has the same underlying driver variables. A transition sub-model can consist of a single land cover transition or a group of transitions that are thought to have the same underlying driver variables. These driver variables are used to model the historical change process. The transition potential maps are obtained by Multilayer Perceptron (MLP) in LCM. The MLP option can run multiple transitions and undertakes the classification of remotely-sensed imagery through the artificial neural network multi-layer perceptron technique. It uses an algorithm to set the number of hidden layer nodes.

MLP automatically evaluates and weights each factor and implicitly takes into account the cor-relations between the explanatory maps (Eastman 2012).

3.2.2 Land Change Simulation: The Estimated Quantities

The Markov transition area matrix is based on land-use changes without drivers that are produced within the Markov chain from two different dates. This matrix results from the multiplication of each column in the transition probability matrix by the number of cells for the corresponding land use in the last image for the year 2013 or 2014.

Markov chain analysis is used to calculate the estimated quantities in 2023 within the urban data for all the scenarios (1972–2013), (1982–2013), (1990–2013), (2002–2013) and (2002–2014) up to 2023.

The MARKOV module computes the transition areas matrix and the transition probability matrix by cross tabulation between LUC categories from two maps (t0 to t1), which reflect data from the calibration stage, to project the estimated changes and persistence at the simulation stage (t1 to T). The estimation to T is based on the number of time periods between t0 and t1 and the number of time periods between t1 and T, respecting in any case the same time units. A more detailed description of the MARKOV matrix can be found in the Idrisi Terrset Help System and also in Mas et al. (2014). The Markov chain analysis is one of the most widely used stochastic approaches in ecological and environmental modeling (Paegelow and Camacho 2008).

Linear regression was used to compare the results of the Markov chain data, i.e. an approach which uses the historical relationship between a dependent variable and one or more independent variables (the year and the population) to predict the future values of the dependent variable, in this case urban areas. The multiple linear statistical regression was used for simulation of the built-up area using the Enter method to enter all variables for the year 2023 at the same time on the basis of the urban area in 1972, 1982, 1990, 2002, 2013 and 2014. The growth rate and other statistics were calculated using Microsoft Excel.

3.2.3 Land Change Simulation: The Scenario

The five scenarios are simulated in a single model to predict the likely urban area in 2023. The LCM model uses an a priori identical Multi-Objective Land Allocation (MOLA) to solve the concurrences between different uses or transitions, in which the MOLA works only once. This process is based on the choice of the most suitable pixels, i.e., those with the greatest change potential in the change potential maps (ranked from high to low). Through the Markov matrix, the MOLA creates a list of host classes (categories that will lose area, in rows) and claimant classes (categories that will gain area, in columns) for each host. The land allocation process is conducted for all the claimant classes in each host class. In this way it solves the conflicts based on a minimum-distance-to-ideal-point rule using the weighted ranks, and the final result is the overlay of each host class reallocation (Eastman et al. 1995; Mas et al. 2014).

4 Results

4.1 Land Change Analysis of Chronological Series of LUC Maps

The results showed a drastic change in land cover and the growth of the urban area between 1972 and 2014, as shown in Fig. 3, when many agricultural areas were urbanized. This has happened in a largely unplanned, somewhat chaotic fashion, so revealing the need for land-use managers and city planners to understand future growth and plan further developments. Over this period urban areas have grown continuously, whereas non-urban (agricultural) areas have shrunk at similar rates, as shown in Table 3.

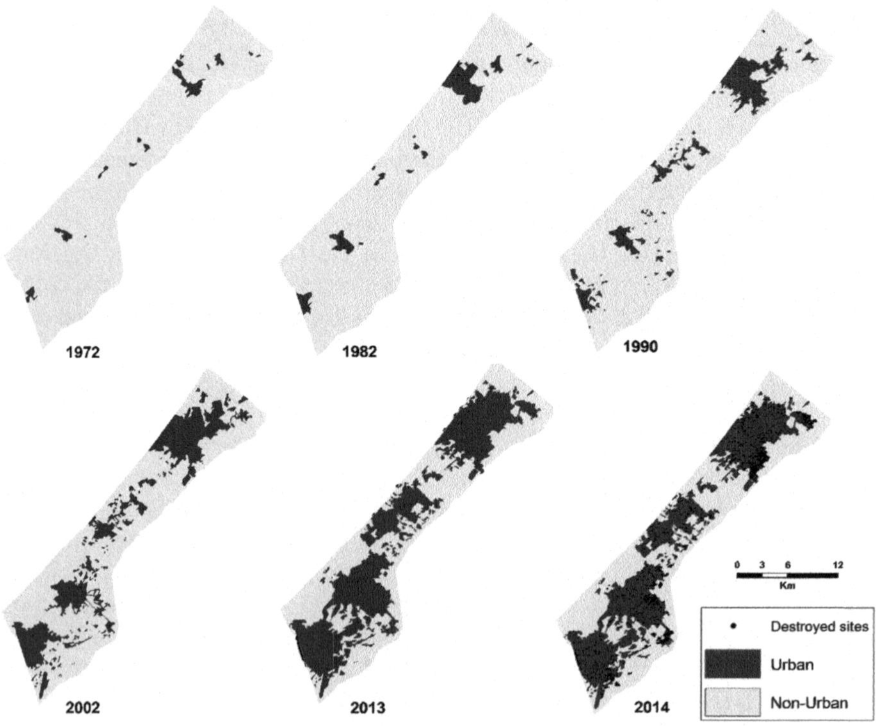

Fig. 3 Urban areas from 1972 to 2014

Table 3 Urban and non-urban areas from 1972 to 2014

Year	Urban, km^2	Urban, %	Non-urban km^2	Non-urban, %
1972	10.94	3.00	349.06	96.96
1982	25.29	7.00	334.71	92.98
1990	46.88	12.80	313.12	86.98
2002	100.23	27.40	259.77	72.16
2013	166.29	46.20	193.71	53.81
2014	164.80	45.78	195.20	54.22

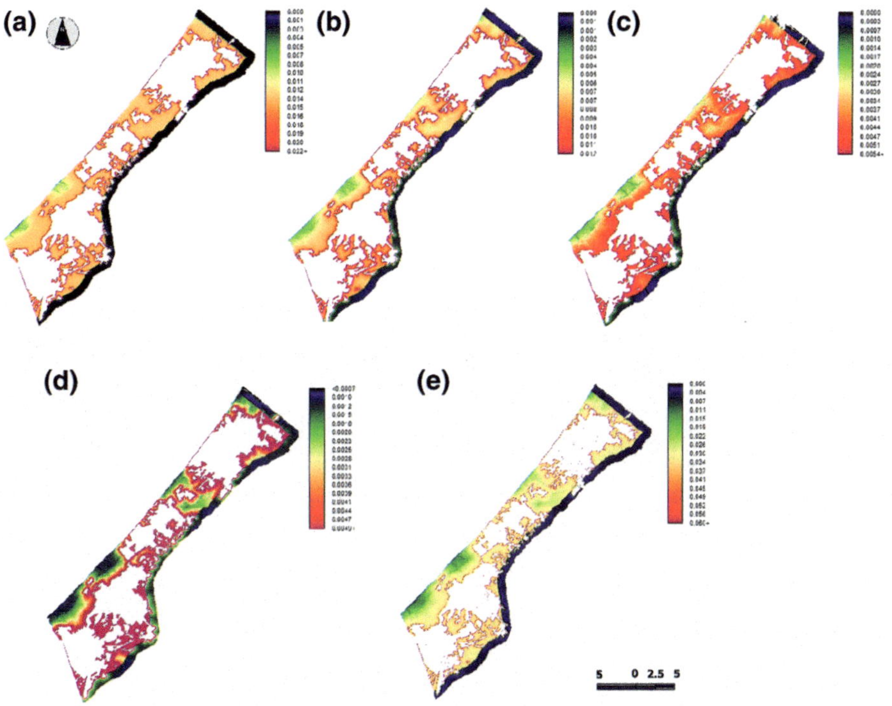

Fig. 4 MLP transition potential map from non-urban to urban area for LCM for **a** (1972–2013), **b** (1982–2013), **c** (1990–2013), **d** (2002–2013), **e** (2002–2014)

4.2 Transition Potential Maps

The MLP Neural network was used to obtain the transition potential map for the transition from Non-Urban to Urban area, as shown in Fig. 4a to Fig. 4e, based on the real transition over the various calibration periods (1972, 1982, 1990, and 2002) to 2013, and (2002) to 2014. The high transition potential values are located around the built-up area with the biggest population density (low distance). Figure 4 shows the transition potential maps for the five scenarios.

4.3 The Estimated Quantities

The Markov transition area matrix (Fig. 5) shows areas (km^2) in which a transition between two classes will have taken place by 2023. Rows represent land use in the calibration period in 2013 or 2014 and columns represent land use in the simulation year 2023, based on the five scenarios.

A multiple regression analysis shows the historical relationship between the urban area, the year and the population (independent variables) to project the future of the urban area for the year 2023, i.e. 240.79 km^2, using the Enter method. The results of the stepwise method show that population growth has had a direct effect on urban expansion. The significant number is around zero and the urban area is 246.5 km^2.

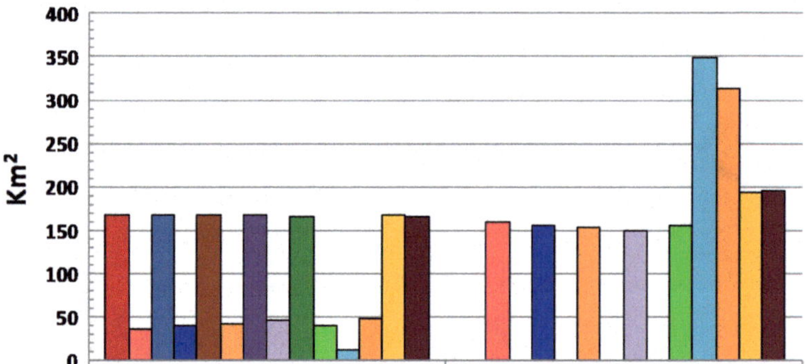

	Urban	Non-Urban
Urban 1972-2013	166.29	0
Non-Urban 1972-2013	36.09	158.56
Urban 1982-2013	166.29	0
Non-Urban 1982-2013	38.61	156.03
Urban 1990-2013	166.29	0
Non-Urban 1990-2013	40.67	153.97
Urban 2002-2013	166.29	0
Non-Urban 2002-2013	46.06	148.59
Urban 2002-2014	165.55	0
Non-Urban 2002-2014	39.18	155.86
1972	10.94	349.06
1990	46.88	313.12
2013	166.29	193.71
2014	164.80	195.20

Fig. 5 Transition area matrix for the estimation of urban areas in the five scenarios by the year 2023, and 1972, 1982, 1990, 2002, 2013 and 2014 in area (km^2)

The adjusted R2 is therefore 0.98, meaning that predicted values statistically demonstrate a high 'goodness of fit'. The stepwise regression equation can be expressed as follows:

$$Y = -1,016.667 + 0.501\mathbf{X_1} + 9.985 * 10^{-5}\mathbf{X_2}$$

The diagram in Fig. 6 shows a comparison between six past trend scenarios. The first five scenarios are the Markov chains from (1972–2013), (1982–2013), (1990–2013), (2002–2013) and (2002–2014) to 2023; and the sixth one is the regression line to 2023 depending on the basic data using the Enter method, which gave areas of 202.35, 204.89, 206.95, 212.32, 204.70 and 240.79 km^2, respectively.

The diagram in Fig. 6 shows a comparison between six past trend scenarios. The first five scenarios are the Markov chains from (1972–2013), (1982–2013), (1990–2013), (2002–2013) and (2002–2014) to 2023; and the sixth one is the regression line to 2023 depending on the basic data using the Enter method, which gave areas of 202.35, 204.89, 206.95, 212.32, 204.70 and 240.79 km^2, respectively.

4.4 *Simulation Maps: Scenario to 2023*

The results of the simulation in the five scenarios varied according to the Markov chains, i.e. (1972–2013), (1982–2013), (1990–2013), (2002–2013) and (2002–2014) were 202.35, 204.89, 206.95, 212.32 and 204.70 km^2, respectively, as shown in Fig. 7.

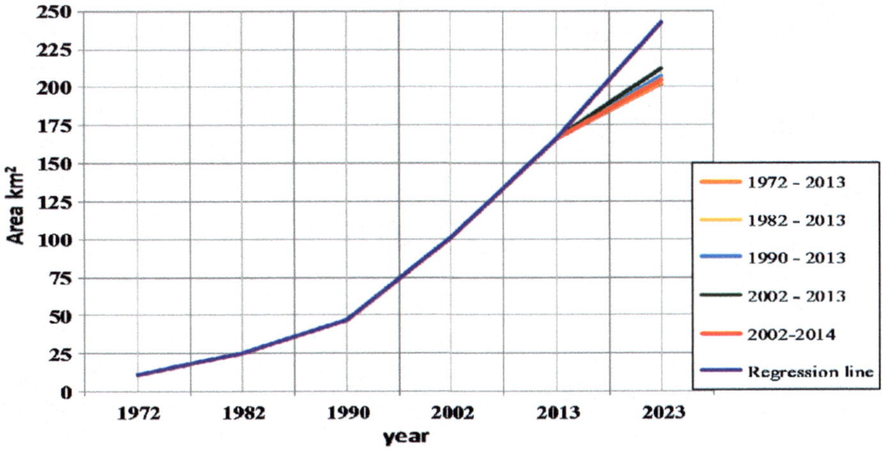

Fig. 6 The plot lines for urban area in km^2 from 1972 to 2023, Markov chain (1972–2013), Markov chain (1982–2013), Markov chain (1990–2013), Markov chain (2002–2013), Markov chain (2002–2014), and regression analysis

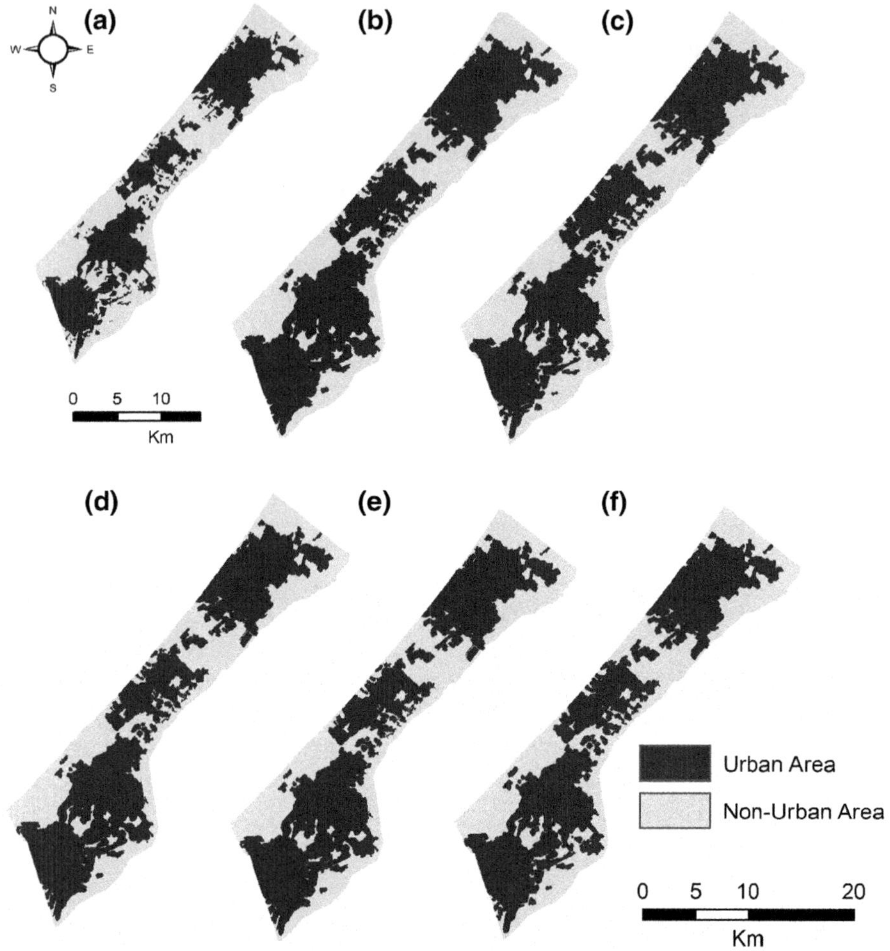

Fig. 7 **a** Real map 2013 and simulated maps for the year 2023 as results of **b** Scenario (1972–2013), **c** Scenario (1982–2013), and **d** Scenario (1990–2013), **e** Scenario (2002–2013), **f** Scenario (2002–2014)

The results of the simulation (individual results for each scenario and the average of all five) are presented in Fig. 8 and Table 4, which show an increase in urban areas and a decrease in non-urban areas between 1972 and 2014. The predicted urban and non-urban areas for 2023 for the periods (1972–2013), (1982–2013), (1990–2013), (2002–2013), and (2002–2014 after the war) are presented in Table 4.

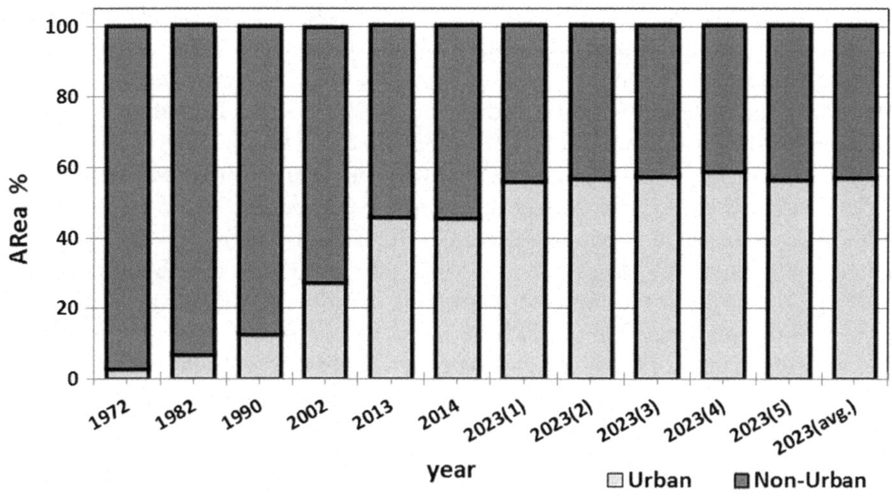

Fig. 8 Increase in urban areas and decrease in non-urban areas from 1972 to 2023

Table 4 Area and percentage of urban and non-urban areas in all scenarios

Scenario	Year	Urban, km^2	Urban, %	Non-urban, km^2	Non-urban, %
1972–2013	2023 (1)	202.37	56.21	157.63	43.79
1982–2013	2023 (2)	204.89	56.91	155.11	43.09
1990–2013	2023 (3)	206.95	57.49	153.05	42.51
2002–2013	2023 (4)	212.34	58.98	147.66	41.02
2002–2014	2023 (5)	204.70	56.86	155.30	43.14
Average	2023 (avg.)	206.25	57.29	153.75	42.71

5 Discussion

The overall results of the five LCM scenarios analyze and simulate land-use changes in the Gaza Strip. The results of the past trend scenarios for spatial distribution per area in 2023 presented both differences and similarities in the allocation of urban area. We discovered an inverse relationship between the predicted area by 2023 and the length of the calibration period, in that the longer the calibration period the smaller the growth in urban area predicted. The urban areas for 2023 predicted by the calibration periods (1972–2013), (1982–2013), (1990–2013) and (2002–2013) were 202.35, 204.89, 206.95 and 212.32 km^2. The calibration period (2002–2014), which showed an increase in urban area to 204.7 km^2 by 2023, is slightly exceptional due to the fact that it includes the 2014 War.

The results for calibration periods 2002–2013 and 2002–2014 have a high "goodness of fit", because they both obtained values close to the regression analysis value (240.79) used to measure statistical best fit values, while the values for the other scenarios were substantially further away from the regression analysis value.

As a percentage of the total area of the Gaza Strip, the scenarios predict that between 56.21 and 58.98% will be urbanized by 2023. The data analysis shows an increase in the urban area from 10.9 (1972) to 25.3 (1982), 46.9 (1990), 100.2 (2002), 166.3 (2013) and 206.24 km^2 in 2023, the average area predicted by the various simulations for the whole Gaza Strip (i.e. around 57.13% of the total). While the decrease in Agricultural areas (Non-Urban Area) was caused by an increase in population growth rate and a lack of management and future planning.

Figure 9 illustrates the increase in the rate of growth in urban area as a percentage of the total area of the Gaza Strip for each time period (1972–1982), (1983–1990), (1991–2002), (2003–2013), 2014 and (2015–2023), with rates of 0.40, 0.7584, 1.35, 1.83, −0.39 and 1.44% from 1972 to 2023, which implies a positive relationship with the rate of population growth.

The population density for the whole of the Gaza Strip will therefore have in-creased from 4,661.5 inhabitants per km^2 in 2013 to 6,704.3 inhabitants per km^2 in 2023. However, in the urban areas the increase will be from 10,231.1 inhabitants per km^2 in 2013 to 11,865.1 inhabitants per km^2 in 2023. Table 5 shows that urban expansion is positively correlated with population growth in the Gaza Strip, which already has one of the highest population densities in the world.

The Palestinian economy in the Gaza strip grew in line with the Israeli economy over the period from 1972 to 2000. There was a dramatic rise in the Palestinian standard of living from 1972 until the eruption of the first Intifada (uprising) in 1987. The main reason for improved living standards was the opening of the rapidly expanding Israeli job market to Palestinian workers (Swirski 2008). The situation continued until the signing of the 1993 Oslo Accords. From 1994 to 2000 there were huge urban projects and a great deal of investment leading to urban expansion (Abuelaish and Camacho 2016).

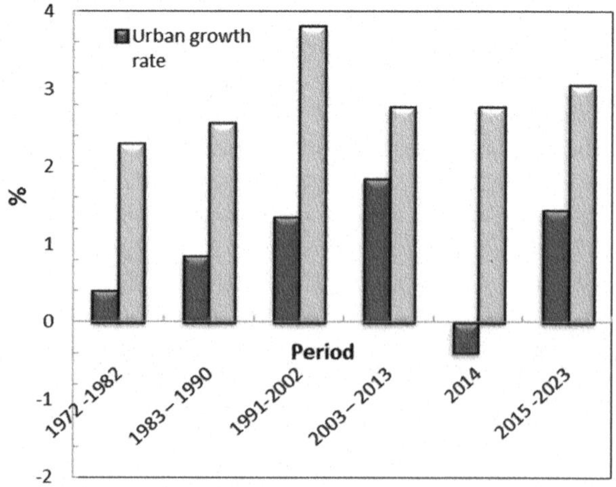

Fig. 9 Population and urban growth rates for the different periods, from 1972 to 2023

Table 5 Increase in urban area, population number and population density from 1972 to 2023

Year	Population No.	Area (km^2)	% Area	Population Density No./area	Actual Pop. Density
1972	393,800	10.9	3	1,078.9	36,128.4
1982	511,115	25.3	7	1,400.3	20,202.2
1990	642,814	46.9	12.8	1,761.1	13,706.1
2002	1,182,908	100.2	27.4	3,240.8	11,805.5
2013	1,701,437	166.3	46.2	4661.5	10231.1
2014	1,760,037	164.80	45.78	4822.0	10679.84
2023	2,447,054	206.24	57.13	6704.3	11865.1

Many of the Palestinian workers in Israel, considered the mainstay of the Palestinian economy have been unemployed since the conflicts in 2000. In 2007 an economic blockade was started around the Gaza Strip, which for a short period prevented urban expansion from continuing at the same rate as before. From 1972 to 1994 urbanization was more vertical than horizontal, a situation that was reversed thereafter (Abuelaish and Camacho 2016).

Effects of the 2104 War

According to the Ministry of Public Works and Housing (MOPWH), an estimated 2,000 tons of cement for residential construction purposes enter the Gaza Strip daily. This would give a monthly figure of around 44,000 tons. Ground floors require about 0.54 tons of cement per m2, while all other floors require 0.21 ton/m^2; the average area of buildings in the Gaza Strip is 150 m^2. Of the 11,000 housing units destroyed during the war, 5,990 had ground floors, and there were 5,189 on other floors. Hence, 648,643.5 tons of cement would be required for reconstructing all the destroyed housing units. At the current supply rate of 44,000 tons a month, reconstruction would therefore take 15 months. This is an ideal scenario in which allowing cement to enter the country and receiving funds to rebuild the residential housing units are essential factors.

According to the temporary agreement for the Gaza Reconstruction Mechanism (GRM) in Shelter Cluster Palestine (February, 2016), 1,107,519 tons of cement have entered the Gaza Strip since October 2014. Around 44% has been used for residential purposes, i.e. 487,308.36 tons. This is enough cement to rebuild 6,016 ground-floor apartments or 15,470 apartments above ground-floor level.

Table 6 shows the completed housing units, those in progress, funded, and awaiting funds from donors as of February 2016, according to Shelter Cluster Palestine. Around 83% of the destroyed housing units are still awaiting funds from donors, which means that a significant amount of the cement must be being used to build new housing units in different places. These construction materials are not only being used to reconstruct destroyed buildings but also to cover normal urban growth and supply building companies everywhere. The black market is playing a major role in selling cement outside the GRM as a result of shortages in the system. This has allowed people to build new houses without a license. Since the war, people prefer to buy housing units in the center of urban areas and to live in

Table 6 Repairs and reconstruction of the housing units damaged and destroyed during the 2014 War

	# Units*	Completed	In progress	Funded	Gap
Totally destroyed	11,000	937	591	3,479	5,993
Severe damage	6,800	2,034	3,027	1,097	642
Major damage	5,700	119	1,075	1,747	2,759
Minor damage	147,500	69,428	9,936	10,060	58,076

Source Shelter Cluster Palestine (February, 2016)

apartment buildings. This is because town centers are considered safer and new building on urban land with planning permission tends to be very expensive.

According to the 2007 Census, there were 241,873 housing units in the Gaza Strip, and according to the projected number of households and housing units in the Gaza Strip using the hypothesis of average number of households per year (PCBS 2009), about 15,529 housing units were required in 2015 and 16,284 in 2016. The amount of cement that entered the Gaza Strip in the previous period was enough for normal urban growth but there were problems for reconstructing the destroyed buildings due to a shortage of donor finance. Around 10–15% of the destroyed housing units were reconstructed in the 15 months between September 2014 and February 2016. International donors at the Cairo Donor Conference on 12th October 2014 pledged over USD 5.4 billion to support the plan to rebuild the Gaza Strip.

Reconstruction of the Gaza Strip is a priority for the Palestinian Authority, and all the destroyed buildings during war 2014 are entitled to financial support. The reconstruction efforts however depend on financial support from the donors, and also on Israel allowing construction materials to enter the Gaza Strip. Since the shortage of building materials due to the blockade affects both reconstruction efforts and natural urban growth, if the blockade ended, the Gaza Strip would return to its previous natural urban growth rate. The fact is however that there is no guarantee that Israel will not repeat past behavior and launch new wars on the Gaza Strip, and there is no expectation of the economic blockade coming to an end soon.

This study tries to answer questions about the future of the Gaza Strip and the impacts on its environment. This information is useful for decision-makers and politicians, who are regularly faced with questions about the complicated situation in the Gaza Strip as a result of its weak economic resources, and the lack of donor support from countries concerned by other conflicts such as in Syria. Most of the houses destroyed during the war belong to poor people who are waiting for financial support to rebuild their houses. Many urban areas were destroyed during the war and their reconstruction would be harder without modeling exercises such as the one presented here.

6 Conclusions

This paper presents, analyses, evaluates, and simulates urban expansion for the years 1972 to 2014 to 2023, using the historical free Landsat data and the free UNITAR/UNOSAT Geodatabase for the areas attacked during the 2014 war. These simulations are based on the continuity of observed past trends and are not exact predictions. Instead they are plausible scenarios of a future state assuming the maintenance of macro-political and social conditions.

The following conclusions were drawn from the results of this research in which we performed a simulation of urban growth in the Gaza Strip for the year 2023 using five scenarios and the Land Change Modeler:

- Around 57.13% of the Gaza Strip will be urban land.
- Around 10–15% of the buildings and infrastructure damaged during the war had been rebuilt and returned to their previous state of natural growth by February 2016.
- The amount of building materials entering the Gaza Strip must be increased and additional support must be given to the people who lost their houses during the 2014 War.
- There is an inverse relationship between the predicted urban area for 2023 and the length of the calibration period.
- Urbanization in the Gaza Strip is increasing dramatically because of natural population growth. This is placing more stress on agricultural areas, causing soil erosion and impairing water quality and quantity.
- Urban planners should take into account that in the near future the three main urban areas will merge into one.
- Urban sprawl increases over time at the expense of agricultural land, above all due to an increase in population.
- The reduction in agricultural land in the Gaza Strip and the pressure placed on natural resources will contribute to local and global climate change.

Acknowledgements This work was supported by the BIA2013-43462-P project funded by the Spanish Ministry of Economy and Competitiveness and by the FEDER European Regional Fund.

References

Abuelaish B, Camacho Olmedo MT (2016) Scenario of land use and land cover change in the Gaza Strip using Remote Sensing and GIS models. J: Arab J Geosci. http://dx.doi.org/10.1007/s12517-015-2292-7

Alcamo J, Kok K, Busch G, Priess J (2008) Searching for the future of land: Scenarios from the local to global scale. In: Alcamo J (ed) Environmental futures: the practice of environmental scenario analysis. Elsevier, Amsterdam, The Netherlands

Curran PJ (1985) Principles of remote sensing. In: Remote sensing today. longman scientific & technical, London

Duh JD, Shandas V, Chang H, George LA (2006) Rates of urbanisation and the resiliency of air and water quality. Sci Total Environ 400(1–3):238–256

Eastman JR (2012) Idrisi Selva Tutorial. Clark University, Worcester, MA, USA

Eastman JR, Jin W, Kyem W, Toledano P (1995) Raster procedures for multi-criteria? Multi-objective decisions. Photogramm Eng Remote Sens 61(5):539–547

Ennab WR (1994) Population and demographic developments in the West Bank and Gaza Strip until 1990. UNCTAD, Geneva

Goodchild MF (2005) GIS and modeling overview. In: Maguire DJ, Batty M, Goodchild MF (eds) GIS, Spatial analysis, and modeling. Redlands, CA: ESRI Press, pp 1–18

Haas J, Ban Y (2014) Urban growth and environmental impacts in Jing-Jin-Ji, the Yangtze River Delta and the Pearl River Delta. Int J Appl Earth Obs Geoinf 30:42–55

Hualou L, Guoping T, Xiubin L, Heilig GK (2007) Socio-economic driving forces of land-use change in Kunshan, the Yangtze River Delta economic area of China. J Environ Manag 83:351–364

Janssen LLF (1993) Methodology for updating terrain object data from remote sensing data: the application of landsat TM Data with respect to agricultural fields. Dissertation Wa-geningen Agricultural University

Mas JF, Kolb M, Paegelow M, Camacho Olmedo MT, Houet T (2014) Inductive pattern-based land use/ cover change models: a comparison of four software packages. Environ Model Softw 51(2014):94–111. doi:10.1016/j.envsoft.2013.09.010

Nakicenovic N, Swart R (eds) (2000) IPCC special report on emission scenarios. Cambridge University Press, Cambridge, UK

Paegelow M, Camacho Olmedo MT (2008) Modelling environmental dynamics. Advances in geomatic simulations. Series environmental science. Springer, Heidelberg. ISBN 978-3-540-68489-3

Paegelow M, Camacho Olmedo MT, Mas JF, Houet T, Pontius RG Jr. (2013) Land Change Modelling: moving beyond projections. Int J Geogr Inf Sci 27(9):1691–1695. Taylor & Francis doi:10.1080/13658816.2013.819104

Palestinian Central Bureau of Statistics (PCBS) (2009) Housing Projections up to 2017

Palestinian Central Bureau of Statistics (PCBS) (2012) Census final results in The Palestinian Territory Summary (Population, Housing and Establishments Census-2007)

Palestinian Central Bureau of Statistics (PCBS) (2015) Local Communities Survey

Palestinian Ministry of Agriculture (MOAg) (2013) Unpublished material and statistical data

Peterson GD, Cumming GS, Carpenter SR (2003) Scenario planning: a tool for Cconservation in an uncertain world. Conserv Biol 17:358–366

Raskin P, Gallopin G, Gutman P, Hammond A, Swart R (1998) Bending the curve: toward global sustainability. a report to the global scenario group. PoleStar Series Report No. 8. Stockholm Environment Institute, Stockholm, Sweden

Ren W, Zhong Y, Meligrana J, Anderson B, Watt WE, Chen J, Leung HL (2003) Urbanization, land use, and water quality in Shanghai: 1947–1996. Environ Int 29(5):649–659

Shelter Cluster Palestine (2016) Palestine report, February 2016. www.ShelterCluster.org Accessed February October 2016

Sleeter BM, Sohl TL, Bouchard M, Reker R, Sleeter RR, Sayler KL (2012) Scenarios of land use and land cover change in the conterminous United States -Utilizing the Special Report on Emission Scenarios at ecoregional scales: global Environmental Change at http://dx.doi.org/10.1016/j.gloenvcha.2012.03.008

Swart RJ, Raskin P, Robinson J (2004) The problem of the future: sustainability Science and scenario analysis. Glob Environ Change 12:137–146

Swirski S (2008) The burden of occupation. The cost of the occupation to israeli society, polity and economy. Adva Center, Tel Aviv

Taubenbock H, Esch T, Felbier A, Wiesner M, Roth A, Dech S (2012) Monitoring urbanization in mega cities from space. Remote Sens Environ 117:162–176

UNEP (2009) Environmental Assessment of the Gaza Strip following the escalation of hostilities in December 2008–January 2009 United Nations Environment Programme
UNEP (2006) Environmental Assessment of the Areas Disengaged by Israel in the Gaza Strip. United Nations Environment Programme
UNITAR/UNOSAT (2014) Impact of the 2014 Conflict in the Gaza Strip, UNOSAT Satellite derived geospatial analysis

Chapter 14
Constraint Cellular Automata for Urban Development Simulation: An Application to the Strasbourg-Kehl Cross-Border Area

J.P. Antoni, V. Judge, G. Vuidel and O. Klein

Abstract Urban sprawl and space consumption have become key issues in sustainable territorial development. Traditional planning approaches are often insufficient to anticipate their complex spatial consequences, especially in cross-border areas. Such complexity requires the use of dynamic spatial simulations and the development of adapted tools like LucSim, a CA-based tool offering solutions for sharing spatial data and simulations among scientists, technicians and stakeholders. Methodologically, this tool allows us to simulate future land use change by first quantifying and then locating the changes. Quantification is based on Markov chains and location on transition rules. The proposed approach is implemented on the Strasbourg-Kehl cross-border area and calibrated with three contrasting prospective scenarios to try to predict cross-border territorial development.

Keywords Cellular automata · Markov chains · Cross-border area · Land use scenarios · Prospective

J.P. Antoni (✉) · V. Judge · G. Vuidel
Laboratoire the MA UMR 6049 CNRS, University of Burgundy,
Franche-Comté, Dijon, France
e-mail: jean-philippe.antoni@u-bourgogne.fr

V. Judge
e-mail: valentine.judge@univ-fcomte.fr

G. Vuidel
e-mail: gilles.vuidel@univ-fcomte.fr

V. Judge · O. Klein
LISER Luxembourg Institute of Socio-Economic Research,
Esch-sur-Alzette, Luxembourg
e-mail: olivier.klein@liser.lu

M.T. Camacho Olmedo et al. (eds.), *Geomatic Approaches for Modeling Land Change Scenarios*, Lecture Notes in Geoinformation and Cartography,
https://doi.org/10.1007/978-3-319-60801-3_14

1 Context and Research Objectives

In the current context of increasing urbanization and daily mobility, urban sprawl and space consumption have become crucial issues for achieving sustainable territorial development (European Environment Agency 2006). This problem is further complicated in the case of cross-border areas where operational procedures on each side of the frontier differ from an administrative, legal and cultural point of view (Stoklosa and Besier 2014). Moreover, open border areas are currently undergoing particular growth dynamics which have given rise to numerous cross-border spatial planning issues (Coplan 2012; Kaiser 2012; Kolossov 2012). In this context, the Strasbourg-Ortenau Eurodistrict Project (French-German cross-border territory) is promoting the development of cross-border initiatives in what is a pilot scheme for the EU. This project is currently supported by local political actions (Antoni 2009) and is widely backed by the European Union. Within this pilot region, we will be focusing specifically on the Strasbourg-Kehl cross-border Area (SKA). SKA is located on the banks of the upper Rhine and covers parts of South-West Germany and North-East France. The area is physically composed of a large plain that is symmetrically organized and delimited by the Vosges and the Black Forest mountains (*graben*). The River Rhine is not only a major fluvial axis running through the middle of the region, but also forms the border between France and Germany, which are linked by bridges with high levels of traffic (Durr and Kayali 2014). Despite its geomorphological consistency, SKA has two different geographic configurations (Fig. 2). The French side is currently highly urbanized around the agglomeration of Strasbourg, while the German part remains predominantly rural. Despite this, people on both sides of the border suffer similar residential housing issues such as urban sprawl, air pollution and congestion. SKA is an interesting case to study for three main reasons that make it quite unique. Firstly, because there is no strong cross-border differential like that between France and Luxembourg or between France and Switzerland (job market, taxes). Secondly, because there is a genuine political will to create a Eurodistrict (defined by the UE as a cross-border administrative and planning institution) and finally because residential mobility from Strasbourg to Kehl and from Germany to France (northern part of the case study) is becoming more and more important.

Nevertheless, despite the cooperation at local and European level, cross-border planning issues remain difficult to manage because many different disciplines (e.g. urban planning, transport, housing, labour market, industrial and commercial investment etc.) and stakeholders are involved. Moreover, trans-national territorial analyses are constrained by the problem of geographical information and data harmonization (i.e. scale, temporality, accuracy of data). Classical planning approaches and methods are therefore often incapable of addressing the complexity of these situations and predicting their spatial implications. This means that spatial planning must look for more collaborative solutions that integrate dynamic and complex spatial analyses in a prospective way. Any strategy to implement sustainability and planning with the available regulatory tools requires planners to imagine the future layout of their territory. Predictions of this kind are however very difficult to make and numerous experiments have shown that a simple trend projection often provides poor spatial extrapolations, disconnected from territorial realities. In this context, spatial simulations are widely viewed as an appropriate tool to help planner stake decisions. Such simulations rely on several kinds of simulations models, among which Cellular Automata (CA) are particularly well designed for managing spatial planning issues.

CA are considered useful tools for modeling and simulating urban development because they allow us to implement simple spatial rules based on empirical knowledge that take into account the role of neighborhood in urban growth processes. They have been widely used to simulate land use changes and scenarios for future urban development in different contexts. The seminal work of Couclelis (1985; 1987), White and Engelen (1993), Batty and Xie (1994) and later Clarke et al. (1997) paved the way for CA to be considered a powerful tool for modeling and simulating spatial phenomena of various types. The research on CA gathered new momentum during the 2000s in a surge in research that coincided with a second wave of faster and cheaper computational capacities (Torrens 2000; Benenson and Torrens 2004; Couclelis 2005; Koomen et al. 2011).

The aim of this paper is to present prospective urban development scenarios for the Strasbourg-Kehl area in the medium term. The methodology (argued in Sect. 2.2) was used to select the year 2038 as a suitable target date for these predictions. This provides a sufficiently long period of time for prospective anticipation and decision making in the field of land planning and regulation policies. Simulations are provided by LucSim (*Land Use Change Simulation*), an open-source operational CA dedicated to geographical analysis and simulations (Antoni 2006). This CA has been developed *from scratch* to offer comprehensive and user-friendly cartographic and mathematical solutions, but also to harmonize and share spatial data and simulations among scientists, territorial and administrative technicians, elected representatives and stakeholders. We use it to construct and simulate cross-border scenarios showing how residential growth in border areas can be planned and controlled by means of comprehensive rules and regulations. We begin by presenting the main assumptions of the CA model, based both on the Markov chains process and the creation of transition rules (Sect. 2), before going on to calibrate three contrasted scenarios for predicting future urban changes (Sect. 3). Results are then presented and discussed in the Conclusion (Sect. 4).

2 Methodology

From a methodological point of view, LucSim can be defined as a constrained cellular automata designed to aid decision-making in urban and land planning. Its main original feature (compared to similar geographical CA) is to simplify the land-use evolution processes into two "fundamental" steps, namely the quantification and location of future land use changes. Land use is assessed within a cellular grid space obtained from the European Corine Land Cover classification.

2.1 Data and Material

As the Strasbourg-Kehl case study takes place on a cross-border field, it is essential to use harmonized data. Indeed, to avoid any mismatch problem between data from France and data from Germany, all aspects of the objects being studied must be defined in exactly the same way on both sides of the border at temporal (collection date), spatial (spatial accuracy and resolution) and thematic levels (the different land use categories). The best way to tackle this issue is to use data created at a higher level within the framework of international cooperation. Corine Land Cover (CLC) is a database designed to that effect. It is a European biophysical land occupation database provided by the European Environment Agency at several dates. With a resolution of 100 m, the database classifies land use into 44 items or categories (Fig. 1) and is used above all to analyze land use change and measure the artificialization of land. For the research presented in this chapter, we reduced the land use classification to 8 main categories, focusing mainly on artificial occupation of land for human activities for two dates: 1990 and 2006 (Fig. 2).

In the cellular space obtained from CLC, each date corresponds to a system defined by N cells in a grid. Cells are associated to one, and only one, land use category. The specific land use of any given cell N_i at time t is referred to as k and the land use of any given cell N_i at time $t + 1$ is called l.

Reclassification (9 Items)			**CLC Classification (44 items)**
Name	*Color*	*Code*	
Urban Fabric		**UR**	Continuous urban fabric; Discontinuous urban fabric
Industrial		**IN**	Industrial or commercial units; Mineral extraction sites; Dump sites; Construction sites
Transport		**TR**	Road and rail networks and associated land; Port areas; Airports
Equipment		**EQ**	Green urban areas; Sport and leisure facilities
Agricultural		**AG**	Non-irrigated arable land; Permanently irrigated land; Rice fields; Pastures; Annual crops associated with permanent crops; Complex cultivation patterns; Land principally occupied by agriculture; Agro-forestry areas
Vine		**VI**	Vineyards; Fruit trees and berry plantations; Olive groves
Forest		**FO**	Broad-leaved forest; Coniferous forest; Mixed forest; Natural grasslands; Moors and heathland; Sclerophyllous vegetation; Transitional woodland-shrub
Water		**WA**	Inland marshes; Peat bogs; Salt marshes; Salines; Intertidal flats; Water courses; Water bodies; Coastal lagoons; Estuaries; Sea and ocean
Natural areas (N/A in SKA)			Beaches, dunes, sands; Bare rocks; Sparsely vegetated areas; Burnt areas; Glaciers and perpetual snow

Fig. 1 Corine land cover reclassification in 8 classes

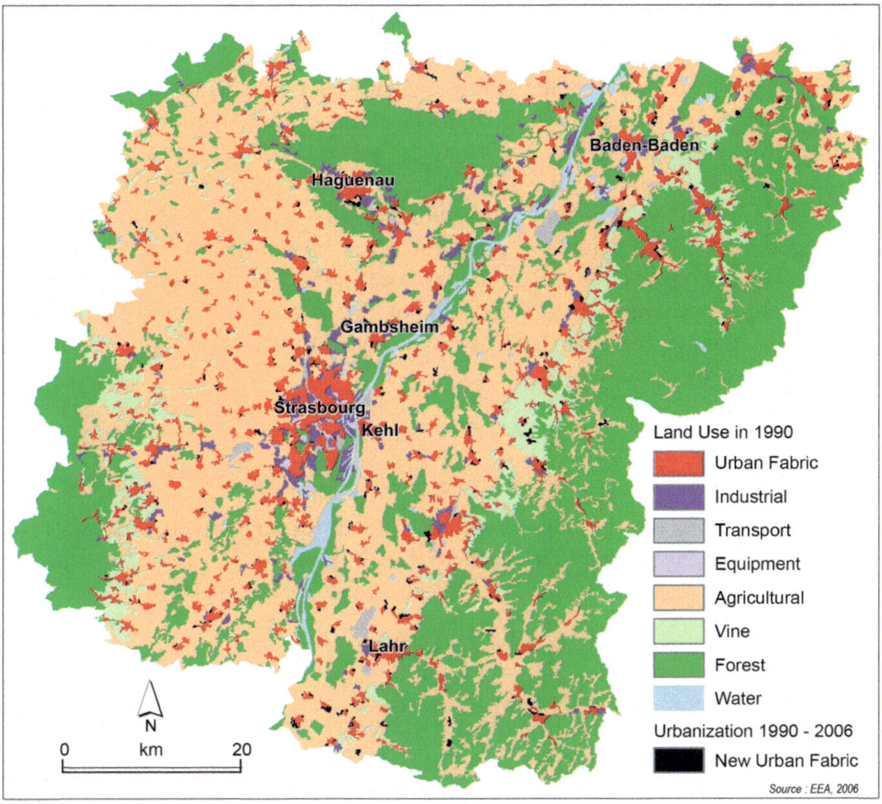

Fig. 2 Land use in the Strasbourg-Kehl area in 1990 and 2006

Table 1 Past land-use vectors

	UR	IN	TR	EQ	FI	VI	FO	WA	$\sum$
1990 (cells)	42,143	10,163	2,628	1,747	242,397	24,410	20,232	6,850	532,659
1990 (%)	8.59	2.07	0.54	0.36	49.42	4.98	41.25	1.40	100
2006 (cells)	44,612	11,977	2,634	2,078	238,826	23,556	202,237	6,739	532,659
2006 (%)	9.14	2.45	0.54	0.43	48.94	4.83	41.44	1.38	100

In the SKA, the quantitative analysis of $N_{i,k}$ and $N_{i,l}$ (1990 and 2006) shows that urbanized cells (UR) expanded by 5.9% between 1990–2006, while natural and agricultural soils (FI) decreased by 1.5%. Land use cover can be summarized more precisely for each date within vectors indicating the proportion of each land use category (Table 1).

2.2 *Quantification of Land Use Changes*

Our first step was to quantify the land use change process. Comparing two static land use images or vectors (1990, 2006) is of little interest in the context of a dynamic simulation, but finding out what happens between each image can enable us to formulate a transition process. By comparing the land use categories date by date and cell by cell, it is possible to determine cellular changes between t and $t + 1$, and identify the land use dynamics. Theoretically, each cell can move from one land use category to another, or remain in the same category. The dynamics of the model can therefore be presented as a series of possible transitions from one land use category k at time t to another land use category l at $t + 1$. For a given cell N_i, a transition Δ can be written as:

$$\Delta \mathrm{N}_{\mathrm{i,kl}} = 1 \text{ if } N_{\mathrm{i,k}}(t) = 1 \text{ and } N_{\mathrm{i,l}}(\mathrm{t} + 1) = 1$$

To simplify the complexity resulting from the high number of cells and possible transitions, changes can be aggregated by land cover categories. The aggregate transition for the complete system is then:

$$\Delta \mathrm{N}_{\mathrm{kl}} = \sum_{\mathrm{i}=1}^{n} \Delta \mathrm{N}_{\mathrm{i,kl}}$$

This formulation allows us to build a contingency matrix indicating the number of cell transitions from a category k to a category l between t and $t + 1$ (i.e. between 1990 and 2006). When associated with the previous vectors, this matrix provides all the elements needed for the construction of a Markov chain (MC). In the literature, a Markov chain is defined as a mathematical process where transition probabilities are conditional on the past, and express the state of a variable at a time t as a function of observations of this variable at $t - 1$ (Feller 1968, Berchtold 1998). It relies on the connection of three items: (i) the description of the relative values associated to an initial state (land occupation visualized as a vector for example); (ii) a transition matrix expressing the transition probabilities of different groups of

Table 2 1990–2006 transition matrix

	UR	IN	TR	EQ	FI	VI	FO	WA	Σ
UR	**0.9893**	0.0063	0.0001		0.0038	0.0003		0.0001	1
IN	0.0093	**0.9539**	0.0006	0.0076	0.0120		0.0100	0.0066	1
TR		0.0030	**0.9844**	0.0118	0.0008				1
EQ	0.0074	0.0137		**0.9788**					1
FI	0.0104	0.0069	0.0001	0.0011	**0.9794**	0.0005	0.0010	0.0006	1
VI	0.0106	0.0016			0.0274	**0.9593**	0.0011		1
FO	0.0001	0.0013			0.0018	0.0001	**0.9962**	0.0005	1
WA		0.0028	0.0001		0.0159		0.0441	**0.9371**	1

observations from one category to another; and (iii) a diachronic transformation by an operator in the form of a matrix multiplication iteration.

If we follow this procedure, land use at time $t + 1$ can be simulated by multiplying the corresponding vector at time t by the corresponding contingency matrix, after the transformation of the latter into transition probabilities from a land use category k to another l. To transform observed contingencies into transition probabilities, we use the following:

$$p_{\mathrm{kl}}(t) = \frac{\Delta \mathrm{N}_{\mathrm{kl}}}{N_k(t)} \text{ and } \sum_{\mathrm{k}=1}^{m} p_{\mathrm{kl}}(t) = 1$$

We then consider the Markov chain as follows:

$$N_i(\mathrm{t} + 1) = \sum_{\mathrm{k}=1}^{m} p_{\mathrm{kl}} . N_k(t)$$

$$\text{where } p_{\mathrm{kl}} = \frac{\Delta \mathrm{N}_{\mathrm{kl}}}{N_k(t)} = \frac{\Delta \mathrm{N}_{\mathrm{kl}}}{\sum_l \Delta \mathrm{N}_{\mathrm{kl}}} \text{ and } \sum_l p_{\mathrm{kl}} = 1$$

According to this formulation, the Markov chain process gives us the chance to prospectively calculate future states from known past states, based on observation of past trends and probabilities. According to the method, this calculation is based on the assumption that future changes will follow the trend of past changes, but as it is based on a matrix calculation, this trend is not necessarily linear. Moreover, the values of the transition matrix can also be modified by users of the model to integrate different parameters for the quantification of future land use changes. In our case, LucSim uses the original transition matrix to calculate the number of cells in each land use category in 2022, 2038, 2054, etc., from 1990 and 2006 land uses (same interval of 16 years between each date). This system gives us a better picture of urban dynamics by calculating land use vectors for each future date, as presented in Table 3.

This table also indicates that the total number of cells that should be urbanized (including UR, IN and EQ categories) by 2038 is:

Table 3 Expected future land-use vectors

	UR	IN	TR	EQ	FI	VI	FO	WA	Σ
2022 (cells)	47,027	13,700	2,640	2,411	235,334	22,735	202,162	6,644	532,653
2022 (%)	9.68	2.82	0.54	0.50	48.46	4.68	41.63	1.37	100
2038 (cells)	49,391	15,339	2,648	2,748	231,921	21,948	202,096	6,566	532,657
2038 (%)	10.22	3.17	0.55	0.57	47.99	4.54	41.82	1.36	100

$$
\begin{array}{c}
[N_{\mathrm{k=UR}}(\mathrm{t}+1)+\mathrm{N}_{\mathrm{k=IN}}(\mathrm{t}+1)+\mathrm{N}_{\mathrm{k=EQ}}(\mathrm{t}+1)] \\
-[N_{\mathrm{k=UR}}(t)+\mathrm{N}_{\mathrm{k=IN}}(t)+\mathrm{N}_{\mathrm{k=EQ}}(t)] \\
8,811
\end{array}
$$

2.3 *Location of Land Use Changes*

The second step was to try to identify the location of land use changes with a method based on Cellular Automata. Developed as a result of the progress of artificial intelligence in computer science, Cellular Automata have the double advantage of being able to determine the land use category of cells according to their neighborhood, and also to integrate the previous Markovian process. By definition, CA are based on the assumption that the class of each cell is determined by its neighborhood, or in our case, by the land use categories of surrounding cells within a given radius:

$$\forall i \in \mathrm{E}, \mathrm{V}_{\mathrm{i,kl}} = \mathrm{f}\left(V_{\mathrm{i,k}}(t), \Omega_i(t)\right)$$

where

$$\Omega_i = \mathrm{f}\left(V^r_{\mathrm{k=1}}, \mathrm{V}^r_{\mathrm{k=2}}, \ldots, \mathrm{V}^r_{\mathrm{k=n}}\right) \text{ and } r \in \{0, \ldots, \infty\}$$

where E is a set of cells that can undergo a transition (non locked), V_i is the land use of the cell i, Ω_i is the neighborhood of the cell i within a radius r (at time t), and C^r_n is the number of cells with a land use S within a radius r at time t.

CA can then be constrained with the results of the Markov chain to produce a model for land use change simulations. This means that the CA transition process from one given category to another is automatically halted when the number of cells given by the MC for each date is reached. This CA transition process is based on transition rules that allow us to consider different configurations. The main problem is then to define relevant rules to simulate realistic scenarios of spatial development, a generalized problem in all modeling and especially in model calibration.

3 Spatial Development Scenarios

After analyzing past transitions (Table 2), we decided to base all our scenarios on the general assumption that new built-up areas can only be developed on agricultural fields (FI). These scenarios present three contrasted configurations for land use changes in 2038: urban sprawl, urban densification and cross-border development based on the bridge connections available on the SKA specific test-field. Although results are calculated at the original 100 meters resolution of the land use cells, they are aggregated and mapped within a larger grid with a resolution of 4,000 meters to improve visualization of the changes.

3.1 Landscape Sprawl

The main idea of the "Landscape Sprawl" (LS) scenario is that future residential preferences will favor natural landscapes and rural amenities, as well as relative proximity to slightly dense urban areas (villages). This means that residential development of new built-up areas is determined by the following transition rules:

- The proportion of UR in a radius of 200 meters must be over 30%.
- The proportion of FI in a radius of 500 meters must be over 50%.
- There must be at least 1 VI cell in a radius of 5 km.
- There must be at least 1 FO cell in a radius of 5 km.
- The total number of new built-up cells is less than 8,811.

The LS scenario (Fig. 3) leads to a gain of 8,976 cells in only 2 CA iterations. This result can be explained by considering spatial configurations that are very generic and numerous in the case of the rules created above. LucSim therefore quickly spots the cells that meet the requirements to be transformed into urban land. A typical example of this process of urbanization can be seen between the "Piémont" area and the high density urban area of Strasbourg. We can also observe a generalized expansion of areas with low urban density (max 200) and a high dispersion of the cells that become urbanized. Nevertheless, this general dispersion is quite homogenous except for a slight concentration around small cities. The urban expansion on the German side appears to be more linear than in France, which is probably due to the topographic features in that area.

Fig. 3 "Landscape sprawl" scenario: land use changes simulation in 2038

3.2 *Urban Densification*

The main idea of the "Urban Densification" (UD) scenario is that future residential preferences will favor dense urban areas, close to urban amenities (e.g. parks, sport and leisure facilities), but relatively far away from industry and related nuisances. Consequently, residential development of new built-up areas will be determined by the following transition rules:

- The proportion of UR in a radius of 200 meters must be over 30%.
- There must be at least 1 EQ cell in a radius of 2 km.
- There must be no IN cells in a radius of 1 km.
- There must be at least 1 IN cell in a radius of 2 km.
- The total number of new built-up cells is less than 8,811.

The UD scenario (Fig. 4) produces a gain of 9,391 cells in 9 iterations. A much higher number of iterations is needed because the rules for this configuration make the transition less likely to happen. Moreover the Markov constraint can only be achieved when newly urbanized cells are taken into account. This explains why the process is slower and more iterations are required to converge toward the solution provided by the set of rules for the UD scenario. In this case new urbanization is concentrated around the bigger cities and expands on the existing urban structure rather than following the area's physical geography features. The fact that the existing urban area is already much larger on the French side favors further urbanization on this side. The urban density is clearly higher than in the LS scenario (max 408).

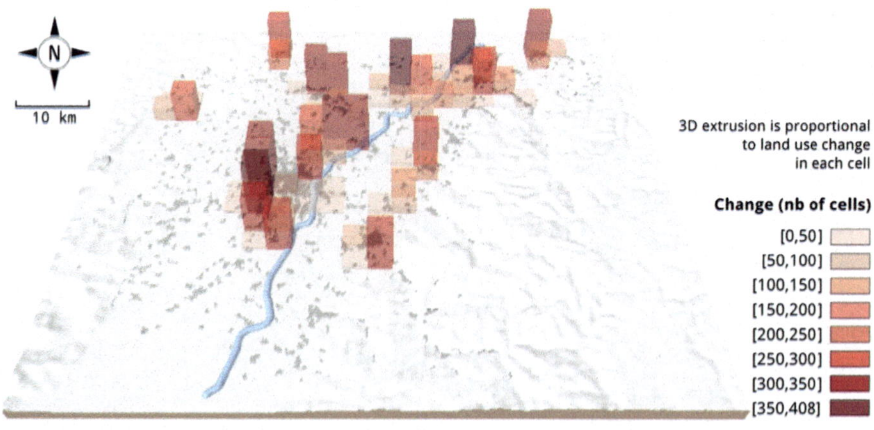

Fig. 4 "Urban densification" scenario: land use changes simulation in 2038

3.3 Bridge Transbordering

The main idea of the "Bridge Transbordering" (BT) scenario is that future residential preferences will favor mixed residential areas (with both LS and UD scenarios), located in quite heavily urbanized areas near the border crossing points. Consequently, residential development of new built-up areas is determined by the following transition rules:

- The proportion of UR in a radius of 200 meters must be over 30%.
- The proportion of FI in a radius of 500 meters must be over 25%.
- There must be at least 1 IN or 1 EQ cell respectively in a radius of 1.5 and 1 km.
- There must be at least one bridge in a radius of 7.5 km.
- The total number of new built-up cells is less than 8,811.

The BT scenario (Fig. 5) leads to gains of 8,852 cells in 10 iterations, roughly the same number as the UD scenario. As in the previous scenario, few spatial configurations are adapted to the transition towards urban land use categories. This situation leads to urban development being highly concentrated in certain places in the study area (max 450), most of which are close to the River Rhine and its crossing points (bridges, ferry). New high density urban development is also predicted around the big cities. Urban development will be essentially linear and more intensive on the French side (especially around the southern part of Strasbourg city, and close to the Gambsheim dam). The three places most affected in the German part are: Lahr, Kehl and around Baden-Baden.

Fig. 5 'Bridge Transbordering' scenarios: simulation of land use changes in 2038

4 Discussion

The three residential development scenarios presented above in succinct form were developed on the basis of expert judgment. Rather than attempting to justify these expert opinions, our aim is to use CA to highlight the compatibility of the language used by experts, decision makers and modelers. To this end the scenarios are expressed verbally and in the form of simple rules that are easy to implement in cellular automata. From a thematic point of view, we have also shown that the scenarios are initially very contrasting and that the resulting CA rules naturally lead to very different configurations in terms of land use changes. However, the results produced by the CA also show some similarities. For example some areas are urbanized whatever the scenario. This convergence clearly shows the areas where the main challenges for future urbanization will lie. It also demonstrates the utility of the tool when taking planning decisions and when debating future regulation policies.

From a scientific point of view, our results have not been validated. Forecasting the future in a complex context is difficult and in the absence of a crystal ball, there is no known technique for validating future urban development results at such a fine scale. Nevertheless the various scenarios involve realistic processes and rules based on accurate expert knowledge to provide images of the future that can be used in debate and decision-making about desirable urban development and land-use changes. The method presents a CA-based tool that, according to its structure and data-feeding, can be widely used on both sides of the border by institutions that aim to merge at some time in the future to form a Euro district. In this context, the objective of the model is not to separate France from Germany by offering independent analyses or forecasts for each one, but to reflect on scenarios for their common future development.

Another way to construct prospective scenarios and define CA rules could involve using a Decision Tree (Judge et al. 2015) or Artificial Neural Networks (Basse et al. 2014). Artificial intelligence helps to automatically determine transition rules based on the analysis of past processes (e.g. 1990–2000–2016). However, such artificial intelligence based solutions only produce a continuation of past trends. In a move to more sustainable forms of development, this should not necessarily be exclusive and other trends and directions that mix these approaches should be included in the simulations. However, forecasting the future in a complex context remains difficult, even if supported by geographic cellular automata models, or indeed any other intelligent methodology for planning and decision making. Anyway, even if they can provide convincing answers that anticipate land use changes, CA remain totally silent on mobility issues (Timmermans 2003). CA can however be combined with residential mobility models, daily mobility or traffic models to simulate the flows generated by land use changes. In this context, a possible extension of this work could involve coupling different models together to propose a more complex LUTI (Land Use and Transport Integrated) model (Wegener and Fürst 1999), in which a CA approach can make an important contribution.

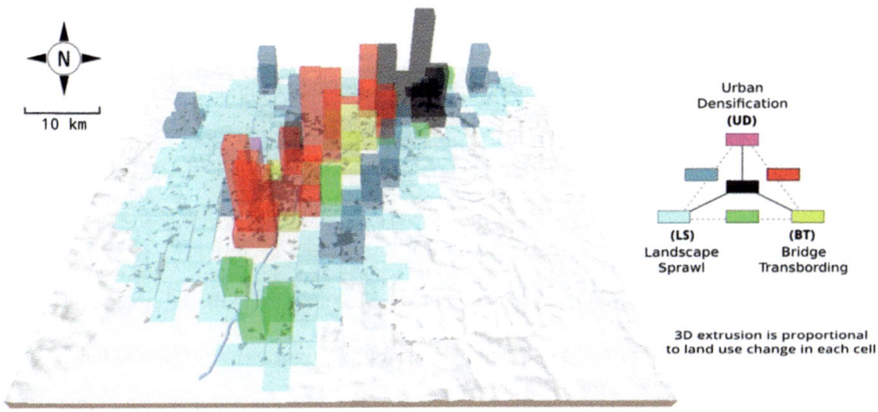

Fig. 6 Land use changes in 2038: combination of the results of the simulated scenarios

5 Conclusion

If we combine the results of the three simulations, areas with differing potential for urbanization emerge. Two specific areas (in black on Fig. 6) systematically appear in all the scenarios; this suggests that these areas have a particularly complex spatial configuration in that they are near the border, close to cities and suburban. Development is therefore likely in these areas irrespective of the preferences associated with each scenario. Some other areas result from the combinations of two of the three scenarios: (i) in green, the "border sprawl", namely a suburbanization along the border but outside the main urban centers; (ii) in red, a densification around the border crossing points in the north of the study area and around Strasbourg; (iii) and finally in blue, a dispersed suburban area away from both large cities and the border area.

By comparing these different scenarios, we can see that this model can assess the impact of single neighborhood rules on urban development. This global modeling enables us to study urban changes easily and efficiently. Breaking down the process into two steps (MC+CA) makes it sufficiently straightforward to be simultaneously understood by all the stakeholders involved in urban planning. LucSim therefore allows a wide range of different points of view to be considered and specific actions to be imagined for territorial development and innovation, within the perspective of more sustainable land and urban planning.

Acknowledgements The research presented in this chapter is part of the Smart. Boundary project supported by the *Fonds National de la Recherche* in Luxembourg and *CNRS* in France (ref. INTER/CNRS/12/02). The authors would like also to thank the Grasp Program of LISER for allowing cross-collaboration between the two teams based in Luxembourg and France.

References

Antoni JP (2006) Calibrer un modèled'évolution de l'occupation du sol urbain. L'exemple de Belfort.Cybergeo : Eur J Geogr. http://cybergeo.revues.org/2436

Antoni JP (2009) Un territoire de projet à co-construire. In: Grossouvre (de) H, Maulin E (eds) Euro-District Strasbourg-Ortenau : la construction de l'Europeréelle. Xénia, pp 25–31

Basse RM, Omrani H, Charif O et al (2014) Land use changes modelling using advanced methods: Cellular automata and artificial neural networks. The spatial and explicit representation of land cover dynamics at the cross-border region scale. Appl Geogr 53:160–171

Batty M, Xie Y (1994) From Cells to Cities. Environ Plan 21:531–538

Benenson I, Torrens PM (2004) Geosimulation: automata-based modeling of urban phenomena, 1st edition. Wiley

Berchtold A (1998) Chaînes de Markov etmodèles de transition. Hermes Science Publications, Paris, Applications aux sciences sociales

Clarke KC, Hoppen S, Gaydos L (1997) A self-modifying cellular automaton model of historical urbanization in the San Francisco Bay Area. Environ Plan 24:247–261

Coplan DB (2012) Border Show Business and Performing States. In: Wilson TM, Donnan H (eds) A companion to border studies. Wiley-Blackwell, pp 507–521

Couclelis H (1985) Cellular Worlds: a framework for modeling micro-macro dynamics. Environ Plann A 20:99–109

Couclelis H (1987) Cellular dynamics: How individual decisions lead to global urban change. Eur J Oper Res 30:344–346

Couclelis H (2005) "Where has the Future Gone?" Rethinking the role of integrated land-use models in spatial planning. Environ Plann A 37:1353–1371

Durr MJ, Kayali ML (2014) Six millions d'habitants pour l'Alsace, Chiffres pour l'Alsace, 50

European Environment Agency (2006) Urban sprawl in Europe: the ignored challenge. EEA Report

Feller W (1968) An introduction to probability theory and its applications, 3rd edn. Wiley

Judge V, Antoni J-P, Klein O (2015) Land use simulation: statistical analysis approaches to calibrate cellular automata

Kaiser RJ (2012) Performativity and the Eventfulness of Bordering Practices. In: Wilson TM, Donnan H (eds) A companion to border studies. Wiley-Blackwell, pp 522–537

Kolossov V (2012) Euroborderscapes: State of the Debate, Report for the 7th Framework Programme of the European Commission

Koomen E, Hilferink M, Borsboom-Van Beurden J (2011) Introducing land use scanner. In: Koomen E, Borsboom-Van Beurden J (eds) Land-use modelling in planning practice. Springer, pp 3–21

Stoklosa K, Besier G (2014) European border regions in comparison: overcoming nationalistic aspects or re-nationalization?, 1st edn. Routledge, New York, London

Timmermans HJP (2003) The Saga of integrated land use-transport modeling: how many more dreams before we wake up? In: 10th International conference on travel behaviour research, Lucerne, Switzerland, p 35

Torrens PM (2000) How cellular models of urban systems work (1. theory). CASA Working Paper Series

Wegener M, Fürst F (1999) Land use transport interaction: state of the art. Transland, Irpud

White R, Engelen G (1993) Cellular automata and fractal urban form: a cellular modelling approach to the evolution of urban land-use patterns. Environ Plann A 25:1175–1199

Chapter 15
Modeling Land-Use Scenarios in Protected Areas of an Urban Region in Spain

M. Gallardo and J. Martínez-Vega

Abstract Land use change due to human activity can have serious, often irreversible effects on the environment. It affects ecosystem functions and the sustainability of protected natural areas. Problems such as fragmentation, low habitat connectivity or a decline in a territory's ability to capture carbon are some of its consequences. By studying past land use trends we can simulate future land uses, and modeling such trends is essential if a preventive approach to the management of protected areas is to be adopted. The aim of this chapter is to simulate different change scenarios in protected natural areas in the urban region of Madrid, from National and Nature Parks to Special Areas of Conservation and Special Protection Areas. To this end we study land use changes both inside and around these protected areas. CORINE Land Cover maps from 1990, 2000 and 2006 are used. Cross-tabulation techniques are applied in order to study trends in land use change. Three scenarios are designed: a baseline or trend scenario, an economic crisis scenario and a green scenario. The CLUE model (based on logistic regression) is used. LCM (based on neural networks) is also used but only in the trend scenario. Biophysical, socio-economic and accessibility factors and incentives and restrictions are considered. FRAGSTATS and GUIDOS are used to analyse the effect of infrastructure and built-up land growth on connectivity and fragmentation. In recent decades, the region of Madrid has experienced intense urban and infrastructure development (48,332 ha). Protected areas have been affected by this urbanization process. Built-up areas have grown at an average annual rate of 5.52% in protected areas and around them. According to the trend scenario, the built-up area will increase by 28,000 ha over the period 2006–2025 to 7.6% of the study area. No fragmentation processes are expected in the National Park. However, fragmentation of agricultural and natural habitats around protected areas is expected to increase by

---

M. Gallardo (✉)
Department of Geography, University of Murcia, Murcia, Spain
e-mail: marta.gallardo@um.es

J. Martínez-Vega
Institute of Economics, Geography and Demography, Spanish National Research Council, Madrid, Spain
e-mail: javier.martinez@cchs.csic.es

M.T. Camacho Olmedo et al. (eds.), *Geomatic Approaches for Modeling Land Change Scenarios*, Lecture Notes in Geoinformation and Cartography,
https://doi.org/10.1007/978-3-319-60801-3_15

7.2% during this period. These findings should alert land use planners and the managers of protected areas to the potential threats.

Keywords Simulations · Land use scenarios · Protected areas · Region of Madrid · Spain

1 Introduction

According to the World Database on Protected Areas, from 2003 to 2014 the number of protected sites increased from 84,577 to 217,294. In 1990, protected areas covered 8.6% of the land area. Since 2012, these areas have grown by 1.6 million km^2 as a result of new declarations. Today, they occupy 15.4% of the land area and of continental and inland waters, 3.4% of the global ocean area, 8.4% of marine areas covered by national jurisdictions and 10.9% of coastal waters (Juffe-Bignoli et al. 2014). In order to reach Aichi Target 11 (Strategic Plan for Biodiversity 2020), the Convention on Biological Diversity recommends that by 2020 at least 17% of terrestrial and inland water surfaces and 10% of coastal and marine areas be protected. In Europe, protected areas occupy 13.6% of the land mass and of continental waters (Deguignet et al. 2014). In Spain, from 1990 to 2013 the number of protected natural areas multiplied by 7 and their surface area tripled (EUROPARC-España 2014). In the world and European contexts, Spain has an important role to play in the conservation of biological diversity. Today, over 27% of the surface occupied by terrestrial ecosystems are protected by national, European or worldwide networks. Within the EU, Spain is the largest contributor to the Natura 2000 network.

In spite of their importance, Protected Areas (PAs) are increasingly under threat from factors such as climate change (Ruiz-Mallén et al. 2015), land use changes (Martínez-Fernández et al. 2015), deforestation (FRA 2010), forest fires (Chuvieco et al. 2013), habitat fragmentation (Dantas de Paula et al. 2015), loss of biodiversity (Sastre et al. 2002), propagation of invasive species (Lei et al. 2014), urban pressure (McDonald 2013) and public use (López Lambas and Ricci 2014).

Land-use change is a matter of concern for the scientific community. Spatio-temporal analysis can be used for a number of purposes (Lambin et al. 2001; Moreira et al. 2001; Améztegui et al. 2010; Viedma et al. 2015): (1) to observe land use changes in the past and explore the factors explaining them, (2) to simulate possible environmental and socio-economic impacts, and (3) to assess the influence of political alternatives in order to improve planning.

A vast number of studies and projects related to Land Use and Cover Change (LUCC) have been carried out. Of importance at a global level is the Land-Use and Land-Cover Change Science/Research Plan (Turner et al. 1995), a core project of the International Geosphere-Biosphere Programme (IGBP) and the International Human Dimension Programme on Global Environmental Change (IHDP). In Europe, one of the most interesting programmes is CORINE Land Cover, CLC

(Feranec et al. 2007). The results of these projects and studies can help managers take decisions and enable the objectives of the aforementioned strategic plan to be achieved.

However, little is known about LUCC trends at different protection levels. Recent studies have focused on analysing changes in protected areas of differing importance and in the unprotected areas around them (Sastre et al. 2002; Romero-Calcerrada and Perry 2004; Ruiz Benito et al. 2010; Hewitt and Escobar 2011; Martínez-Fernández et al. 2015; Martinuzzi et al. 2015). It is important to simulate future land-use scenarios so that a dual approach can be adopted in preventive planning for protected areas and their surroundings (Martinuzzi et al. 2015). Such scenarios are important firstly for predicting threats associated with increased built-up land and the risk of forest fires stemming from growth in the Wildland-Urban Interface. They may also be a source of opportunities. New naturalised areas resulting from the abandonment of agricultural land might be included in buffer zones or ecological corridors that would improve connectivity among PA cores.

In short, although the perceptions of scientists and manager differ (Rodríguez-Rodríguez and Martínez-Vega 2016), LUCC would seem to be a basic component for evaluating the efficiency of PAs (Rodríguez-Rodríguez and Martínez-Vega 2012).

Our research is in line with previous approaches. The simulated scenarios and initial knowledge of their consequences for landscape structure could be a good starting-point for discussion and for reaching agreements between local communities and managers of protected areas.

The main goal of this research is to simulate land use in 2025 in PAs and their surrounding areas in the region of Madrid using two simulators, one based on logistic regression and the other on artificial neural networks. A secondary goal is to analyse the LUCC that took place between 1990 and 2006 and the changes expected by 2025 in order to determine trends and threats arising inside and around PAs. A third goal is to analyse the changes that have taken place or are expected in landscape structure and in a selection of landscape ecology indices.

2 Test Area and Data Sets

The Madrid region covers an area of 8,027 km^2 and in 2015 had a population of 6,436,996 inhabitants.[1] It is the most densely populated region in Spain with about 800 inhabitants/km^2. The region has a continental Mediterranean climate. Forest and semi-natural areas occupy about 48% of the total area, agricultural land 37%, built-up areas 14% and wetlands and water bodies 0.84%, according to CORINE Land Cover 2006 (CLC06).

[1]http://www.madrid.org/iestadis (last accessed March 3, 2016).

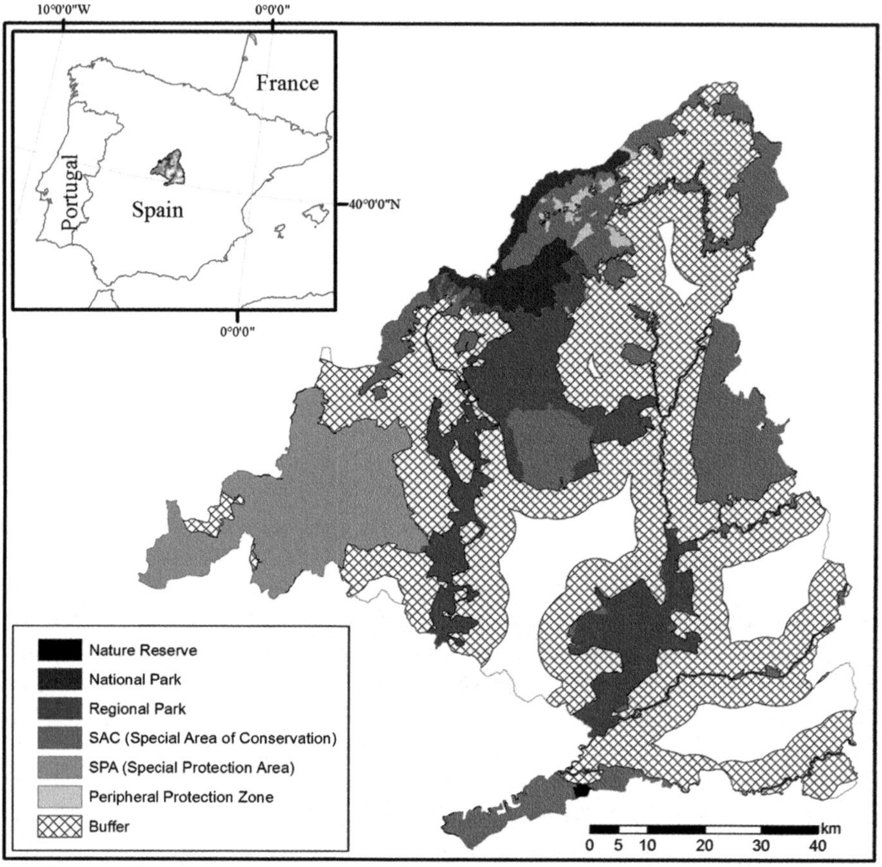

Fig. 1 Test area: Madrid region, Spain

In the region of Madrid, PAs occupy 329,164 ha, equivalent to 41% of the region's total surface area (Fig. 1). Table 1 shows their main characteristics, listing them in order of protection—from greatest to least. 15% of the Madrid Region is protected in SACs (Special Areas of Conservation), 12% in Regional Parks (RP), 10% in an SPA (Special Protection Area), about 3% belongs to a National Park (NP) and the remaining 1% is occupied by the Peripheral Protection Zone (PPZ) around this National Park and by a Nature Reserve (NR). All the PAs studied contain terrestrial ecosystems typical of the Mediterranean biogeographic region.

We also took into account a 5 km buffer zone around all the PAs in the region, which has no protection from the point of view of biodiversity. It occupies 372,865 km^2 equivalent to 46% of the region's area. Its aim is to mitigate threats to the PAs and as such it plays a strategic role in the conservation of biodiversity. About 13% of the region's land surface falls outside the scope of the study. Most of it is occupied by the city of Madrid and by other towns within the metropolitan area.

Table 1 Main characteristics of the protected areas considered in the study

Protected area	Designation year	Designation target	Main ecosystems
El Regajal-Mar de Ontígola Nature Reserve	1994	Fauna (Lepidoptera; birds), botanical	Scrub; semi-natural wetland (dam)
Sierra de Guadarrama National Park	2013	Ecological, geomorphology, landscape, scientific, cultural, education	Montane scrub and alpine grasslands; pine forests (*P. sylvestris*); deciduous forests (*Q. pyrenaica*); wetlands; peatlands
Cuenca Alta del Manzanares Regional Park	1985	Environmental, cultural, agricultural and landscape, ecologicalcorridor	Montane; deciduous forests (*Q. pyrenaica*); evergreen forests (*Q. rotundifolia, P. sylvestris*); pasturelands
Sureste Regional Park	1994	Ecological, palaeontological and archaeological	Unirrigated cropland; pine forests (*P. halepensis*); riparian forests; artificial wetlands; scrub
Curso medio del río Guadarrama Regional Park	1999	Natural and cultural, water ecosystems, landscape, ecological corridor, tourism	Evergreen forests (*Q. rotundifolia, P. pinea*); riparian forests; scrub; unirrigated cropland
Cuenca del río Lozoya y Sierra Norte SAC	1998/ 2014[a]	Ecological, habitats	Montane; deciduous forests (*Q. pyrenaica*); evergreen forests (*T. baccata*); scrub (*G. purgans*)
Cuenca del río Manzanares SAC	1998/ 2014[a]	Ecological, fauna, hábitats	Evergreen forests (*Q. ilex, Q. rotundifolia*); riparian forests (*Salix and Populus alba*); Sclerophillous grazed forests (dehesas); substeppes (*Thero-Brachypodietea*)
Cuenca del río Guadalix SAC	1998/ 2014[a]	Ecological, fauna, hábitats	Evergreen forests (*Q. ilex, Q. rotundifolia*); Arborescent matorral with *Juniperusspp*; riparian forests (*Salix and Populus alba*); Sclerophillous grazed forests (dehesas); substeppes
Cuencas de los ríos Jarama y Henares SAC	1998/ 2011[a]	Ecological, fauna, hábitats	Cereal steppes; riparian forests (*Salix and Populus alba*); Arborescentmatorral with *Juniperus spp.*
Vegas, Cuestas y Páramos del Sureste de Madrid SAC	1998/ 2014[a]	Ecological, fauna, hábitats	Wetlands; salt and gypsum inland steppes; riparian forests (*Salix and Populus alba*)

(continued)

Table 1 (continued)

Protected area	Designation year	Designation target	Main ecosystems
Encinares de los ríos Alberche y Cofio SPA	1990	Ecological, fauna, hábitats	Evergreen forests (*Q. ilex, Q. rotundifolia, P. pinea, P. pinaster*); Sclerophillous grazed forests (dehesas); scrub
Peripheral Protection Zone Guadarrama National Park	2013	Ecological	Montane; pine forests (*P. sylvestris*); pasturelands

[a]For the SACs, two dates are given in the "Designation year" field. The first refers to the year when the regional government proposed to the EU that the area be declared an SAC. This marked the beginning of their commitment to preventive protection in order to conserve the biodiversity of the area's habitats. The second date is the actual date of the declaration, after which the corresponding management plans were approved

We have selected two sets of geographical data. First, we downloaded all the updated perimeters and their corresponding attributes for the Nationally Designated Protected areas (NDP) in the Madrid region.[2] We also downloaded the Natura 2000 Network areas (Nn2000).[3]

This information comes from the Nature Data Bank of the Spanish Ministry for the Environment, the national contact point. In order to find the dates for final approval of the SACs, we linked the cartography with the Common Database on Designated Areas (CDDA) of the European Environment Agency.[4] We then downloaded land use/land cover maps from the CLC project for the years 1990, 2000 and 2006.[5] We did not consider the most recent map (CLC 2012) because it is still under review.

Finally, we took into account a collection of auxiliary geographic data in order to map the driving factors and the restrictive and incentive factors during design of future land use scenarios. We considered roads, rivers and railway stations (Numerical Cartographic Base 1:100,000, obtained from the Spanish National Geographical Institute) when drawing up accessibility maps that take into account the cost of transport and distances to the city of Madrid, to other cities, to the airport and to the roads themselves. We used a Digital Elevation Model (raster 30 m GMES RDA, EU-DEM) to generate altitude and slope maps, the lithological map

[2]http://www.magrama.gob.es/es/biodiversidad/servicios/banco-datos-naturaleza/informacion-disponible/cartografia_informacion_disp.aspx (last accessed March 21, 2016).

[3]http://www.magrama.gob.es/es/biodiversidad/servicios/banco-datos-naturaleza/informacion-disponible/red_natura_2000_inf_disp.aspx (last accessed March 21, 2016).

[4]http://www.eea.europa.eu/data-and-maps/data/natura-6#tab-european-data (last accessed March 21, 2016).

[5]http://centrodedescargas.cnig.es/CentroDescargas/buscadorCatalogo.do?codFamilia=02113 (last accessed March 21, 2016).

of Madrid, the map of public-utility forest areas (Regional Government of Madrid), PA zoning in the region (Autonomous Body for National Parks) and specific legislation on land and territorial planning (General Urban Land Plan for Madrid for 1997, Law 9/2001 of 17 July on land in the Region of Madrid, Law 9/1995 of 28 March on measures for territorial policy, land and planning, and Law 3/1991 of 7 March on roads in the Region of Madrid).

3 Methodology and Practical Application to the Data Sets

Our research follows the workflow shown in Fig. 2. We used ArcGIS v10.3 (ESRI Inc.) for vector processing of the downloaded data and to draw up the buffer. For LUCC analyses, we used IDRISI-Selva (Eastman 2012). We also used CLUE (Verburg and Overmars 2007) and the IDRISI-Selva Land Change Modeller (LCM) for designing the scenarios. Finally, we used GUIDOS-MSPA (Soille and Vogt 2009) to analyse the spatial landscape pattern, and FRAGSTATS 4.0 (McGarigal et al. 2002) to evaluate trends in landscape metrics depending on the type of PA and trends in their surrounding areas.

Firstly, we selected the PAs to be considered in the study. Areas where several types of protection overlap are classified as areas of greatest protection. In descending order, the level of priority is as follows: (1) Nature Reserve,

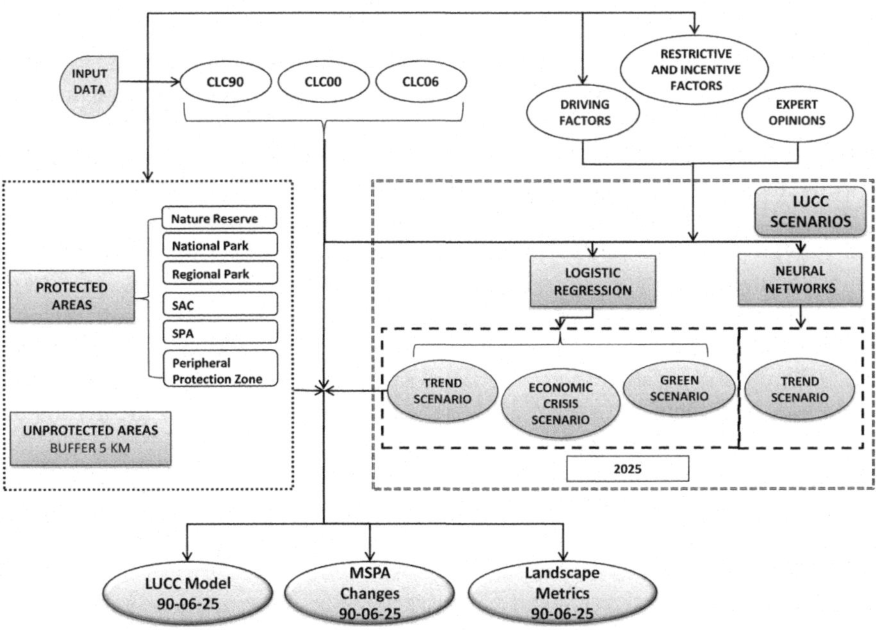

Fig. 2 Framework

(2) National Park, (3) Regional Park, (4) SAC, (5) SPA, (6) Peripheral Protection Zone in *Sierra de Guadarrama* National Park.

We then established an unprotected area around each PA, joining up areas that are adjacent to each other. From this buffer we excluded land that might be protected for other reasons (public-utility forest, public waters, roads, etc.). In line with the literature, we began using a 10-km buffer (Martínez-Fernández et al. 2015; Martinuzzi et al. 2015). However, in the end we opted for a 5-km buffer which is more appropriate for the characteristics of this triangular urban region. This means that 13% of the region is outside the PAs and the buffers analyzed. Their ecosystems are very different to those inside the PAs.

Second, we transformed the CLC vector maps to a raster format with a 50 × 50 m pixel size. From the CLC legend we made three different groupings. The first is a simplification of level 3 in seven categories: (1) urban fabric, (2) industrial and commercial, (3) arable land and permanent crops, (4) heterogeneous agricultural areas, (5) forests, (6) shrubs and herbaceous vegetation, and (7) others (open spaces with little vegetation, wetlands and water bodies). We used this grouping to simulate future scenarios and to analyse temporal trends in landscape metrics. In the second we grouped land uses into three types: (1) built-up surfaces, (2) agricultural areas and (3) natural areas. This classification (Martinuzzi et al. 2015) was used to evaluate land-use changes according to the type of PA and in their surrounding areas. Finally, in order to assess the dynamics of landscape structure we took level 1 of the CLC legend into account. We reclassified the maps in binary format. We considered class 1 as background, and combined classes 2, 3, 4 and 5 into a single target category (agricultural and natural areas) linked to the habitats represented in PAs in the region of Madrid.

Third, using CLUE we simulated land use in 2025 in three different scenarios: (a) trend scenario, (b) economic crisis scenario and (c) green scenario. The first one, the trend scenario or "business as usual" shows what would happen if the past trend in 1990–2000–2006 were to continue until 2025. The crisis scenario shows what would happen if the economic crisis in Spain and the region of Madrid were to continue until 2025. To draw up this scenario, we carried out 117 surveys with experts (scientists, land and protected area managers, ecologists and representatives of non-governmental organisations, neighbourhood associations, etc.); they were asked about how much the different land use types could grow or decrease under an economic crisis scenario and where these land use changes could preferentially be located (see Gallardo 2014; Gallardo et al. 2016). Finally, the green scenario shows what would happen if there were more active reforestation policies and if greater importance were placed on the natural environment. It does, however, take into account that Madrid is an urban region and that built-up areas will continue to grow. This means on the one hand, that greater protection is offered to natural uses than in the past and, on the other, that greater growth is assigned to built-up land. We used LCM to design the trend scenario in order to compare its results with those of CLUE.

In the models drawn up with CLUE, we related land use and driving factors by means of logistic regressions (LR). In the model simulated with LCM, we used a multi-layer perceptron neural network (MLP). Previously, we performed a

Pearson's correlation analysis to observe the correlations between the selected variables. We eliminated highly correlated variables as they did not make a significant contribution (see Gallardo 2014; Gallardo et al. 2016). In CLUE, we assigned the demand for each land use specifying the number of hectares for each land use in 2025, based on what had happened in previous years. In LCM, this was determined by a transition matrix indicating the probability of change from one use to another (see Gallardo 2014).

We calibrated the model in order to improve the scenario results. Taking the sequence of maps 1990–2000 as a base, we simulated a land-use model in 2006 and compared it with the real map for 2006. We varied the amount of land-use change, the driving factors used and/or the size or weight of the neighbourhood in order to obtain a better result. For validation, we carried out comparisons in terms of quantity and location; the former considered the proportion of each category of land use appearing on one map and whether this was similar to the proportion for that same category on the other; the latter compared the position of each category on the two maps. We used Kappa statistics, K Location (location) and K Histogram (quantity) (Pontius 2000; Van Vliet 2009) and we compared them with a null model and a random model. We obtained values and maps of hits, misses and false alarms (Eastman 2012; Sangermano et al. 2012).

Fourth, we drew up cross-tabulation matrices (Pontius et al. 2004) to obtain values and maps of changes between 1990–2006 and 2006–2025. We then compared the results with the protected areas depending on their level of priority and with the 5-km buffer. The aim was to find some of the main processes of land-use change that had already taken place and that could be expected in different scenarios: urbanisation, naturalisation and disturbances and exchanges in natural areas (Stellmes et al. 2013; Martínez-Fernández et al. 2015).

Fifth, we calculated an index for fragmentation of agricultural and natural habitats and for temporal variations in terms of their size and spatial pattern. The MSPA algorithm in the GUIDOS software (Soille and Vogt 2009) classified each pixel by its geometric position on the matrix being analysed, distinguishing between seven entities: (1) cores, (2) islets, (3) perforations, (4) edges, (5) loops, (6) bridges and (7) branches. We took into account the following parameters: analysis of pixel connectivity in 8 directions (cardinal and diagonal) = 1; transition pixels = 1; distinction between external and internal edges (perforations) in the core class. We calculated a Habitat Fragmentation Index (Chuvieco et al. 2013), in our case the sum of agricultural and natural habitats (HFI) in each type of PA and in its corresponding buffer. This goes from 1 (greatest fragmentation) to 2 (least fragmentation). It assigns a different weight to each of the entities mapped in terms of the relations between resilience and spatial coherence (Opdam et al. 2003, 2006). There is a constant gradation from the core (greatest weight) to the islets (least weight). The index relates the number of pixels in each category or fragmentation entity to their weights.

Finally, we calculated temporal variations in some landscape metrics. Following the recommendations of Aguilera and Botequilha-Leitão (2012) for a Mediterranean region with similar processes to those of Madrid, we selected six

FRAGSTATS indices that give us an estimate of fragmentation (NP) of the landscape patches, their size (LPI and AREA_CV), the complexity of their shape (FRAC_AM), their closeness (CONTIG_AM) and their isolation (ENN_AM).

4 Results

Table 2 shows the LUCC that took place between 1990 and 2006 by zone (types of PA and their surroundings), a period of intense change. Globally, the built-up surface expanded by 41,000 ha, equivalent to 11.23% of the total study area. Over these 16 years, more than 9% of the buffer land and 2% of PAs were sealed. There are large differences depending on the degree of protection enjoyed by the different PAs. The Nature Reserve and Regional Parks were the most affected by land-use changes. In general, agricultural areas contributed most to the growth of urban areas. Although in relative terms persistence is very high inside the PAs, the increase in built-up area is a worrying process from an ecological point of view. In short, almost half the surface area that changed its use inside the PAs in the region was urbanized. Naturalisation of abandoned agricultural land is less worrying from the ecological and surface area points of view. Revegetation affected over 10,000 ha, about 3% of the area studied. Both processes occurred with greater intensity in the areas around the PAs.

Map CLC06, reclassified in 7 categories, and the three scenarios are represented in Fig. 3. The result obtained with LCM for the trend scenario is not shown because the result was fairly similar to that obtained with CLUE.

The trend scenario (Fig. 3b) shows that extensive growth of urban, industrial and commercial areas can be expected in the region. In both CLUE and LCM, these areas will grow by more than 30% compared to 2006 levels. Urban areas will spread in a compact way around the metropolis, especially to the south and south-east along the main transport routes. Industrial and commercial areas will spread towards the south and south-east of the capital. Heterogeneous agricultural areas and forests will remain stable, with slight gains of less than 0.1% over 2006.

Table 2 Land use cover change that took place between 1990 and 2006 in protected areas and in their surroundings

	LUCC CLC90-CLC06													
	NR		NP		RP		SAC		SPA		PPZ		BUFFER	
LUCC	ha	%	ha	%	ha	%	Ha	%	ha	%	ha	%	ha	%
FBA	36	5.8	0	0	1,486	1.54	440	0.4	678	0.8	3	0.1	11,156	3.0
ABA	10	1.6	0	0	1,775	1.84	1,460	1.2	533	0.6	21	0.4	23,502	6.3
AFA	1	0.2	0	0	1,870	1.93	1,124	0.9	789	0.9	4	0.1	6,291	1.7

NR Nature Reserve, *NP* National Park, *RP* Regional Park, *SAC* Special Area of Conservation, *SPA* Special Protection Area, *PPZ* Peripheral Protection Zone, *FBA* Forest to built-up areas, *ABA* Agricultural to built-up areas, *AFA* Agricultural to forest areas

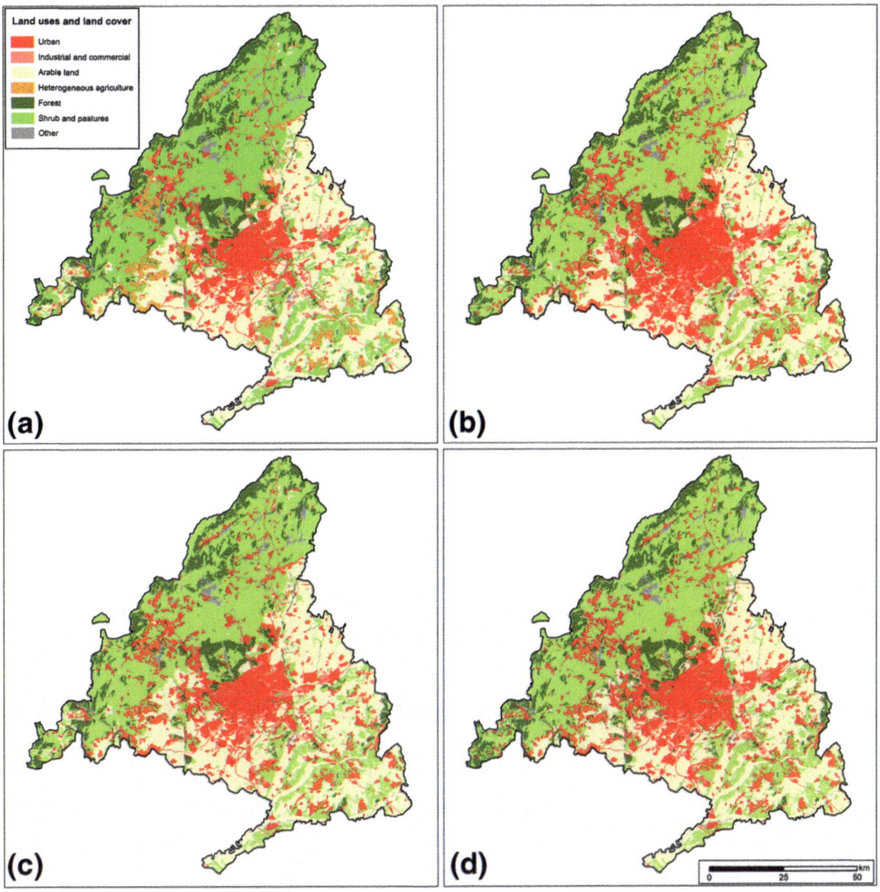

Fig. 3 LUCC trend in the region of Madrid between 2006 and 2025: **a** CLC map for 2006; the other quadrants show the scenarios modelled using CLUE in 2025: **b** trend scenario, **c** economic crisis scenario, **d** green scenario

The greatest losses will be in arable land and permanent crops. In short, the current processes will be reinforced, that is, anthropization of natural habitats and to a lesser extent naturalisation of agricultural habitats.

In the economic crisis scenario (Fig. 3c), growth in built-up areas will be much more restrained, 6 times less than in the trend scenario. It would be located to the south-east of Madrid, where the large urban patch would spread in a compact way. Forests, shrubs, and herbaceous vegetation will remain stable.

In the green scenario (Fig. 3d), growth in built-up areas will be half that forecast in the trend scenario. While in the previous scenarios, shrub and pastures record losses, here there will be a slight gain. Following the same trend, forests will see marked growth compared to 2006 (+13.72%).

Table 3 Projected Change 2006–2025 in protected areas and their buffers

	NR		NP		RP		SAC		SPA		PPZ		BUFFER	
LUCC CLC06-SCEN25 TREND														
LUCC	ha	%	ha	%	ha	%	ha	%	ha	%	ha	%	ha	%
FBA	3	0.5	0	0	820	0.9	53	0.1	14	0.1	0	0	6,341	1.7
ABA	5	0.8	0	0	218	0.2	88	0.1	1	0.0	0	0	20,828	5.6
AFA	0	0	0	0	1	0.0	55	0.1	4,372	5.3	0	0	169	0.1
LUCC CLC06-SCEN25 CRISIS														
LUCC	ha	%	ha	%	ha	%	ha	%	ha	%	ha	%	ha	%
FBA	3	0.5	0	0	364	0.4	21	0.0	13	0.0	0	0	2,417	0.7
ABA	5	0.8	0	0	120	0.1	24	0.0	0	0	0	0	2,021	0.5
AFA	0	0	0	0	2	0.0	60	0.1	2,050	2.5	1	0.0	63	0.0
LUCC CLC06-SCEN25 GREEN														
LUCC	ha	%	ha	%	ha	%	ha	%	ha	%	ha	%	ha	%
FBA	3	0.5	0	0	193	0.2	26	0.0	14	0.0	0	0	2,897	0.8
ABA	5	0.8	0	0	211	0.2	106	0.1	0	0	0	0	11,283	3.0
AFA	0	0	0	0	1,788	1.9	455	0.4	5,656	6.8	9	0.2	5,914	1.6

NR Nature Reserve, *NP* National Park, *RP* Regional Park, *SAC* Special Area of Conservation, *SPA* Special Protection Area, *PPZ* Peripheral Protection Zone. *FBA* Forest to built-up areas, *ABA* Agricultural to built-up areas, *AFA* Agricultural to forest areas

Table 3 summarises the projected change between 2006 and 2025 by scenario and zone (types of PA and their surroundings). Although in the region as a whole a growing trend in soil sealing will continue, this will be slightly mitigated in comparison with the first period. Overall, in the trend scenario the built-up area will grow by 28,000 ha (7.6% of the study area). The buffer will increase its built-up land by 7.29%, and PAs by 0.36%.

The latter result is to some extent the product of the design of the three scenarios and takes into account the restrictions set out in the management plans for the different natural areas. The Regional Parks and the Nature Reserve will continue to be the most affected by this process. In the economic crisis scenario, the expansion of new urban zones will drop sharply in the buffer (+1.19%) and inside PAs (+0.16%). In the green scenario the built-up area will increase in the buffer (+4%) but only very slightly in the PAs (+0.16%). As in the first period, most of the new urban zones will be developed on abandoned agricultural land. The process of naturalisation will take place in the SW of the region, in the *Encinares de los ríos Alberche y Cofio* SPA. In this scenario, 60% of the area affected by land use change in PAs will be naturalised.

Figure 4 shows a representative window of what will happen in the centre-south of the region, around the city of Madrid and its metropolitan area. It represents the processes of urbanisation and naturalisation that will affect the PAs and the buffer, in the three scenarios. If restrictions on the construction of new urban and industrial buildings and of new infrastructure are complied with, most of the PAs will be unaffected by urbanisation.

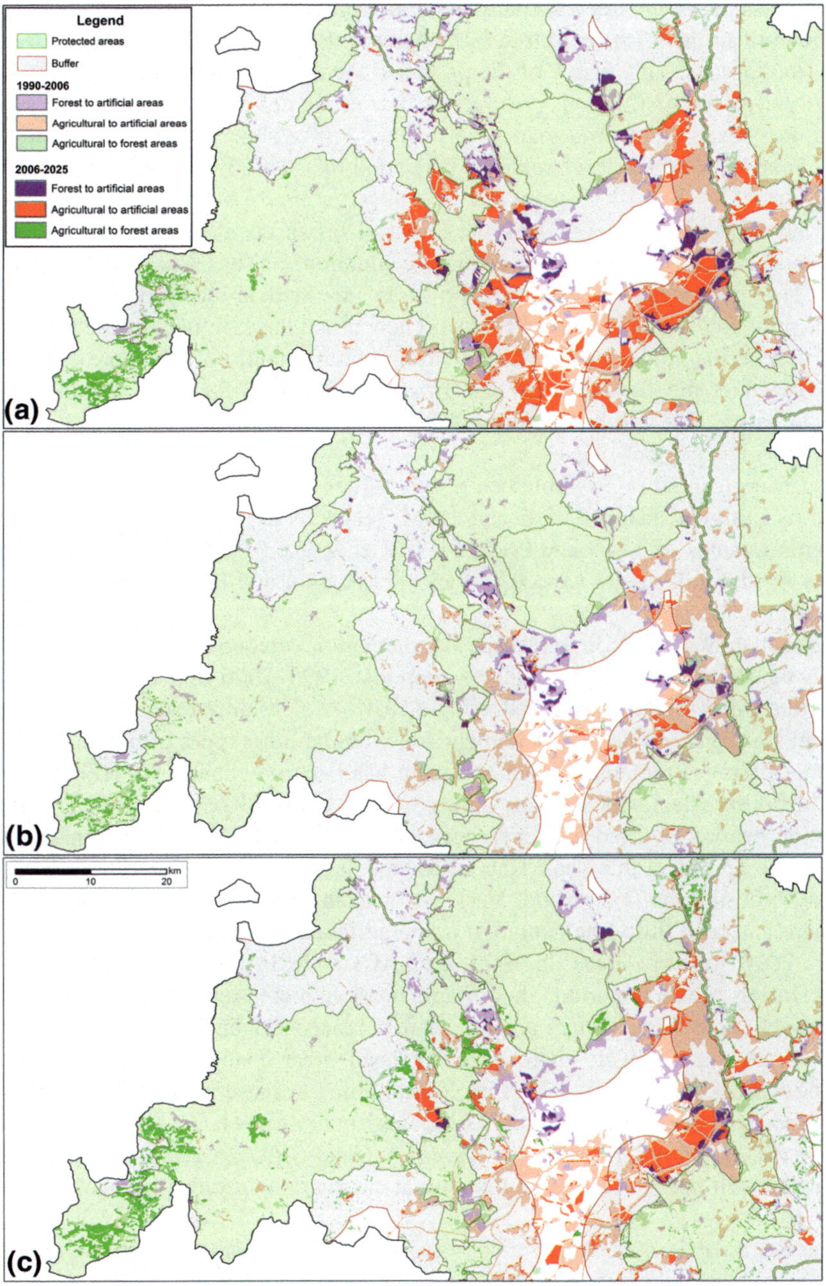

Fig. 4 Processes of urbanisation and naturalisation that took place in the centre-south sector of the region of Madrid during the period 1990–2006 (*light colours*) and projected changes 2006–2025 (*dark colours*), for **a** the trend scenario, **b** the economic crisis scenario and **c** the green scenario

All the same, in the trend scenario (Fig. 4a), medium and small residential areas will be built in the *Cuenca Alta del Manzanares* Regional Park along the A6 and M607 motorways. This is one of the protected areas that suffered most from urban sprawl during the first period. Most of the new urban developments will take place in the unprotected territorial matrix outside the buffer zone and, especially, in the buffer zones around the three regional parks and the SACs in the east and south of the region.

In the crisis scenario (Fig. 4b), the threat of urban development will be much more moderate as a result of the economic situation that has affected Spain since 2008. The green scenario (Fig. 4c) shows an intermediate situation. Soil sealing will be mitigated in the buffers of PAs, and new urban development inside protected areas will be insignificant. Forestation of agricultural land will be more extensive and will affect the *Encinares de los ríos Alberche y Cofío* SPA and the three regional parks. This will be the response to the incentives for revegetation included in the PA management plans.

Regarding validation, we obtained Kappa values of 0.868 and 0.892 for the trend scenarios designed using CLUE and LCM, respectively, and K Location and K Histogram values of 0.869 and 0.998 for CLUE and of 0.927 and 0.962 for LCM. Values for the null model were 0.879 and K Location and K Histogram values of 0.951 and 0.925.

Table 4 shows trends in landscape fragmentation categories in each type of PA and in their respective buffers, in two periods (1990–2006 and 2006–2025) and taking the trend scenario designed using CLUE. A clear difference exists between the National Park and its PPZ in comparison with the other types of protection. The National Park has remained intact and there has been no change. Its HFI was still 2.00 in 2006. The forest habitats survive today and will persist in their current condition bearing in mind the severe restrictions imposed by the land management plans to be approved this year. Habitat fragmentation in the buffer zone increased by 0.40% during the first period. Very minor changes are expected in the future.

There has been little fragmentation in the agricultural and natural habitats in the Natura 2000 Network areas. In 1990, the SACs and SPAs studied had an HFI of 1.98. During the first period, their fragmentation increased by 2.34 and 1.39% respectively, and these figures are expected to reach 2.51 and 2.01% by 2025.

Although quantitatively there have not been great changes, there has been a striking loss of core zones of high ecological value because of the construction of roads and new associated urban areas. Ecologists are especially worried about the case of the *Encinares de los ríos Alberche y Cofio* SPA. 26 years after its declaration, it is still not covered by any plan that clearly and specifically regulates land use.

The Regional Parks are also a source of worry. Although they are covered by plans, management has not been as efficient as expected. In 2006, the fragmentation index was 1.89, almost 5% greater than 16 years before, and 1.6% lower than the

Table 4 Trend over time of the spatial landscape pattern of PAs in and around Madrid

CLC90

	Backgr	Core	Islet	Perfor	Edge	Loop	Bridge	Branch	Total	HFI90
NR	45	466	0	2	89	0	18	0	620	1.81
NP	2	21,572	0	0	32	0	0	0	21,606	2.00
RP	3571	89,576		1514	1434	294	90	183	96,662	1.94
SAC	1542	117,988	2	1258	626	91	37	44	121,588	1.98
SPA	1095	80,406	0	1026	45	18	0	39	82,629	1.98
PPZ	20	5461	0	60	0	1	0	4	5546	1.99
Buffer	39,024	313,985	27	8725	7874	1577	447	1080	372,739	1.86

CLC06

	Backgr	Core	Islet	Perfor	Edge	Loop	Bridge	Branch	Total	HFI06
NR	66	430	0	19	102	0	3	0	620	1.75
NP	2	21,571	0	1	32	0	0	0	21,606	2.00
RP	6767	84,077	8	1853	3106	248	261	342	96,662	1.89
SAC	3429	114,758	14	1916	1153	159	34	125	121,588	1.95
SPA	1860	79,095	0	1524	45	36	0	69	82,629	1.96
PPZ	23	5431	0	85	0	3	0	4	5546	1.98
Buffer	72,045	274,367	131	9349	12,411	1598	830	2008	372,739	1.76

SCEN25-TREND

	Backgr	Core	Islet	Perfor	Edge	Loop	Bridge	Branch	Total	HFI25
NR	95	401	0	0	110	2	10	2	0	1.71
NP	3	21,582	0	4	33	0	3	0	125	2.00
RP	8158	82,446	51	1564	3635	134	250	428	28	1.87
SAC	3524	114,486	8	1792	1383	155	88	141	442	1.95
SPA	2264	78,584	1	1644	61	21	0	93	115	1.96
PPZ	44	5370		99	0	4	0	15	14	1.98
Buffer	99,525	246,424	995	9011	10,425	2162	1215	2695	645	1.68

expected figure for 2025. The *Sureste* and *Curso medio del río Guadarrama* Regional Parks have been crossed by new motorways and occupied by new urban zones, which have increased the background category and the edges associated with perforations.

The Nature Reserve is a particularly striking case. Even though it falls under the category for maximum protection, in 1990 it was the most fragmented zone in the region (HFI of 1.81). Between 1990 and 2006, its fragmentation index grew by 5.6% and is expected to reach an accumulated figure of 10% by 2025. A new motorway (R4), crosses the reserve parallel to a previous one (A4) so encouraging urban growth around the historic town of Aranjuez. It is true, however, that this is a small protected area covering less than 0.2% of the total PAs in the region.

Unexpectedly, there are no great differences between this nature reserve and the extensive buffer (46% of the regional surface area) around all the protected areas in the Madrid Region. This area is the second most affected by the general process of built-up land growth over this short period (average annual increase of 0.63% in the fragmentation index). In 2006, its HFI was 1.76 and this process of landscape fragmentation is expected to reach 1.67 by 2025.

Comparison of the left and right parts of Fig. 5 shows an increase in the background and in perforation in the core of agricultural and forest habitats in the region of Madrid during the period of most intensive growth in built-up land (1990–2006). The urban areas and new corridors opened up by motorways and railway lines are perfectly visible. At the other extreme and as already stated, there has been revegetation associated with the abandonment of agricultural lands in the *Encinares de los ríos Alberche y Cofio* SPA. However, this phenomenon does not compensate for the loss of habitats in the buffer, the nature reserve and the regional parks that are closest to the capital.

The landscape metrics reinforce the key ideas expressed above. In the buffer the number of urban patches (NP) increased by 26% between 1990 and 2006, and is expected to rise over the base year by 142% by 2025. The percentage occupied by the largest urban patches is also increasing. During the first period, the largest patch index (LPI) doubled and is expected to quadruple by 2025. With the increase in the number and surface area of urban patches, their contiguity is almost at maximum level (ENN_AM = 0.93). A similar, albeit less intensive, process has taken place in industrial and commercial uses.

In the Nature Reserve, the increase in the number and surface area of built-up patches stems from the widening of roads, as stated above, and from new urban and industrial developments linked to improved accessibility. A similar progression is expected up to 2025 which will be reflected in increased contiguity of patches with this type of land use.

In the Regional Parks, the number of urban patches grew by 60% between 1990 and 2006, and additional growth of 36% is expected by 2025. The index for the largest patch within this category is very low but it doubled during the first period and quadrupled during the second. The average distance between urban patches (ENN_AM) has also grown. This may indicate the isolation of such zones among large forest patches in the search for more scenic landscape. This has already happened in the *Manzanares* and *Guadarrama* Regional Parks. Great changes are not apparent in the metrics of other categories, probably because of internal exchanges between the forest and agricultural classes.

Nor are there great changes in forest areas within SACs. In the SPA there is an incipient process of regeneration of arboreal vegetation. In 2025 the number of forest patches will be three times greater than in 1990. Built-up land growth has had no effect on the National Park and its PPZ, with no appreciable change in indices.

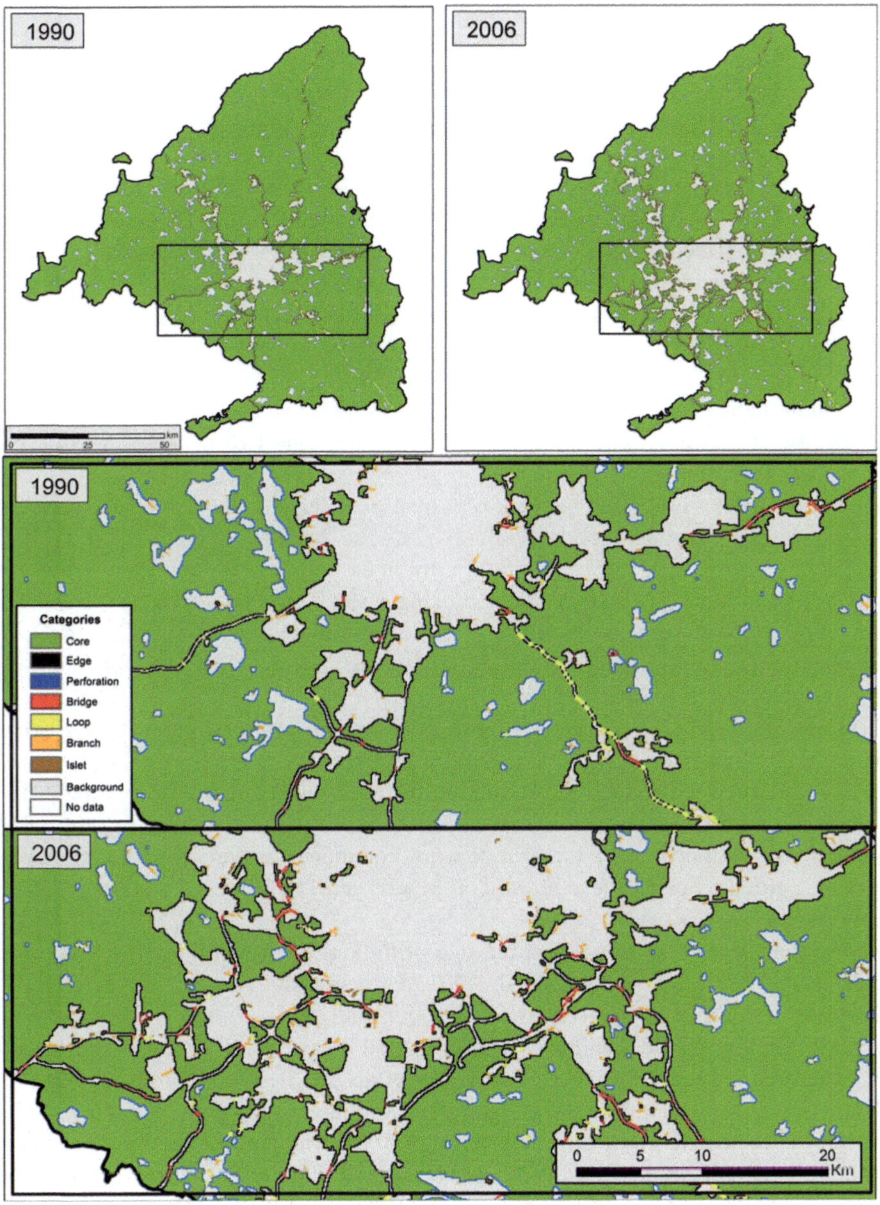

Fig. 5 Trends in fragmentation of agricultural and natural habitats in the region of Madrid since 1990 (*left*) and 2006 (*right*). Expanded view of a window of the southern part of the city of Madrid and its metropolitan area

5 Discussion of Results

We consider CLC to be a valid source of information for this research. This cartography is available throughout Europe, so the study could be replicated in other areas. The scale 1:100,000 is appropriate for studying regional and national PA networks. In order to update our study, it would be very useful to have access to CLC 2012 but, as already stated, it will be some months before the errors detected in it can be corrected. In fact, in our study site errors were also found in the previous CLC, especially in CLC00 (Catalá Mateo et al. 2008) and CLC06 (Hewitt and Escobar 2011; Díaz-Pacheco and Gutiérrez 2013; Martínez-Fernández et al. 2015; Gallardo et al. 2016). A lot of effort was made to correct these errors to avoid generating false land use change values.

A more detailed scale could be used for this type of study at the level of PAs or of specific ecosystems. The Information System on Land Cover in Spain (SIOSE), on a scale of 1:25,000, might be a good alternative. However, its complex legend including mixed classes and the lack of a historical series make it inappropriate for this research. Another good alternative might be the Spanish Forest Map (MFE2012) on a scale of 1:50,000. It combines with the SIGPAC covering the agricultural surface area and is updated using photointerpretation of SPOT images and with support from the National Plan for Aerial Orthophotography (PNOA). However, the coverage for 1990 does not have the same level of detail to enable us to analyse changes in land use.

Going further back in time, an effort needs to be made to interpret the aerial photographs made in 1956–57 and build an earlier land use map to start the time series. However, experts in the simulation of future use scenarios recommend that the initial and final maps be built on data from similar sources and using the same methods. In addition, such a long series would include some very different and even opposing trends. For all these reasons, it is advisable to use a more recent, shorter time period (Candau et al. 2000).

Another topic for discussion is the size of the buffer. A width of 10 km is often used in the literature, (Bruner et al. 2001; Figueroa and Sánchez-Cordero 2008; Martinuzzi et al. 2015). In the USA, Mexico and other countries this might be suitable because of the smaller size of protected areas. But a 10 km buffer would include 94% of Spain (Martínez-Fernández et al. 2015). We must remember that Spain plays an important role in biodiversity conservation and that 27% of its territory is protected. In the case of the region of Madrid, a 10 km buffer would be a complex solution because, with the territorial distribution of its PAs, much of the regional surface area would be within that buffer and it would include ecosystems that are very different to those represented in the PAs that were urbanised many decades ago. The buffer would therefore no longer be an effective means of controlling what happens inside and outside the PAs.

Regarding the design of future scenarios, in spite of the variety of simulation techniques, we opted for logistic regression because it is easy to use. And although there are technical differences between CLUE and LCM, the results obtained in the trend scenarios with both models are fairly similar.

Regarding validation of the results, the goodness of fit of the models depends on whether these values are due to good prediction or to the fact that there is high persistence in the study area (Pontius and Millones 2011). This phenomenon occurs with the K Histogram in CLUE. Its fit is almost perfect because real values for land use demand are taken.

The results obtained in our research are in line with the findings of previous studies on land use change in similar or nearby areas (Ruiz Benito et al. 2010, Hewitt and Escobar 2011; Díaz-Pacheco and Gutiérrez 2013; Díaz-Pacheco and García-Palomares 2014; Gallardo and Martínez-Vega 2016). They are also in line with the results of future scenarios in protected areas and their surroundings in the region of Madrid (Ruiz-Benito et al. 2010) and in the USA (Martinuzzi et al. 2015). The results on habitat fragmentation in regional parks and in the nature reserve are also in line with the findings of Rodríguez-Rodríguez and Martínez-Vega (2013).

Finally, although new episodes of built-up land growth are not expected inside Madrid's protected areas, threats to their peripheral zones are still a matter of concern. Those in charge of preserving biodiversity should remain on the alert for breaches of management plans or land use changes inside PAs approved on the basis of, for example, considerations of general interest. This type of reasoning and the impotence of managers were behind the high rates of built-up land growth and the increase in habitat fragmentation that took place in the period 1990–2006. We propose an exercise in collective reflection, comparing the results of the three scenarios proposed with a new trend scenario in which there are no restrictions on use changes in PAs and no incentives. The graphic and statistical results indicate clearly what might happen if the authorities were to sit back and allow economic agents to adopt an aggressive attitude.

6 Conclusion

Clearly land use changes linked to processes of anthropization and soil sealing are amongst the main threats to biodiversity, the preservation of natural resources and the production of environmental goods and services. For this reason, the analysis of land use changes during recent periods and the simulation of future scenarios can facilitate effective preventive planning for protected areas. Sustainable development can only be achieved when we understand the full implications of land use changes.

In urban areas such as the Madrid region the spill-over effect of protected areas should be monitored. It is clear that they attract urban developments to less protected areas around them. Transformation of their agricultural and natural habitats may have irreversible effects on biodiversity. Fragmentation brings with it longer exterior and interior edges. It can also create external threats for protected areas such as invasion by exotic species or the propagation of forest fires. These threats increase the potential ecological vulnerability of these spaces.

Acknowledgements This research received funding from the Spanish National R&D project "DISESGLOB: Design of a Methodology for Monitoring and Assessing the Overall Sustainability of National Parks and the Influence of Expected Changes of Use" (CSO2013-42421-P). Marta Gallardo was sponsored by a JAE-Predoc Grant from the Spanish National Research Council (CSIC). We specially thank Pilar Echavarría (CSIC) for her assistance in designing the cartographic figures.

References

Aguilera F, Botequilha-Leitão A (2012) Selección de métricas de paisaje mediante análisis de componentes principales para la descripción de los cambios de uso y cobertura del suelo del Algarve, Portugal. GeoFocus 12:93–121

Améztegui A, Brotons L, Coll L (2010) Land-use changes as major drivers of mountain pine (Pinusuncinata Ram.) expansion in the Pyrenees. Glob Ecol Biogeogr 19(5):632–641. doi:10.1111/j.1466-8238.2010.00550.x

Bruner AG, Gullison RE, Rice RE, da Fonseca GAB (2001) Effectiveness of parks in protecting tropical biodiversity. Science 291:125–128

Candau J, Rasmussen S, Clarke KC (2000) A coupled cellular automaton model for land use/land cover dynamics. 4thInternational Conference on Integrating GIS and Environmental Modeling (GIS/EM4): problems, prospects and research needs Banff. http://www.geog.ucsb.edu/~kclarke/ucime/banff2000/529-jc-paper.htm

Catalá Mateo R, Bosque Sendra J, Plata Rocha W (2008) Análisis de posibles errores en la base de datos Corine Land Cover (1990–2000) en la Comunidad de Madrid. Estudios Geográficos 264:81–104. doi:10.3989/egeogr.2008.i264.80

Chuvieco E, Martínez S, Román MV, Hantson S, Pettinari L (2013) Integration of ecological and socio-economic factors to assess global vulnerability to wildfire. Glob Ecol Biogeogr 23(2):245–258. doi:10.1111/geb.12095

Dantas de Paula M, Groeneveld J, Huth A (2015) Tropical forest degradation and recovery in fragmented landscapes. Simulating changes in tree community, forest hidrology and carbon balance. Glob Ecol Conserv 3:664–677. doi:10.1016/j.gecco.2015.03.004

Deguignet M, Juffe-Bignoli D, Harrison J, MacSharry B, Burgess N, Kingston N (2014) 2014 United Nations list of protected areas. UNEP-WCMC, Cambridge

Díaz-Pacheco J, Gutiérrez J (2013) Exploring the limitations of CORINE Land Cover for monitoring urban land-use dynamics in metropolitan areas. J Land Use Sci. doi:10.1080/1747423X.2012.761736

Díaz-Pacheco J, García-Palomares JC (2014) A highly detailed land-use vector map for Madrid region based on photo-interpretation. J Maps 10(3):424–433. doi:10.1080/17445647.2014.882798

Eastman JR (2012) IDRISI Selva. Clark University, Worcester, Guía para SIG y procesamiento de imágenes

EUROPARC-España (2014) Anuario 2013 del estado de las áreas protegidas en España. Fundación Fernando González Bernáldez, Madrid

FRA (2010) Global ForestResourcesAssessment 2010. FAO, Rome http://www.fao.org/docrep/013/i1757e/i1757e.pdf. Accessed 22 March 2016

Feranec J, Hazeu G, Christensen S, Jaffrain G (2007) CORINE land cover change detection in Europe (case studies of the Netherlands and Slovakia). Land Use Policy 24:234–247. doi:10.1016/j.landusepol.2006.02.002

Figueroa F, Sánchez-Cordero V (2008) Effectiveness of natural protected areas to prevent land use and land cover change in Mexico. Biodivers Conserv 17:3223–3240. doi:10.1007/s10531-008-9423-3

Gallardo M (2014) Cambios de usos del suelo y simulación de escenarios en la Comunidad de Madrid (Changes in land use and simulation scenarios in the region of Madrid). Universidad Complutense, Madrid. http://eprints.ucm.es/25253/. Accessed 22 March 2016

Gallardo M, Martínez-Vega J (2016) Three decades of land-use changes in the region of Madrid and how they relate to territorial planning. Eur Plan Stud 24(5):1016–1033 doi.10.1080/09654313.2016.1139059

Gallardo M, Gómez I, Vilar L, Martínez-Vega J, Martín MP (2016) Impacts of future land use/land cover on wildfire occurrence in the Madrid region. Reg Environ Change 16(4):1047–1061. doi:10.1007/s10113-015-0819-9

Hewitt R, Escobar F (2011) The territorial dynamics of fast-growing regions: unsustainable land use change and future policy challenges in Madrid. Spain. Appl Geogr 31(2):650–667. doi:10.1016/j.apgeog.2010.11.00

Juffe-Bignoli D, Burgess ND, Bingham H, Belle EMS, de Lima MG, Deguignet M, Bertzky B, Milam AN, Martinez-Lopez J, Lewis E, Eassom A, Wicander S, Geldmann J, van Soesbergen A, Arnell AP, O'Connor B, Park S, Shi YN, Danks FS, MacSharry B, Kingston N (2014) Protected Planet Report 2014. UNEP-WCMC, Cambridge. https://portals.iucn.org/library/sites/library/files/documents/2014-043.pdf. Accessed 22 March 2016

Lambin EF, Turner BLI, Geist HJ, Agbola SB, Angelsen A, Bruce JW, Coomes OT, Dirzo R, Fischer G, Folke C, George PS, Homewood K, Imbernon J, Leemans R, Li X, Moran EF, Mortimore M, Ramakrishnan PS, Richards JF, Skänes H, Steffen W, Stone GD, Svedin U, Veldkamp TA, Vogel C, Xu J (2001) The causes of land-use and land-cover change: moving beyond the myths. Glob Environ Change 11:261–269. doi:10.1016/S0959-3780(01)00007-3

Lei C, Lin Z, Zhang Q (2014) The spreading front of invasive species in favorable habitat or unfavorable habitat. J Differ Equ 257:145–166. doi:10.1016/j.jde.2014.03.015

López Lambas ME, Ricci S (2014) Planning and management of mobility in natural protected areas. Procedia Soc Behav Sci 162:320–329. doi:10.1016/j.sbspro.2014.12.213

Martínez-Fernández J, Ruiz-Benito P, Zavala MA (2015) Recent land cover changes in Spain across biogeographical regions and protection levels: Implications for conservation policies. Land Use Policy 44:62–75. doi:10.1016/j.landusepol.2014.11.021

Martinuzzi S, Radeloff VC, Joppa LN, Hamilton CM, Helmers DP, Plantinga AJ, Lewis DJ (2015) Scenarios of future land use change around United States' protected areas. Biol Conserv 184:446–455. doi:10.1016/j.biocon.2015.02.015

McDonald RI (2013) Implications of Urbanization for Conservation and Biodiversity Protection. In: Levin SA (ed) Encyclopedia of biodiversity, 2nd edn. Academic Press, Amsterdam, pp 231–244

McGarigal K, Cushman SA, Neel MC, Ene E (2002) FRAGSTATS: spatial pattern analysis program for categorical maps. University of Massachusetts, Amherst. http://www.umass.edu/landeco/research/fragstats/fragstats.html. Accessed 22 March 2016

Moreira F, Rego FC, Ferreira PG (2001) Temporal (1958–1995) pattern of change in a cultural landscape of northwestern Portugal: implications for fire occurrence. Landsc Ecol 16(6):557–567

Opdam P, Verboom J, Pouwels R (2003) Landscape cohesion: an index for the conservation potential of landscapes for biodiversity. Landsc Ecol 18:113–126

Opdam P, Steingrover E, Rooij SV (2006) Ecological networks: a spatial concept for multi-actor planning of sustainable landscapes. Landsc Urban Plan 75(3–4):322–332. doi:10.1016/j.landurbplan.2005.02.015

Pontius RG Jr (2000) Quantification error versus location error in comparison of categorical maps. Photogramm Eng Remote Sens 66:1011–1016

Pontius RG Jr, Shusas E, McEachern M (2004) Detecting important categorical land changes while accounting for persistence. Agr Ecosyst Environ 101:251–268. doi:10.1016/j.agee.2003.09.008

Pontius RG Jr, Millones M (2011) Death to Kappa: Birth of quantity disagreement and allocation disagreement for accuracy assessment. Int J Remote Sens 32:4407–4429. doi:10.1080/01431161.2011.552923

Rodríguez-Rodríguez D, Martínez-Vega J (2012) Proposal of a system for the integrated and comparative assessment of protected areas. Ecol Ind 23:566–572. doi:10.1016/j.ecolind.2012.05.009

Rodríguez-Rodríguez D, Martínez-Vega J (2013) Results of the implementation of the System for the Integrated Assessment of Protected Areas (SIAPA) to the protected areas of the Autonomous Region of Madrid (Spain). Ecol Ind 34:210–222. doi:10.1016/j.ecolind.2013.04.019

Rodríguez-Rodríguez D, Martínez-Vega J (2016) What should be evaluated from a manager's perspective? Developing a salient protected area effectiveness evaluation system by managers and scientists. Ecol Ind 64:289–296. doi:10.1016/j.ecolind.2016.01.011

Romero-Calcerrada R, Perry GLW (2004) The role of land abandonment in landscape dynamics in the SPA Encinares del rio Alberche y Cofio, Central Spain, 1984–1999. Landsc Urban Plan 66:217–232. doi:10.1016/S0169-2046(03)00112-9

Ruiz Benito P, Cuevas JA, Bravo de la Parra R, Prieto F, García del Barrio JM, Zavala MA (2010) Land use change in a Mediterranean metropolitan region and its periphery: assessment of conservation policies through CORINE land cover data and Markov models. For Syst 13 (3):315–328

Ruiz-Mallén I, Corbera E, Calvo-Boyero D, Reyes-García D (2015) Participatory scenarios to explore local adaptation to global change in biosphere reserves: experiences from Bolivia and Mexico. Environ Sci Policy 54:398–408. doi:10.1016/j.envsci.2015.07.027

Sangermano F, Toledano J, Eastman JR (2012) Land cover change in the Bolivian Amazon and its implications for REDD + and endemic biodiversity. Landsc Ecol 27:571–584. doi:10.1007/s10980-012-9710-y

Sastre P, Lucio JV, Martínez V (2002) Modelos de conectividad del paisaje a distintas escalas. Ejemplos de aplicación en la Comunidad de Madrid. Ecosistemas 11(2). http://revistaecosistemas.net/index.php/ecosistemas/article/viewFile/288/283. Accessed 22 March 2016

Soille P, Vogt P (2009) Morphological segmentation of binary patterns. Pattern Recogn Lett 30 (4):456–459. doi:10.1016/j.patrec.2008.10.015

Stellmes M, Röder A, Udelhoven T, Hill J (2013) Mapping syndromes of land change in Spain with remote sensing time series, demographic and climatic data. Land Use Policy 30(1):685–702. doi:10.1016/j.landusepol.2012.05.007

Turner BL, Skole D, Sanderson S, Fischer G, Fresco L, Leemans R (1995) Land-use and Land-cover change: science/research plan. International Geosphere-Biosphere Programme, Stockholm

Verburg PH, Overmars KP (2007) Dynamic simulation of land-use change trajectories with the CLUE-s model. In: Koomen E, Stillwell J, Bakema A, Scholten HJ (eds) Modelling land-use change. Progress and applications. Springer, Heidelberg, pp 321–335. doi:10.1007/978-1-4020-5648-2_18

Van Vliet J (2009) Assessing the accuracy of changes in spatial explicit land use change models. In: 12th AGILE international conference on geographic information science, Leibniz Universität, Hannover, 2–5 June 2009

Viedma O, Moity N, Moreno JM (2015) Changes in landscape fire-hazard during the second half of the 20th century: agriculture abandonment and the changing role of driving factors. Agr Ecosyst Environ 207:126–140. doi:10.1016/j.agee.2015.04.011

Chapter 16
Navigating the Future: Land Redevelopment Scenarios and Broader Impact Assessment in Southern California

J.H. Kim, J.R. Hipp and V. Basolo

Abstract While land use and cover change (LUCC) modeling and simulation technologies have been widely disseminated in urban planning and other public decision-making domains, their application to site redevelopment is still limited. This chapter presents a case study in which land use change simulation and impact assessment models are employed to facilitate public dialogue for reuse of a decommissioned air force base site (known as the Orange County Great Park) in Southern California. Emphasis is on the uniqueness of site renewal in an urban context that requires special attention in modeling, impact assessment and decision support. It is also suggested that both relevance and coherence are crucial to the success of LUCC applications.

Keywords Site renewal · Land use change simulation · Impact assessment · Southern california

1 Introduction

Land use decisions often become the subject of public deliberation. The presence of externalities–either negative or positive–makes it implausible to allow individual property owners to make these decisions purely out of personal interest even in a static world. In reality, land use may also need to be aligned with community

J.H. Kim (✉) · V. Basolo
Department of Planning, Policy and Design, University of California, 206 Social Ecology I, Irvine, CA 92697, USA
e-mail: jaehk6@uci.edu

V. Basolo
e-mail: basolo@uci.edu

J.R. Hipp
Department of Criminology, Law and Society, University of California, 3311 Social Ecology II, Irvine, CA 92697, USA
e-mail: hippj@uci.edu

M.T. Camacho Olmedo et al. (eds.), *Geomatic Approaches for Modeling Land Change Scenarios*, Lecture Notes in Geoinformation and Cartography,
https://doi.org/10.1007/978-3-319-60801-3_16

visions or to be managed from a long-term perspective. This is especially true in a highly urbanized area where potential land use conflicts can emerge in a complicated, dynamic fashion, and the demand for systematic land management is high (see e.g. Plotkin 1987; Kaiser et al. 1995; Taleai et al. 2007; Kim 2011).

Public interest can be taken into account in many forms, and there is no single ideal way to accomplish this important task. While some claim that expert knowledge is essential, others argue for a more participatory approach to attain "efficiency and effectiveness, currency, relevance, responsiveness and their supposed low cost … [and to] foster a sense of ownership of a plan and commitment to its implementation" (McCall and Dunn 2012, p. 81). Moreover, it is debatable to what extent price signals and other attributes of a market mechanism need to be employed in determining an appropriate use of land rather than relying on a political process (see e.g. Lee 1981; Pennington 1999; Cheshire and Sheppard 2005).

In any case, land use decisions do require analytical support to enable us to navigate the future with careful consideration of the tradeoffs associated with the decisions. What is likely to happen? What if we implement an alternative action to modify the trajectory? Will it lead to a better future (from social, economic, and/or environmental perspectives)? Who gains, who loses?

This chapter provides a case study using land use and cover change (LUCC) simulation and impact assessment models to provide the analytical support for a land use decision. Specifically, we consider the reuse of a decommissioned air force base site, known as the Orange County Great Park area, located in Southern California. LUCC simulation techniques have been increasingly employed in spatial planning and other public decision-making domains (see e.g., Koomen et al. 2008; Koomen and Borsboom-van Beurden 2011). However, their real-world applications have typically focused on cases of urban growth and physical expansion. Relatively little attention has been paid to urban decline and/or renewal, while the demand for decision support systems for these issues has been growing rapidly (Zheng et al. 2015). Our case study seeks to fill this gap in the literature and provide some meaningful lessons on ways of using LUCC simulation and associated tools for a broader scope of community and regional planning tasks.

The remainder of this chapter is structured as follows. Section 2 provides a description of the study area and discusses some unique characteristics of site renewal projects which need to be considered carefully in devising an analytical framework for decision making. Section 3 presents our methodologies, namely the land use change and impact assessment models, while the results and validation outcomes are reported in Sects. 4 and 5, respectively. We conclude with a discussion of our case study findings in Sect. 6.

2 Test Area and Data Sets

Our study area covers the former Marine Corps Air Station (MCAS) El Toro and surrounding land parcels, located in Orange County, California (Fig. 1). Over the second half of the twentieth century, the site was an air station that served "as a training facility in peacetime and a staging area for support of overseas military missions in times of conflict" (Orange County Great Park, n.d.). However, as a result of the 1993 Defense Base Closure and Realignment decision, the MCAS El Toro was closed in 1999, and reuse of the site (approximately five thousand acres of land) became an important item on the planning agenda for the City of Irvine, which annexed the site in 2003, and for nearby communities within the county.

At the very beginning, "*[i]n November 1994, … Measure A was passed by Orange County voters, designating MCAS El Toro for commercial aviation use. The Orange County Board of Supervisors, supported by the John Wayne Airport neighbors [i.e., those living around an existing airport in the area], hoped to*

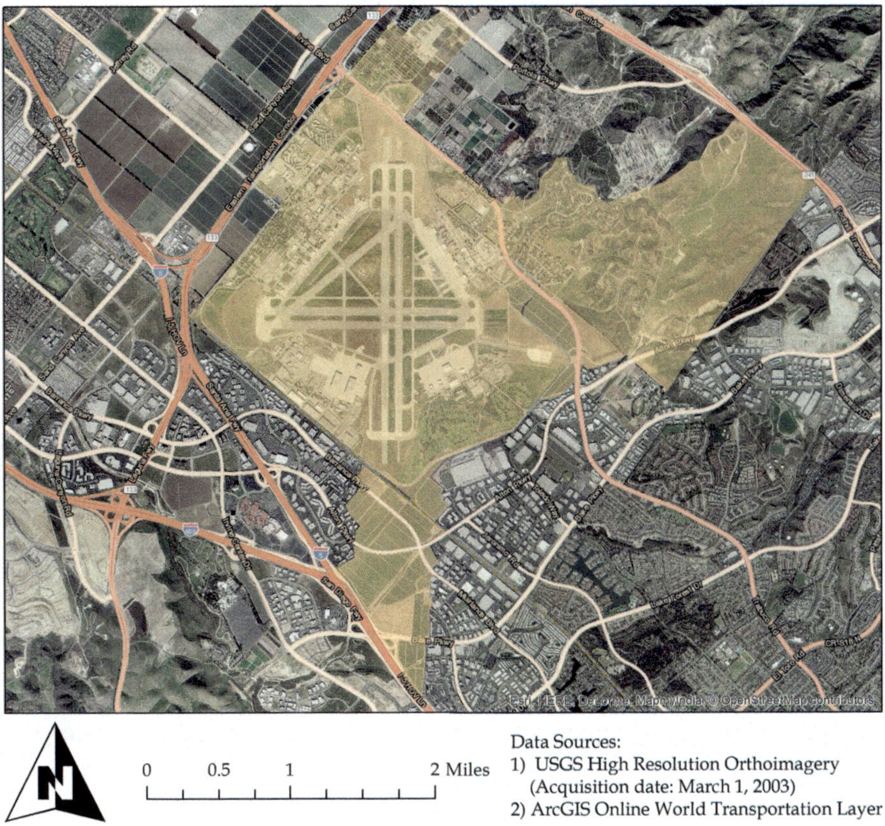

Fig. 1 Study area–aerial view in 2003

develop a large commercial airport that would serve 38 million passengers annually, and eventually replace John Wayne as Orange County's airport. As plans for the El Toro airport project were made public, the communities surrounding El Toro organized to oppose it and developed a competing plan, the Orange County Central Park and Nature Initiative. The initiative supported the development of a 1,300 acre public space that would include a sports park, botanical garden, and cultural terrace. After an intense grass-roots campaign, the initiative was placed on the ballot as Measure W and passed by a 58 percent to 42 percent vote on March 5, 2002. The next day, the U.S. Navy and the City of Irvine announced plans for the development of the Orange County Great Park" (Lamb 2009–Guide to the UCI's special collection on the development of the El Toro Airport, 1992–2003). But, the detailed plan/layout of the site reuse was not finalized immediately. Rather, it has been the subject of lengthy analyses, plan revisions, and debates among various groups of local stakeholders (Kranser 2005; Stockstill 2014). Recently, a modified project plan, which incorporates the ideas of mixed-use community creation and transit oriented development, has gone through a comprehensive impact analysis to meet the California Environmental Quality Act requirements (City of Irvine 2012). In its current general plan, the City of Irvine creates a separate land use category, called "Orange County Great Park" with the following definition: "*The development of regionally significant conservation and open space, parks and recreation, educational facilities, and other public-oriented land uses, integrated with privately developed multi-use, residential, commercial, and industrial properties, at the former MCAS El Toro site*."

Although unique in many respects, our study area presents some key attributes of site renewal projects that require special considerations in modeling/simulation, impact assessment, and decision support. Among others, given that site renewal opportunities often arise in highly urbanized areas, the land use decision is likely to involve tensions among various community groups affected (either positively or negatively) by the detailed renewal plan. It is not unusual for local politics to come into play. Sometimes, consideration has to be given to the interests of nearby jurisdictions.[1] Existing policies in and outside of the jurisdiction can be a barrier to renewal, and thus may need to be modified through systematic cooperation among policy authorities. Furthermore, to be successful, large-scale site renewal projects often require strong public support, which can be gained from consensus building or other forms of collaborative planning.

To deal with this complicated situation effectively, a comprehensive impact assessment needs to be conducted, covering not only immediate traffic and environmental impacts but also long-term socioeconomic consequences. Multiple relevant scenarios may also need to be explored in a coherent manner for communication and informed decision-making. When a LUCC modeling/simulation approach is employed, the models need to be designed in a way that

[1]Our study site, although annexed into the City of Irvine in 2003, is surrounded by multiple jurisdictions, such as Lake Forest and Laguna Hills.

can reflect the detailed site reuse pattern at an appropriate scale. In other words, a binary or crude land use/cover classification is less likely to be precise. A detailed urban land use categorization based on land parcels is more likely to be appropriate for this purpose, while a grid cell-based technique can also provide meaningful decision support.

For our study site and other parts of the metropolitan area, the Southern California Association of Governments (SCAG) provides detailed parcel-level land use information, dating back to 1990. The parcel data file is based upon SCAG's land use coding system with over 100 categories, ranging from low-density single-family to duplexes, low-rise, medium-rise, and high-rise apartments as well as detailed commercial, industrial, institutional and open space designations. This dataset enables us to investigate the dynamics of land use change within the region at a finer scale, as it reveals the evolution of land use over the last fifteen years (1990–2005) in a consistent format.

In addition to the parcel-level land use, we combine a variety of spatially-explicit data, needed for the investigation of causes and consequences of land use change. These include elevation/slope, transportation infrastructure, and locations of key attractors (including the shoreline) within the Southern California metropolitan area, made up of Los Angeles, Orange, Riverside, San Bernardino, and Ventura counties. We also use a range of neighborhood-level socioeconomic data, derived from Census products and other sources of information. For instance, to investigate how land use changes can influence surrounding neighborhoods, we gather Zipcode Business Patterns data provided by the US Census Bureau and the data on average loan values and average household income of in-movers coming from the Federal Financial Institutions Examination Council (FFIEC), which collects the information under the Home Mortgage Disclosure Act (HMDA).

3 Methodology and Practical Application to the Data Sets

In an attempt to support more informed decision making, we conduct a baseline land use change simulation and scenario-based impact assessment. For the baseline simulation, we employ a multinomial logit model, which has been widely used for empirical investigations of urban land use change (see e.g., Zhou and Kockelman 2008; Fragkias and Geoghegan 2010; Kim et al. 2017) and can be briefly expressed as follows

$$p_{ij} = \frac{\exp(X\beta_{ij})}{\sum_m \exp(X\beta_{im})}$$

where p_{ij} indicates the probability of parcel-level land use conversion from i to j; X and β_{ij} represent land use change factors and the estimable coefficients, designed to capture their effects on the i-to-j probability, respectively.

More specifically, our model is constructed with the following 10 land use categories, obtained through an aggregation of the SCAG's coding system, to avoid certain drawbacks of a highly disaggregated scheme that can emerge in multinomial logistic regression (particularly due to an insufficient number of land parcels for a certain category of land use transition in the dataset)—0: No development, 1: Single-family residential, 2: Multi-family residential, 3: Other residential, 4: Commercial & Services, 5: Industrial, 6: Transportation, Communication, & Utilities, 7: Public facilities, 8: Mixed use, and 9: Open space & Recreational (see Table 1). The model is calibrated using the region-wide data (all parcel observations with valid land use information within the Southern California metropolitan area, including unincorporated areas) for 1990–2005 with consideration of a range of potential land use change factors, including each parcel's physical/ecological attributes (e.g., parcel size, shape, slope), accessibility measures (e.g., proximity to employment centers and transportation infrastructure), and neighborhood characteristics (e.g., socio-demographic variables and surrounding land uses).

Our main focus in the baseline simulation is to reveal how (in terms of land use) the decommissioned MCAS El Toro site will be transformed in the future, if no specific action is taken. In other words, we attempt to use a calibrated model, based on past development patterns in the larger region, to generate a baseline reference about what is likely to occur in this particular site, if the parcels in this site follow the past trend (or market forces) in the metropolitan area. The simulation outcomes are expected to inform the stakeholders involved, particularly to help them understand the gap between their desires and the probable future with no specific actions beyond the status quo. The outcomes, when presented and delivered effectively, can also contribute to building consensus about the need for actions toward a more desirable future.

Although useful, the baseline simulation alone does not enable us to determine what actions are needed or how the site needs to be redeveloped. Therefore, to provide decision support and facilitate planning dialogue more effectively, we develop a set of alternative site reuse scenarios and conduct an impact assessment for each of the scenarios. More specifically, we consider the following five possible ways of reusing the site (for each scenario, we set the percentage park area to 20% where the existing Great Park is located).

- Housing-heavy development: housing at 80%, others at 0%
- Industrial-heavy development: housing at 50%, industrial at 30%
- Retail-heavy development: housing at 50%, retail at 30%
- Office-heavy development: housing at 50%, office at 30%
- Mixed development: 50% housing, combined with 30% industrial, retail, and offices (i.e., 10% each)

These scenarios acknowledge the desire to use this site to accommodate the increasing need for housing within the City of Irvine (in each scenario, at least 50% of the site is allocated to housing–single-family, multi-family, and other residential combined). More importantly, this set of scenarios roughly represents stakeholders'

Table 1 Land use classification scheme

No.	Category	Detailed land uses	Remark
0	No development	Vacant; Agricultural	
1	Single-family residential	Low density and high density single-family residential units	
2	Multi-family residential	Duplexes; Triplexes; 2- or 3-unit condominiums and townhouses; Low-rise, medium-rise, and high-rise apartments	
3	Other residential	Trailer parks; Mobile home courts; Mixed residential	
4	Commercial & Services	Low-rise, medium-rise, and high-rise office buildings; Skyscrapers; Shopping centers; Modern and older strip development; Commercial storage; Hotels and motels	Attended pay public parking facilities are included in this category
5	Industrial	Manufacturing, assembly, and industrial space; Motion picture and television studio lots; Packing houses and grain elevators; Petroleum refining and processing; Open storage; Metal and chemical processing; Mineral extraction facilities; Wholesaling and warehousing units	
6	Transportation, Communication, & Utilities	Airports; Railroads; Freeways and major roads; Park-and-ride lots; Bus and truck terminals; Harbor facilities; Communication facilities; Electrical power and other energy generation facilities; Solid/liquid waste disposal sites; Water, natural gas, petroleum facilities and maintenance yards	
7	Public facilities	Government offices; Police and sheriff stations; Fire stations; Public health care facilities; Religious facilities; Correctional facilities; Special care facilities; Educational institutions ranging from pre-schools and day care centers to colleges and universities	Non-attended public parking facilities are included in this category.
8	Mixed use	Various types of mixed urban uses	
9	Open space & Recreational	Golf courses; Local/regional parks; Cemeteries; Wildlife preserves; Specimen gardens and arboreta; Beach parks	

(conflicting) aspirations to have more space for diverse commercial/industrial purposes, although they do not articulate the detailed configuration within the site. The scenarios can also enable stakeholders to easily comprehend the impacts of an increase in a certain type of land use within the site and the associated tradeoffs (e.g., what if we decide to reuse more space for housing rather than industrial, retail, or office uses).

With these scenarios, we conduct an impact assessment that focuses on the following key socio-economic indicators: (1) average home sales price, (2) unemployment rate, (3) change in three types of jobs (white collar; blue collar; retail), (4) average household income of in-movers, and (5) average home loan values.[2] Although these indicators are considered important in the decision-making process, they have not been systematically analyzed. Our impact assessment focuses on these variables to fill the information gaps left behind by the previous impact analyses which mainly focused on the environmental and transportation implications of the proposed redevelopment plans.

More specifically, for these variables, we estimate neighborhood change models in which measures at one point in time are used to project the level of the measure of interest during the subsequent years, having the following general form.

$$\Delta y_{k,t} = \alpha + \beta \cdot y_{k,t-1} + \theta \cdot X_{k,t-1} + \rho \cdot WX_{k,t-1} + \tau \cdot T + \varepsilon_{k,t}$$

where $\Delta y_{k,t}$ represents the change in a socio-economic variable of interest in neighborhood k between year t-1 and t; $X_{k,t-1}$ and $WX_{k,t-1}$ indicate a set of covariates, including land use variables, and their spatial lags, respectively; T represents a collection of binary variables included to capture the fixed effects for years; $\varepsilon_{k,t}$ is an error term assumed to have a normal distribution; $\alpha, \beta, \theta, \rho$ and τ are the parameters to be estimated.[3]

For instance, to account for the (spatio-temporal) complex nature of job growth, we include a broad range of potential predictors as well as land use composition metrics in each job change model. Specifically, to explain the annual increase or decrease in jobs at a zip code area scale, we consider several measures of the number of jobs in the spatial area around a zip code area, such as the number of jobs of the same type within 1 mile, from 1–5 miles, and from 5–10 miles and similar spatial measures showing how the number of such jobs changed in the previous

[2]Given the data availability, the first three variables are analyzed at the zipcode area level, while our analysis of the remaining two are carried out at the census tract level. Census tracts have certain advantages over zipcode areas in that they are smaller and typically considered more representative of "neighborhoods", even though tracts do not always work perfectly in delineating neighborhoods (Chaskin 1998; Hipp 2007). However, loan amounts may not be an ideal measure of home prices in a neighborhood, and therefore we use data aggregated to zipcode areas that captures sales price information obtained from the RAND Corporation's statistics service as well as the tract-level average home loan values. Analyzing these two variables—i.e., zipcode area-level sales price and tract-level loan amounts—enables us to check the possible scale sensitivity of the analysis outcomes.

[3]For the job projection models we also include the change in jobs in the previous year in the models as this adds significantly to the model fit. This measure is not included in the other models.

two years.[4] These are meant to capture both potential agglomeration economies and diseconomies (i.e., centripetal and centrifugal forces that can largely shape business location patterns). To assess the possible cross-sectoral effects, in modeling the growth in one type of job, we include both neighborhood and nearby measures of other jobs (e.g., blue-collar and retail jobs considered in the white-collar job change model).

The neighborhood change models are estimated using the annual data from 1990 to the most recent year for the outcome variables. We use the coefficient estimates from those models for our forward simulations. We then substitute various values for the land use measures in the key zipcode areas or tracts of interest in the Great Park based on the five scenarios. We then project forward in time based on the models to compute predicted probabilities of home prices, unemployment, jobs, and income in the area.

4 Results

We present our baseline land use change simulation results in the following two ways: (1) the most likely development of the parcels ***if no restrictions are enforced*** (Fig. 2, left) and (2) the most likely development of the parcels, ***when "no development" is not an option*** (Fig. 2, right). The first presentation shows what is likely to happen over the next fifteen years using our land use change model calibrated with the data for 1990–2005, if the parcels simply follow the region-wide trend of parcel-level land use change. In other words, it provides an answer for "what will happen if no conscious plans or actions are implemented?" Basically, the second presentation also assumes that no specific actions will be taken, but it reflects the possibility of high demand for redevelopment in this area and shows the most likely development of individual parcels within the site. It should be made clear that the model estimates are contingent upon the assumption that the Great Park parcel characteristics are given. Changing the size and shape, as well as the grade, of the parcels would affect the model results.

As shown in the Fig. 2 (left), when no specific actions are taken, our simulation indicates that no development would be the most likely outcome for a majority of the parcels. Under this baseline scenario, a handful of large land parcels are projected to be reused for urban open space and recreational purposes. It is also expected that approximately 22 acres of the land, mostly small parcels, will be transformed into single-family residential units.

If we assume that all of the parcels are to be reused for urban purposes, open space and recreational uses would occupy over 85% of the site (in terms of land

---

[4]For all of these spatial buffers, we compute the measures with an inverse distance decay function. This essentially means that neighborhoods closer to the neighborhood of interest have a stronger effect than neighborhoods further away.

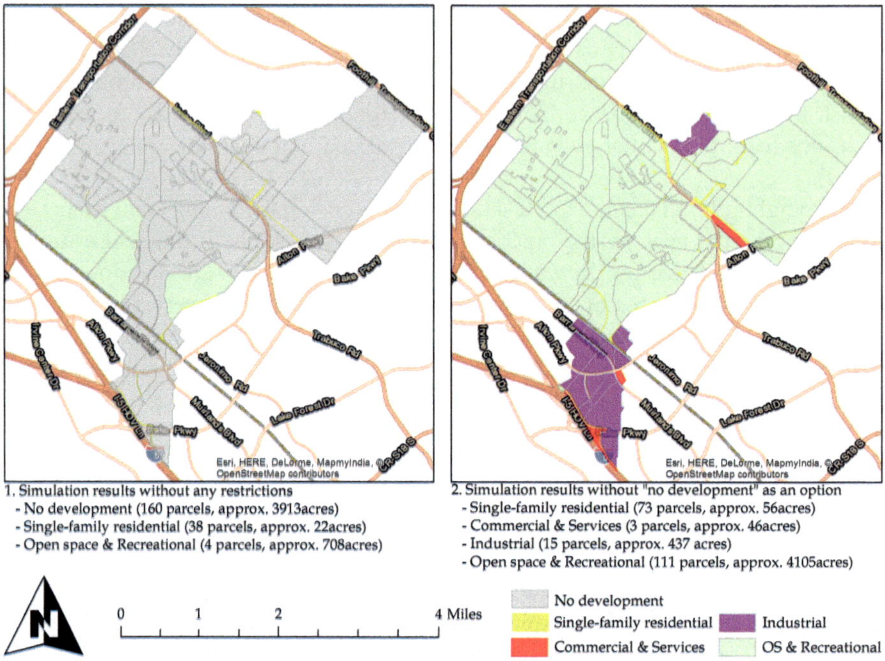

Fig. 2 Baseline simulation

area), as demonstrated in Fig. 2 (right). Single-family residential units could also expand, while no multi-family units are expected to be built. In this case, some parcels near the interstate highway (I–5) would be developed into commercial or industrial space. This is because their large size and proximity to the freeway significantly increase the probability of development for such purposes.

These presentations of the baseline simulation are useful in the sense that they show what is likely to happen in the future if no specific actions are taken and help to figure out to what extent the outcome meets the community's expectations. However, these two presentations are not enough to facilitate the dialogue for devising a better land use layout with careful consideration of the forces behind the outcome. The detailed probability patterns provide a basis for a more fruitful conversation in which we can discuss ways of achieving the objective of reusing the site more intensively for a variety of urban purposes. Figures 3, 4, and 5 present the probability distributions for each type of possible development.

Figure 3 demonstrates that small parcels tend to have a higher score for residential development, particularly single-family residential purposes, as our land use change model captures a negative association between parcel size and the probability of this type of development. Non-residential development shows quite distinct patterns, as small parcel areas would not be appropriate for these purposes, unless they were combined. For instance, there is a higher probability of commercial and service development (Fig. 4) in medium-sized parcels next to arterial roads.

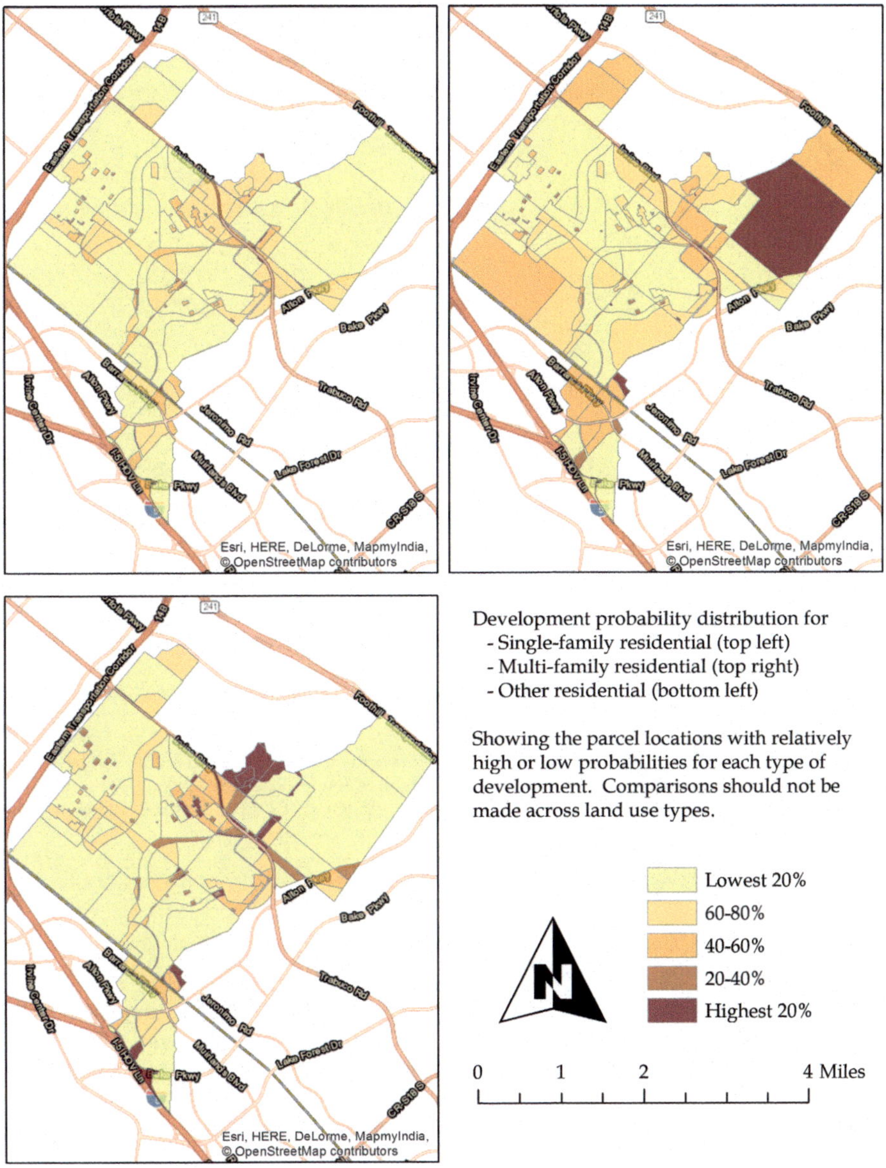

Fig. 3 Development probability distribution—Part 1

Industrial development, transportation, communications and utilities are more likely in large parcels, especially those close to the interstate highway and exiting industrial lands. Public facilities (Fig. 5) have a high score in the areas where open space and recreational purposes are found to be feasible, suggesting that these two land uses are likely to compete with each other.

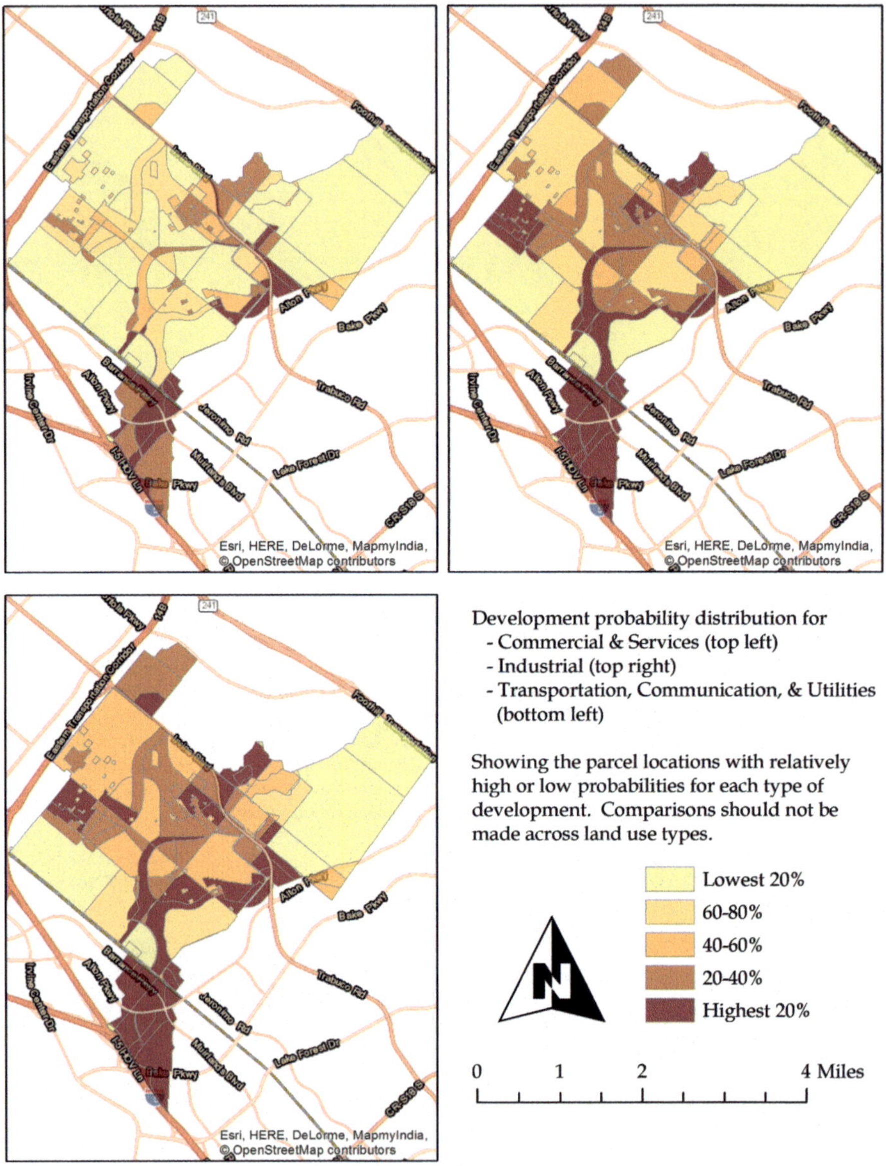

Fig. 4 Development probability distribution—Part 2

As explained in the previous section, for our impact assessment, we test five land use scenarios that roughly reflect stakeholders' varying desires: (1) Housing-heavy development, (2) Industrial-heavy development, (3) Retail-heavy development, (4) Office-heavy development, and (5) Mixed development. The results for each of the scenario impact analyses are summarized in Table 2. We take Scenario 1

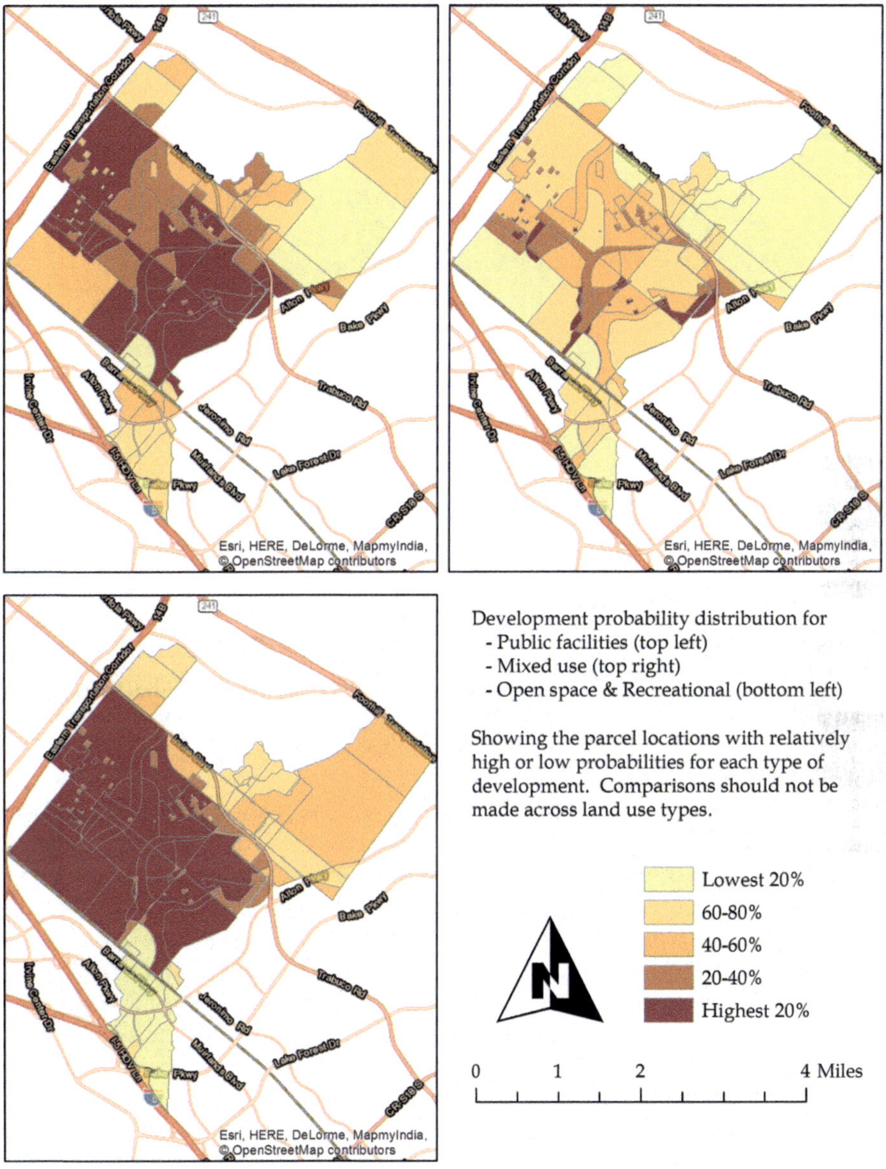

Fig. 5 Development probability distribution—Part 3

(housing-heavy development) as the benchmark with which to compare the other scenarios. The results set out in the table therefore compare the expected change in the various measures for each scenario compared to the change under this initial scenario of 80% housing and 20% land for the park area.

Table 2 Impact assessment summary

Category	Items	Scenario1. Housing-heavy	Scenario 2. Industrial-heavy	Scenario 3. Retail-heavy	Scenario 4. Office-heavy	Scenario 5. Mixed development
Land use mix	Housing	80%	50%	50%	50%	50%
	Industrial	0%	30%	0%	0%	10%
	Retail	0%	0%	30%	0%	10%
	Office	0%	0%	0%	30%	10%
	Park	20%	20%	20%	20%	20%
Impacts	Avg. home sales price	–	+5.5%	+12.3%	−15.3%	+1.2%
	Unemp. rate	–	+0.32	+0.50	+0.45	+0.34
	White collar jobs	–	+7.8%	−1.7%	+74.5%	+29.2%
	Retail. jobs	–	−14.5%	+20.2%	−28.7%	−9.5%
	Blue collar jobs	–	+33.3%	−6.8%	+23.7%	+18.0%
	Avg. hh. income	–	+2.0%	−0.1%	+3.1%	−3.0%
	Avg. home loan values	–	+1.9%	+1.2%	−3.9%	−3.1%

In the first scenario, a housing-dominant development is generally expected to result in a relatively slower pace of job growth in the area. For instance, this scenario would generate a much smaller number of white-collar jobs compared to Scenarios 4 and 5, while the growth rate for white-collar jobs is even lower under the third scenario (i.e., retail-heavy development). The growth rate for blue-collar jobs would also be lower in this scenario than in most of the other scenarios. However, it appears that the housing-heavy development would reduce unemployment in the area, even though its contribution to job creation in the broader region is questionable.

Scenarios 2 through 4 suggest that job creation can be achieved more effectively by devoting a certain proportion of the land to non-residential uses. This does not mean that all types of jobs can be equally generated through the provision of non-residential space. In fact, we find the mechanism to be quite complex, since each type of non-residential land use has not only direct effects on white-collar, retail, and blue-collar jobs but also indirect effects via the linkages between job types (e.g., a negative effect of white-collar job change on retail job growth). Industrial development (Scenario 2), for example, would induce a large increase in blue-collar jobs and a moderate increase in white-collar jobs, while retail job growth could be dampened. The office-heavy development scenario (Scenario 4) could create an even larger number of white-collar jobs, but again at the expense of retail job opportunities.

A mixed development of housing, retail, offices, and industrial sites (Scenario 5) could reduce the conflicts among the three types of jobs. It is projected that this mix of land use could increase white-collar and blue-collar jobs by +29% and +18%

respectively, while the retail job growth rate is not as high as that for housing-heavy development. It is important to note that this mixed development scenario had both the lowest average household income of new residents and the lowest average home loan values.

In sum, we see complex tradeoffs that need to be taken into account in the collective land use decision-making process. One of the tradeoffs is the tension between job growth (considered important from a regional perspective) and the creation of a residential community that can attract residents with a lower level of unemployment (often supported by local stakeholders). The impact assessment also reveals that the land use decision is closely associated with detailed job change patterns in the area and that tradeoffs exist between white/blue-collar and retail jobs.

5 Model Validation

We validate our statistical models through the following steps. First, we estimate the model for the entire region from 1990 to 2001 and obtain the coefficient estimates. Second, we compute the land use change that actually occurred in census tracts or zipcode areas from 2001–2005. Third, we multiply those land use changes by the coefficients from the model. We also multiply the values of the other exogenous variables in the model by the estimated coefficients and compute the predicted value of the outcome variable of interest in 2005. We then compute the predicted values for each subsequent year by multiplying the coefficients by the values of the exogenous variables and the predicted value of the outcome variable of interest from the previous year.

This approach may cause our model projections to diverge from real values further into the future. For example, whereas the correlation between the value predicted by the model and the actual sales price ranges from 0.92 to 0.97 from 1992–2001 (when real data are being used to estimate the model), the correlations fall to 0.64 to 0.67 during 2002–2006 (when the data are outside the range of the model, but we do not use the predicted values for the previous year sales price to compute new predicted values, and use real values instead).[5] The key question then is how the model does when projecting time points beyond the data (after 2006, when we are using the predicted values for the sales price for the previous year to compute new predicted values).

[5]For the unemployment models in zip code areas, the correlations in the earlier years are above 0.98 from 1992–2001, and from 0.87 to 0.99 from 2002–2006. For the models for average loan values using data aggregated to tracts, the earlier year correlations range from 0.57 to 0.92 from 1991–2001 and about 0.91 to 0.92 during 2002–2006. For the average income level of new residents the earlier year correlations range from 0.34 to 0.91 from 1991–2001, and about 0.86 to 0.89 during 2002–2006.

Our validation checks suggest that for the average sales price models, the correlations between our predicted values and actual values are 0.51 in 2007, 0.45 in 2008, 0.43 in 2009, 0.41 in 2010, 0.40 in 2011 and 0.40 in 2012. The validation checks for the unemployment models using data aggregated to zipcode areas show correlations of 0.66 in 2007, 0.53 in 2008, 0.46 in 2009, 0.39 in 2010, 0.35 in 2011, and 0.31 in 2012. For the models predicting types of jobs, validation checks show correlations for each type of job are found to be over 0.95 from 2007 to 2010. For the models for average loan values using data aggregated to tracts, the validation checks show correlations of 0.82 or 0.83. Similarly, for the average income level of new residents in tracts, the correlations range from 0.75 to 0.82.

These validation checks assess how well our model does in explaining the neighborhood change trajectories compared to what actually occurred and indicate that our models produce good projections for the future number of jobs, the average loan values and the average income of new residents. The models are basically satisfactory, although less effective, in projecting average sales prices and unemployment rates in zip code areas. It may be that these larger units negatively affect the performance of the models, although we cannot be certain without a more rigorous investigation beyond the scope of this study.

6 Conclusion and Outlook

Urban site renewal has huge potential as a means of curbing unchecked urban expansion (generating serious social, fiscal, and environmental problems–see e.g., Ewing 1996; Johnson 2001; Burchell et al. 2005) and preventing abandonment of core areas, thus enabling more sustainable urban development. However, in reality, site renewal projects in urban areas have often been impeded not only by real estate market uncertainties but also by many regulatory and political barriers (Farris 2001). Difficulties also exist in building consensus and garnering public support, particularly when the projects are expected to produce large impacts on nearby areas.

In this chapter, we provide a way to utilize LUCC simulation and impact assessment models to support site renewal and associated decision-making processes. As discussed above, these tools can help us understand what is likely to happen and test various alternative scenarios in a coherent manner. Moreover, they can provide an opportunity to complement traditional environmental or transportation impact analysis techniques and fill the information gaps in a way that can facilitate dialogue among various stakeholders as well as planning professionals. Ideas for the future of the site can be effectively explained to the public through land use visualization, land use-based scenario development, and relevant socio-economic projections or by bringing stakeholders into the process of LUCC modeling and simulation (Voinov and Bousquet 2010; Pettit et al. 2011; Voinov et al. 2016). However, LUCC simulation technology does not always guarantee success. Intrinsic nonlinearities, scale dependency and other sensitivities in LUCC simulation are challenging problems that must be overcome (Kim 2013). Stakeholders also often fail to understand various assumptions behind simulation models and differentiate the model results from reality (Becu et al. 2008). Understanding how to convey the model information (including the model specification, assumptions, inputs and outputs) is crucial to realizing the great potential of LUCC simulation and impact assessment models.

It is also important to put model results into context. While this type of model-generated information is generally valuable in the planning process, it must be tempered with an understanding of the effects of land use decisions over time on communities situated in a particular planning context. Land use decisions can have a cumulative effect on a community, often with unintended consequences. For example, in our case, years of undersupply in the residential market has resulted in high rents and home purchase prices. In fact, the Orange County Business Council, in its 2015 Housing Scorecard, asserts that "Insufficient planning for, and provision of, workforce housing supply will impede Orange County's growth potential and continue to perpetuate the region as 'desirable but unattainable' for recent graduates, many new families, and workforce talent that might otherwise move to the county" (Orange County Business Council, 2015, p. 43). Land use distributions, therefore, must be assessed, not merely on the basis of high job creation in the past and high future housing values, but also on whether past growth has laid the foundations for a healthy economy and vibrant communities in the future.

Acknowledgements This material is based upon work supported by the Metropolitan Futures Initiative in the School of Social Ecology at the University of California, Irvine. The authors thank Harya S. Dillon, Hiroshi Ishikawa, Asiya Natekal, and Amrita Singh for their excellent research assistance.

References

Becu N, Neef A, Schreinemachers P, Angkapitux C (2008) Participatory computer simulation to support collective decision-making: potential and limits of stakeholder involvement. Land Use Policy 25(4):498–509

Burchell RW, Downs A, McCann B, Mukherji S (2005) Sprawl costs: economic impacts of unchecked development. Island Press, Washington, DC

Chaskin RJ (1998) Neighborhood as a unit of planning and action: a heuristic approach. J Plan Lit 13(1):11–30

Cheshire P, Sheppard S (2005) The introduction of price signals into land use planning decision-making: a proposal. Urban Stud 42(4):647–663

City of Irvine (2012) Heritage Fields Project 2012 GPA/ZC Second Supplemental Environmental Impact Report. Retried from https://legacy.cityofirvine.org/cityhall/cd/planningactivities/pda/heritage_fields_revision/great_park_neighborhoods_eir/

Ewing RH (1996) Characteristics, causes, and effects of sprawl: a literature review. Environ Urban Stud. 21:1–15

Farris JT (2001) The barriers to using urban infill development to achieve smart growth. Hous Policy Debate 12(1):1–30

Fragkias M, Geoghegan J (2010) Commercial and industrial land use change, job decentralization and growth controls: a spatially explicit analysis. J Land Use Sci 5(1):45–66

Hipp JR (2007) Block, tract, and levels of aggregation: Neighborhood structure and crime and disorder as a case in point. Am Sociol Rev 72(5):659–680

Johnson MP (2001) Environmental impacts of urban sprawl: A survey of the literature and proposed research agenda. Environ Plan A 33(4):717–735

Kaiser EJ, Godschalk DR, Chapin FS (1995) Urban land use planning. University of Illinois Press, Urbana

Kim JH (2011) Linking land use planning and regulation to economic development: a literature review. J Plan Lit 26(1):35–47

Kim JH (2013) Spatiotemporal scale dependency and other sensitivities in dynamic land-use change simulations. Int J Geogr Inf Sci 27(9):1782–1803

Kim JH, Hipp JR, Basolo V, Dillon HS (2017) Land use change dynamics in Southern California: does geographic elasticity matter? J Planning Educ Res. doi:10.1177/0739456X166882

Koomen E, Rietveld P, de Nijs T (2008) Modelling land-use change for spatial planning support. Ann Reg Sci 42(1):1–10

Koomen E, Borsboom-van Beurden J (eds) (2011) Land-use modelling in planning practice. Springer, Heidelberg

Kranser L (2005) The grounding of El Toro. The orange county business Journal. Retried from http://www.eltoroairport.org/issues/grounding.html

Lamb J (2009) Guide to the collection on the development of the El Toro airport. MS-R141. Special collections and archives, The UC Irvine libraries, Irvine, California. Retrieved from http://pdf.oac.cdlib.org/pdf/uci/spcoll/r141.pdf

Lee DB Jr (1981) Land use planning as a response to market failure. In: de Neufville JI (ed) The land use policy debate in the United States. Plenum Press, New York, pp 149–164

McCall MK, Dunn CE (2012) Geo-information tools for participatory spatial planning: fulfilling the criteria for 'good' governance? Geoforum 43(1):81–94

Orange county business council (2015) Orange county workforce housing Scorecard. Retrieved from http://www.ocbc.org/research/workforce-housing-scorecard/

Orange County Great Park (nd) History of the Land. Retrieved from http://www.ocgp.org/learn/history/

Pennington M (1999) Free market environmentalism and the limits of land use planning. J Environ Plan Policy Manag 1(1):43–59

Pettit CJ, Raymond CM, Bryan BA, Lewis H (2011) Identifying strengths and weaknesses of landscape visualisation for effective communication of future alternatives. Landsc Urban Plan 100(3):231–241

Plotkin S (1987) Property, policy and politics: towards a theory of urban land-use conflict. Int J Urban Reg Res 11(3):382–404

Stockstill M (2014) Too big to fail? Planning 2014(April):20–24

Taleai M, Sharifi A, Sliuzas R, Mesgari M (2007) Evaluating the compatibility of multi-functional and intensive urban land uses. Int J Appl Earth Obs Geoinf 9(4):375–391

Voinov A, Bousquet F (2010) Modelling with stakeholders. Environ Model Softw 25(11): 1268–1281

Voinov A, Kolagani N et al (2016) Modelling with stakeholders–next generation. Environ Model Softw 77:196–220

Zheng HW, Shen GQ, Wang H, Hong J (2015) Simulating land use change in urban renewal areas: A case study in Hong Kong. Habitat Int 46:23–34

Zhou B, Kockelman KM (2008) Neighborhood impacts on land use change: a multinomial logit model of spatial relationships. Ann Reg Sci 42(2):321–340

Chapter 17
Modeling the Future Evolution of Chilean Forests to Guide Current Practices. Native Forest and Industrial Timber Plantations in Southern Chile

N. Maestripieri, M. Paegelow and G. Selleron

Abstract Scientific research builds projects and seeks to achieve specific goals that refer to the principles of scientific inference: deduction, induction and abduction. These inferences correspond to the time path of the prediction—which belongs to the world of rationality and accuracy—and scenarios—which transcribe the uncertain nature of the studied process and can describe, in some cases, a probable future, desirable or not. Because the conclusion of deductive inference stems from premises, predictive simulation must be the result of past observations. Optimization of these results requires a rigorous calibration of the model, in order to reproduce a known situation (past or present). Scenarios are not predictions. For exploratory scenarios (*forecasting*), plausible hypotheses are built from observed processes and can only be verified a posteriori. The scenario begins with a given situation in the present and moves forward into the future, responding to the question "*What may happen if …?*" The normative scenario (inductive inference) describes a probable or desirable (or undesirable) future and then moves backwards to the present, i.e. retrospectively. The attitude is proactive towards the future and responds to the question "*How can a specific target be reached?*". These inferences give rise to specific approaches in terms of modeling and simulation. By focusing on forest dynamics in the south of Chile, this paper presents an expert approach (multi-criteria evaluation with Markovian chains) to map predictive and exploratory scenarios. The results open up various interesting lines of discussion in terms of resource management and clearly show the importance of model calibration (choice of data and configuration) upstream of the simulation process.

N. Maestripieri (✉) · M. Paegelow · G. Selleron
GEODE, University of Toulouse Jean Jaurès, Toulouse, France
e-mail: nicolas.maestripieri@univ-tlse2.fr

M. Paegelow
e-mail: paegelow@univ-tlse2.fr

G. Selleron
e-mail: gilles.selleron@univ-tlse2.fr

M.T. Camacho Olmedo et al. (eds.), *Geomatic Approaches for Modeling Land Change Scenarios*, Lecture Notes in Geoinformation and Cartography,
https://doi.org/10.1007/978-3-319-60801-3_17

Keywords Modeling · Prediction · Exploratory scenarios · Multi-criteria evaluation · Southern chile

1 Introduction

Southern Chile has witnessed severe change in its landscape over the past 40 years. The application of Decree Law 701 by the military government of Augusto Pinochet in 1974 intensified forestry activity in many parts of the region, by encouraging the planting of forests on suitable land, devoid of tree cover and in most cases heavily eroded. It offered subsidies of between 75 and 90% of the costs of planting the trees on land classified as suitable for forest, which was declared exempt from taxation (Barrué Pastor 2004). Business production criteria quickly supplanted what was initially a soil protection strategy. Between 1975 and 2007 an average of 95,141 hectares of timber forest (afforestation and reforestation) were planted each year nationwide, reaching a total of 2.2 M hectares in 2007 (infor 2008). By 2011 this area had reached 2.8 M according to the update of the Cadastre of Native Vegetation Resources of 2011.[1] The “conquest” of new spaces and the concentration /monopolization of land holdings in the hands of a small group of actors has created a complex set of relationships between different components of the biophysical and human worlds. The spatiotemporal dynamics of industrial timber forest plantations (pine and eucalyptus) is a multifaceted process with major environmental and socioeconomic consequences (Donoso and Otero 2005; Altamirano and Lara 2010; Zamorano-Elgueta et al. 2015). In addition, the medium and long term vision necessary for the sustainable management of forest resources has been negatively affected by an excessively short-term approach.

The stakeholders (direct and indirect) in the Chilean forestry sector face a lack of clarity (in terms of land availability) regarding the future evolution of timber plantations. There is an essential need for a long-term vision of land use, to optimize the definition of a sustainable forestry policy and overcome the problems at local level. This gives rise to two main tasks. The first, based on a retrospective approach (Maestripieri and Paegelow 2013; Maestripieri et al. 2015), involves assessing how the industrial timber plantations evolved (rhythmicity and underlying factors) and their spatial trajectories and the second requires the use of prospective modeling to determine their spatiotemporal dynamics in the medium term. Prospective modeling can provide a useful framework for analysis as it embraces spatial concepts, temporality (past-present-future) and the intentionality of the stakeholders. It therefore focuses on the complexity of the interactions that affect or are affected by the landscape.

The main goal is to develop a dynamic and spatially explicit model based on an exploratory time path in order to simulate exploratory prospective and predictive

[1]http://www.conaf.cl/nuestros-bosques/plantaciones-forestales/.

scenarios. Models like these are reflexive support tools which by understanding and acting on the spatiotemporal dynamics of the different types of land use can mitigate their impacts on the environment, the economy and society in general.

2 Test Areas and Data Sets

We modeled land use change in the municipality of San Juan de la Costa in the Los Lagos Region of southern Chile (Fig. 1) (latitude 40°14–40°44S and longitude 73° 18–73°48 W). The municipality covers an area of approximately 1520 km^2 and has two of the characteristic landscape units of southern Chile. The pre-coastal range (east) with an average altitude of 300 m.a.s.l., and the Coastal Range (west) with peaks that reach 800 m.a.s.l. and receive abundant rainfall (3500 mm /year). With more than 96,000 ha in 2008, the native forest covers almost all of the Coastal Range and fragmented parts of the pre-coastal range, although its area has decreased by 13% since 1986. Industrial timber plantations however are located exclusively in the pre-coastal range and have followed a quite different growth trajectory, increasing quickly from a mere 387 ha in 1986 to 9888 ha in 2008 (+2455%).

According to the latest census (INE 2002), the population of San Juan de la Costa is 8831 inhabitants, a high proportion of whom live in rural areas (7929 inhabitants or 90%). Ethnically speaking, most of the people are of Mapuche-Huilliche origin (60% of the total population), and of the 62 indigenous communities identified in 2000, 44 were located on the pre-coastal range (Maestripieri and Paegelow 2013).

In order to understand the past dynamics of land-use and cover changes (LUCC) at multiple scales, we used thirteen Landsat and Spot satellite images, although only three Landsat images were used for the calibration and simulation of our models (Thematic Mapper from 1986 and 2008; Enhanced Thematic Mapper from 1999). The vesting period for the Landsat images corresponds to the months of September and December, the spring season in the southern hemisphere. The temporal coverage is interesting because it began two years after the promulgation of the DL 701 and ended with the first field surveys in 2008. The spatial resolution of all the images is 30 m.

Additional digitized data were also collected. These data (shapefile) are from several state agencies (Department of *Prospección Sectorial* of the Corporación Nacional Forestal (CONAF), the Instituto Geográfico Militar (IGM), and the Corporación Nacional de Desarrollo Indígena (CONADI) and are combined in a Geographic Information System.

Digital cadastral data from 1999 (generated by the CIREN) were obtained by a forest engineer from the CONADI. This information was initially provided by the Servicio de Impuestos Internos (SII), and considers all changes in property boundaries—including subdivision and/or merger of various properties.

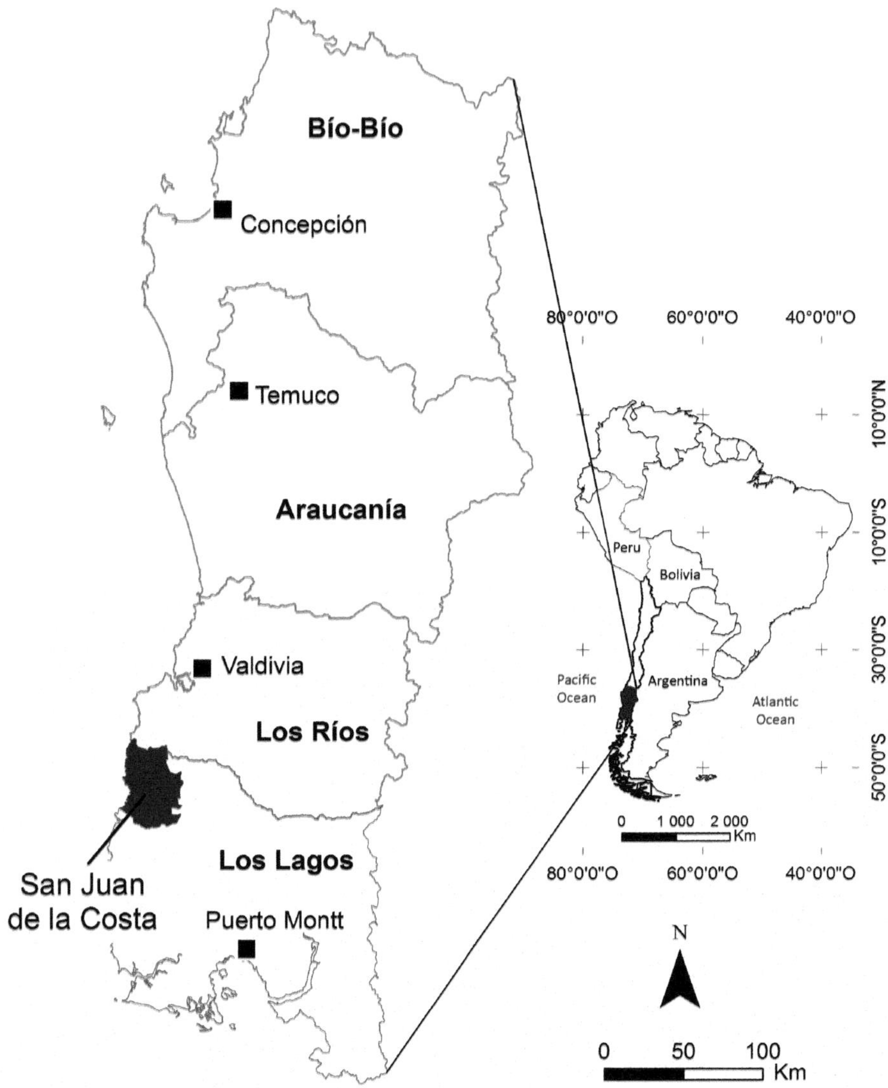

Fig. 1 Municipality of San Juan de la Costa

The information includes the name of the town, the name of the property and of the owner, the ROL SII (the property's identification number) and the total area.

We completed our study using Google Earth, a free application with a 3D Geographical Information System accessible online. It consists of satellite images and high (and very high) resolution aerial photographs (Quickbird). This system covers our entire study area although the spatial resolution varies from one part to the next. The application allows us to integrate various georeferenced pieces of information, such as raw and processed satellite images (MODIS-Landsat). Browsing is made easier with a Time Slider, which enables the user to estimate the evolution of the vegetation.

3 Methodology and Practical Application to the Data Sets

Before choosing the right model, we have to present the hypothesis on which each scenario is based.

3.1 *Prospective Scenarios*

Two exploratory scenarios are developed. The first is a *business-as-usual* scenario (the time horizon is 2017). This scenario is a prediction and not a prospective model, even if we used scenario-building techniques. The second is a 'sustainable development' scenario (the time horizon is 2035). Both are non-participatory models and are added to two normative scenarios (eco-centric and intensive) (Fig. 2) that were presented in previous research (Maestripieri et al. 2015).

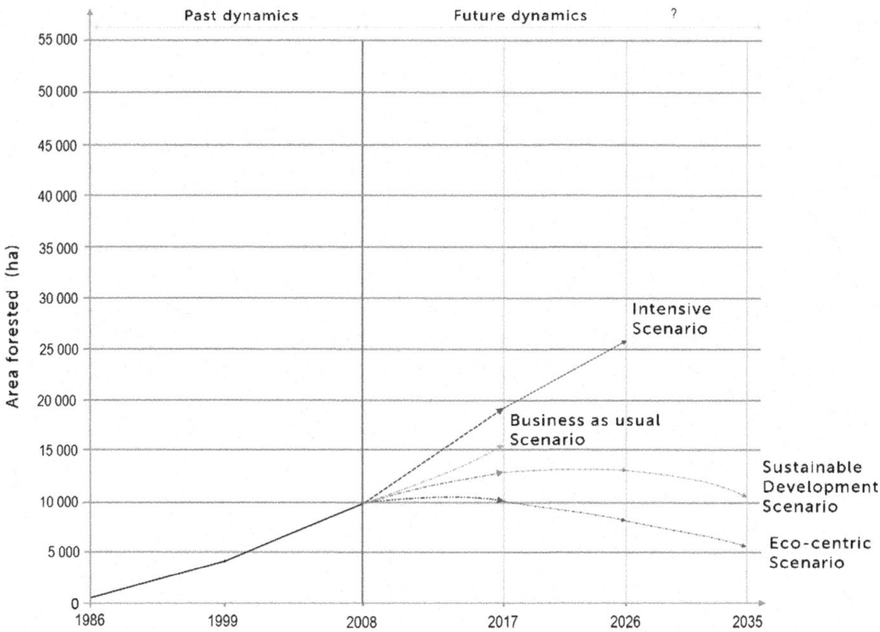

Fig. 2 Representation and evolution of scenarios between 2008 and 2035

3.2 Business-as-Usual Predictive Scenario and "Sustainable Development" Exploratory Scenario

The business-as-usual scenario is designed to predict future developments on the basis of "the logical consequences of prior assumptions or trends". We hypothesize that the growth in industrial plantations in the municipality will remain constant. In other words we do not foresee either a slow-down in the rate of growth or, conversely, large-scale land purchases.

Sustainable development at a regional level requires a project that promotes the implementation of sustainable forestry for both native forests and exotic plantations (Cruz and Schmidt 2007, pp. 290–293). This must take into account the economic situation of the study area, the possible environmental consequences and the expectations of local people. In this case, "expectations" must be viewed within a context of autonomous decision-making vis-à-vis the central government. Under this scenario, native forest assets will increase slightly between 2008 and 2035 while timber plantation will stagnate. The main hypotheses of both scenarios are listed in Table 1.

Before modeling these scenarios we need to understand how the model works and how the driving forces can be integrated into this model. The model has been calibrated with known past dates (1986–1999–2008) so as to assess how it works and how the drivers "react".

3.3 Calibration

Two models were calibrated (Maestripieri and Paegelow 2013): Land Change Modeler (Artificial Intelligence) and CA-Markov (expert approach). The latter, presented here, combines a Markovian procedure (Markovian probability maps) and a multi-criteria evaluation (MCE) approach for the spatial allocation of future LUCC.

For the prediction, the quantity of change depends on Markovian chains and takes into account the rate of transition between 1999 (t1) and 2008 (t2). The latest image is the one for 2008, which can be used for the first simulation test, calibrated by two earlier dates (t0 = 1986 and t1 = 1999) (Markov chain of order 2). Using the images for 1986 and 1999 we can extrapolate the future quantity of change. As for the exploratory scenario, the quantity of change also depends on Markovian chains, but the modeler changes the rates according to the scenario. The spatial allocation also varies because of the changes in the weight factors.

Table 1 Overview of the hypotheses on which the scenarios are based

Hypothesis	Comments
Increase of production capacity	Seguel (2010) notes that one of the ways that forestry conglomerates have to develop their forest resources and thus to increase production, is to rely on smallholders
Land availability	Moguillansky and Silva (2001) explain that "it is known that for large enterprises, the land available today in Chile has lower quality and higher prices, which makes it economically unprofitable"
Updating of DL 701	The changes seek to involve small and medium landowners and indigenous communities so as to increase the rate of afforestation in Chile
Wood energy	According to PricewaterhouseCoopers International Limited (PwC 2011), bioenergy has become the new corporate strategy for hedging against market fluctuations and meeting the requirements of sustainable development
Territorial governance	The centralization of power—characteristic of Chilean decision-making—gradually gives way to a decentralization process, to new forms of collective territorial organizations, independent of the government of local administrative units (Leloup 2005)
Mapu Lahual	The Land Mapu Lahual Project (Red de Parques Comunitarios Mapu Lahual) falls within this context of territorial enhancement. This is a conservation and ecotourism initiative that is part of an overall development strategy led by indigenous organizations
Agroforestry	Although the only potential use of the land in the municipality appears to be in forestry (Santana 2004), forest grazing could also be an option. Financial returns from planted and native forests are one of the most important factors driving forest management, conservation, and investments throughout the world (Cubbage et al. 2007). Grazing of livestock could provide regular income to the owners while they wait for their first harvest of wood (pruning, thinning)
Increase the value of the native forest resource by increasing sales of firewood	The economic attractiveness of native forest can be increased by reducing afforestation costs in line with those for exotic plants Cubbage et al. (2007) demonstrated that indigenous native forest management can contribute to positive financial returns, one of the most important factors driving forest management, conservation and investment around the world

3.3.1 Modeled Variables

Using the method of supervised classification we defined four land cover categories to classify the three Landsat scenes: (i) timber plantations, (ii) native forest (iii) other (non-forest land cover), and (iv) water (not included in the modeling process). By merging all non-forest land use into one category, we are simplifying the prospective modeling process and highlighting the timber plantation dynamic.

The process of selection of physical, natural, social and economic drivers is detailed in Maestripieri et al. (2015). Briefly, we used GIS and satellite data (from Landsat imagery) such as Land Use (LU), Distance from Existing Land Cover features (DExLC), Slope (SLP), Altitude (ALT), Coastal Range (CoR) and Pre-Coastal Range (Pre-CoR), Distance from Road Network (DRoNet), Land Tenure (LdTen), Distance from Coastal Road (DCoRo). Because the selection of criteria depends on the availability of the data, the MCE does not claim to provide completeness or optimum precision.

3.3.2 Methods for Estimating Quantity, Allocation and Calibration Outputs

Mapping scenario hypotheses depends on the MCE procedure and more specifically on the weight attributed to each driving factor in the modeling. The objective of the MCE is to generate suitability or probability maps by integrating a set of measurable and mappable criteria. These maps (hard-classified maps) can be used to develop specific land use strategies. The modeler controls the process by identifying and characterizing the driving forces with an expert approach (i.e. using his/her expert knowledge). We then discussed these driving forces with the interviewees, who determined the weight that should be given to each one. In addition, we analyzed the interactions between the changes observed (for instance in the shift from native forest to timber plantation) and the driving forces (land tenure, proximity to roads, and so on) (Maestripieri and Paegelow 2013). All the standardization process, weighting procedure and technique compensation between the factors and the level of risk-taking are presented in research by Maestripieri et al. (2015).

3.4 *Validation of the Model*

In order to ensure the optimization of the results and the accuracy of the predictive model, rigorous calibration is essential. Calibration is performed using empirical historical data and seeks to replicate a known situation. The validation of a model usually involves comparison with a real situation (Fig. 3). Although this step is essential for the modeler, it does not meet the expectations of planners and policy makers, who are looking for more specific results about future land use. The validation stage is important for the planners too however because, as Pontius and

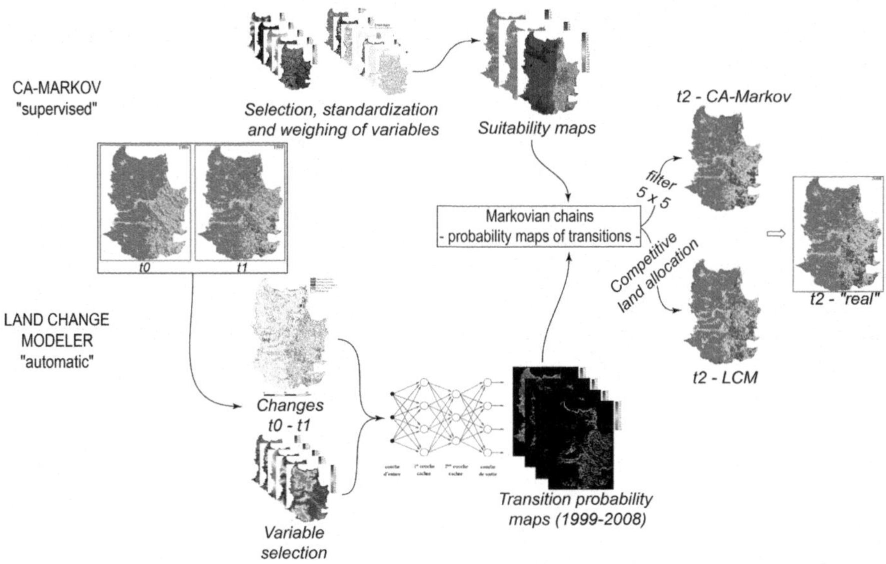

Fig. 3 Organigram of calibration process (CA_Markov and LCM)

Spencer (2005) point out, "one important purpose of the validation exercise is to allow the modeler and decision-maker to understand the appropriate level of confidence to have in the model as it extrapolates to points in time that are not known, for example, the future".

Unlike predictive models, when we create a model based on a scenario it is impossible to estimate its quality and accuracy. The success of a scenario must be assessed retrospectively by comparing it with the current situation. Hulme and Dessai (2008) explain that the success of a scenario is not so much a question of its (retrospective) accuracy or the (retrospective) efficiency of the decision, "but more on establishing an enabling condition for 'good' (robust) decisions to be made; i.e., in which a wide range of relevant uncertainties have been considered". The retrospective allows us a posteriori to reflect on the failures of a scenario and construct a sound base for the development of new decisions. "Retrospective helps prevent us from making the same mistakes, by helping to develop our knowledge of the content and implementation of methods" (Van Der Helm 2002, in Houet 2006).

Scenarios can be evaluated and validated by an expert and by stakeholders (Leclerc et al. 2010). Finally, Houet (2006) explains that "the evaluation may also focus on methods used in the construction of scenarios (models, probability …) as well as the prospective scenario verifying compliance with the four fundamentals of scenario building: relevance, coherence, plausibility and transparency".

3.4.1 Qualitative Validation

Given the temporal coverage of high-resolution images provided by Google Earth in the municipality (18/01/2005 to 19/02/2011), the 2008 classification is the only one that has been validated. The availability of these images is not synchronous to the entire area, as the timing gradient shortens the coverage period (which now runs from 26/01/2006 to 02/09/2010) to eliminate certain blurred images. Errors are located by importing the classification under GE and adjusting the opacity, so allowing the simultaneous visualization of the data.

We also compared each land cover of our supervised classification (2008) with the digitized and updated Cadastre of Native Vegetation Resources (CONAF, 2006).

3.4.2 Quantitative Validation

Although visual examination by comparing the reference map (2008) and the simulation (2008) provides an initial estimate of the quality of the prediction, it fails to locate and accurately quantify the errors and the correctly predicted areas. In order to overcome the subjectivity of the modeler, Pontius et al. (2004) proposed a statistical comparison between these maps. Here is a summary presentation of the three methods we used to assess and validate the models (each method will be detailed in the following pages):

- For hard-classified maps: LUCC-budgets (Pontius et al. 2004)
- By comparing two land cover maps at two different dates (*t1-t2*), the budget method highlights the components of the dynamic—*dominant signals of land change*—The aim is to compare two-LUCC budgets (*t1-t2* and *t1-t2 predicted*) in order to characterize the errors.
- For soft-classified maps: ROC (Pontius and Schneider 2001)
- Although it does not separate the errors due to amount from the errors due to location, the Relative Operating Characteristic evaluates the quality of prediction in terms of location. This is done by comparing a binary map (land use) with a suitability map.
- Budgeting of errors/accuracies (Chen and Pontius 2010)
- This method allows us to quantify and locate errors/accuracies in LUCC by crossing two reference maps (t1 and t2) and a prediction map (t2'). Comparison of the observed and predicted changes generates four categories of pixels: *null successes, false alarms, hits* and *misses* (Maestripieri and Paegelow 2013).

4 Results

4.1 Calibration Results

> The problem is that scientists usually already know that persistence dominates the landscape. Scientists want to identify the dominant signals of land change. [...] identify the signals of change separately from any given level of persistence (Pontius et al. 2004).

The swap measures changes in spatial allocation by subtracting the balance (gains minus losses) from the total change (gains plus losses). Let us imagine for instance that a land use category recorded a gain of 30 ha between two dates and a loss of 20 ha over the same period. The total change, calculated by adding these two amounts together, is 50 ha, the balance (net change) is 10 ha and the swap is 40 ha. Although the area covered by this land use increased by only 10 ha, this low figure hides significant spatiotemporal dynamics, as land use changed on 50 ha.

In our study, timber plantations represented 2.7% of the landscape in 1999 and 6.4% in 2008. Figure 4 shows that all the changes in this category were gains (3.7%). No losses were observed, which also implies a swap of 0. The figure shows that the CA-Markov model perfectly simulates the dynamic.

The native forest is the second largest category in terms of total change (7.9%—swap 3% and net change: 4.8%). The model clearly underestimated total change

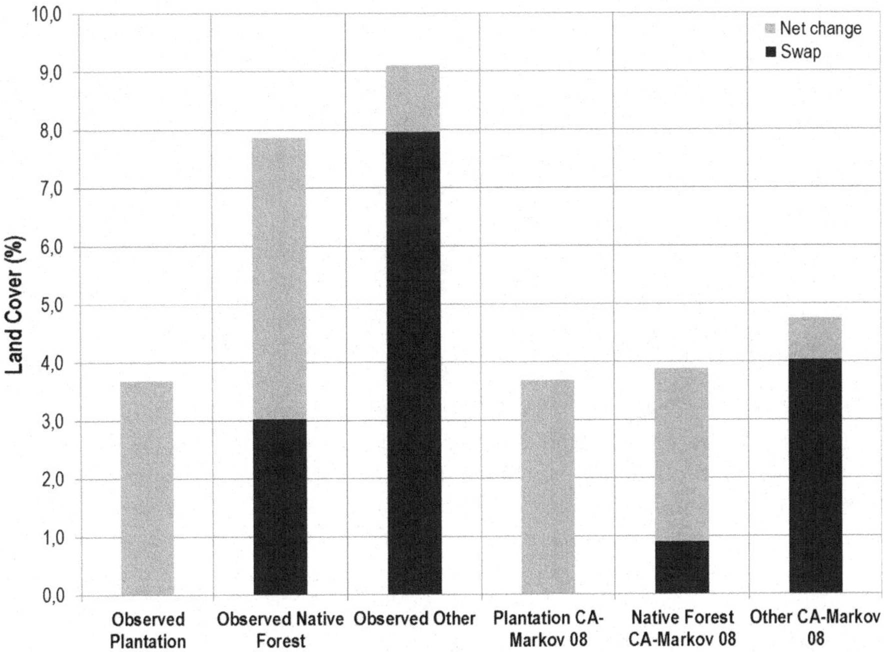

Fig. 4 Comparison between LUCC-budget (1999–2008) and simulated LUCC-budget (1999–2008 CA_Markov). Net change + swap equals total change (%)

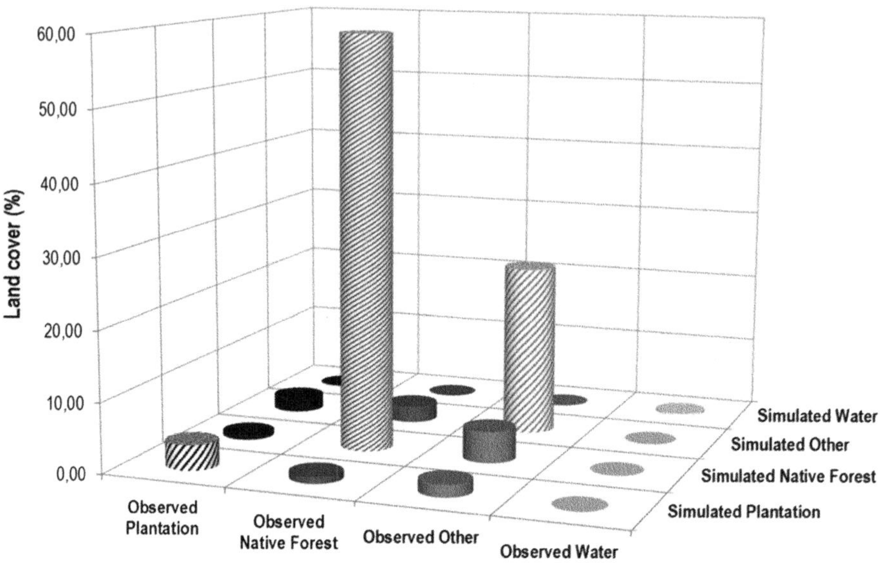

Fig. 5 Residues between observed and simulated (CA-Markov) land cover. Cylinders with stripes represent concordances

(3.9%), which is mainly due to a minimization of loss (3.4% against 6.4%) and higher persistence (64.1% against 61.2%). Nevertheless the share between swap and net change is correctly simulated.

The "other land use" category shows the highest total change with 9.1%. Although the net changes were only 1.2%, there was a swap of 7.9%, reflecting a loss in a given location (afforestation and natural recolonization—4%), offset by gains elsewhere (deforestation—5.1%). Although the 4.7% figure predicted by the model for total change was an underestimation (it simulated 2% for deforestation against the real figure of 5.1%), the ratio between net change and swap is correct.

In order to characterize under (and over) estimation vis-à-vis the observed map, i.e. to understand where the mistakes (or residues) come from, we compared the land cover map in 2008 to the CA-Markov simulation for 2008.

This comparison revealed two important facts: (i) the share of spatial concordance (striped cylinders—Fig. 5) and (ii) residues that represent the simulated types of land use that do not correspond to the real land cover in 2008.

Of the 6.4% of plantations simulated in 2008, 3.7% are consistent with observation. The remaining 2.73% (0.99% of native forest and 1.74% of "other land use") are the residues and are located, for the most part, near the plantations. Conversely, the residues for the other categories (2.02% for "other land use" and 0.71% for native forest) translate simulation errors in which the model fails because it does not predict the appearance of new plantations.

For timber plantations, ROC classifies their suitability classes in descending order with thresholds defined by the modeler. The occurrence of each resulting class

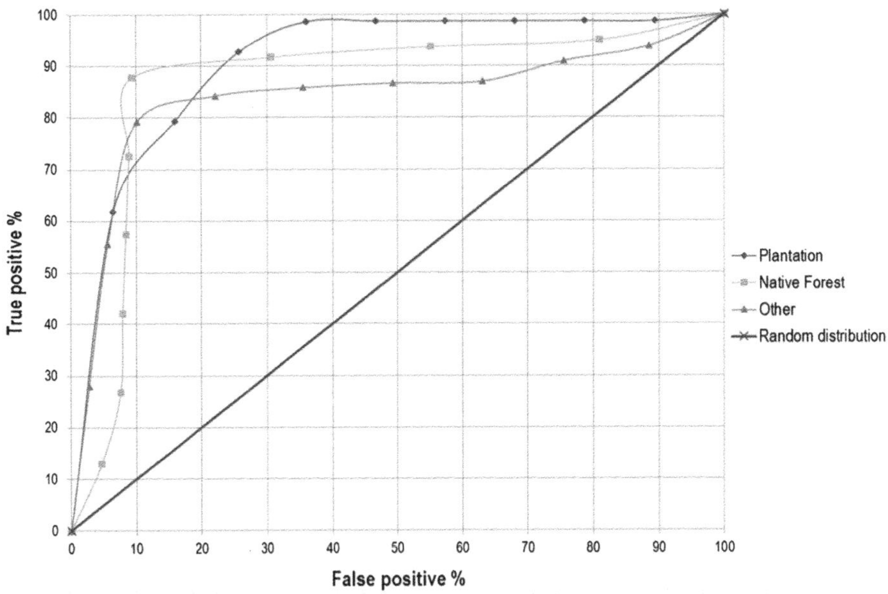

Fig. 6 ROC for CA_Markov model

is compared to the real location map (Paegelow and Camacho Olmedo 2008) to determine whether it actually corresponds to plantations (positive true) or non-plantations (false positive).

Performance is measured by the Area under the Curve (AUC). If the suitability values for land use correspond perfectly to their location on the map, then the ROC will equal 1. Pérez Vega et al. (2012) argue that "a highly predictive model will produce a curve that rises rapidly from the lower left to a point near the upper left corner and then moves slowly near the upper edge of the graph to reach the upper right hand corner".

If the suitability values were randomly distributed between plantation and non-plantation for example, the ROC would be 0.5 (random distribution in Fig. 6). The AUC for plantations simulated by CA-Markov is 0.90, 0.87 for native forest and 0.84 for the "other land use" category, which shows that the model took the suitability values for each category into account to ensure their location.

As we explained above, the values were discussed with the interviewees who compared the factors to determine their relative importance. These values were integrated into a pairwise comparison matrix (Table 2). The majority of the variables presented in each scenario are non-spatial, so we had to translate them into spatial variables and/or "play" with spatially explicit variables in order to come closer to our hypotheses.

The assumptions mentioned for both scenarios (decree law, agroforestry, etc.) cannot be directly introduced into the model because they are non-spatial data. The best way to include these assumptions is by adjusting the weight factors to take

Table 2 Pairwise comparison matrix—sustainable development scenario (objective PLANTATION)

	LU	LdTen	CoR	Pre-CoR	DExLC	DRoNet	SLP	ALT
LU	1							
LdTen	9	1						
CoR	2	1/6	1					
Pre-CoR	6	1/2	2	1				
DExLC	5	1/3	4	1	1			
DRoNet	1/2	1/4	3	1	1/3	1		
SLP	3	1/3	2	1/2	1/3	2	1	
ALT	1/4	1/7	1/2	1/4	1/4	1/3	1/3	1
Eigenvalue	0.06	0.33	0.06	0.15	0.18	0.08	0.10	0.03
Consistency ratio: 0.08 (acceptable)								

them into account. For instance, the updating of the DL 701 in 2011, just like its earlier amendment, will have no impact (in the short-term) on indigenous communities, and only affects the medium and (very) large landowners. As part of the multi-criteria evaluation, we therefore give large landowners a high weight of 255, while the weight for small landowners is very low (20).

In the MCE, the optimistic strategy (Strategy # 4) gives too much importance to pixels with high suitability at the expense of criteria with lower values, and therefore does not take account of the laws offering incentives to small landowners to plant forests (low suitability). Strategy # 2 (low risk taking and low levels of compensation—pessimistic strategy) gives a greater weight to factors with a lower degree of suitability.

This scenario sees the creation of new plantations and the reforestation of all existing areas in which trees are felled. The calculation of Markovian transition probabilities between 2008 and 2017 produces a total of 15,590 ha of timber plantations in the municipality.

4.1.1 "Sustainable Development" Exploratory Scenario

This scenario is halfway between the sustainability of exotic plantations (economic stability), which implies a slight drop in forest production, and the respect for indigenous claims resulting in the recovery of their ancestral lands and the establishment of development policies. This shift does not reduce the likelihood of further plantations on the foothills. According to Estades and Escobar (2005) "although pine plantations are no longer interesting, because of its great capacity for natural regeneration, its total eradication is extremely difficult (at least with existing techniques to date). This suggests that pine plantations are an artificial ecosystem that will dominate much of the landscape of the CC of Chile for a long time".

Thus, all the landowners obtain a maximum suitability score for the objective PLANTATION. The evolution of exotic plantations is modulated by estimating their coverage (ha) in 2035 in the Markovian matrix (10,690 ha). The strategy (for plantations) is identical to the business-as-usual scenario. Finally, and in order to promote the emergence of native forest in the "other land use" category, the suitability of this category is also maximum (255).

4.1.2 Results for the Different Scenarios

Established timber plantations remain stable between 2008 and 2017, making small gains from native forests (1926 ha, or 2% of native forests in 2008). Other land use also increases by 3776 ha, or 8%, to a total of 15,590 ha. These areas are illustrated in Fig. 7, which shows that all these changes take place in the foothills of the Coastal Range (to the east) near the plantations detected in 2008. If we compare these results with the transitions observed between 1999 and 2008, we can see that the direct substitution (native forest to new plantation) process is growing very slightly (+43 ha), while the rate of change from "other land use" to "plantation" has fallen.

The only category in which total area decreases is native forest, although losses were lower than in the previous period (6673 ha against 7928 ha). The trend suggests that these losses will continue beyond 2017. Most of this area (almost 4750 ha) was lost to the "other land use" category. The deforested areas are concentrated on the foothills and in the center of the Coastal Cordillera near the Ruta U-40.

Established timber plantations remain virtually stable between 2008 and 2035 with the only new spaces captured from the "other land use" category (801.9 ha, or 2%)

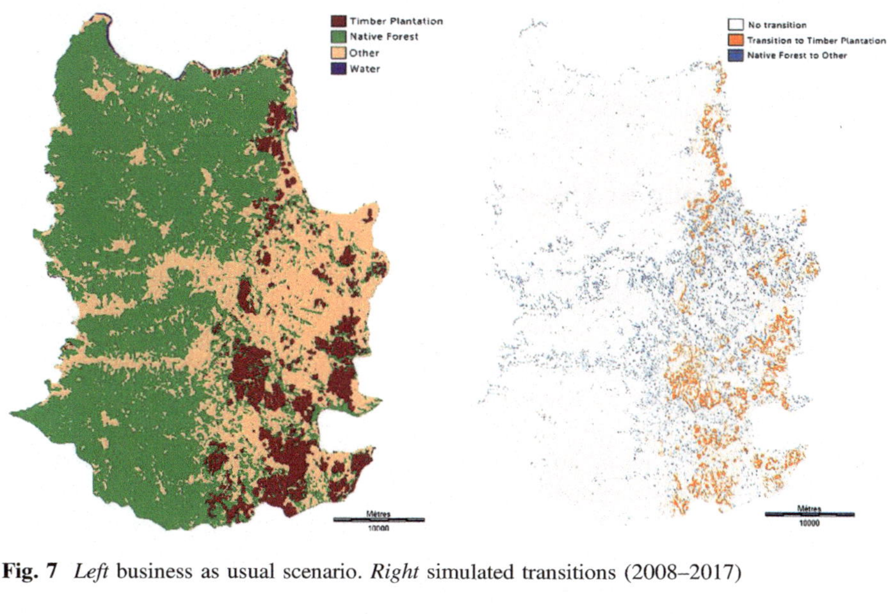

Fig. 7 *Left* business as usual scenario. *Right* simulated transitions (2008–2017)

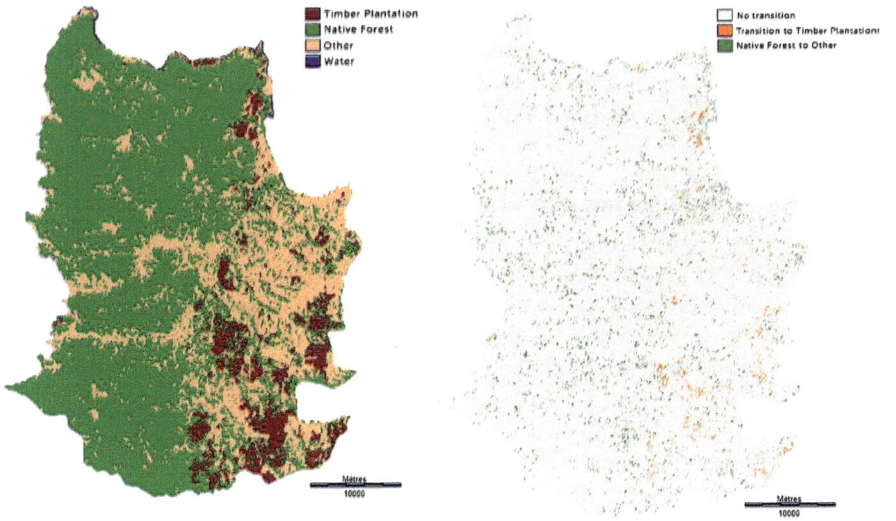

Fig. 8 *Left* exploratory prospective scenario. *Right* simulated transitions (2008–2035)

increasing the total to 10,690 ha. This scenario marks the end of direct substitution. Figure 8 shows that all of these areas are in the foothills near the plantations detected in 2008.

Our results for this scenario indicate that native forest will recolonize "other land use" areas in a diffuse and heterogeneous way, and will not suffer any losses. In 2008 native forest covered an area of 96,857 ha (63% of the landscape). In 2035 this area is expected to have expanded by about 3000 ha to around 99,000 (65% of the landscape). In addition to this natural 'reconquest', it is likely that native forests will be planted in the foothills, on managed forest land, according to a management plan.

5 Validation and/or Discussion of Results

The approach combining multi-criteria evaluation with Markovian chains offers good flexibility for modeling prospective scenarios. It is proving to be more effective than LCM for the calibration process (Maestripieri and Paegelow 2013) and makes the mapping of normative scenarios easier thanks to the weight attributed to each driving factor (Maestripieri et al. 2015).

However, there is room for improvement in the process by which the factors are weighted and then adjusted to take the non-spatially explicit assumptions into account. This step has a certain degree of subjectivity. This subjectivity is often wrongly perceived as a defect of the criteria selection process. It is also sometimes considered synonymous with inaccuracy or uncertainty (Joerin 1997).

The multi-criteria approach has an element of subjectivity because of the states of consciousness of the researcher, but also and above all because of his/her knowledge of the fieldwork. Belton and Stewart (2002) consider that subjectivity is inherent in all decision making, especially in the choice of the criteria on which the decision is based, and the relative weight given to these criteria.

There are also many issues surrounding decision making. Scenario planning has been defined as 'a process of positing several informed, plausible and imagined alternative future environments in which decisions about the future may be played out, for the purpose of changing current thinking, improving decision making, enhancing human and organization learning and improving performance' (Chermack and Lynham 2002). Even if we developed non-participatory models, these scenarios could only be built with the involvement of stakeholders. These models must be seen as a tool for reflection and debate, and not as a turnkey solution (Maestripieri et al. 2015). The question then arises as to the position of the stakeholders in this decision-making process. Godet (1993) argues that "a scenario approach can only be credible and useful if it complies with our prerequisites - relevance, consistency, likelihood and transparency". So, the next challenge is to present the scenarios (including the driving forces of each scenario and their weight) to local stakeholders to find out whether they consider them to be realistic or not.

6 Conclusion and Outlook

After failing to fully integrate the local population in the decision-making process, policymakers should question the productive model taking into account the environmental and socio-economic impacts generated by timber plantations dynamic and native forest fragmentation. Our retrospective analysis matches the conclusions of Zamorano-Elgueta et al. (2015), who found that "deforestation and native forest fragmentation in the Coastal Range of Región de Los Ríos was found to be less intensive than in other regions of Chile. […] Nevertheless, the continuing expansion of exotic tree plantations and loss and fragmentation of native forest may lead to microclimatic changes at the forest edges that may facilitate the spread of exotic species towards the interior of the forest fragments."

Government policy on forest resources (timber plantations and native forest) and their management at local and national level has an important role to play. There is also a need to strengthen public policies for the conservation of native forest outside protected areas (Miranda et al. 2015). The DL 701 (promulgated under the military government of Augusto Pinochet) should be reconsidered, even though the economic conditions of the forestry sector make timber plantations a highly profitable land use (Manuschevich and Beier 2016).

In this way, the exploratory scenarios (and prospective scenarios in general) may only include specialist expert opinion if—and only if—the knowledge of local stakeholders is taken into account (Kleiche-Dray and Waast 2016).

Acknowledgements The authors acknowledge with gratitude the ECOS-Sud Chile project n_ C07H03 entitled "La forêt de la cordillère côtière continentale dans le sud du Chili: Dynamiques contemporaines et modélisations prospectives".

References

Altamirano A, Lara A (2010) Deforestation in temperate ecosystems of pre-Andean range of south-central Chile. Bosque 31(1):53–64. doi:10.4067/S0717-92002010000100007

Barrué Pastor M (2004) Forêts et développement durable au Chili. Presse Universitaire du Mirail, Toulouse, Indianité Mapuche et mondialisation

Belton V, Stewart TJ (2002) Multiple criteria decision analysis: an integrated approach. Kluwers Academic Publishers, Norwell-Massachusetts

Chen H, Pontius RG (2010) Diagnostic tools to evaluate a spatial land change projection along a gradient of an explanatory variable. Landsc Ecol 25:1319–1331. doi:10.1007/s10980-010-9519-5

Chermack T, Lynham SA (2002) Definitions and outcome variables of scenario planning. Hum Resour Dev Rev 1(3):366–383

Cruz G, Schmidt H, (2007) Silvicultura en bosques natives. In: Hernandez PJ et al (eds) Biodiversidad: Manejo y conservación de recursos forestales, Santiago de Chile, Edición Universitaria, pp 279–307

Cubbage et al (2007) Timber investment returns for selected plantations and native forests in South America and the Southern United States. New For 33(3):237–255. doi:10.1007/s11056-006-9025-4

Donoso PJ, Otero LA (2005) Hacia una definición de país forestal: ¿Dónde se sitúa Chile? Bosque 26(3):5–18. doi:10.4067/S0717-92002005000300002

Estades CF, Escobar MA (2005) Los ecosistemas de las plantaciones de pino de la Cordillera de la Costa. In: Smith-Ramírez C et al (eds) Historia, biodiversidad y ecología de los bosques costeros de Chile. Edición Universitaria Bosque Nativo, Santiago de Chile, pp 600–616

Godet M (1993) Anticipation and scenarios are not synonymous. Futur 25(3):350–351. doi:10.1016/0016-3287(93)90143-H

Houet T (2006) Occupation des sols et gestion de l'eau: Modélisation prospective en paysage agricole fragmenté (Application au SAGE Blavet). Thèse de doctorat, Université de Rennes 2 - Haute Bretagne

Hulme M, Dessai S (2008) Predicting, deciding, learning: can one evaluate the 'success' of national climate scenarios? Manuscript for special Focus Issue of Environmental Research Letterson the topic 'Where next with global environmental scenarios?'

INE (2002) Estadísticas Sociales de los pueblos indígenas en Chile. Censo 2002, Instituto Nacional de Estadísticas / Programa Orígenes (MIDEPLAN / BID)

INFOR (2008) Anuario Forestal 2008. Centro de Información Forestal

Joerin F (1997) Proposition d'une approche par utilisation de SIG et de méthodes d'analyse multicritère. Département de génie rural, Ecole Polytechnique, Lausanne

Kleiche-Dray M, Waast R (2016) Indigenous knowledge in Mexico: between environmentalism and rural development. In De Castro F et al (eds) Environmental governance in Latin America, Palgrave Macmillan UK, pp 86–110. doi:10.1007/978-1-137-50572-9

Leclerc G et al (2010) Elaboration participative de modèles et de scénarios: une entrée pour analyser la coévolution des systèmes d'élevage extensif et des territoires. Cahiers Agricoles 19 (2):152–159

Leloup F (2005) La gouvernance territoriale comme nouveau mode de coordination territoriale? GéographieÉconomieSociété7:321–332. doi:10.3166/ges.7.321-331

Maestripieri N, Paegelow M (2013) Validation spatiale de deux modèles de simulation: l'exemple des plantations industrielles au Chili. Cybergéo—Eur J Geogr. http://cybergeo.revues.org/26042

Maestripieri N et al (2015). Explanatory prospective modelling of normative scenarios. The case of timber plantations dynamics in southern Chile. Futures (in press). http://dx.doi.org/10.1016/j.futures.2015.10.013

Manuschevich D, Beier CM (2016) Simulating land use changes under alternative policy scenarios for conservation of native forests in south-central Chile. Land Use Policy 51:350–362. doi:10.1016/j.landusepol.2015.08.032

Miranda et al (2015) Different times, same story: native forest loss and landscape homogenization in three physiographical areas of south-central of Chile. Appl Geogr 60:20–28. doi:10.1016/j.apgeog.2015.02.016

Moguillansky G, Silva V (2001) Estrategias empresariales y políticas públicas: el futuro del complejo forestal en Chile. In: Muñoz O (ed) Más allá del bosque: transformar el modelo exportador. Editorial FLASCO, Santiago de Chile, pp 107–144

Paegelow M, Camacho Olmedo MT (2008) Modelling environmental dynamics: advances in geomatic solutions. Springer, Berlin, Heidelberg

Pérez Vega et al (2012) Comparing two approaches to land use/cover change modeling and their implications for the assessment of biodiversity loss in a deciduous tropical forest. Environ Model Softw 29:11–23. doi:10.1016/j.envsoft.2011.09.011

Pontius RG, Schneider LC (2001) Land-cover change model validation by an ROC method for the Ipswich watershed, Massachusetts, USA. Agr Ecosyst Environ 85:239–248. doi:10.1016/S0167-8809(01)00187-6

Pontius RG et al (2004) Useful techniques of validation for spatially explicit land change models. Ecol Model 179:445–461. doi:10.1016/j.ecolmodel.2004.05.010

Pontius RG, Spencer J (2005) Uncertainty in extrapolations of predictive land-change models. Environ Plan 32:211–230. doi:10.1068/b31152

PwC (2011) Growing the future. www.pwc.com/fpp

Santana R (2004) Exploitation de la forêt originelle et reboisements dans la commune de San Juan de la Costa. Enjeux du développement durable à l'échelon local. In: Barrué-Pastor M. (dir), Forêts et développement durable au Chili. Indianité Mapuche et mondialisation. Presse Universitaire du Mirail, Toulouse, p 115–146

Seguel A (2010) Conflicto público de tierras y Recursos naturales: Expansión forestal y territorialidad Mapuche (Chile)

Van der Helm (2002) Comment faut-il appécier la prospective? Vers une évaluation ex post de la Vision Mondiale de l'eau, Rapport de DEA, ENGREF, p96

Zamorano-Elgueta C et al (2015) Native forest replacement by exotic plantations in southern Chile (1985–2011) and partial compensation by natural regeneration. For Ecol Manag 345:10–20. doi:10.1016/j.foreco.2015.02.025

Chapter 18
Urban Transportation Scenarios in a LUCC Model: A Case Study in Bogota, Colombia

D. Páez and F. Escobar

Abstract In this chapter, we present a practical implementation of a LUCC simulation based on transport scenarios. The model, called the Bogota Land Development model or BoLD, was built on Metronamica to address information gaps in decision-making. Using BoLD we modeled and compared two types of transport infrastructure: a highway-based transport network and a suburban rail system. These transport scenarios were combined with options to expand the city into green areas currently protected as nature reserves. Customized geospatial analyses were developed for calculating accessibility distance decay factors (ADDF) based on a methodology developed in this research called Over-Time Spatial Decay Calculation (OSDC). Results of the scenarios are presented graphically in what we call a Mobility Circle, a key contribution of this research. Validation of the results obtained suggests that both OSDC and the Mobility Circle appear to enhance the information available to decision-makers when evaluating urban scenarios driven by transport projects. In any case, those working in this field should approach LUCC based primarily on changes in transport systems with caution, as they provide a narrow view of future scenarios without clearly considering important aspects such as changes in land demand.

Keywords Transport-land-use interaction · Scenario planning · Indicators

D. Páez (✉)
Department of Civil and Environmental Engineering, University of Los Andes, Bogotá, Colombia
e-mail: dpaez@uniandes.edu.co

F. Escobar
Department of Geology, Geography and Environmental Sciences, University of Alcalá, Alcalá de Henares, Spain
e-mail: francisco.escobar@uah.es

M.T. Camacho Olmedo et al. (eds.), *Geomatic Approaches for Modeling Land Change Scenarios*, Lecture Notes in Geoinformation and Cartography,
https://doi.org/10.1007/978-3-319-60801-3_18

1 Introduction

Building scenarios in a LUCC model requires a thorough knowledge of the study region, as described in previous chapters of this book. This is particularly important when the scenarios explore changes in the transport infrastructure. The LUCC application presented here is unique as the scenarios focus on a specific transport technology instead of evaluating overall infrastructure solutions.

Political debates about the future of a city can center on specific transport infrastructure projects for specific areas (Al-Kodmany and Ali 2013). In some cases, particularly in developing cities, the focus is narrowed to projects involving one specific technology (for example, a new metro or tramway). Projects like these are gaining momentum in democracies around the world, at the expense of holistic or strategic infrastructure plans, and are often key priorities in the campaign programs of national and local politicians.

Transport models have been used for over 30 years. They are particularly important for designing infrastructure to meet expected demand, such as train or bus station capacity. They also provide valuable information for designing operational aspects, such as fleet size and frequencies (Ortuzar and Willumsen 2005).

New approaches to infrastructure financing, based on land development, are increasingly common today. Concepts like "transport oriented development" are used in urban planning in conjunction with sustainable development (Suzuki et al. 2014).

Given these new financial approaches to infrastructure, the planning and development of transport infrastructure requires important decision-making information that cannot be obtained using the traditional models which do not provide information about land-use cover changes (LUCC).

Although previous researchers have combined traditional transport models with LUCC models (Zhao and Peng 2012), little work has been done on the use of LUCC models as an overarching tool for urban planning in which specific transport technology alternatives are evaluated not only from a demand or operational perspective, but also as a driver of urban development.

In this chapter we present a LUCC model developed for the city of Bogota to evaluate transport alternatives in the growth areas to the west of the city. The model, which is known as the Bogota Land Development model or BoLD uses Metronamica software. Advanced spatial analyses on a Geographic Information System (GIS) were used to determine accessibility indexes for each type of infrastructure and land use. These analyses, combined with those produced for neighboring interactions between land-uses, allowed specific indicators to be developed to evaluate the proposed transport alternatives.

In order to understand land-use transport interaction, the scenarios evaluated with BoLD included the possibility of regulatory changes permitting development in an environmentally protected area in the north of the city.

2 Case Study: Bogota

In this section we present the case study developed for Bogota. We begin with an introduction describing the general urban geography of the city to provide the reader with a context for the model. We then explain the LUCC modeling needs in Bogota, which were used as the design parameters for the model. This section ends with an explanation of current debates in Bogota about land development and transport proposals. We used these proposals as the basis for the formulation of scenarios for the BoLD model.

2.1 General Context of Bogota

Bogota (Fig. 1) is the capital of Colombia. The city has a population of 8 million people, over 20% of the entire population of the country (Munoz-Raskin 2010). A large proportion of Colombia's business activity takes place in the city, which generates over 25% of GDP (DANE 2015).

From an administrative perspective, Bogota has the political status of "capital district". This status, awarded under the Colombian Constitution of 1991, allows it to be governed as an independent state and not as a local municipality. Bogota has significant autonomy in terms of land-use planning, taxation and infrastructure development and management.

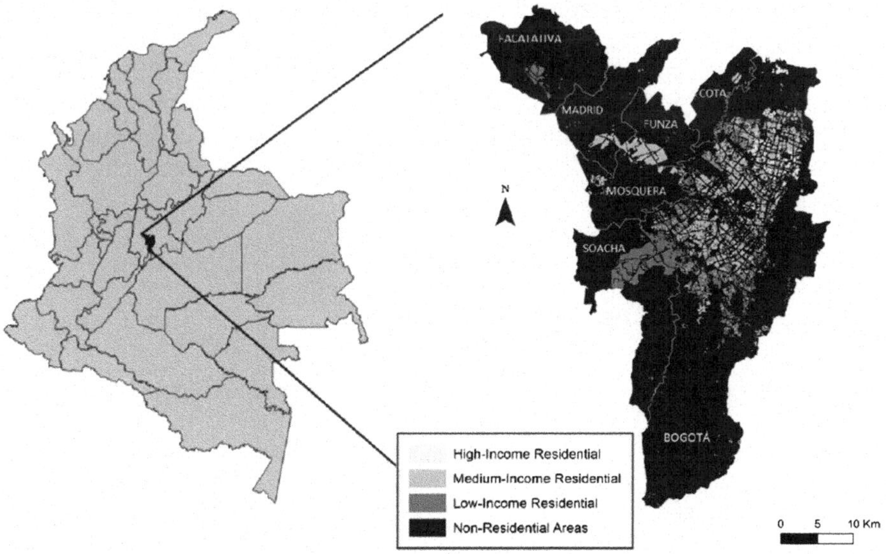

Fig. 1 Bogota and its surrounding municipalities, within the context of Colombia, and location of residential areas

The population of Bogota has grown from around 5 million in 1990 to almost 7.2 million in 2010. Population projections by the National Administrative Department of Statistics (DANE) estimate that Bogota will grow at an average rate of 1.1% per year in the next 25 years (DANE 2011).

The city's urban area extends across 355 square kilometers (Bocarejo et al. 2013) with an average population density of 20,500 inhabitants per square kilometer (Bocarejo et al. 2013). It is divided into 19 urban communities which are in turn divided into 112 planning zones. The purpose of the planning zones is to define and regulate urban land-use and management at a detailed scale (Bocarejo et al. 2013).

Urban growth trends in Bogota today show a concentration of low-income households in the south and west of the city, while the high-income households are located in the north and in the center (Bocarejo et al. 2013). Lack of available land for expansion and high land prices have forced low-income earners out to peripheral areas of the city (Oviedo Hernandez and Davila 2016). As available land within Bogota's jurisdiction runs out, most of the city's urban growth is likely to occur outside the municipality of Bogota. This is why the BoLD model incorporates six municipalities to the west of Bogota, namely Funza, Mosquera, Madrid, Facatativá, Cota and Soacha (Fig. 1). Most of their inhabitants are either low or middle-income. Cota, as the smallest town in this group, has a population of around 25,000 (DANE 2011), while Soacha has the highest population with more than 500,000 inhabitants (DANE 2011). The others have populations ranging be-tween 75,000 and 132,000 inhabitants.

In terms of density, the peripheral areas of Bogota have the lowest formal employment densities and the highest population densities. High population densities in these areas are due to minimal public spaces (including streets and parks) and small areas of privately owned land per inhabitant. They are not due to high-rise development (Bocarejo et al. 2013). In general these areas have low-rise housing with large numbers of occupants, whereas in the high-income areas there are high-rise developments with large flats with few occupants, so creating lower population densities. There is also a high concentration of formal employment (Bocarejo et al. 2013).

2.2 LUCC Modeling Needs in Bogota

BoLD was conceived to address the need to understand the global impacts of transport infrastructure projects on urban development. This is particularly important for Bogota, a city whose political agenda is driven by transport projects. In 1998, the local government created Transmilenio, a Bus Rapid Transit (BRT) system. It was developed as a response to deteriorating mobility in the city demonstrated by increasing travel times. Another objective of the new BRT system was to reduce greenhouse gas emissions by renovating the city's aging bus fleet (Suzuki et al. 2014).

The system increased accessibility in many areas of the city resulting in the creation of new urban growth centers (Bocarejo et al. 2014). The initial plan was for a network of 388 km, which was to be completed in 5 phases over 18 years (Suzuki et al. 2014). Currently it is in Phase 3 with 113 km constructed. In 2015, demand for the system reached 2,345,803 trips a day (Transmilenio 2015).

Understanding how transport infrastructure changes land use has become increasingly important for Bogota city planners, as international funding organizations (bankers and donors) often require information about the contribution of specific transport projects to the sustainable development of the city.

Additionally, previous research has identified significant urban development challenges caused by a lack of proper planning (Paez et al. 2014). These challenges include; unequal distribution of transport infrastructure and services between the most disadvantaged, the middle class and the most prosperous. To improve outcomes, Bogota needs to include these challenges as part of the decision-making process.

BoLD is therefore intended to address the lack of technical tools to assess specific transport infrastructure proposals from a sustainable development perspective.

2.3 Infrastructure and Land Development Proposals for Bogota

Although other areas of the city have had significant debates about increased urban development (for example, the south for low income population and the north for the rich), stakeholder workshops as well as current administration priorities turned our attention to the west growth area (Universidad de los Andes 2015). Transport proposals for the western part of the city aim either to increase current road infrastructure for buses (the existing transport option) or to create a new suburban rail service running on the existing freight infrastructure (Regiotram 2014).

The Light Rail Transit (LRT) is planned to be developed as a public-private partnership between private investors; the city of Bogota and the State of Cundinamarca. All the municipalities to the west of Bogota belong to the state of Cundinamarca. The objective of the LRT is to supply a fast, environmentally-friendly, safe, integrated transport option for the west. It is intended to provide users with an alternative to the current road-based public transport (Regiotram 2014). The idea is for the LRT to operate as a commuter train in the inter-urban areas outside Bogota, and as a tramway in Bogota's urban areas, reaching speeds of 110 km/h and 60 km/h respectively (Regiotram 2014).

As an alternative to the LRT, the road proposal includes road improvement schemes in the western and northern parts of the city, with the construction of urban highways to replace some of the roads connecting Bogota CBD to other town centers.

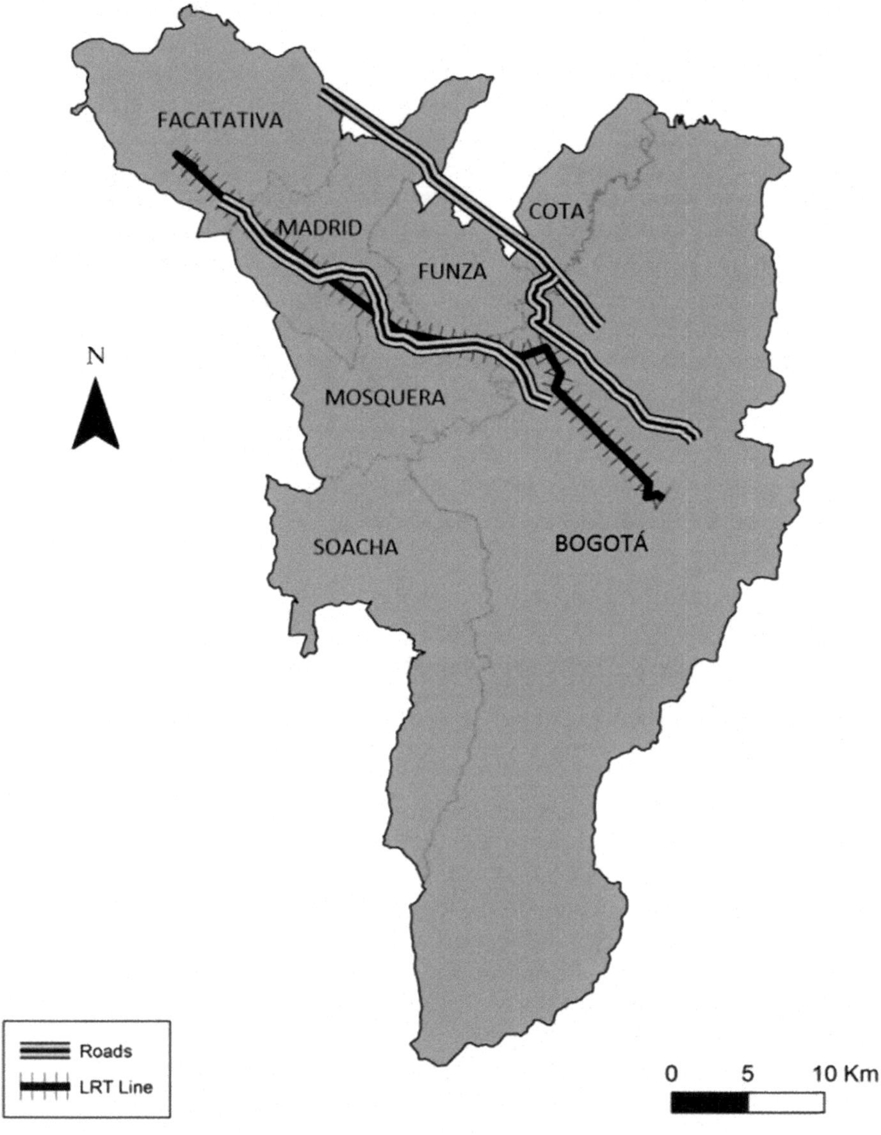

Fig. 2 Proposed transport projects for Bogota and the growth area in the west

Figure 2 shows the proposed road and rail alternatives currently being debated for Bogota and its western corridor.

A challenging aspect of modeling specific transport infrastructure projects is including a standard infrastructure plan for the years being modeled.

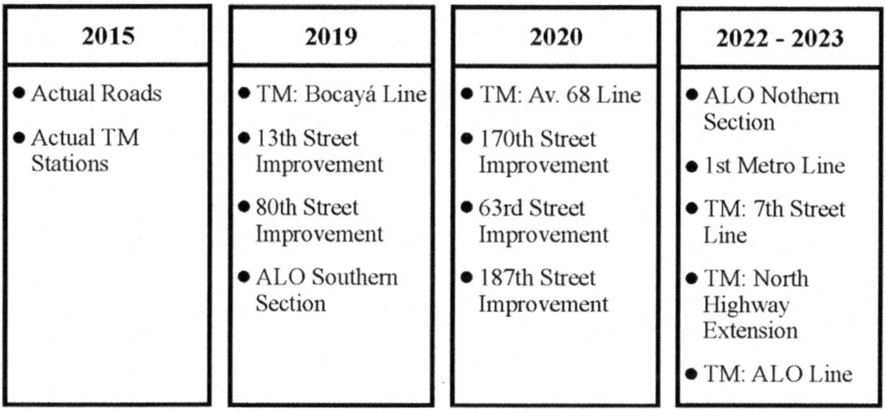

Fig. 3 Temporal changes in infrastructure for the BoLD Model

Figure 3 shows all the roads improvements on the Transmilenio BRT network that were taken into account for this model as the general infrastructure plan for the rest of the city. Although stakeholders in Bogota were extensively consulted about this general plan, many of these projects, and their year of becoming operational, have been a source of controversy.

To complement these infrastructure scenarios in which either improved roads or a train service were constructed, land regulation options in Bogota likely to have an impact on the west were also modeled. In particular, we included options for the Van Der Hammen reserve (VDH), a 1400-hectare nature reserve to the north of Bogota, in the analysis (Fig. 4). Supporters of developing the reserve argue that increasing the available land for residential developments close to existing commercial and industrial areas in Bogota would provide shorter travel distances for Bogota residents. If combined with sensitive urban development, these benefits could potentially surpass those obtained from maintaining the land as an environmental reserve (El Tiempo 2016).

The options for preserving the nature reserve or not were combined with the proposed transport infrastructures in the west to produce four scenarios to be modeled in BoLD. These scenarios are detailed in Table 1.

In the next section we present the methodology used for developing BoLD and its technical parameters.

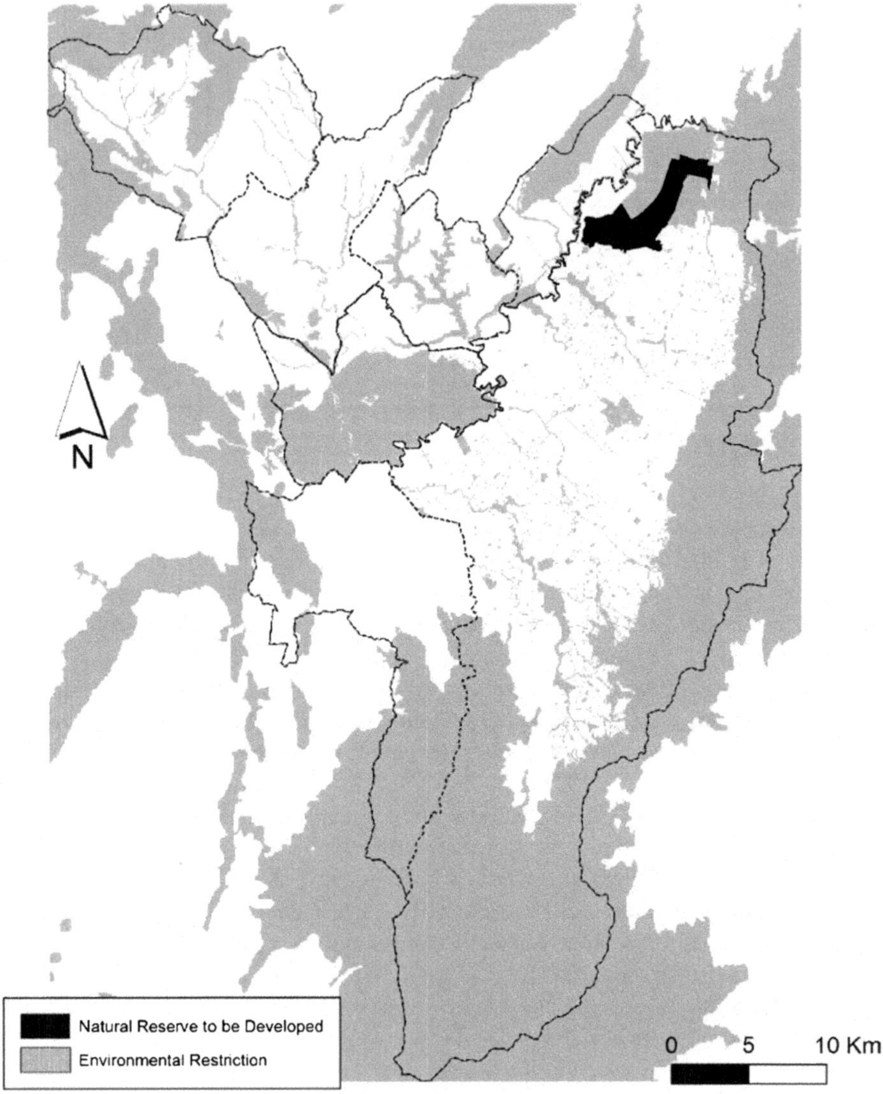

Fig. 4 Area of the nature reserve under consideration for development

Table 1 Scenario narratives in BoLD

	Road infrastructure	Suburban train infrastructure
Nature reserve maintained	Scenario 1: Road infrastructure continues to be the main form of transport for growth areas in the west. New roads allow additional connections between outlying municipalities and Bogota. No changes to existing restrictions on urbanization in the VDH reserve	Scenario 2: Existing freight rail infrastructure upgraded to provide a suburban passenger-rail service for Bogota and the municipalities in the west. New road constructions or upgrades will only take place in areas where no rail infrastructure currently exists
Nature reserve urbanized	Scenario 3: As in Scenario 1, roads are upgraded to provide accessibility in the west. However, land regulations are changed to allow the VDH reserve to be urbanized, so enabling additional road infrastructure and BTR services	Scenario 4: As in Scenario 2, a new train service is developed for the west. However, land regulations are changed to allow the VDH reserve to be urbanized, so enabling additional road infrastructure and BTR services

3 Methodology and Practical Application of the LUCC Model

BoLD was conceived as part of a technical cooperation project between the University of Los Andes in Colombia and the French Development Agency (AFD in French). The project had the following objectives and principles:

- To design and develop technical tools to aid decision-making for local authorities in Bogota and nearby municipalities, particularly in terms of sustainable development.
- These tools should be practical and replicable under different scenarios.
- There should be active participation of Bogota's governmental institutions, particularly of the planning and infrastructure departments.
- Indicators must be created to measure the impact of transportation problems on the sustainable development of the city.

BoLD was developed according to a series of general steps: Diagnosis, Model architecture, Model development and Results and Indicators.

It is important to note that significant stakeholder engagement was conducted. This occurred as part of the steering committee and technical committee workshops set up as part of the project. Government officials and representatives of key non-governmental entities participated in the workshops. At least one technical committee meeting was held in each phase of the process. In total, the project had three steering committees with representatives from local authorities, NGOs, the AFD and the University of Los Andes.

Community meetings are often part of the process of developing LUCC models (Escobar et al. 2015). BoLD, however, was conceived as a strategic tool to resolve some of the key discussions already happening in the community.

Therefore, our intention was from day one to engage with key decision-makers that are likely to use the findings of the model. It was also to engage with the civil servants in Bogotá City Council with responsibilities in this field who would benefit from using the tool to support their decision-making processes.

We will now describe each step in the methodology in detail. For each step, we outline the technical parameters designed and developed for use in BoLD.

3.1 Step 1: Diagnosis

The first step was to conduct a diagnosis of current decision-making practices in key government institutions of Bogota and to understand the specific needs of key decision-makers. Interviews with stakeholders in both public and non-governmental organizations were conducted. The aim was to understand how modeling is currently used to inform decision-making relating to major transport infrastructure projects in the Bogota region and to find out where there is room for improvement.

One of the first issues we had to address was that strategic technical tools were not used in most decision-making, a problem that previous research into decision-making processes in Bogota had also encountered (Ardila 2004).

As part of the assessment, we conducted an extensive review of the literature on the latest developments in LUCC modeling. During this, we explored the options of programming a specific tool or of using existing software.

Although previous experiences in Bogota have shown the advantages of creating one's "own tool" (Guzman and Gómez-Gélvez 2014), we decided that existing commercial software would be more appropriate for achieving the objectives set out in the cooperation project. In particular, if we wanted to continue modeling scenarios beyond this first run of BoLD, we would need continuous support for the tool as well as a platform for training the government officials who were likely to use it in their work.

After comparing the different software tools on the market, we selected Metronamica as the most suitable in terms of the data demands and results it provided, as well as the large number of modeling exercises previously conducted with it, some of them in developing countries (Barredo et al. 2004). Our previous experience with this software had also proved successful (Hewitt et al. 2014; Escobar et al. 2015).

Metronamica is a cellular automata-based tool for modeling LUCC scenarios developed by RIKS (http://www.riks.nl/). The advantages of the model are the freedom to run different scenarios for the future, its capacity to enable the definition of very complex functions, the facility to "learn" the characteristics of a particular area, the ability to link to GIS and the easy incorporation of raster-based spatial data (Linke 2008). The transition rules affecting cell mutation from one land use category to another throughout the model run are computed using five main parameters; neighboring effect, zoning regulations, suitability linked to bio-physical factors, accessibility to communication infrastructure and a stochastic factor. The model is described in detail in RIKS (2007) and Hewitt et al. (2014).

Previous experiences with LUCC models have shown some limitations, particularly in relation to the need for a similar set of input data and the need to recalibrate and revalidate when the model is moved to other locations (Research Institute for Knowledge Systems BV 2007). Metronamica is no exception.

In order to address these limitations we conducted specific stakeholder interactions along with customized spatial analyses.

A LUCC model implemented with Metronamica is calibrated by providing two different datasets for the study area (Straatman 2004). Generally, these datasets need to be separated in time by around 10 years to meet the dynamics of LUCC (Robert et al. 2004) and are both considered as baseline land-use datasets. The calibration process involves replicating the LU map for the second date with a sufficient level of similarity to the actual LU map. The goodness of the model can be assessed using both qualitative and quantitative methods (Hewitt et al. 2014). The model is then extended to the final simulation date.

When it came to calibrating BoLD, instead of using the traditional calibration process, we decided to apply the neighboring rules and the accessibility analysis detailed below. Although this decision could be contested, the literature shows that calibration assessment is still a challenge and that indices such as kappa and others have certain disadvantages (Pontius and Millones 2011). As a result, visual assessment of calibration (van Vliet 2012) can be considered as effective as quantitative methods. In addition, the perceived value of models of this kind is shifting away from their very debatable predicting capacity to their usefulness as a tool for shared learning throughout the modeling process (Hewitt et al. 2014).

In the Bogota model, after adjusting the model parameters (neighboring relations, accessibility, zoning, suitability and stochasticity) in an iterative process, we found the observed and estimated land use maps for 2014 to be visually close enough. However, the estimated map was rejected as it substantially incremented the number of individual cells of different land use classes scattered within the area (salt and pepper effect) and causing abnormal results in the prospective land use map. The direct application of the model parameters to the observed land use map for 2014 produced satisfactory results as it was plausible and met all the criteria (land demands were met). More importantly, it showed the model's capacity and aroused so much interest amongst stakeholders that Bogota City Council requested that it be implemented for a larger area with more diverse scenarios.

Multiple datasets for land-use coverage were explored for Bogota and its surrounding area to the west. Cadastral or planning datasets containing complete land-use cover in two different years were only obtainable for the Bogota municipality. The information about the municipalities in the west was incomplete, so we decided to use a combination of datasets to produce a complete dataset of land-use cover for 2005 and 2014. Although the time lapse was only 9 years, just below what is recommended in the literature, the rapid growth experienced in Bogota over this period has caused more than enough LUCC to enable us to calibrate the model properly.

Table 2 Datasets used in the BoLD Model

Dataset	Description	Application in BoLD
2014 cadastral dataset for Bogota	Parcel-based cadaster dataset for Bogota that includes land-use coverage for every land parcel and the rate able value of each one	Calibration of land-use coverage areas in Bogota
2005–2011 planning zones	Planning zones for areas outside Bogota municipality with their intended or authorized land-use coverage	Calibration of land-use coverage areas in Bogota by detecting vacant zones and most likely land-use based on regulatory restrictions
2005 and 2014 water body inventory	Official dataset of rivers, lakes and other water bodies in the area	Identification of areas covered by water not always identifiable by Landsat images
2005 and 2014 national and regional parks and reserves	Official dataset from national government describing land with protected status in the study area	Separation of parkland from agricultural lands, and identification of forest reserves

The 2005 and 2014 datasets were formed on the basis of satellite images from Landsat. However, these images had significant limitations including:

- Errors in the images that resulted in incomplete pictures of some areas
- Spectral signature similarities meant that not all land-uses could be discriminated from each other.
- The images do not provide information about the income levels within the different residential areas, which became an important aspect, as explained in the next section.

To address these issues, we sourced other datasets, particularly in vector form. Table 2 contains a description of these datasets and how they were used to improve information from the Landsat images.

3.2 *Step 2: Model Architecture*

The architecture of BoLD was developed on the basis of the diagnosis conducted in the first step of the process. In this stage, we focused particularly on the land-uses to be modeled and on the extent and resolution of the model.

3.2.1 Land-Uses Modeled

Within Metronamica, land-use classes are divided into vacant, feature and function categories. Feature land-uses are defined as those that are not supposed to change during the simulation (for example, water bodies, roads, wetlands etc.) whereas vacant land-uses are those that lose land in favor of the more active land-use classes, referred to as function. The latter are typically residential, industrial or commercial (Research Institute for Knowledge Systems BV 2007). A fundamental aspect of the model architecture is the decision as to which land-uses should be classified as either vacant, feature or function. As part of this classification process, we conducted a specific division of residential land cover based on population income. This division was based on information obtained in interactions with stakeholders and previous research into urban layout in Bogota (Munoz-Raskin 2010).

Colombia has one of the highest levels of income inequality in the world, resulting in a GINI index of 0.5, which at a regional level is exceeded only by Honduras (The World Bank 2013). It is hardly surprising therefore that its capital, Bogota, also has significant economic inequalities, which are expressed in its urban geography (Aliaga-Linares and Álvarez-Rivadulla 2010). Residential land-use is heavily spatially clustered on the basis of the income of the residents. Generally, in Bogota the poorest residents are located to the south and west, while the rich are concentrated in the north.

Bogota has a system that classifies residential properties (and therefore their occupiers) from 1 to 6, where 6 represents those earning the highest incomes and 1 the lowest. The purpose of the stratification system is to calculate land taxes and many other public services, including subsidies for utility services and social welfare benefits.

In a general sense, the price of land per square meter for properties in strata 6 (the highest income band) is estimated to be 10 times higher than in strata 1 (ALCALDE MAYOR DE BOGOTÁ, D. C. 2014).

The following graph (Fig. 5) shows the different prices per square meter in each strata according to government taxation system calculations.

Table 3 shows the land-uses included in BoLD together with their description and classification under the Metronamica model. This was developed using the stratification system in Bogota as a basis and then simplifying it into the three categories outlined above.

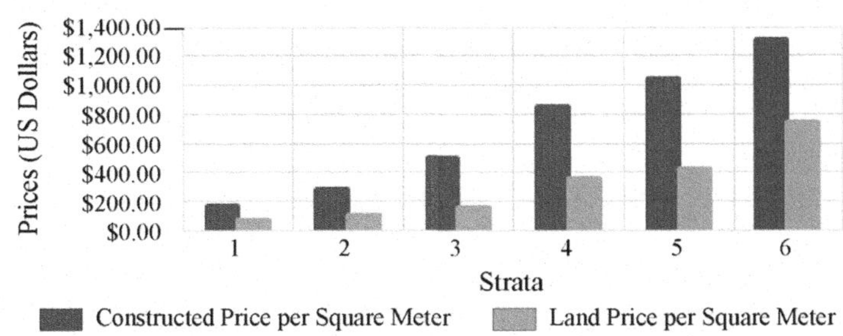

Fig. 5 Prices per square meter. Us Dollars (*Source* Secretaría de Hacienda)

Table 3 Land-use in BoLD model

Land-use	Classification	Description
Available for change	Vacant	Vacant land or areas reserved for urban growth
Agriculture	Vacant	Land used primarily for agricultural activities that could change to a different use in the future
High-income residential	Function	Land covered with residential properties in strata 5 and 6.
Medium-income residential	Function	Land covered with residential properties in strata 3 and 4.
Low-income residential	Function	Land covered with residential properties in strata 1 and 2.
Commercial	Function	Land covered with large commercial properties. It includes shopping complexes or malls but excludes local shops or small retail activities at ground level in residential or industrial areas
Industrial	Function	Land covered with industrial activities of significant size. It includes factories and agro-industrial complexes but excludes other small industrial activities such as informal car garages or workshops located in predominantly retail complexes
Urban facilities	Feature	Land covered by institutional buildings, both private and public, that provide fundamental services to the community such as schools, hospitals, sports facilities, etc.
Roads	Feature	Land covered by roads and other transport infrastructure, such as railways and stations
Water bodies	Feature	Land covered by water bodies such as lakes, swamps and rivers
Other	Feature	Land covered with uses different to those described previously that are likely to remain unchanged over time. For example, government buildings, museums, parks, historical sites, etc.

It is important to note that significant amounts of commercial and industrial activity in Bogota occur on residential land or land not specifically allocated for such purposes (Izquierdo and Horta 2013).

In view of the strategic nature assigned to BoLD, modeling small industrial or commercial areas was considered to be beyond its scope.

3.2.2 Model Extent, Timeframe and Resolution

Metronamica is a spatially-explicit cellular automata LUCC model and therefore requires a specific geographic area to be modeled. It also requires a specific resolution or cell size. These are fundamental parameters. Results of the model can vary significantly depending on the extent of the area and the cell size. Boundary effects could also distort results (Linke 2008). The entire municipality of Bogota was included as well as those to the west of it. In view of the current growth patterns in low-income population in the south (Botero and Gakenheimer 1999), we also decided to include the municipality of Soacha.

It is important to acknowledge that the extent we selected has certain limitations:

- It does not include all potential growth areas outside Bogota, particularly those in the far north (north zones inside the municipality of Bogota are included). This was primarily due to a lack of resources and interest on the part of local authorities. Although we acknowledge this as a limitation of the model, its impact on the overall result is minimal as most of the population growth in the city is happening in the south and southwest. The north is traditionally a growth area for the most prosperous section of society. Although as this group makes up less than 5% of the population of the city, the impact of their demands, in terms of land use changes is minimal in comparison with the land use changes caused to fulfil the needs of middle and low income population in the south and southwest.
- Although the municipalities west of Bogota are quite big and have significant vacant areas for future growth, their interaction with other municipalities outside the study area is not included in the model. In particular, Facatativa, which is not a commuter town for Bogota, but acts as a regional center for rural activities in the broader context.

The modeling timeframe for BoLD was from 2014 to 2040. This is in line with standard practice for LUCC models, which are normally used for strategic decision-making, to cover modeling periods of about 30 years (Y.Sato et al. 2003). This timeframe appears to be suitable not only for the scenarios proposed but also to fill the gap in terms of decision-making information identified for Bogota.

The resolution of the BoLD model was set with a cell size of 100 $\times$ 100 meters, so each cell covers an area of 1 hectare. This resolution was based on previous experiences of LUCC models where a cellular automata approach was used in cities with similar characteristics to Bogota (Escobar et al. 2015). This resolution was also appropriate to the level of information from the Landsat images and improved with local datasets.

3.3 Step 3: Model Development

In a general sense, a LUCC model based on cellular automata requires inputs (of both information and parameters) in the following areas:

- future land demands
- current and future land zoning changes and suitability conditions
- neighboring relationships between land-uses, and
- accessibility analysis based on transport infrastructure

3.3.1 Future Land Demands

Future land demand inputs were developed by forecasting current growth trends. Demands for the three income levels of residential land-uses were based on population growth forecasted by the National Statistics Department (DANE 2016). The other two key land-uses, commercial and industrial, were projected on the basis of GDP forecasts for Bogota and its surrounding region (DANE 2016, DANE (2) 2016). According to reports from the Bank of the Republic, industry is projected to grow by 2.7% annually over the next 2 years. FENALCO (National Federation of Retailers) have estimated that for the next two to three years, commercial activity is likely to grow at around 3% a year (FENALCO 2014). In both cases, future residential, commercial and industrial land use demands were estimated in direct proportion to the growth forecasted by DANE and FENALCO respectively.

Our review of the literature revealed that there are two main sources of information used to estimate the future land demand of a city. While in some models the information was provided by a national entity (Mancosu et al. 2015; Aljoufie 2014), in other cases land demand was constructed by the researchers themselves (Hewitt et al. 2012). We opted for the second method, obtaining the projections for population growth by mathematical extrapolation (Table 4). Using this method, the following values were obtained for the city's population and GDP for trade and industry (in millions of COP- Colombian Pesos):

Based on this expected growth, we calculated the expected increase in future land demand in Bogota (in hectares or cells in the model), as observed in Table 5.

For the case of Bogota, increase in land demand is shown in Fig. 6:

When calculating the data shown in Table 5 and Fig. 6, we made the following assumptions:

1. High-income residential land-use will always represent 5% of total residential land-use.
2. Poverty will decrease by 5% in 9 years and the medium income group will increase at the same rate over the same period.
3. Population densities remain constant over time.

Table 4 Projection of growth in population and GDP in Bogota

Year	2005	2014	2023	2032	2040
Total population (people)	7,556,515	8,661,781	9,737,843	10,808,780	11,760,724
Total commercial GDP (mill COP)	69,324	152,931	201,478	262,883	333,012
Total industrial GDP (mill COP)	29,119	51,215	61,807	73,865	86,544

Table 5 Estimated land demand

Year	2014	2023	2032	2040	% Cells per land-use for 2040
Residential high income	1359	1411	1566	1704	4.22
Residential medium income	11607	15519	18530	21582	53.41
Residential low income	7949	6584	6003	5112	12.65
Commercial	1146	1510	1970	2495	6.18
Industrial	5633	6798	8124	9519	23.55
Total cells	27694	31821	36193	40412	100.00
Cells increment (%)		13%	12%	10%	

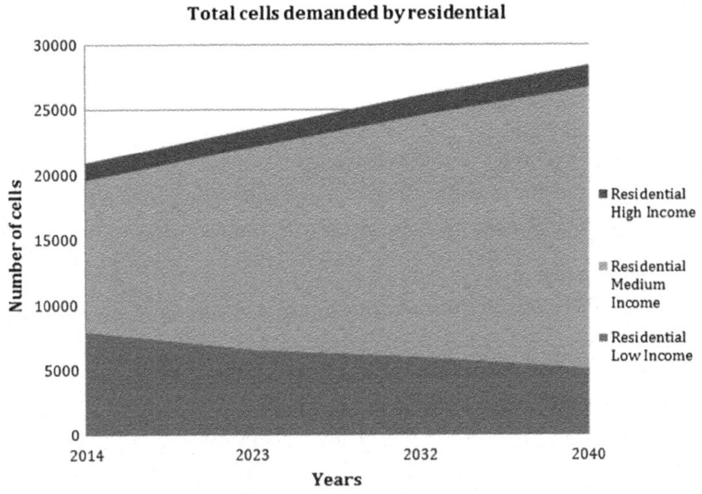

Fig. 6 Population growth for Bogota and its region

3.3.2 Current and Future Land Zoning Changes and Suitability Conditions

As has happened in other developing cities (Lombard 2014; Heinrichs and Bernet 2014), residential growth has not been strictly confined to authorized areas. Illegal settlements are still common today. In Bogota in particular, construction outside authorized areas remains a significant problem for low-income residential land-use (Escobedo et al. 2015).

Bogota and its neighboring municipalities do not have an integrated land planning system. This leads to restrictions and municipality-based land-use plans that are often inconsistent with those of neighboring municipalities. After reviewing all the land-use planning zones in all the municipalities, the following zoning categories were included in BoLD:

- Archaeological: Refers to all areas, lands, buildings, spaces or facilities that have archeological features or value.
- Heritage: Refers to all historical and cultural areas, land, buildings, spaces or facilities, which must be preserved and protected as part of the essential fabric of each culture or nation. This category includes museums, heritage and historic preservation areas. A good example of this is the historical center of Bogota, La Candelaria.
- Environmental Restriction: Refers to all areas, spaces and/or ecosystems that have a strategic role in biological processes and contribute to biological diversity; as well as the provision of basic resources for human subsistence.
- Industrial Use: Refers to all areas, land, buildings, spaces or facilities that currently have high industrial activity.
- Road network: Refers to all areas and corridors through which traffic (private, public, haulage, etc.) flows.
- Environmental slightly restricted: Refers to all areas, spaces and/or ecosystems that have a strategic role in biological processes and contribute to biological diversity. These areas, however, are not protected or are already fragmented or affected by human activity. All of the land in this category falls within the municipality of Mosquera, Cundinamarca.
- Airport: Refers to the whole airport area and other adjacent land specially designated for aeronautical or aerospace logistics activities.

Suitability was included in BoLD using a similar procedure to zoning. Suitability refers to the natural conditions under which land-uses can develop and to this end we evaluated the risk of Landslides, Flood Zones and Heavy Rainstorms. The information we used came from the regional risk management authority.

Finally, we also studied geographic areas which suffer from the ponding effect of precipitation. The ponding effect is the product of progressive increases in the flow of puddles or ponds caused by heavy rainfall. When the ponds join, they generate torrential water flows that can damage or destroy everything in their path, affecting the most vulnerable people in the city. This phenomenon occurs mainly in highly sloping areas and near rivers and mountain streams.

3.3.3 Neighboring Relationships Between Land-Uses

Neighboring interactions are fundamental in a LUCC model. They define how different land-uses affect the development of their surrounding land-uses. To represent neighboring interactions between land-use types, we used a methodology based on spatial analysis and Laplace probability concepts (Hansen 1993). This methodology was taken from Laplace's rule and adapted for the BoLD model using

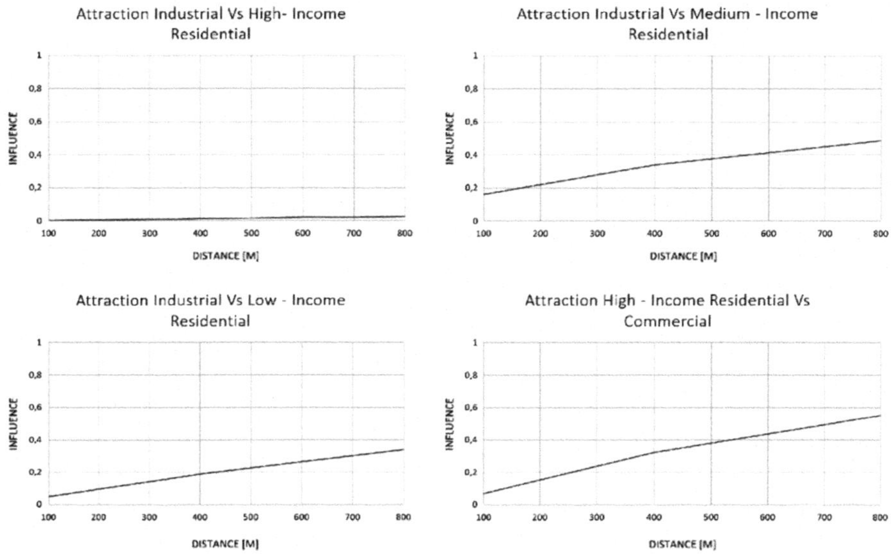

Fig. 7 Neighboring interactions between land-use categories (*Source* Rubio et al. 2015)

ArcGIS to calculate distances between land-uses and their relation to land-use cells in a defined searching radius.

Attractiveness Index

$$X_{AB} = \frac{\sum_{i=1}^{n} \overline{D_{AB}}}{R * n} \tag{1}$$

where,
A = Land-use A
B = Land-use B
D_AB = Average distance between A and B
R = Searching radius
n = Number of cells in A
X_AB = Attractiveness Index of A to B

Examples of results from these analyses are shown in Fig. 7.

The probabilities of land type locations and the relationships between them were calculated for all function land uses. As an example, the above graphs show that in our study area, high-income residential has a very strong attraction or desire to locate close to industrial land. The inclusion of neighboring relationships based on historical probability and spatial analysis strengthens the simulated results, as these curves were used in all future scenarios of BoLD.

3.3.4 Accessibility Analysis

Accessibility refers to the closeness of a particular land use to the transport service provided in a specific area. For LUCC models, accessibility refers to the preference of most function land-uses to locate closer to transport services. A highly accessible location is more likely to be developed than another location with the same conditions but with limited access to roads or public transport.

Mathematically, accessibility in a LUCC model based on cellular automata can be expressed as (RIKS 2007):

$$A_{c,y,s} = \frac{a_{y,s}}{{}^{t}d_{c,y} + a_{y,s}} \tag{2}$$

where:

- $A_{c,y,s}$ is the accessibility of cell, c, in relation to a certain type of node or transport link, y, (for example, a main road or train station) for a specific land-use, s
- $a_{y,s}$ is the accessibility distance decay factor (ADDF) which varies depending on the type of infrastructure, y, and is individual for each land-use, s
- ${}^{t}d_{c,y}$ is the distance from the specific cell being analyzed to the infrastructure, y, at a specific time, t

The result of this equation will have a value of between 0 and 1 for each cell. The inclusion of ADDF in the equation enables us to assess the degree to which accessibility is affected by the distance to the nearest transport infrastructure. This factor varies for each type of infrastructure and for each land-use. As distance to the nearest transport infrastructure is an important parameter in the scenarios being modeled, assigning the ADDF for each type of infrastructure and for each land-use is a key task for the modeler.

Although the calculation of ADDF is commonly based on empirical experiences (Furtado 2009), the significance of these factors for the BoLD model required us to explore advanced technical methods for calculating ADDF. This aspect became a focal point of the investigation and we believe it to be the main contribution that BoLD makes to LUCC modeling.

A methodology based on advanced spatial analysis implemented on a GIS was used to determine ADDF. We call this methodology OSDC (Over-time Spatial Decay Calculation).

OSDC is based on three principles. The first is that ADDF factors determined on the basis of past information can be used for modeling future scenarios. For example, if low-income residential land-use is more attracted to public transport nodes compared to high-income residential uses, this relationship would be maintained in the future.

The second assumption in OSDC is that the ADDF values for each type of infrastructure and for each land-use are proportional to each other. In other words, if two ADDF for two different infrastructures are equal (for example 1) they make the same contribution to the overall attractiveness of the cells in the model. Consequently, and considering that OSDC creates ADDF with values of between 0 and 1, if a specific transport infrastructure, y, for particular land-uses, s, has a low proportional accessibility, the OSDC would assign a value of 0.

The third principle assumes that the average distance (up to a maximum of 2 km) between the cells for a particular land-use and the transport infrastructure is a good indicator of the ADDF. The relationship is inversely proportional, the larger the average distance to the transport hub, the smaller the decay factor value.

These three principles are applied to OSDC with the following equation:

$$ADDF_{y,s} = \frac{\frac{\delta_{y,s}}{\delta_{maxy}}}{max\frac{\delta_{y,s}}{\delta_{maxy}}} \tag{3}$$

In which:

- $ADDF_{y,s}$ is the accessibility distance decay factor for the infrastructure, y, (for example, main roads or train stations) in a specific land-use, s, (for example commercial or industrial land-uses)
- $\delta_{y,s}$ is the average distance to the infrastructure, *y,* of all land-use cells of type,, *s* that are no more than 2 km away from it
- $\delta_{max\,y}$ is the maximum for all $\delta_{y,s}$, values for infrastructure, *y*.

OSDC produces ADDF values of between 0 and 1. Additionally, and considering that transport infrastructure can be modeled as lines or points in a GIS system, two normalization procedures must be conducted, one for each type of infrastructure (the numerator in the previous equation). The denominator of the equation is applied to normalize the ADDF again. This second normalization is needed to generate a proportionality between all the land-uses and infrastructure types in the model.

OSDC was applied using the processed datasets for 2005 and 2014 in the BoLD model, and the results were compared. Table 6 shows the results obtained and the differences between the two years.

Ideally, there should be no difference between the ADDF calculated for 2014 and 2005. Apart from some differences in roads for high-income residential land-use, in general OSDC is consistent (it has a standard deviation of less than 0.13).

4 Indicators and Results

The final step of the methodology involved developing decision-making indicators for mobility. We adapted previous work by James (2015) that presented a circle of sustainability as a representation of the sustainability of a territory. The circle is divided into four domains and each domain into various sub-domains (James 2015).

Table 6 ADDF calculated in BoLD using OSDC

Land-use	2005 dataset	2014 dataset	Differences (2014–2005)						
	BRT line	Station	Roads	BRT line	Station	Roads	BRT line	Station	Roads
Residential high-income	0.70	0.00	0.77	0.77	0.00	0.48	0.07	0.00	−0.29
Residential-middle income	0.49	0.33	0.85	0.40	0.28	0.62	−0.10	−0.04	−0.23
Residential low income	0.00	0.30	0.00	0.00	0.28	0.00	0.00	−0.02	0.00
Commercial	0.75	0.46	1.00	0.89	0.66	1.00	0.13	0.20	0.00
Industrial	0.43	0.38	0.10	0.32	0.51	0.17	−0.11	0.13	0.06

Table 7 Indicators developed for the Mobility Circle

Domain	Indicator	Description
Urbanism	Average distance to work	Approximates the distance from a cell with residential land-use to the land-uses related to work
	Average distance to downtown	Average distance from a cell with residential land-use to Bolivar Plaza in the city's downtown area
	Average distance to large parklands	Average distance from a cell with residential land-use to the main parks in Bogota
Equity	Average distance to public transport	Average distance from a cell with residential land-use to any type of public transport (metro or Transmilenio)
	Access rate to transportation infrastructure	Number of residential use cells within a certain distance of roads or public transport stations. The distances were set at 500 meters to public transport and 1 km to roads.
	Distance to major activity zone	Average distance from a residential land-use cell to any type of public transport (metro or Transmilenio)
Environment	Access ratio for commercial use	Ratio between average distance from a cell with commercial land-use to roads and average distance to public transport
	Habitat fragmentation	Indicates biodiversity according to the 'Probability of Occurrence' and is based on the degree of fragmentation.
	Expansion of urban areas	Ratio indicating how urban areas have appeared and disappeared during the simulation
Risk	Flood risk	Percentage of cells with a residential land-use in areas considered at risk of flooding.
	Landslide risk	Percentage of cells with a residential land-use in areas considered at risk of landslides.
	Torrential rain risk	Percentage of cells with a residential land-use in areas considered at risk of torrential rain.

The general concept is that all the mobility impacts of each scenario can be summarized in a single graphic. We adapted a version of the circle of sustainability based on mobility indicators. Our mobility circle is designed to have three domains and each domain to have three subdomains or indicators. Table 7 shows the individual scenarios that make up the circle with an interpretation for each.

4.1 *Results and Discussion*

After implementing BoLD using the parameters described in the previous section, including the ODDC for accessibility analysis, we obtained results for all the scenarios. Map results for the baseline year (2014) and the simulated result for 2040 for each scenario are presented in Fig. 8.

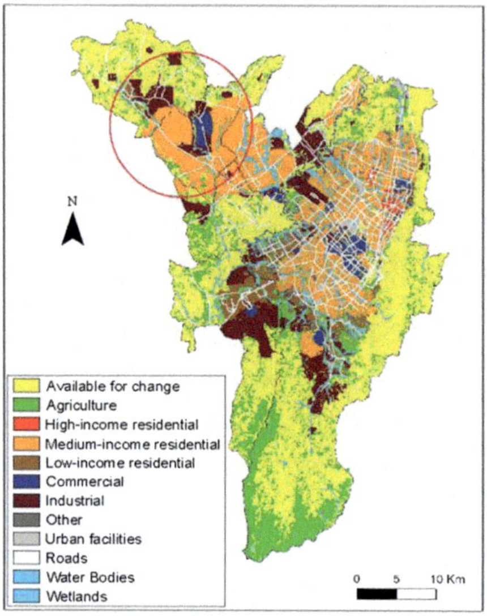

Scenario 1: Highway development and VDHR restricted

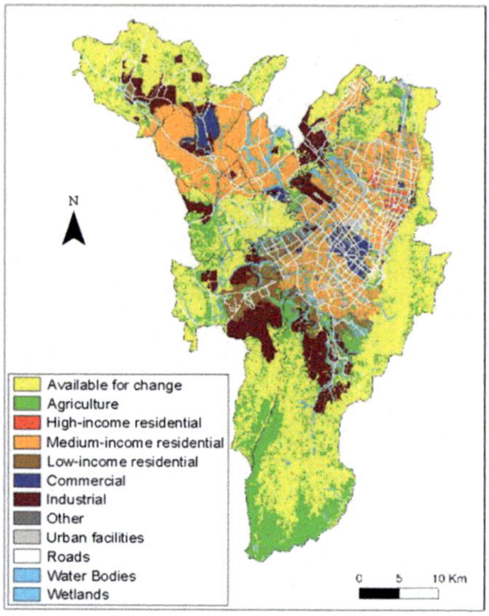

Scenario 2: Train and VDHR restricted

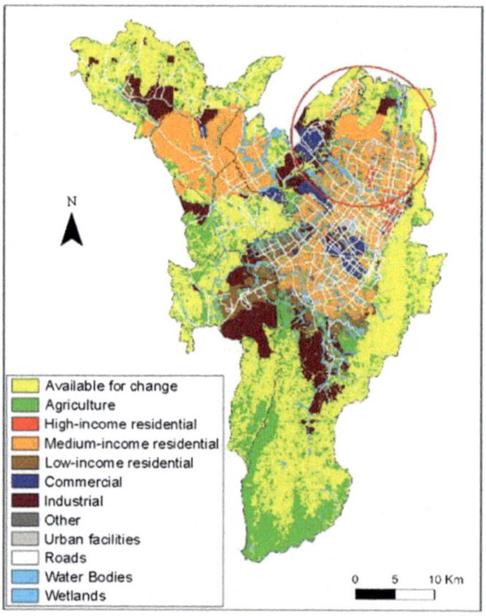

Scenario 3: Highway development and VDHR unrestricted

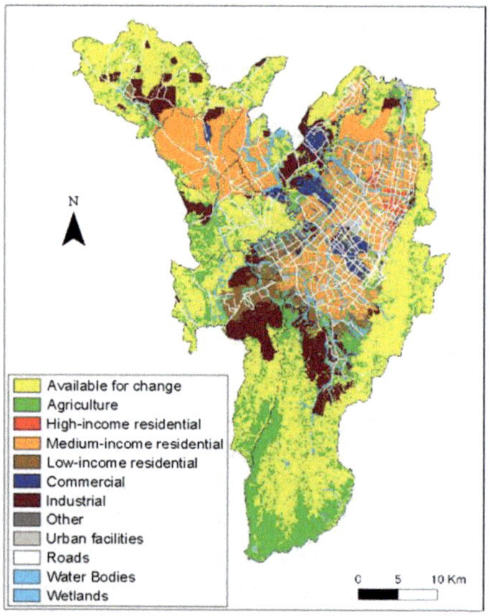

Scenario 4: Train and VDHR unrestricted

Fig. 8 Land-use maps for 2040 scenarios

General patterns of development are maintained in all scenarios. This is probably due to the fact that Bogota is a mature city in which similar trends have been in place for many years. However, in some specific areas there are differences between the scenarios.

For the first scenario (freeway development in the west with conservation of the nature reserve), increased commercial development along the proposed road can be observed along with additional industrial growth in the surrounding areas. Industrial areas appear in the far west on what was previously agricultural land. These results were expected as additional road capacity is particularly attractive for commercial and residential development and the current trend of industrialization in the west is maintained.

For the scenario in which a passenger rail system is built instead of additional roads (while maintaining the nature reserve undeveloped), residential and commercial development concentrates alongside the proposed stations. This is particularly obvious in border areas between Bogota and the municipalities.

In the scenarios in which the nature reserve is open for development (Scenarios 3 and 4) the influence of transport infrastructure in land patterns is very similar to those described for Scenarios 1 and 2. However, the additional land availability in the north creates a concentration of medium-income earners inside Bogota while at the same time promoting low-income development on the outskirts of the study area. This result confirms a pattern already occurring in Bogota where low income population are forced to live in high density locations a long way from the city centre due to land prices. As new transport infrastructures are created, land values will increase, with the result that the only available land for low income people will be even further out.

In order to provide information that can support decision-making and identify differences between all four maps, we produced mobility circles for all the scenarios. These circles can help users understand the implications, in terms of sustainability indicators, of each scenario (Fig. 9).

There are clear differences between Scenario 1 (a highway-based development with the VDHR restricted) and Scenario 2 (a train-based development with the same restrictions for the nature reserve). Average distance to downtown is approximately 1% higher in Scenario 2 than in Scenario 1 causing a change in level inside the circle. The access rate for commercial land-use is 2% lower in Scenario 2 producing the same effect inside the circle.

When Scenario 3 is compared to Scenario 2, significant changes can be ob-served. Average distance to downtown is approximately 1% higher in Scenario 3 causing a change in level inside the circle. The access rate for commercial land-use is about 10% higher in Scenario 3 causing a change in level inside the circle.

As regards the other indicators, they show the same rates across all scenarios. Expansion of urban areas, landslide risk and torrential rain risk have rates of between 90 and 100% in all scenarios. Average distance to work, average distance to significant parklands and flood risk are between 60 and 80% in all scenarios. However, average distance to public transport remains under 30% in all four scenarios.

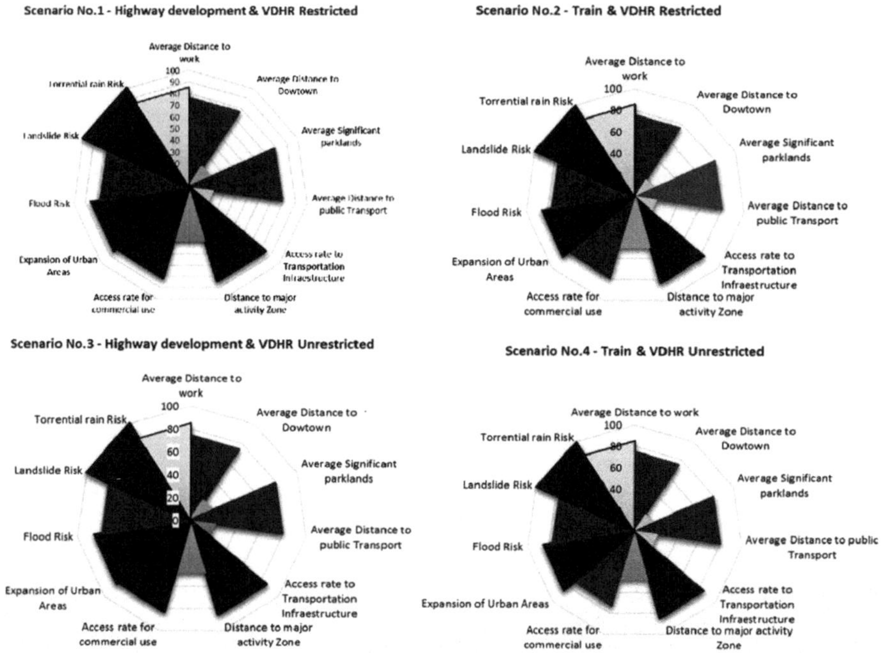

Fig. 9 Circles of mobility for all four scenarios

The most striking changes in the first comparison between Scenario 1 and Scenario 2 are in terms of average distance to downtown. The average distance is lower with a highway scenario in which more residential land-use is developed near the city, while the train-based scenario promotes development in the nearby municipalities. As commercial land-use tends to allocate near new roads, an increase in this land-use was likely to happen in Scenario 1, as indeed occurred.

The comparison between Scenario 3 and 4 indicated greater changes. New development of commercial land-use is higher in Scenario 3 because of the lifting of restrictions on development in the nature reserve. With VDHR unrestricted, residential land-use tends to allocate in the new unrestricted area while commercial land-use tends to allocate to areas previously dominated by low-income residential. Average distance to downtown also increases due to the house-building process in the reserve.

4.1.1 LUCC Changes

As the location and type of transport infrastructure varies in each scenario, different changes will take place in different locations. The matrix below (Table 8) shows the total area for each land-use type in the baseline year (2014) and the forecast year (2040).

Table 8 Contingency table resulting from cross tabulation of 2014 and 2040 (Scenario 1) land-use maps (hectares)

		2014												
	Land use	AFC	Agricul.	High Res.	Med. Res.	Low Res.	Commer.	Indus.	Other	U. Faci.	Roads.	W. bod.	Wetl.	TOTAL
2040	AFC	60264	77	120	2711	2318	587	3434	0	0	0	0	0	69511
	Agricul	0	23859	61	2357	1489	343	2763	0	0	0	0	0	30872
	High Res.	0	211	1047	36	25	21	19	0	0	0	0	0	1359
	Med. Res.	0	1416	300	9775	17	8	91	0	0	0	0	0	11607
	Low Res.	0	723	13	3230	2966	1014	3	0	0	0	0	0	7949
	Commer.	345	3	154	310	0	331	3	0	0	0	0	0	1146
	Indus.	0	768	9	1459	0	191	3206	0	0	0	0	0	5633
	Other	0	0	0	0	0	0	0	3162	0	0	0	0	3162
	U. Faci.	0	0	0	0	0	0	0	0	2340	0	0	0	2340
	Roads.	0	0	0	0	0	0	0	0	0	13628	0	0	13628
	W. bod.	0	0	0	0	0	0	0	0	0	0	6784	0	6784
	Wetl.	0	0	0	0	0	0	0	0	0	0	0	1497	1497
	TOTAL	60609	27057	1704	19878	6815	2495	9519	3162	2340	13628	6784	1497	155488

As expected, most of the new development occurs by converting land considered available for this purpose (both agricultural land and land reserved for expansion). According to the model, 3121 hectares of agricultural land would be converted into urban areas. This result was also expected as most of the expansion of Bogota over the last 30 years has taken place in this way.

The model results also suggest that the current trend in Bogota in which industrial areas are converted into residential would continue in the future, especially if they are close to middle income residential areas. It is expected that 19 hectares of industrial land would be transformed into high-income residential and 91 hectares into medium-income residential from 2014 to 2040.

A shift from commercial to low-income residential areas could also be ob-served from the matrix (total of 1014 hectares).

Some unexpected changes can also be observed. For instance, the conversion of commercial and industrial land into agricultural land does not, in principle, make much sense. This is probably due to over calibration of the attraction and repulsion between function classes (residential, industry and commercial) and low land claims for these land use classes, making the land free for vacant classes such as agriculture.

5 Conclusion and Outlook

In this chapter, we present BoLD, a LUCC model for the city of Bogota in Colombia. The main objective of BoLD is to assist decision-making processes by providing LUCC information based on scenarios. The different development scenarios for Bogota involve specific transport infrastructure projects, which is why BoLD was designed and implemented with this focus.

The main objective of BoLD has been achieved. We have developed a spatially-explicit model that can assist decision-making processes. Results with four scenarios have shown that the model is capable of producing technical results in relation to the positive and negative effects of the proposed transport infrastructures. In the application of the LUCC model we were also able to incorporate land management policies such as the urbanization of nature reserves, which were included in the analysis.

In this research we also developed the ODDS, an advanced spatial methodology for calculating ADDF using a formula. ODDS improves LUCC modeling by filling a gap in the literature in which the influence of accessibility was often modeled using empirical experiences.

Limitations inherent to this kind of exercise in general, and our case study in particular, have been highlighted throughout the text. They can be summarized as follows:

- We have assumed that the impacts on land-use of a particular transport infrastructure are not influenced by other transport projects. Although this assumption facilitates a clearer differentiation between the options, possible synergies between transport alternatives are not considered, something that could be significant for other case studies outside Bogota.
- Users should approach LUCC simulations that only consider transport changes with caution, as they provide a narrow view of future scenarios without clearly considering important aspects such as changes in land demand.

Considering the limitations in a developing city and the fact that the complex urban patterns of Bogota have not been studied very well, the results we obtained appear to be similar to other previous studies in which LUCC simulations have been used. While other studies have taken the complex path of developing a parallel transport model (for example Aljoufie 2014), we have found an alternative for situations in which limited information is available and the impact of transport on urban patterns is not well understood.

The ODDS geospatial analysis method appears to be a viable option when LUCC simulations are being developed for scenarios based on transport infra-structure proposals. Additionally, the mobility circle offers a graphic representation of the results that can also facilitate decision-making.

Results from the different scenarios reflect the different impact of a highway and a train-based scenario in Bogotá. The indicators in the different mobility circles show that highways promoted the allocation of residential land-use. The majority of indicators were based on distance calculations. Removing restrictions on the VDHR allowed residential land-use to allocate there while commercial tended to occupy low-income residential areas.

Based on this experience some recommendations for the future, both in land-use planning and in modeling procedures and transport network models are:

- Interactions between traditional transport models and BoLD can provide more accurate information on development in the area.
- Mobility indicators using GIS based methods have great potential. These methods can include several network analyses as seen in our review of the literature.
- Information about land-use demand can be improved. It was found that population growth estimates can be improved using several techniques including transportation variables.

Acknowledgements This research was supported by the French Development Agency – AFD and by the project BIA201343462-P (Spanish Ministry of Economy and Competitiveness and European Regional Development Fund FEDER).

References

Alcalde Mayor de Bogotá DC (2014) Decreto número 586 2014. Bogotá

Alcaldía de Bogota (17 de 12 de 2013) Decreto 577 de 2013. Recuperado el 10 de 07 de 2015, de Normas: http://www.alcaldiabogota.gov.co/sisjur/normas/Norma1.jsp?i=55975

Aliaga-Linares L, Álvarez-Rivadulla MJ (2010) Residential Segregation in Bogota across Time and Scales. Lincoln Institute of Land Policy

Aljoufie M (2014) Toward integrated land use and transport planning in fast-growing cities: the case of Jeddah, Saudi Arabia. Habitat Int 41:205–215

Al-Kodmany K, Ali MM (2013) The future of the city: tall buildings and Urban Design. WIT Press, Southampton

Ardila A (27 de October de 2004) Transit planning in Curitiba and Bogotá: roles in interaction, risk, and change. Dspace MIT: https://dspace.mit.edu/handle/1721.1/28791#files-area. Accessed 29 Mar 2016

Banco de La república de Colombia (2016) http://www.banrep.gov.co/es/movil/comunicado. Obtained de http://www.banrep.gov.co/es/movil/comunicado

Barredo JI, Demicheli L, Lavalle C, Kasanko M, McCormick N (2004) Modelling future urban scenarios in developing countries: an application case study in Lagos, Nigeria. Environ Plan 32:65–84

Bocarejo JP, Escobar D, Oviedo Hernández D, Garlaza D (2014) Accessibility analysis of the integrated transit system of Bogota. Int J Sustain Transp

Bocarejo JP, Portilla I, Pardo MA (2013) Impact of Transmilenio on density, land use, and land value in Bogotá. Res Transp Econ 78–86

Botero Bd, Gakenheimer, Ca (1999) Land use expansion in Bogota: can it be controlled? Land Lines 11(2)

DANE (2) (03/ 03/ 2016) Cuentas Nacionales - Departamentales. Obtenido de PIB por ramas de actividad/ 2000–2014 preliminary: http://www.dane.gov.co/index.php/cuentas-economicas/cuentas-departamentales

DANE (2011) Estimaciones de población 1985–2005 y proyecciones de población 2005–2020 total municipal por área. DANE, Bogota

DANE (2015) Producto Interno Bruto de Bogotá DC Bogota: DANE

DANE (2016) Estratificación socioeconómica. de DANE: http://www.dane.gov.co/index.php/estratificacion-socioeconomica/generalidades. Accessed 30 Mar 2016

Dinero (4/ 09/ 2014) El renacimiento de la ALO

El Tiempo. (13/ 02/ 2016). Las claves para entender la discusión sobre la reserva Van der Hammen

Escobar F, Hewitt R, Hernandez V (2015) Usos del suelo en los parques nacionales españoles. Evolución y modelado participativo, Proyectos de investigación en parques nacionales: 2010–2013. Madrid: Organismo Autónomo Parques Nacionales, Ministerio de Agricultura y Medio Ambiente pp 175–211

Escobedo FJ, Clerici N, Staudhammer CL, Tovar Corzo G (2015) Socio-ecological dynamics and inequality in Bogotá, Colombia's public urban forests and their ecosystem services. Urban For Urban Green 14:1040–1053

FENALCO (2014) Comercio: Un sector en expansión y con potencial de crecimiento. Bogotá

Furtado BA (2009) Modeling social heterogeneity, neighborhoods and local influences on urban real estate prices. Neth Geogr Stud 385

Guzman LA, Gómez-Gélvez JA (2014) The projection of travel demand: a dynamic model approach considering the feedback mechanisms of accessibility. Bogotá DC

Hansen PB (1993) Parallel cellular automata: a model program for computational. Concurr Pract Exp 5:425–448

Heinrichs D, Bernet JS (2014) Public transport and accessibility in informal settlements: aerial cable cars in Medellín, Colombia. Transp Res Procedia 4:55–67

Hewitt R, Hernández-Jiménez V, Encinas MA, Escobar F (2012) Land use modelling and the role of stakeholders in natural protected areas: the case of Doñana, Spain. Paper, Madrid. Obtenido de http://www.iemss.org/sites/iemss2012//proceedings/E2_0403_Hewitt_et_al.pdf
Hewitt R, van Delden H, Escobar F (2014) Participatory land use modelling, pathways to an integrated approach. Environ Model Softw 52:149–165
IDIGER (Marzo de 2016) Panorama general de riesgos: Inundación. Obtenido de http://www.idiger.gov.co/inundacion
Izquierdo JP, Horta MP (2013) Informalidad empresarial y laboral en pequeños comerciantes: nueva evidencia para Colombia. Obtenido de Ministerio de Hacienda: http://www.minhacienda.gov.co/portal/page/portal/HomeMinhacienda/politicafiscal/reportesmacroeconomicos/NotasFiscales/21%20Informalidad%20empresarial%20y%20laboral%20en%20peque%F1os%20comerciantes.pdf
James P (2015) Urban sustainability in theory and practice: circles of sustainability. Routledge, New York
Linke SC (2008) Local level application of the dynamic land use model METRONAMICA. Technical University Berlin
Lombard M (2014) Constructing ordinary places: Place-making in urban informal settlements in México. Prog Plan 94:1–53
Mancosu E, Gago-Silva A, Barbosa A, de Bono A, Ivanov E, Lehmann A, Fons J (2015) Future land-use change scenarios for the Black Sea catchment. Environ Sci Policy 46:26–36
Munoz-Raskin, R. (2010). Walking accessibility to bus rapid transit: does it affect property values? The case of Bogotá, Colombia. Transport Policy, 72–84
Ortuzar Jd, Willumsen LG (2005) Modelling transport. WILEY, Chichester
Oviedo Hernandez D, Davila J (2016) Transport, urban development and the peripheral poor in Colombia—Placing splintering urbanism in the context of transport networks. J Transp Geogr 180–192
Paez D, Bocarejo J, Guzmán L (2014) To densify or not densify? mobility and urban life in developing countries. Panam, Santander, España
Peñalosa E (02 de 02 de 2016) Así es el proyecto de vivienda "Ciudad Paz" de Peñalosa. Hora 20. (D Calderón, Entrevistador) Bogotá. Obtenido de http://caracol.com.co/emisora/2016/02/01/bogota/1454364058_248991.html
Pontius R, Millones M (2011) Death to Kappa: birth of quantity disagreement and allocation disagreement for accuracy assessment. Int J Remote Sens 32(15):4407–4429
Regiotram (2014) Implantación de un LRT en Bogotá. Secretaria de Movilidad, Bogotá
Research Institute for Knowledge Systems BV (sf) Metronamica documentation. Maastricht: RIKS BV
Riks (2007) METRONAMICA model description. RIKS BV, Maastricht, The Netherlands
Robert GP, Huffaker D, Denman K (2004) Useful techniques of validation for spatially explicit land-change models. Science Direct, Ecol Model 179(4):445–461
Rubio L, Páez D, Escobar F (2015) Caracterización de las relaciones de vecindad entre usos del suelo a través de herramientas SIG. Estudio de caso, Bogotá D.C http://repositorio.uniandes.edu.co/xmlui/handle/1992/7189?show=full
Sato Y, Breiling M, Yoshikawa N (2003) Comparative model evaluation in LUCC research. Laboratory of Land Resource Management, Bogor, Indonesia
Straatman BR (2004) Towards an automatic calibration procedure for constrained cellular automata. Comput Environ Urban Syst 28(1–2):149–170
Suzuki H, Cevero R, Luchi K (2014) Transformando las ciudades con el transporte público. World Bank, Washington DC
The World Bank (2013) The World Bank. de GINI index (World Bank estimate): http://data.worldbank.org/indicator/SI.POV.GINI?order=wbapi_data_value_2013+wbapi_data_value+wbapi_data_value-last&sort=desc. Accessed 30 Mar 2016
Transmilenio (2015) Estadísticas de oferta y demanda del Sistema Integrado de Transporte Público SITP. Transmilenio, Bogota

Universidad de los Andes (2015) Segundo Informe de Avance. Universidad de los Andes, Cooperación AFD, Universidad de los Andes y Bogotá DC Bogotá

Van Vliet J, Hurkens J, White R, van Delden H (2012) An activity based cellular automaton model to simulate land-use dynamics. Environ Plan B 39(2):198–212

Zhao L. Peng Z-R (2012) LandSys: an agent-based Cellular Automata model of land use change developed for transportation analysis. J Transp Geogr 25:35–49

Chapter 19
Integrating Econometric and Spatially Explicit Dynamic Models to Simulate Land Use Transitions in the Cerrado Biome

T. Carvalho Lima, S. Carvalho Ribeiro and B. Soares-Filho

Abstract Land use changes in Brazil have broad implications within environ-mental, socio-economic, and policy contexts. Despite extensive research on the topic, there are still significant gaps, namely in modeling the nature of drivers of land use change across Brazil's large biomes. We aim to fill this gap by coupling econometric with spatially explicit models to explore future trends in land use change in the Cerrado biome. Cerrado savannas are considered a biodiversity hotspot, occupying 24% of Brazil's territory. Nevertheless, the native vegetation in this region is under mounting pressure due to agricultural expansion. The econometric model we developed determines gross rates of deforestation and regrowth in each municipality within the Cerrado biome from 2002 to 2009. We used GEODA and agricultural Census data (IBGE 1995, 2006) to develop an auto-regression spatial model. This model was coupled with a spatially explicit model developed using Dinamica EGO software. Simulations from 2009 to 2050 resulted in a loss of 14.2 Mha of native vegetation and regrowth of 18.5 Mha, showing that complex land use dynamics are in place. Our results are in line with other studies that show lower probabilities of deforestation inside protected areas and indigenous lands. There is a high probability however of deforestation in some of the buffer zones around these protected areas, which must therefore be continuously monitored. We conclude that there is a need for a consistent monitoring framework, built upon the work of different governmental and non-governmental initiatives, in order to design and implement effective conservation actions in this important Brazilian biome.

T. Carvalho Lima · S. Carvalho Ribeiro (✉) · B. Soares-Filho
Centro de Sensoriamento Remoto, Universidade Federal de Minas Gerais (UFMG), Belo Horizonte, Brazil
e-mail: sonia.carvalhoribeiro@gmail.com

T. Carvalho Lima
e-mail: carvalholimatcl@gmail.com

B. Soares-Filho
e-mail: britaldo@csr.ufmg.br

M.T. Camacho Olmedo et al. (eds.), *Geomatic Approaches for Modeling Land Change Scenarios*, Lecture Notes in Geoinformation and Cartography,
https://doi.org/10.1007/978-3-319-60801-3_19

Keywords Deforestation · Land-use change · Dinamica EGO · Spatial lag regression

1 Introduction

Land use planning in Brazil faces enormous challenges in reconciling economic growth and environmental conservation in a country of continental dimensions (Soares et al. 2014; Soares-Filho et al. 2016). Persistent land-use land-cover changes (LULCC) from a certain class (e.g. forest) to another (e.g. agriculture) are referred to as land use transition, which are linked to complex processes by which human activities transform the landscape (Lambin and Meyfroidt 2010; Basse et al. 2014).

A major land use transition, such as deforestation, entails impacts on carbon and climate cycles at scales from local to global (Pielke et al. 2011; Naudts et al. 2016), as well as on multiple ecosystem goods and services (Lawler et al. 2014). Assessing trends in land use transitions is thus of great importance.

Spatially explicit models and scenarios are important tools to help planners and policy makers understand the drivers of land-use change (Soares-Filho et al. 2006; Lambin et al. 2014; Soares-Filho et al. 2016). In Brazil, the Amazon and the Atlantic Forest biomes have been permanently monitored since the 1980s by PRODES (Monitoring Brazilian Amazon Forest by Satellite) and SOS Mata Atlântica, respectively. Land use change has not however been tracked in the Cerrado (Beuchle et al. 2015), nor has there been any research on the nature of the drivers behind such change (Bürgi et al. 2004).

We combined econometric and spatially explicit models to explore land use transitions in the Cerrado Biome in Brazil. The land use system in this region is distinctive for many reasons (Klink and Machado 2005). It is considered a biodiversity hotspot, occupying 24% of the territory of Brazil (Embrapa_Cerrados 2008) and is the second largest biome in Brazil (after the Amazon), hosting important endemic species and ecosystem services (Klink and Machado 2005). However, the Cerrado also has the most coveted land for producing a wide array of agricultural products (soy, beef etc.). Despite its importance, there is no clear evidence on the land-use trends in this biome. While the Brazilian Ministry of Environment (Ministério do MeioAmbiente-MMA) reports that deforestation rates in Cerrado decreased from 14.2 thousand km^2yr^{-1} between 2002 and 2008 to 6.4 thousand km^2yr^{-1} between 2009 and 2010, other sources report that reduction in deforestation has been less substantial than the government claims (Ferreira 2009; Beuchle et al. 2015). It is therefore vital to adopt a common framework for monitoring and assessing Cerrado land-use trends. This requires consistency in the monitoring systems implemented by governmental institutions and programs. The monitoring procedure must focus on the key variables of the system that are useful for forecasting the system dynamics and their impacts on key issues, such as biodiversity (Pereira et al. 2013). The data obtained from monitoring can be used as an input for spatially explicit modeling (Skidmore et al. 2015). Land-use transitions in Cerrado are driven by a range of environmental and socioeconomic variables. In this article, we contribute to the understanding of the dynamics behind these transitions by combining an econometric model with a spatially explicit model based on data for the 2002–2009 period on losses (deforestation) and gains (re-growth) in native vegetation at the municipality scale.

2 Test Areas and Data Sets

2.1 *Study Area*

The study area encompasses the entire Brazilian Cerrado, comprising the states of Goiás, Tocantins, Bahia, São Paulo, Minas Gerais, Mato Grosso, Mato Grosso do Sul, Piauí, Maranhão as well as part of Paraná (Fig. 1). The Cerrado borders all other Brazilian biomes except for Pampa in the south of the country. In these ecotone areas, there is a broad range of ecosystem services (Embrapa Cerrados 2008).

Cunha et al. (1994) described four deforestation frontiers in Cerrado. The first is a consolidated agribusiness frontier running across western Minas Gerais, central and southeast Goiás, southern Mato Grosso do Sul and southeast Mato Grosso (Zone I). Another agricultural expansion frontier situates in northern Goiás, Distrito Federal (DF) and western Minas Gerais (Zone II). The agricultural expansion during the nineties followed a different deforestation pattern. This included a small area in northern Mato Grosso, northeastern Goiás, Tocantins and Western Bahia (Zone III). A final agriculture frontier is emerging in the central and southern part of Mato Grosso, northeast Tocantins, southern Maranhão and southwest Piauí (Zone IV) (Cunha et al. 1994).

2.2 *Dataset*

Despite its extensive area, there is a lack of systematic data on land use dynamics in Cerrado. The three major sources are: the Brazilian Statistical Office (Instituto Brasileiro de Geografia e Estatística–IBGE), the Monitoring Program on the Deforestation of Brazilian biomes, known as PROBIO (Monitoring Program organized by the Ministry of the Environment–MMA) and the Systematic Monitoring of Deforestation in Cerrado (SIAD-LAPIG) (Table 1). While these monitoring systems have provided data on deforestation, spatially explicit data on regrowth rates are still missing. Even the most recent assessment by Satellite Monitoring of Deforestation in the Brazilian biomes (Programa Monitoramento Desmatamento dos Biomas Brasileitros–PMDBBS) focuses only on gross vegetation losses and there are no data available on annual deforestation rates prior to 2002 (Beuchle et al. 2015).

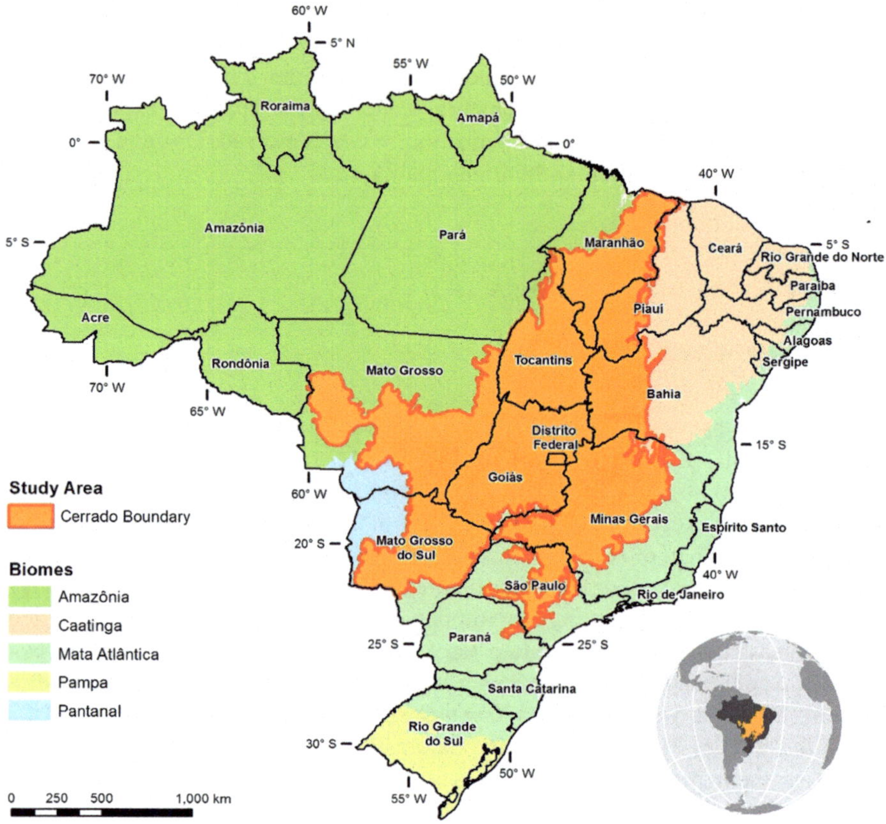

Fig. 1 Major Brazilian biomes

3 Methodology

3.1 Econometric Model

Econometric models apply mathematical and statistical techniques for data estimation and inference. The econometric model we developed determines the gross rates of deforestation and regrowth in each municipality in the Cerrado biome from 2002 to 2009.

The econometric model with spatial dependence (Le Sage and Pace 2009) is as follows:

$$Y_{t+v} = \xi WY + \beta_0 + \beta 1_{X1} + \beta 2_{X2} + \ldots + \beta_{nXn} + \varepsilon \qquad \text{(i)}$$

Table 1 Cerrado datasets

Source	Institution	Time	Data	Data source	Resolution
Agricultural census	Instituto Brasileiro de Geografia e Estatística–IBGE	1995 and 2006	Interviews with landowners	(1)	Municipality
Demographic, economic and geographical data	Instituto Brasileiro de Pesquisa Econômica Aplicada–IPEA	1995 and 2006	Surveys	(2)	Municipality
Satellite Monitoring deforestation in the Brazilian biomes PMDBBS	Ministério do Meio Ambiente–MMA/Instituto Brasileiro do Meio Ambiente e dos Recursos Naturais Renováveis–IBAMA	2002 and 2008	Remote sensing	(3)	1:250,000
		2008 and 2009			
		2009 and 2010			
Systematic Monitoring of Deforestation in Cerrado–SIAD LAPIG	LAPIG, Conservation International and The Nature Conservancy (TNC) (LAPIG, 2012)	2003 to 2009	Remote sensing	(4)	1:250,000
Project for Conservation and Sustainable Use of Brazilian Biological Diversity–PROBIO	Ministério do MeioAmbiente - MMA emparceria com o ConselhoNacional de DesenvolvimentoCientífico e Tecnológico – CNPq	2002	Remote sensing	(5)	1:250,000

(1) http://www.sidra.ibge.gov.br/
(2) http://www.ipeadata.gov.br/
(3) http://siscom.ibama.gov.br/monitora_biomas/PMDBBS%20-%20CERRADO.html
(4) https://www.lapig.iesa.ufg.br/lapig/index.php/produtos/dados-geograficos
(5) http://mapas.mma.gov.br/mapas/aplic/probio/datadownload.htm?/cerrado/mapas_pdf/vegetacao/250000/index.html

where:
Y = dependent variable
X = independent variable
β = coefficient
ε, μ = error terms
ξ = spatial autocorrelation parameter
λ = spatial autocorrelation parameter in the error term
W = weight according to neighborhood matrix

Input variables were computed for the 1204 municipalities comprising data from both the IBGE (IBGE 1995, 2006) and from the Instituto Brasileiro de Pesquisa Econômica Aplicada (IPEA). In addition to this secondary data, we also developed spatial variables such as the influence of urban centers and slope using Dinamica EGO.

The dependent variable (gross rate = Y) was calculated by subtracting the area of native vegetation for 1995 from the area for 2006 (ii). Negative values indicate loss of native vegetation (deforestation), while positive values represent gains (vegetation regrowth) as follows:

$$\mathrm{Tx} = (\mathrm{Ft} + \mathrm{n} - \mathrm{Ft}) \tag{ii}$$

where
Tx = rate of variation of native vegetation cover (Gross_Rate Y)
Ft = initial forest area (t)
Ft + n = forest area in (t + n)

We began the model by analyzing a total of 108 possible explanatory variables. Variables were standardized according to the area of the municipality. Linear stepwise regression was developed using SPSS Statistics 17.0 software. We then excluded all the municipalities with null values. From the initial 1204, a subset of 1192 municipalities was then included in the econometric model.

The model was controlled for outliers by assigning binary values (0, 1) to two new variables named as positive and negative outliers. In the positive outlier variable, municipalities in which residuals were more than two standard deviations above the mean were assigned a value of 1, while the others were assigned a value of 0. For the negative outlier variable, municipalities with residuals below two standard deviations were assigned a value of 1, while all the others were assigned a value of 0. In this way, we were able to separate the influence of outliers without removing them from the analysis (Soares Filho et al. 2008). The variables were selected according to their statistical significance and a spatial auto-regression model was performed using the Geoda 0.9.9.1 software.[1] Once these procedures had been completed, the econometric model was integrated into a spatially explicit simulation model in Dinamica EGO.[2]

3.2 *Spatially Explicit Model*

The econometric model calculates the deforestation and regrowth rates, which are then passed on to the spatially explicit model for allocation.

[1]https://spatial.uchicago.edu/software.

[2]http://csr.ufmg.br/dinamica/ (there is a detailed description of Dinamica EGO in Part I of this book).

3.3 Spatial Model Calibration

The Weights of Evidence model (WOFE) (Bonham-Carter 1994) was used to derive the transition probability map (Fig. 2).

The WOFE model assesses the relationships between a group of explanatory variables and the spatial probability of a transition, in this case the probability of a loss or a gain in native vegetation (Fig. 2). It does this by computing changes inside and outside a certain spatial pattern (Bonham-Carter 1994; Soares et al. 2013). The larger the value of the coefficient of the Weight of Evidence W+, the stronger the association between the explanatory variable and the change. By the same logic, negative coefficients indicate an inhibitory effect, and values close to zero are consistent with no association. Variables used in the model include (i) distance to previously deforested areas (or regrowth) (ii) distance to roads and railways (Soares Filho et al. 2004), (iii) distance to rivers, (iv) elevation, (v) slope, and (vi) distance to croplands (soy). The effect of each spatial variable is calculated independently of a combined solution and the only assumption that must be made is that the explanatory variables are spatially independent, which can be checked using pair-wise tests for categorical maps, such as Cramer's Coefficient, Contingency Coefficient, and Joint Information Uncertainty. Continuous variables also need to be categorized (Bonham-Carter 1994).

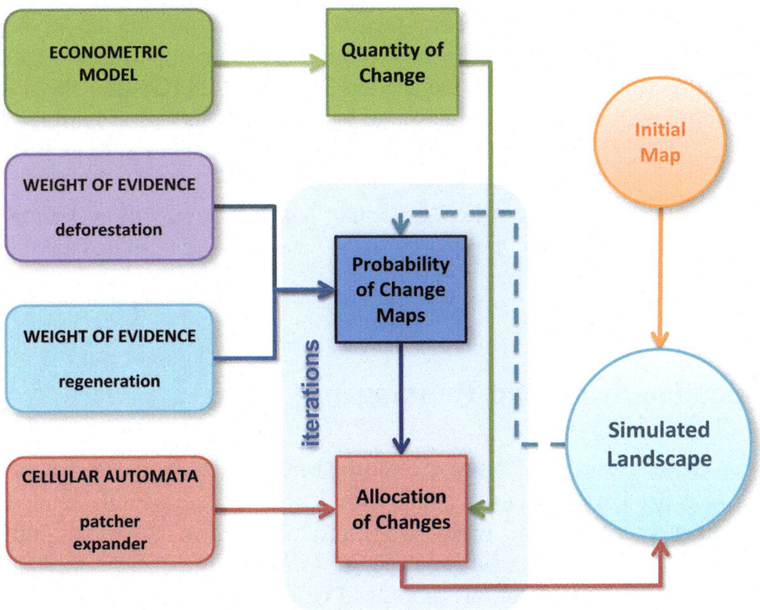

Fig. 2 Model components

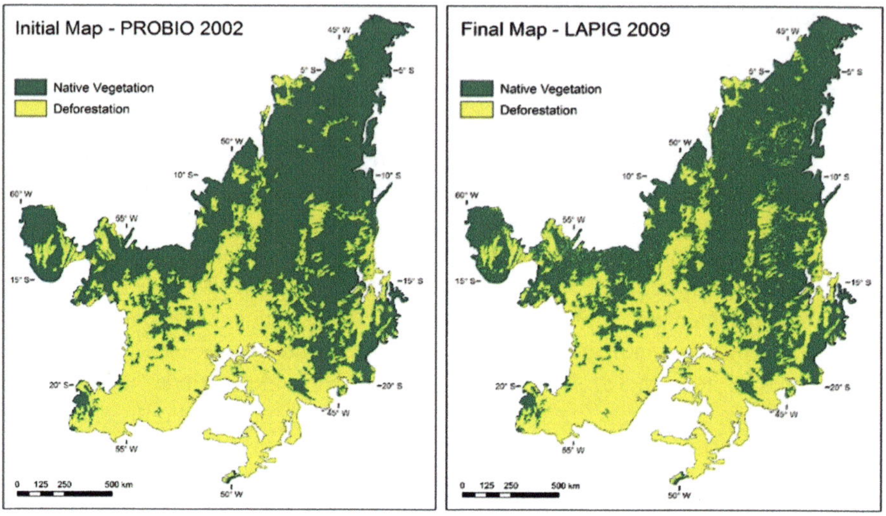

Fig. 3 Initial (*left*) and final (*right*) land cover maps. Up to 2002 only PROBIO assessed Cerrado deforestation. By 2009 a more refined dataset was made available by SIAD-LAPIG. We used the best data available in the different years as input for the model

As the quantity of change was output by the econometric model on the basis of census data, it may overestimate the spatial data mapped by LAPIG. In order to address this issue, we used a correction factor of 0.19 to adjust the model output to the LAPIG data. This correction factor was obtained by comparing the simulated annual rate of deforestation (from the spatial data from LAPIG) with the observed rate (in IBGE census data) for the same period (2002–2009) (Figure 3). After calibrating the model with data for 2006 to 2009, the model simulates annual losses and gains (regrowth) of Cerrado native vegetation from 2009 to 2050. The calibration goal is to capture the "rules" governing land cover transitions (gain/loss of vegetation). The validation procedure then assesses the robustness of the simulation.

4 Simulation Setup and Running

Dinamica EGO (see part V, and www.csr.ufmg.br/dinamica for a description of the software) was used to run land change simulations (Soares Filho et al. 2009). Annual time-step simulation maps were produced from 2006 to 2009. To approximate the simulated landscape structure to the real one, we tested different settings for the transition functions employed in Dinamica: the Expander and the Patcher (Soares Filho et al. 2009; Soares et al. 2013). These functions incorporate cellular automata local rules designed to mimic the neighborhood influence on the transition of a cell state (Soares Filho et al. 2009). The job of the Expander is to expand or contract previous patches of a certain land-use and land-cover class, while the Patcher is designed to form new patches through a seedling mechanism (Soares Filho et al. 2009; Soares et al. 2013).

5 Validation

The model was validated using three procedures: (i) comparison of the rates estimated by the econometric model and those observed by the Brazilian statistical office (IBGE) for the same period (1995-2006), (ii) Quantitative validation of the simulated rates and those from LAPIG for 2009, (iii) spatial allocation comparison using fuzzy logic (Soares et al. 2013).

6 Results

6.1 Econometric Model

The selected variables are displayed in Table 2. There is a direct relationship between cropland area (soy, maize, sugar cane and cotton) and the loss of Cerrado's native vegetation. The same is true for the remaining vegetation area. Cattle rearing is positively associated with vegetation loss. Proximity to urban areas is another important factor. By contrast, higher elevations have a negative association with the loss of vegetation in Cerrado.

Table 2 Significant variables selected in the econometric model and their coefficients

Variable	Coefficient	Significance
Constant	0.114368777	0.000
Cropland area	–0.08114414	0.000
Remaining vegetation	–0.603871017	0.000
Head of cattle	–0.006072578	0.231
Elevation	3.08378E-05	0.000
Urban influence	–2.23467E-09	0.000
Negative Outlier	0.198747832	0.000
Positive Outlier	–0.178983079	0.000
R^2 = 0.609		

6.2 Spatially Explicit Model

6.2.1 Spatial Determinants for Loss/Gain of Native Vegetation in Cerrado

By analyzing the WOFE coefficients we can identify the influence of spatial determinants on the two transitions, namely loss and gain of native vegetation in Cerrado. The analysis of the variable slope shows that flat areas (<7°) have a positive association with loss of native vegetation while steep slopes, by contrast, constrain deforestation (Figs. 4 and 5).

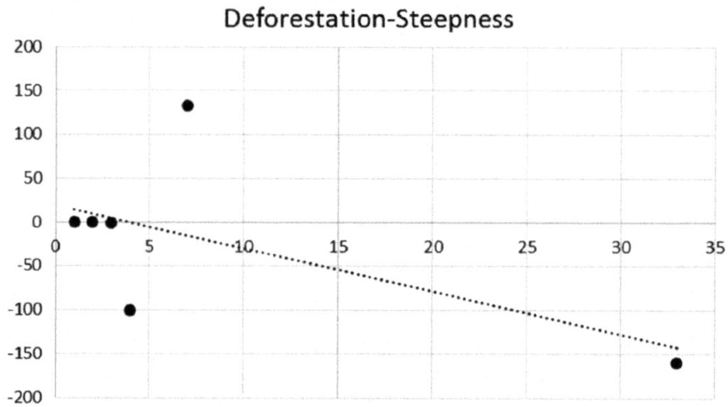

Fig. 4 Loss of Cerrado vegetation and its association with slope. The X axis shows the steepness in degrees and Y the weight of evidence

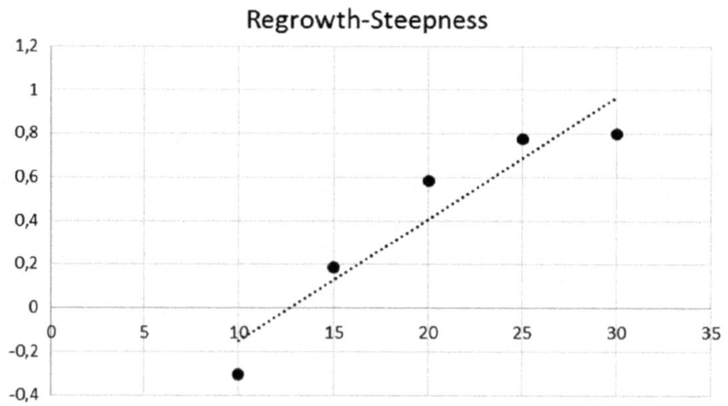

Fig. 5 Gains in native vegetation and their association with slope. The X axis shows the steepness in degrees and Y the weight of evidence

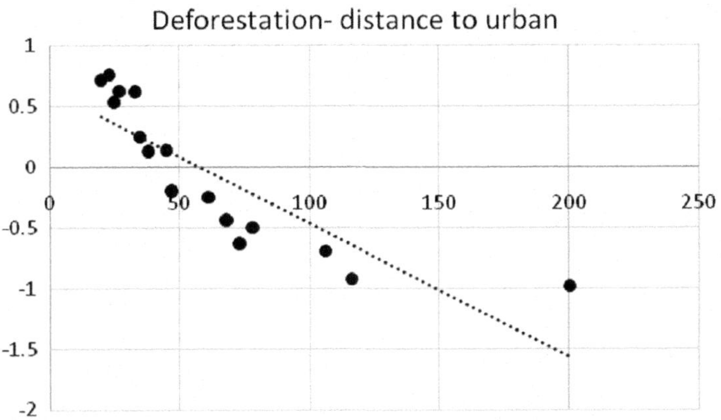

Fig. 6 Deforestation and its association with distance to urban. X axis shows the distance to urban in km and Y the weight of evidence

Another important spatial determinant of transition is elevation. Lower elevation areas have a positive association with the loss of native vegetation in Cerrado, while higher elevations have a negative association with deforestation and a positive association with gains in vegetation.

Proximity to urban areas is also an important determinant for transition in this region. The closer to urban areas, the higher the likelihood of deforestation. Plots of vegetation located near urban centers (within a radius of 50 km) are thus more likely to become deforested (Fig. 6).

Other important spatial determinants for loss/gain of native vegetation are protected areas and proximity to regrowth areas. Regarding the former, we found a negative association between protected areas (indigenous and strictly protected lands) and loss of native vegetation, while proximity to regrowth areas (gains in natural vegetation) is associated with regrowth.

The above results are compatible with those of similar research in the Amazon (Soares et al. 2010). However, the 'distance to roads and railways' variable produces results that at first sight seem to contradict these findings. As shown in Fig. 7, for the period under study, our results show that the areas with greatest deforestation are located farther away from roads. These results might be explained by the fact that the major dynamics in Cerrado deforestation occurred before the study period (2002–2009), when major roads such as the BR 153 (built 1959) and the BR 364 (built 1960) were constructed. Roads and railways were not therefore important spatial determinants of deforestation over the period of study.

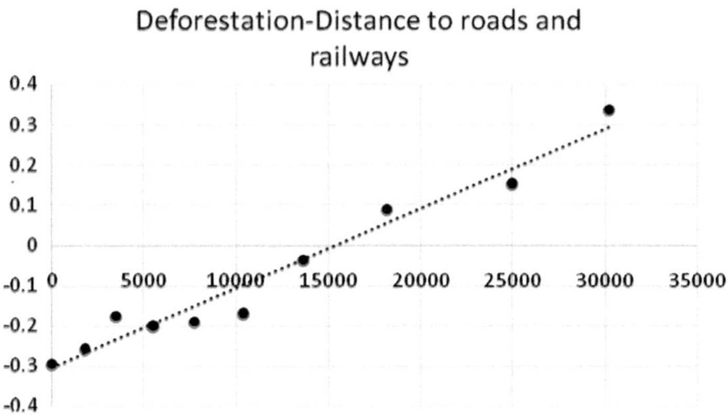

Fig. 7 Deforestation and its association with distance to roads and railways. X axis shows the distance to roads/railways in meters and Y the weight of evidence

6.2.2 Spatial Allocation of Transitions

Figure 8 shows a kernel map of the deforestation. According to Fig. 8 deforestation is concentrated in the central area of Mato Grosso, southern Maranhão, southeastern Piauí and western Bahia.

These results are in line with those produced by Cunha et al. (1994). Figure 8 shows that over the study period (2002 to 2009) the highest rates of deforestation were in western Bahia, southeastern Piauí and southern Maranhão (Zones III and IV of Cunha et al.1994). Figure 8 shows lower rates of deforestation in Zones I and II, where agriculture and husbandry developed earlier, in the 1990s.

7 Calibration and Validation

While the rate of variation in native vegetation (Tx- gross rate) obtained by the census data is 0.21%, the gross rate from the econometric model was 0.29%. The fact that the two estimates are quite similar and that both are positive reveals that regrowth is higher than deforestation. These results are in line with official records by the Ministry of Environment showing, for the period under analysis, that in 2002–2008 the annual deforestation rate was 0.7%. This estimate was further reduced to 0.37% between 2009–2009 and to 0.3 for the period between 2009 and 2010.

Protected areas, such as the Jalapão State Park and the Araguaia National Park, in the State of Tocantins, the Chapada dos Veadeiros National Park, on the border between the States of Minas Gerais and Bahia, amongst others, have a lower probability of deforestation (Fig. 9). The same is true for indigenous areas located in Mato Grosso in ecotone areas (areas of transition between biomes) with the

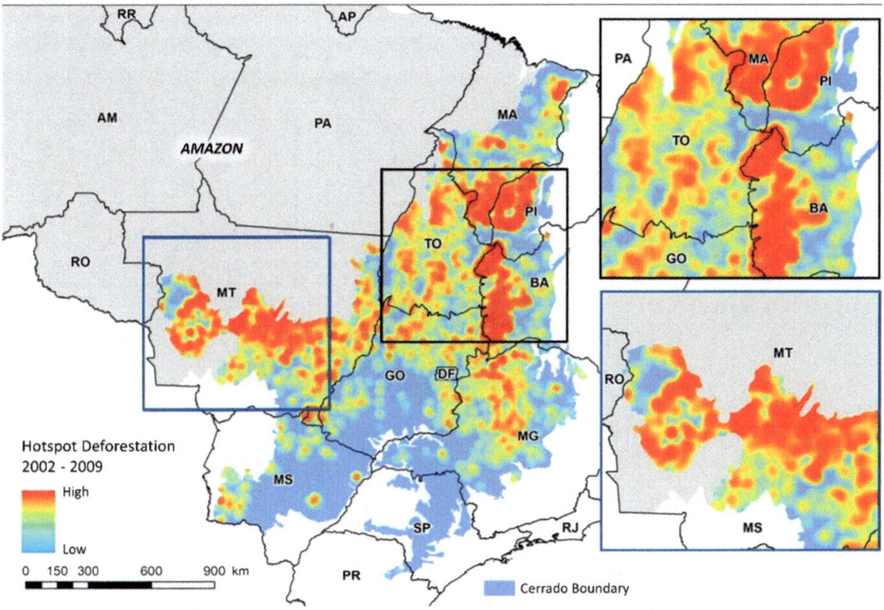

Fig. 8 Hotspots for deforestation in Cerrado 2002–2009 using kernel statistics. Kernel statistics were computed in Dinamica EGO, using a window size of 30 km radius

Amazon. However, as shown in Fig. 10, there is a high probability of deforestation (in red) in some of the buffer zones around protected areas.

Using the method of reciprocal similarity (Soares et al. 2013), we compared the simulated map by using the weights of evidence with a map of equal probability of change for the whole Cerrado region. This procedure checks the consistency of allocations based on spatial determinants with the randomly generated ones.

The simulated model, which used the probability resulting from the weights of evidence, achieved a similarity of 50% for a spatial comparison window of 15 km, while the simulation based on a constant probability (null model) achieved a similarity of 12.5% (Fig. 11). This shows that the model performed better in spatially predicting deforestation than a null (random with same transition rates) model.

8 Simulation

The model produced a loss of 14.2 Mha (6.0% over the whole period at an annual rate of 0.16%) over the period 2009 to 2050 and a gain of 18.5 Mha (annual regrowth rate of 0.79%) (Fig. 12). This means that the rate of deforestation is similar to the mean rate of 0.17% found by Ferreira et al. (2012) for the period 2010–2050. Deforestation is concentrated in the central part of Mato Grosso–ecotone with Amazon–southern

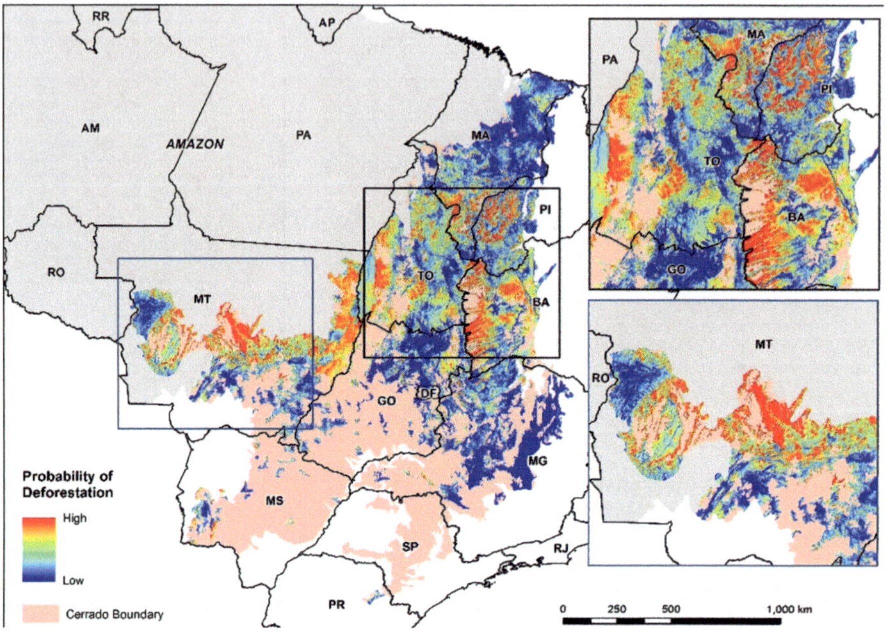

Fig. 9 Map showing the probability of deforestation (based on the weights of evidence for the spatial determinants analyzed)

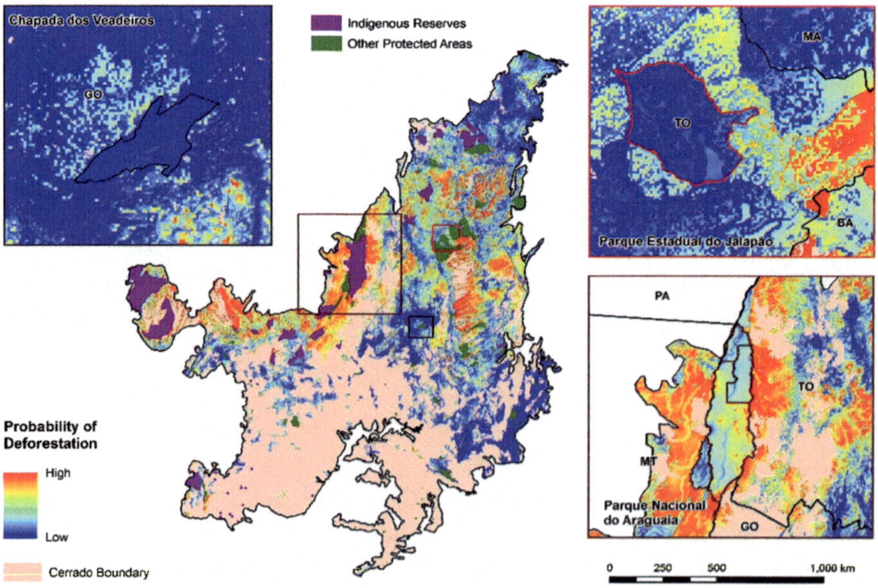

Fig. 10 Deforestation probability map highlighting protected areas

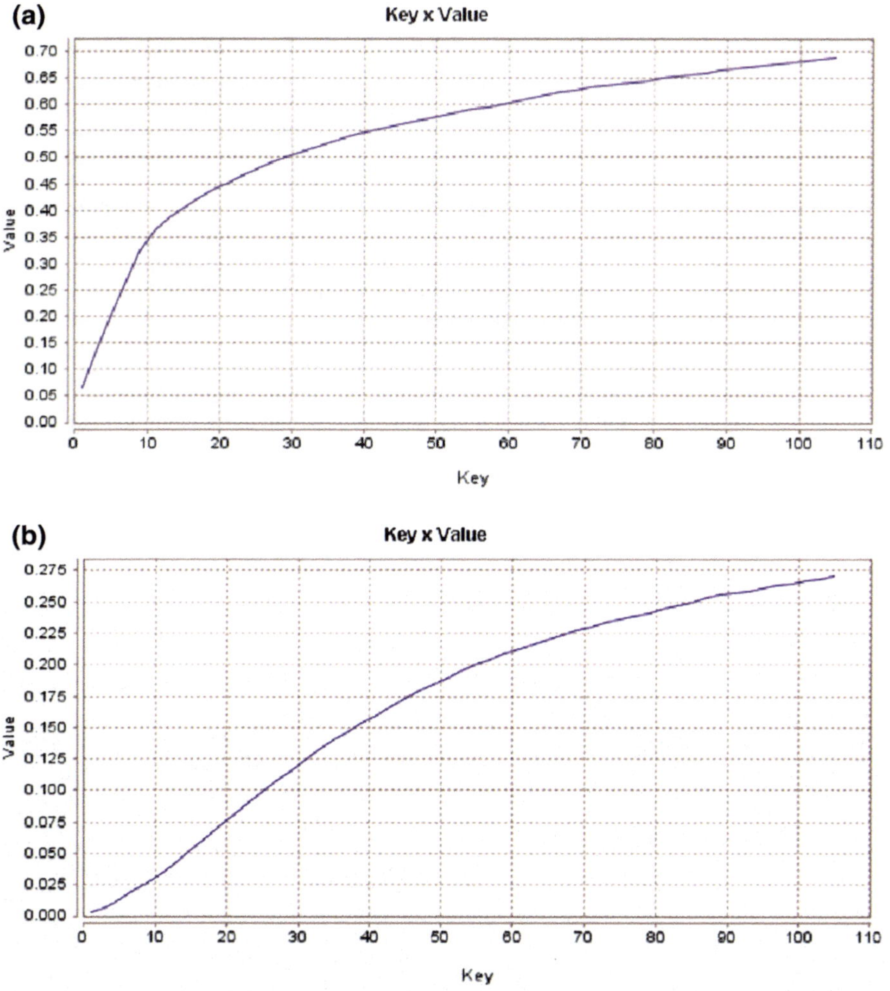

Fig. 11 Similarity indexes for spatial allocation considering window size (pixel). **a** similarity using the WOFE, **b** similarity based on a constant probability map

Maranhão, southeastern Piauí, western Bahia and western Tocantins (frontier between forest and agriculture) (Fig. 12).

As there is no monitoring of regrowth in the Cerrado biome, we used the weights of evidence from Teixeira et al. (2009). These authors show that regrowth occurs near water streams (Fig. 13), in high elevations and on steep slopes. The majority of these areas are classified as Permanent Protected Areas (APPs).

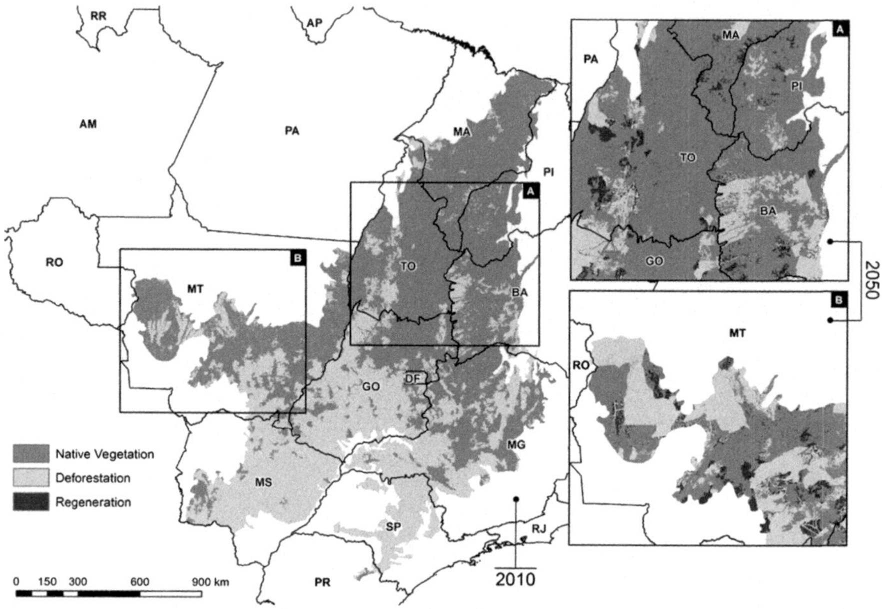

Fig. 12 Simulated spatial allocation of native vegetation, deforestation and regeneration areas in 2010 and 2050

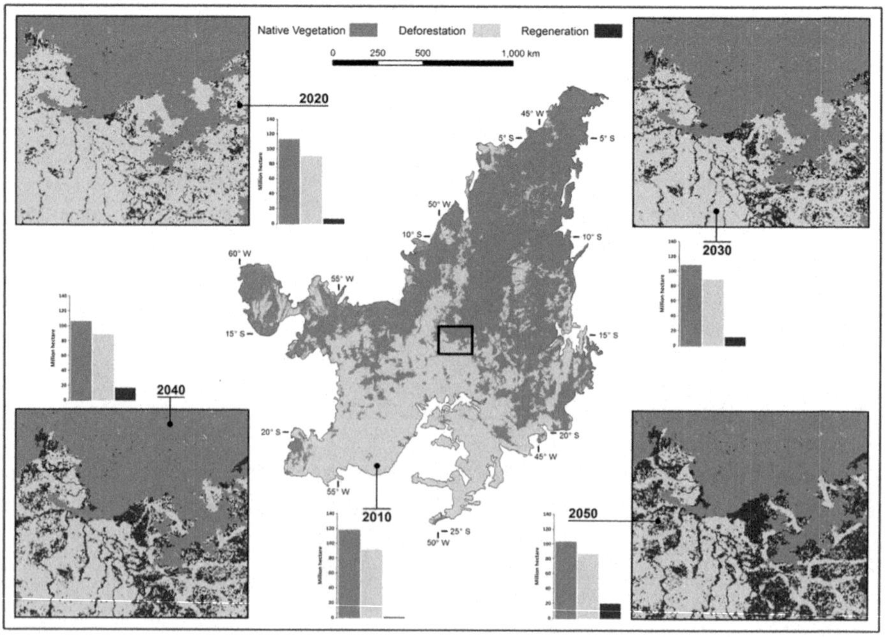

Fig. 13 Simulation of the spatial allocation of native vegetation, deforested and regrowth areas in the years 2010, 2020, 2040 and 2050. Detail is given to the spatial allocation of regrowth in Permanent Protection Areas (APP)

9 Conclusion and Outlook

Spatially explicit land-use models simulate the patterns of change in the landscape in response to both human and ecological dynamics. As there is growing awareness that ongoing land use transitions are linked to major environmental issues (Grecchi et al. 2014; Aguiar et al. 2016), considerable research efforts have been devoted to modeling LULCC in a spatially explicit way in order to inform policy makers.

In line with studies on the Amazon, our results for the Cerrado biome show the lowest probability of deforestation inside protected areas and indigenous lands. However, only 8.21% of this region has protected status–2.85% -full protection and 5.36% under sustainable use. Our results also show that there is a high probability of deforestation in some of the buffer zones surrounding the protected areas, so highlighting the level of threat. These buffer zones must therefore be monitored more carefully.

Several problems can arise when building a spatially explicit land-use model, one of which is the lack of appropriate data to represent the system being studied, thus limiting model calibration and validation (Soares-Filho et al. 2006, 2009, 2013). In this research, we tried to model the dynamics of land use change in the entire Cerrado biome making use of the best data available. By using appropriate statistical techniques to deal with uncertainties and gaps (e.g. spatial data on Cerrado regrowth) in the data, we were able to build a model whose outputs broadly agree with recent estimates of land cover change (Grecchi et al. 2014; Beuchle et al. 2015) and official reports (MMA/IBAMA/PNUD, 2010). The model was able to mimic the deforestation and regrowth spatial patterns (Fig. 11) for the entire Cerrado region. Our simulation indicates those areas that need to be continuously monitored to avoid illegal deforestation.

Although the results of this work are illustrative of the land use dynamics in the Cerrado region, it is important to acknowledge that there are limitations in the data, above all due to the differences between the agricultural census and the monitoring data, and the lack of spatial data for monitoring regrowth. These limitations in the model approach reveal the need for monitoring systems that can capture land use trends in the Cerrado biome in a detailed scale. Another important issue is the adoption of a common monitoring framework that can compare and integrate different monitoring systems. Since the Cerrado comprises an enormous area, there is a wide socioeconomic and environmental heterogeneity across the biome. There is therefore a need to understand how different dynamics take place in specific socioeconomic contexts. One possibility for future research is to regionalize the simulation model to address the specificities of the various frontiers, for example those identified by Cunha et al. (1994). The study of the spatial determinants leading to deforestation in Cerrado at regional or local scale is likely to produce new insights into the social dynamics of deforestation, although these go beyond the scope of this research.

Finally, we need to improve the monitoring systems by mapping the multiple transitions that take place in the Cerrado region. This will certainly enhance the capability of modeling approaches, and will allows us to design better tools for planning the conservation of this important Brazilian biome.

References

Aguiar APD, Vieira ICG, Assis TO, Dalla-Nora EL, Toledo PM, Oliveira Santos-Junior RA, Batistella M, Coelho AS, Savaget EK, Aragão LEOC, Nobre CA, Ometto JPH (2016) Land use change emission scenarios: anticipating a forest transition process in the Brazilian Amazon. Glob Change Biol

Basse RM, Omrani H, Charif O, Gerber P, Bódis K (2014) Land use changes modelling using advanced methods: cellular automata and artificial neural networks. The spatial and explicit representation of land cover dynamics at the cross-border region scale. Appl Geogr 53:160–171

Beuchle R, Grecchi RC, Shimabukuro YE, Seliger R, Eva HD, Sano E, Achard F (2015) Land cover changes in the Brazilian Cerrado and Caatinga biomes from 1990 to 2010 based on a systematic remote sensing sampling approach. Appl Geogr 58:116–127

Bonham-Carter G (1994) Geographic information systems for geoscientists: modelling with GIS. Pergamon, Oxford, UK

Bürgi M, Hersperger AM, Schneeberger N, (2004) Driving forces of landscape change–current and new directions. Landsc Ecol 19(8):857–868

Cunha A, Mueller CC, Alves E, Silva J (1994) Uma Avaliação da Sustentabilidade da Agriculturanos Cerrados. Brasília, IPEA

Embrapa Cerrados (2008) IV Plano Diretor da Embrapa Cerrados 2008–2011–2023/Embrapa Cerrados–Planaltina, DF: Embrapa Cerrados 42 p

Ferreira ME (2009) Modelagem da Paisagem do Cerrado. Universidade Federal de Goiás

Grecchi RC, Gwyn QHJ, Bénié GB, Formaggio AR, Fahl FC (2014) Land use and land cover changes in the Brazilian Cerrado: a multidisciplinary approach to assess the impacts of agricultural expansion. Appl Geogr 55:300–312

IBGE (1995) Censo Agropecuario 1995/96. Rio de Janeiro, Brazil: IBGE. 1998. http://biblioteca.ibge.gov.br/index.php/biblioteca-catalogo?view=detalhes&id=748. Accessed Feb 2017

IBGE (2006) Censo Agropecuario 1995/96. Rio de Janeiro, Brazil: IBGE. http://www.biblioteca.ibge.gov.br/index.php/biblioteca-catalogo?view=detalhes&id=750. Accessed Feb 2017

Klink C, Machado RB (2005) Conservation of the Brazilian Cerrado. Conserv Biol 19:707–713

Lambin EF, Meyfroidt P (2010) Land use transitions: socio-ecological feedback versus socio-economic change. Land Use Policy 27:108–118

Lambin EF, Meyfroidt P, Rueda X, Blackman A, Börner J, Cerutti PO, Dietsch T, Jungmann L, Lamarque P, Lister J, Walker NF, Wunder S (2014) Effectiveness and synergies of policy instruments for land use governance in tropical regions. Glob Environ Change 28:129–140

LAPIG (2012) Laboratório de Processamento de Imagens e Geoprocessamento do Instituto de Estudos Sócio-Ambientais. LAPIG-IESA. Disponívelem: http://www.lapig.iesa.ufg.br/lapig/. Accessed Apr 2016

Lawler JJ, Lewis DJ, Nelson E, Plantinga AJ, Polasky S, Withey JC, Helmers DP, Martinuzzi S, Pennington D, Radeloff VC (2014) Projected land-use change impacts on ecosystem services in the United States. Proc Natl Acad Sci 111:7492–7497

Le Sage JP, Pace RK (2009) Introduction to spatial econometrics. New York: Taylor & Francis Group, pp 321

MMA/IBAMA/PNUD (2010) Relatório Técnico de Monitoramento de Desmatamento do Bioma Cerrado IBAMA. 2008 a 2009. Disponívelem: http://siscom.ibama.gov.br/monitorabiomas/cerrado/RELATORIO_CERRADO_2008–2009.pdf

Naudts K, Chen Y, McGrath MJ, Ryder J, Valade A, Otto J, Luyssaert S (2016) Europe's forest management did not mitigate climate warming. Science 351:597–600

Pereira HM, Ferrier S, Walters M, Geller GN, Jongman RHG, Scholes RJ, Bruford MW et al (2013) Essential biodiversity variables. Science 339:277–278

Pielke RA, Pitman A, Niyogi D, Mahmood R, McAlpine C, Hossain F, Goldewijk KK, Nair U, Betts R, Fall S, Reichstein M, Kabat P, de Noblet N (2011) Land use/land cover changes and climate: modeling analysis and observational evidence. Wiley Interdiscip Rev Clim Change 2:828–850

Skidmore AK, Pettorelli N, Coops NC, Geller GN, Hansen M, Lucas R, Mücher CA, O'Connor B, Paganini M, Pereira, HMM (2015) Environmental science: agree on biodiversity metrics to track from space. Nature 523

Soares-Filho B et al (2013) OTIMIZAGRO: Uma Plataforma Integrada de Modelagem de Uso e Mudanças no Uso da Terra para o Brasil. Secretaria de Assuntos Estratégicos da Presidênciado Brasil. Tech, Report

Soares-Filho B et al (2006) Modeling conservation in the Amazon basin. Nature 440:520–523

Soares-Filho B, Rajão R, Merry F, Rodrigues H, Davis J, Lima L, Macedo, M, Coe M, Carneiro A, Santiago L (2016) Brazil's Market for trading forest certificates. Plos One. doi:10.1371/journal.pone.0152311 6 April 2016

Soares B, Moutinho P, Nepstad D, Anderson A, Rodrigues H, Garcia R, Dietzsch L, Merry F, Bowman M, Hissa L, Silvestrini R, Maretti C (2010) Role of Brazilian Amazon protected areas in climate change mitigation. P Natl Acad Sci USA 107:10821–10826

Soares B, Rajao R, Macedo M, Carneiro A, Costa W, Coe M, Rodrigues H, Alencar A (2014) LAND USE cracking Brazil's forest code. Science 344:363–364

Soares B, Rodrigues H, Follador M (2013) A hybrid analytical-heuristic method for calibrating land-use change models. Environ Model Softw 43:80–87

Soares Filho B, Alencar A, Nepstad D, Cerqueira G, Diaz M, Rivero S, Solorzanos L, Voll E (2004) Simulating the response of land-cover changes to roadpaving and governance along a major Amazon highway: the Santare´m–Cuiaba´corridor. Glob Chang Biolo 10

Soares Filho B, Garcia RA, Rodrigues H, Moro S, Nepstad D (2008) Nexos entre as dimensõessocioeconômicas e o desmatamento da Amazônia: a caminho de um modelo integrado. In: Amazônia: Natureza e Sociedade em Transformação, p 304

Soares Filho B, Rodrigues H, Costa,W (2009) Modeling environmental dynamics with dinamica ego. In: Soares Filho B. (ed) Guidebook. Belo Horizonte, Brazil. http://csr.ufmg.br/dinamica/dokuwiki/doku.php?id=tutorial:start. Accessed 20 Nov 2015

Soares Filho B, Rodrigues H, Follador M (2013) A hybrid analytical-heuristic method for calibrating land-use change models. Environ Model Softw. doi:10.1016/j.envsoft.2013.01.010

Part IV
Technical Notes

Chapter 20
Cellular Automaton

J.F. Mas and H. Rodrigues

Abstract Cellular Automaton (CA) is widely used in land change modeling. In this technical note, we describe two CA: the Game of Life and the CA used in the software package DINAMICA EGO.

Keywords Simulation · Neighborhood effects · Spatial patterns

1 Short Description of Interest

A cellular automaton (CA) consists of a discrete cell space of any dimension. Each cell presents one state among a finite number of states, and changes it according to a set of rules that determines the new state depending on the cell neighborhood. The rules for updating the state of cells are iteratively applied to generate a new grid from the previous one.

See Chap. 3 about simulation and the short presentations in Part V of this book about CA_MARKOV, Dinamica EGO, Metronamica and APoLUS.

J.F. Mas (✉)
Centro de Investigaciones en Geografía Ambiental, Universidad Nacional Autónoma de México (UNAM), Morelia, Michoacan, Mexico
e-mail: jfmas@ciga.unam.mx

H. Rodrigues
Centro de Sensoriamento Remoto, Universidade Federal de Minas Gerais, Belo Horizonte (UFMG), Belo Horizonte, Brazil
e-mail: hermann@csr.ufmg.br

M.T. Camacho Olmedo et al. (eds.), *Geomatic Approaches for Modeling Land Change Scenarios*, Lecture Notes in Geoinformation and Cartography,
https://doi.org/10.1007/978-3-319-60801-3_20

2 Technical Details

The most famous cellular automaton is undoubtedly the Game of Life. It consists of a two-dimensional grid of cells, each of which is in one of two possible states: "populated" or "unpopulated" (black and white cells respectively in Fig. 1). The state of each cell changes according to a few mathematical rules taking into account its eight immediate neighbors:

- A populated cell with one or no neighbors becomes unpopulated.
- A populated cell with four or more neighbors becomes unpopulated.
- A populated cell with two or three neighbors remains populated.
- An unpopulated cell with three populated neighbors becomes populated.

Cellular automata models were first applied in geography by Tobler (1979) and widely applied to land change modeling, particularly for urban growth simulation due to their simplicity, flexibility and intuitiveness, and their ability to represent spatio-temporal processes (White and Engelen 1993; Santé et al. 2010). The use of CA to simulate land change is based on the assumption that landscape spatial configuration affects future patterns of change through local interactions among land uses. For example, Besussi et al. (1998) used a CA which modifies the density of residential cells as a function of the presence of commercial cells in the neighborhood.

Some models such as CA_MARKOV (see the technical note about Cellular Automata in CA_MARKOV in Part IV of this book) and Dinamica EGO use a neighborhood filtering referred to as CA to simulate a proximity effect that makes changes occur in the form of patches in order to mimic landscape patterns and to avoid a salt and pepper effect. Dinamica EGO patcher CA is designed to generate new patches through a seeding mechanism. The user can set parameters to control the mean patch size and the patch size variance (Soares-Filho et al. 2002). Figure 2 illustrates the CA behavior in a simplified way. First a patch seed (S) is selected using an approach which selects one cell from amongst those with the highest probabilities but without restricting the selection to these cells.

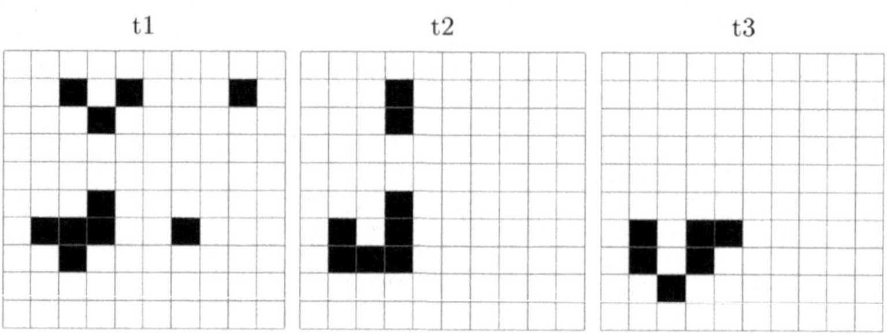

Fig. 1 Evolution of the grid of cells during three iterations according to the rules of the Game of Life

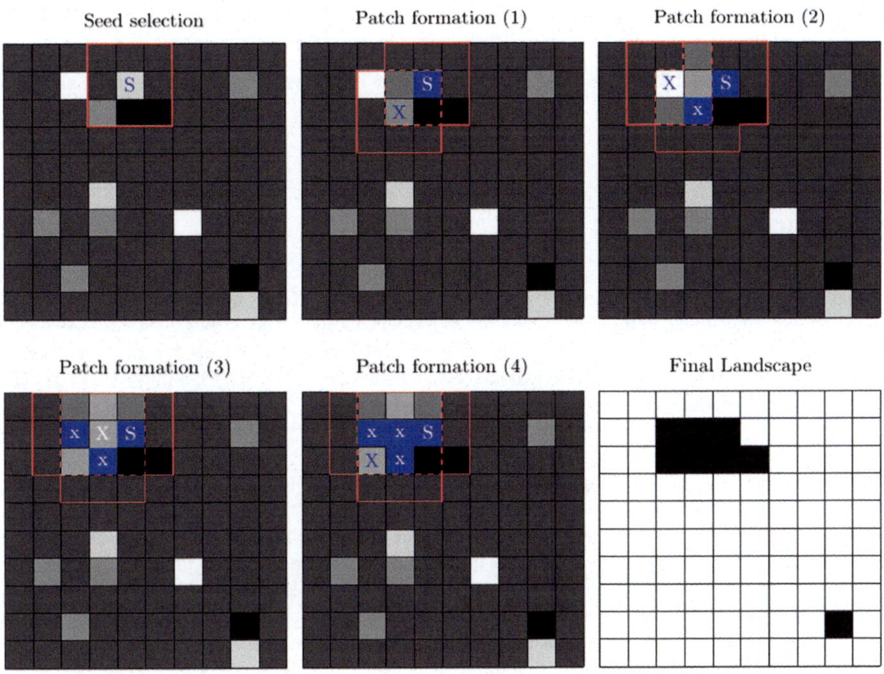

Fig. 2 Procedure of patch creation in Dinamica patcher

Then, seed neighbors are selected using a window (red box) and all neighbors in which transition is possible are collected and placed on a "patch formation pool". A cell is selected from that pool using the same approach used to select a seed cell. The selected cell (X) is used as part of the patch and its neighbors are collected and placed in the patch formation pool. If a cell is already in the pool (dashed red box), its probability is increased. This process continues until the expected number of cells for that patch is reached. The number of cells in a patch is chosen as a random number from a normal distribution based on the mean and variance patch sizes defined by the user. In the example in Fig. 2, patch size was five cells, and the patch formation process stopped when this size was reached.

Dinamica EGO also has a second CA known as the expander, which deals exclusively with the expansion of previously formed patches. It works in the same way as the patcher but only takes into account the cells in proximity to existing areas.

CA constraint-based models use an approach closer to the Game of Life as the transition rules are neighborhood-based. A common way of defining transition rules is to calculate the probability of change while taking the neighborhood into account. As this basic CA formalism is too simplified and fails to represent real landscape, it is often extended by adding elements such as suitability, zoning, accessibility and random perturbation (Zhao and Murayama 2007). For example, Metronamica computes the transition potential by combining neighborhood effect, accessibility, zoning and suitability.

References

Besussi E, Cecchini A, Rinaldi E (1998) The diffused city of the Italian north-east: identification of urban dynamics using cellular automata urban models. Comput Env Urban Syst 22(5):497–523

Santé I, García AM, Miranda D, Crecente R (2010) Cellular automata models for the simulation of real-world urban process: a review and analysis. Landscape and Urban Planning 96(2):108–122

Soares-Filho BS, Coutinho Cerqueira G, Lopes Pennachin C (2002) DINAMICA–a stochastic cellular automata model designed to simulate the landscape dynamics in an Amazonian colonization frontier. Ecol Model 154:217–235

Tobler WR (1979) Cellular geography. In: Gale S, Olsson G (eds) Philosophy in geography, pp 279–386

White R, Engelen G (1993) Cellular automata and fractal urban form: a cellular modelling approach to the evolution of urban land-use patterns. Environ Plann A 25(8):1175–1199

Zhao Y, Murayama Y (2007) A constrained CA model to simulate Urban growth of the Tokyo metropolitan area. In: Proceedings of the 9th international conference on geocomputation, national centre for geocomputation, National University of Ireland, Maynooth Eire, 3–5 September 2007. http://www.geocomputation.org/2007/7B-Cellular_Automata/7B3.pdf

Chapter 21
Cellular Automata in CA_MARKOV

M.T. Camacho Olmedo and J.F. Mas

Abstract In this technical note we present the Cellular Automata (CA) incorporated by default into CA_MARKOV (TerrSet software), that produces important effects in the simulation step. After a short description of interest, the technical details are showed followed by an example applying and ignoring the CA.

Keywords Simulation · Cellular Automata · Neighborhood filter · Suitability · Land change

1 Short Description of Interest

The Cellular Automata (CA) incorporated by default into CA_MARKOV (Eastman 2015) produces important effects in the simulation step. This filter up-weights the suitability of pixels that are contiguous to existing LUC pixels and down-weights the suitability of pixels that are not. As a result this filter homogenizes the simulated map through spatial aggregation and increases the probability of change in pixels that are both suitable and close to the existing LUC, producing a dilation effect around existing patches.

See Chap. 3 about simulation and the short presentations in Part V of this book about CA_MARKOV.

M.T. Camacho Olmedo (✉)
Departamento de Análisis Geográfico Regional y Geografía Física,
Universidad de Granada, Granada, Spain
e-mail: camacho@ugr.es

J.F. Mas
Centro de Investigaciones en Geografía Ambiental, Universidad Nacional
Autónoma de México (UNAM), Morelia, Michoacán, Mexico
e-mail: jfmas@ciga.unam.mx

M.T. Camacho Olmedo et al. (eds.), *Geomatic Approaches for Modeling Land Change Scenarios*, Lecture Notes in Geoinformation and Cartography,
https://doi.org/10.1007/978-3-319-60801-3_21

2 Technical Details

In CA_MARKOV, the CA reduces the probability of change away from existing areas of that category, using a 5 × 5 filter. The changes are simulated in various steps, so favoring the aggregation of cells where a change occurs and creates a "patchy" pattern. Figure 1 illustrates this procedure in a simplified way.

In Fig. 1 (top), the simulated changes are obtained by allocating the change (10 cells) to the cells with the highest probability of change without any CA. The bottom part of the figure illustrates the neighborhood filtering approach: in the first step, the probability of change is reduced for the cells furthest away from existing areas of that category using a 3 × 3 window. In step 1 only part of the change (five cells) is allocated to higher change probability. Probability is again decreased away from existing areas including newly created ones. In the second step, the rest of the change is allocated taking into account this modified change probability map, which increases the probability of obtaining clusters of simulated change cells. Note that this is a simplified example (only two steps for change allocation and changes in probability) and that different maps of simulated change can be obtained because many cells have the same probability of change.

3 Example

Figure 2 shows an example of the CA in CA_MARKOV and ignoring this CA. The filter incorporated by default in CA_MARKOV (Fig. 2, above) is a simple contiguity 5 × 5 filter (00100; 01110; 11111; 01110; 00100) normalized to force a

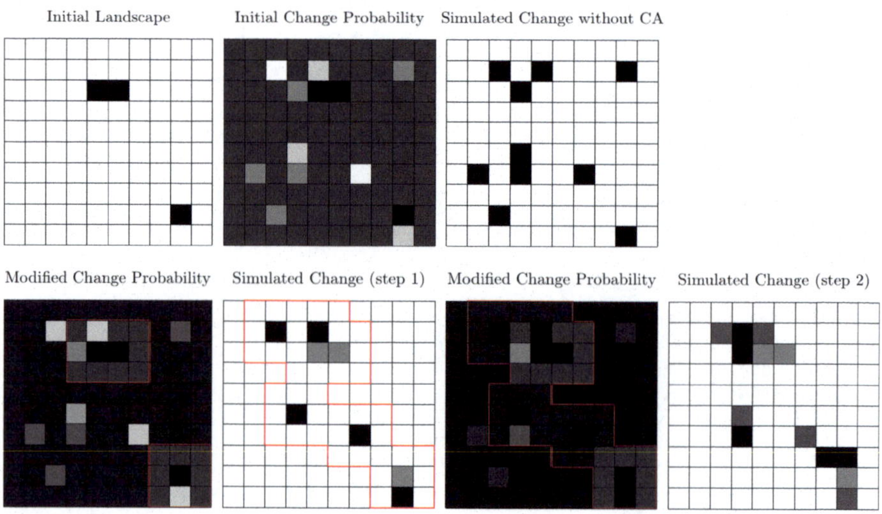

Fig. 1 Procedure for neighborhood filtering to simulate change patches

sum of 1. The CA is applied to a Boolean image for each class of the LUC at t1 (initial date) for each iteration. After that a value of 0.1111 is added to each filtered map to create the set of weighted images with values from 0.1111 to 1.1111. This ensures that the filter will always find suitable areas. These images multiply the suitability maps with original values from 0 to 1, giving results that can vary from 0 (null original suitability, e.g. constraints) to 1.1111 (maximum suitability) that will be stretched to 0–255 values. The down-weighting never exceeds 90% of the original value.

A CA_MARKOV simulation can be carried out using a simple filter (000; 010; 000) to ignore the CA (Fig. 2, below), that is, to eliminate the effects of down and up-weighting contiguity (Camacho Olmedo et al. 2013, 2015). By adding 0.1111 we create a set of Boolean weight images with a value of 0.1111 for pixels that are not the existing LUC at t1 and with a value of 1.1111 for pixels that are the existing LUC at t1. This enables us to ignore the CA, because the effect of contiguity disappears and only the existing LUC at t1 receives the higher value.

In the simulation maps (Fig. 3), the use of the Cellular Automata in the standard CA_MARKOV produces a buildup effect around existing patches and partially avoids the "salt-and-pepper" effect (left). In the CA_MARKOV simulation that ignores the Cellular Automata (right), new patches with spatial artifacts from the used factors are simulated and the "salt-and-pepper" effect is visible. Errors and

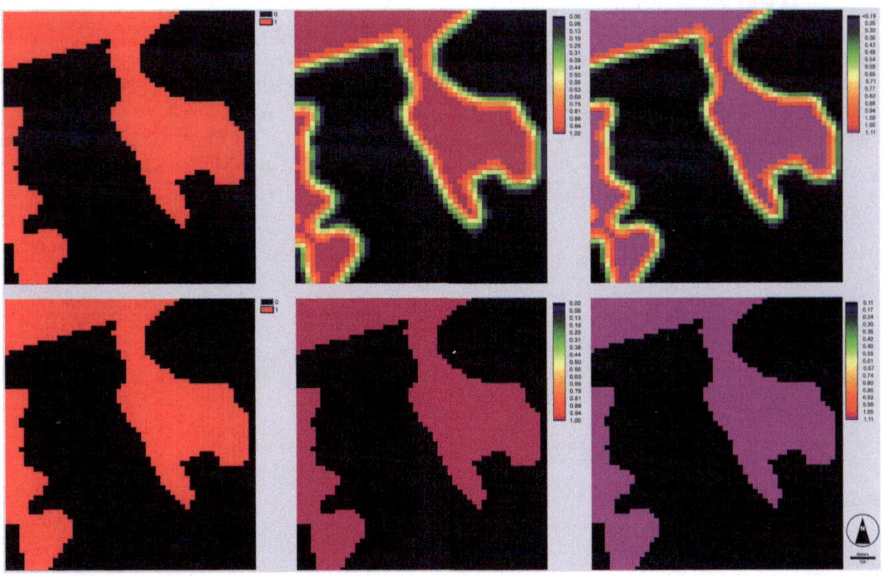

Fig. 2 *Above* CA default in CA_MARKOV. Boolean LUC map at t1 (*left*); after one iteration of CA default filter normalized (*middle*); and after adding 0.1111 (*right*). *Below* Ignoring CA in CA_MARKOV. Boolean LUC map at t1 (*left*); after one iteration of a user-defined filter (*middle*); and after adding 0.1111 (*right*)

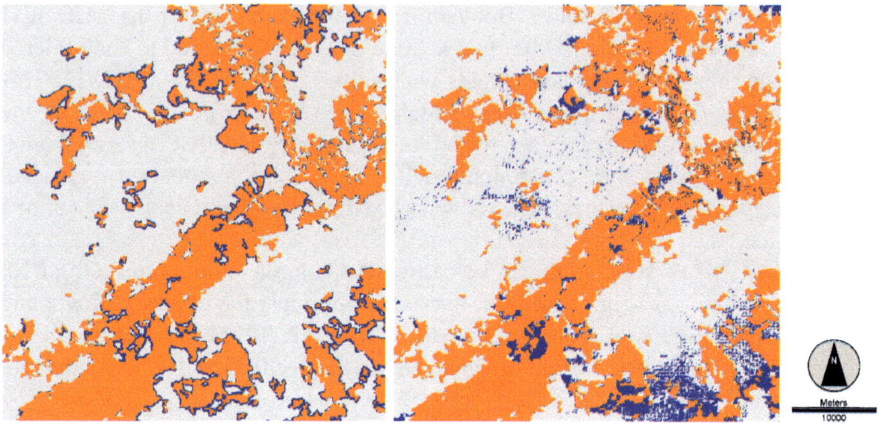

Fig. 3 Irrigated at 2000 (*in orange*), gain of Irrigated during 2000–2006 (*in blue*) for CA_MARKOV (*left*) and CA_MARKOV ignoring filter (*right*)

correct predictions from the two modeling approaches correspond to different allocation procedures that draw different effects on the simulation maps.

References

Camacho Olmedo MT, Paegelow M, Mas JF (2013) Interest of intermediate soft-classified maps in land change model validation: suitability versus transition potential. Int J Geogr Inf Sci 27 (12):2343–2361

Camacho Olmedo MT, Pontius RG Jr, Paegelow M, Mas JF (2015) Comparison of simulation models in terms of quantity and allocation of land change. Environ Model Softw 69:214–221

Eastman JR (1987–2015) TerrSet, Manual. Clark University 394 p. © 1987–2015. http://www.clarklabs.org/

Chapter 22
Fuzzy Coincidence

J.F. Mas

Abstract Fuzzy logic provides techniques to deal with inaccuracies or ambiguities in both the attribute and the geometry of spatial data. In this technical note, the fuzzy approach used to assess the spatial coincidence between a modeled map and an observed (true) map is presented.

Keywords Validation · Fuzzy logic · Map overlay · Map comparison

1 Short Description of Interest

In order to assess the model, the simulated map is often compared with an observed map of the simulated event. However, when this comparison is done pixel-by-pixel, the simulated event (red cells in Fig. 1) is considered as correctly predicted only when it coincides perfectly with the observed event (blue cells).

As shown in Fig. 1, this strict requirement of perfect coincidence prevents us from assessing and comparing model performance well. In order to avoid this problem, the maps can be compared using the concept of fuzziness of location, in which spatial coincidence is not restricted to a strict, pixel-by-pixel overlay, but also includes the cells in a neighborhood (Hagen 2003).

See Chap. 4 about validation.

J.F. Mas (✉)
Centro de Investigaciones en Geografía Ambiental,
Universidad Nacional Autónoma de México (UNAM), Morelia, Michoacán, Mexico
e-mail: jfmas@ciga.unam.mx

M.T. Camacho Olmedo et al. (eds.), *Geomatic Approaches for Modeling Land Change Scenarios*, Lecture Notes in Geoinformation and Cartography,
https://doi.org/10.1007/978-3-319-60801-3_22

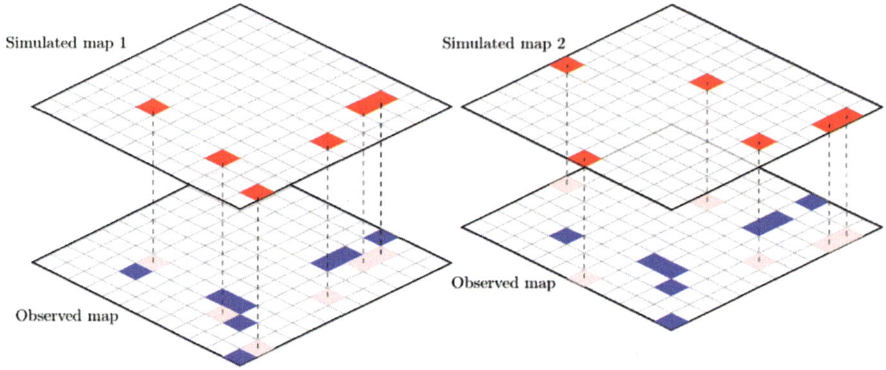

Fig. 1 There is no coincidence with the observed map in either of the simulated maps, although the first simulated map allocates the simulated events much more accurately than the second

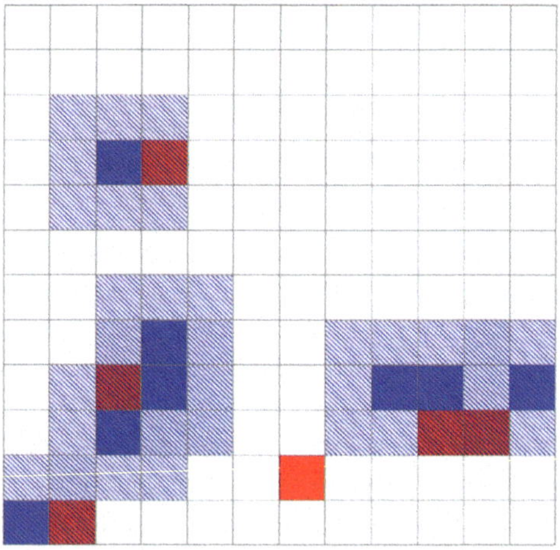

Fig. 2 Fuzzy tolerance of one cell width around the observed changes (*blue shaded cells*). Using this tolerance, five cells of simulated change (*red cells*) are considered as coincident with observed change (*blue cells*)

2 Technical Details

In raster format, fuzziness of location is represented by a tolerance region around event cells obtained by applying a kernel window to the image. For example in Fig. 2, shaded blue areas represent a fuzzy tolerance region around the blue cells obtained by a window of 3 × 3 pixels. Based on this fuzzy tolerance, five of the six simulated (red) cells that we considered coincide with the observed (blue) cells. A larger tolerance region can be obtained using larger windows. Two-way comparisons can be obtained by applying the fuzziness to the simulated or to the observed maps of the event alternatively. As simulated maps with scattered small patches tend to score higher, because they produce a large tolerance area, the minimum coincidence value from the two-way comparison is used in order to obtain a conservative assessment of the model (Almeida et al. 2008). An exponential decay function can be used to weight the similarity using the distance from the center of the window.

3 Example

Cuevas and Mas (2008) modeled land use/cover in a dry tropical region of Mexico (area about 2000 km^2) using two scenarios that encompass a plausible range of future trajectories of deforestation. The first one assumes that the past observed trends will continue and the second assumes that deforestation rates will increase due to cattle raising.

These two simulations were compared with the observed land use/cover map using the fuzzy coincidence method with windows from 1 to 2000 m. The coincidence value was graphed as a function of the window size. Results showed that no scenario was able to predict the exact position: no coincidence was found when the assessment was based on a strict (no fuzzy) evaluation. However, when increasing the window size and thus the tolerance to positional error, the coincidence augments notably, indicating that the model was able to identify the location of change coarsely. The second scenario presented a coincidence value of 0.7 with a tolerance distance (half window size) of about one kilometer, which indicates that it identified most of the small regions where deforestation took place.

References

Almeida C, Gleriani J, Castejon E, Soares-Filho B (2008) Neural networks and cellular automata for modeling intra-urban land use dynamics. Int J Geogr Inf Sci 22(9):943–963

Cuevas G, Mas J-F (2008) Land use scenarios: a communication tool with local communities. In: Paegelow M, Camacho Olmedo MT (eds) Modelling environmental dynamics. Springer, Berlin, pp 223–246

Hagen A (2003) Fuzzy set approach to assessing similarity of categorical maps. Int J Geogr Inf Sci 17(3):235–249

Chapter 23
LUCC Based Validation Indices: Figure of Merit, Producer's Accuracy and User's Accuracy

M. Paegelow

Abstract This technical note presents the method of LUCC based validation indices commonly used during the validation step and including techniques such as of figure of merit, producer's and user's accuracy. We present first the interest and the technical details before giving an example.

Keywords Validation · LUCC · Accuracy · Figure of Merit · Producer's Accuracy · User's Accuracy

1 Short Description of Interest

There are various map comparison techniques based on the LUCC-budget approach. One of the best known and most frequently employed is a comparison between the last model-known LUC map (t_0 observed), the projected LUC map (t_1 simulated) and the model-unknown, observed LUC at t1 (t_1 observed).

Comparisons of observed and simulated land change produce various possible results: correct prediction (change or persistence correctly predicted), erroneous prediction (observed persistence predicted as change—commission error, observed change predicted as persistence—omission error and observed change predicted as wrong gaining category change). The combination of these comparison categories produced the following validation indices proposed by Pontius et al. (2008):

- Figure of Merit, which expresses the overlap between observed and simulated change

See Chap. 4 about validation.

M. Paegelow (✉)
Département de Géographie, Aménagement, Environnement, Université Toulouse Jean Jaurès, Toulouse, France
e-mail: paegelow@univ-tlse2.fr

M.T. Camacho Olmedo et al. (eds.), *Geomatic Approaches for Modeling Land Change Scenarios*, Lecture Notes in Geoinformation and Cartography,
https://doi.org/10.1007/978-3-319-60801-3_23

- Producer's Accuracy, which expresses "the proportion of pixels that the model predicts accurately as change, given that the reference maps indicate observed change" (Pontius et al. 2008)
- User's Accuracy, which measures the proportion of pixels accurately predicted as change when all model-predicted changes are included.

These indices, focusing on land change, are simple measures of the accuracy of simulated LUCC.

2 Technical Details

Continuing the work done by Perica and Foufoula-Georgiou (1996), Pontius et al. (2008) distinguished the map comparison results between anterior (t0, last model known) LUC and posterior (t1 observed and t1 predicted) LUC, by placing them in the following categories:

A = *Error owing to observed change predicted as persistence*
B = *Correct. Observed change predicted as change with the same, correct, gaining category*
C = *Error owing to observed change predicted as change but with a wrong gaining category*
D = *Error due to observed persistence predicted as change*

By combining the quantitative proportions of these comparison categories, we obtain the three validation indexes listed above and illustrated in Fig. 1:

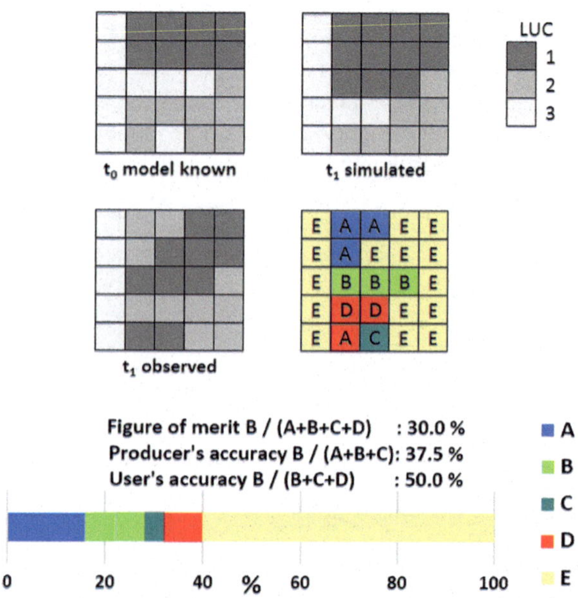

Fig. 1 Fictional LUC at t0, simulated LUC at t1, observed LUC at t1 and comparison results between observed and simulated LUC at t1. The graph *below* quantifies these comparisons

- Figure of Merit—B/(A + B + C + D): Figure of Merit expresses the correspondence between observed and predicted change.
- Producer's Accuracy—B/(A + B + C): "the proportion of pixels that the model predicts accurately as change, given that the reference maps indicate observed change" (Pontius et al. 2008).
- User's Accuracy—B/(B + C+D): the proportion of area that the model predicts accurately as change when all model-predicted changes are given.

3 Example

Figure 2 shows land use in the Murcia region of Spain in 2000 (top left), 2006 (top right) and simulated LUC for 2006 (down left). The fourth map (bottom, right) shows the comparison between these three maps and computed validation indices. For more details about LUCC research in this area, see Camacho Olmedo et al. (2015).

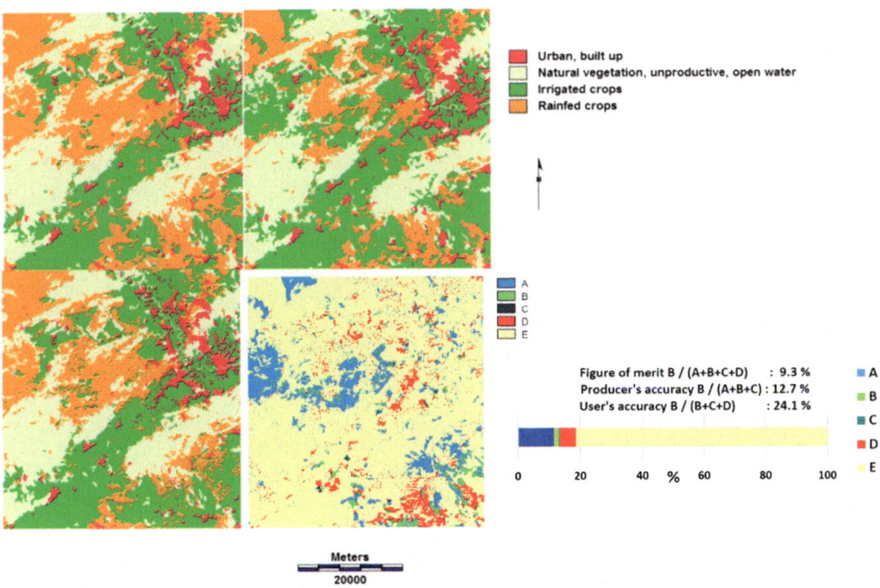

Fig. 2 *Top* LUC in Murcia region in 2000 (*left*) and 2006 (*right*). *Bottom* Simulated LUC in 2006 (*left*) and comparison between observed and simulated LUC in 2006 (*right*). The bar graph shows the proportions of comparison categories and various validation indices

References

Camacho Olmedo MT, Pontius RG Jr, Paegelow M, Mas JF (2015) Comparison of simulation mo-dels in terms of quantity and allocation of land change. Environ Model Softw 69:214–221

Perica S, Foufoula-Georgiou E (1996) Model for multiscale disaggregation of spatial rainfall based on coupling meteorological and scaling descriptions. J Geophys Res 101(D21):26347–26361

Pontius RG Jr, Boersma RW, Castella JC, Clarke K, de Nijs T, Dietzel C, Duan Z, Fotsing E, Goldstein N, Kok K, Koomen E, Lippitt CD, McConnell W, Sood AM, Pijankowski B, Pidhadia S, Sweeney S, Trung TN, Veldkamp AT, Verburg PH (2008) Comparing the input, output, and validation maps for several models of land change. Ann Reg Sci 42(1):11–27

Chapter 24
LUCC Budget

M. Paegelow

Abstract This technical note presents the technique of LUCC budget that is commonly used during the modeling process in both the calibration and the validation stage. First we present the interest and technical details of this technique before illustrating the technique by an example.

Keywords Calibration · Validation · LUCC budget

1 Short Description of Interest

Land Use/Cover Change (LUCC)-budget is a map comparison technique comparing LUC maps at two different dates. Focusing on changes in time and space, this cross tabulation procedure synthesized by Pontius (2000) and Pontius et al. (2004a, b) allows us to characterize land change by quantifying the following components:

- Gains
- Losses
- Net change (balance between gains and losses)
- Swap (changes balanced by equal amount of gains and losses)
- Total change

In both the calibration and validation stages, LUCC-budget provides useful information by comparing observed LUCC to simulated LUCC, particularly in terms of the amount of expected change and the proportion of swap and net change. Predominant net change means that land change for the category in question is simple extension or regression, while predominant swap is an indication of more

See Chap. 4 about validation.

M. Paegelow (✉)
Département de Géographie, Aménagement, Environnement,
Université Toulouse Jean Jaurès, Toulouse, France
e-mail: paegelow@univ-tlse2.fr

M.T. Camacho Olmedo et al. (eds.), *Geomatic Approaches for Modeling Land Change Scenarios*, Lecture Notes in Geoinformation and Cartography,
https://doi.org/10.1007/978-3-319-60801-3_24

complex land change processes involving multiple categories. To illustrate this, we analyzed western European mountain areas that have experienced economic and demographic decline followed by land abandonment. While recently abandoned crops become bushy, earlier abandoned cropland became bushy and later woody. As a result, LUCC budget for bushy land may not show any net gains or losses, despite showing a considerable amount of swap. The simple balance for changes in bushy land (gains minus losses) at the two compared dates may be close to zero, so disguising land change processes. Our studies show that swap is the most difficult component for modeling software to gauge.

2 Technical Details

If we consider the following two binary LUC maps at t0 and t1 for forest and non-forest, we can see the following changes (Fig. 1):

Gain: 1 pixel
Loss: 2 pixels
Net change: −1 pixel (absolute value of net change: 1)
Swap: 2 pixels (1 gain balanced by 1 loss)
Total change: 3 pixels changed between t0 and t1

The following matrix presents primary LUCC-budget components: gains and losses

		t1					
		Class A	Class B	Class C	…Class n	Total t0	Losses
t0	Class A	**1A**	1B	1C	1n	$\sum 1$	$\sum 1-1A$
	Class B	2A	**2B**	2C	2n	$\sum 2$	$\sum 2-2B$
	Class C	3A	3B	**3C**	3n	$\sum 3$	$\sum 3-3C$
	…Class N	NA	NB	NC	**Nn**	$\sum N$	$\sum N-Nn$
	Total t1	$\sum A$	$\sum B$	$\sum C$	$\sum n$		
	Gains	$\sum A-1A$	$\sum B-2B$	$\sum C-3A$	$\sum n-Nn$		

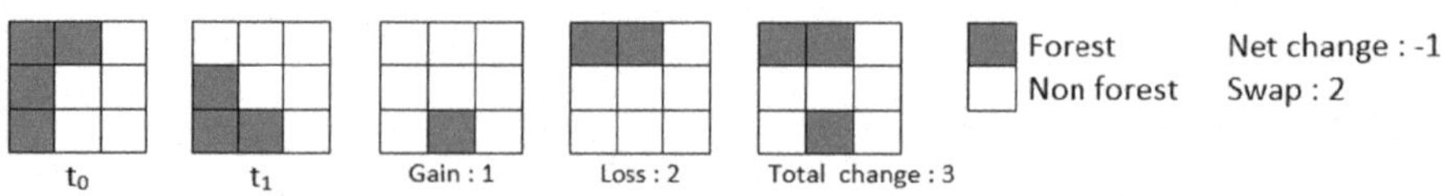

Fig. 1 Binary LUC maps for t0 and t1 and resulting LUCC budget components as gain, loss and total change. In addition: net change and swap

- Gain—sum of changes towards a specific LUC category.
- Loss: sum of changes from a specific LUC category.
- Total change—expresses the overall change (gains and losses) between two LUC maps (dates).
- Absolute net change—the absolute balance of the sum of gains and losses for each LUC category (e.g. 2% gain and 4% loss results in an absolute net change of 2%).
- Swap—the difference between total change and absolute net change that expresses a change of allocation without a change of quantity.

3 Example

Paegelow et al. (2014) compared the results of 3 LUCC models (CA_MARKOV, LCM, Dinamica EGO) with observed land change in a study area in the Eastern Pyrenees. Figure 2 shows the proportion of net change and swap as components of total change. The graphic gives useful information to answer the following questions:

- How close is modeled land change to reality?
- Does the software model the complexity of land change (swap component) well?

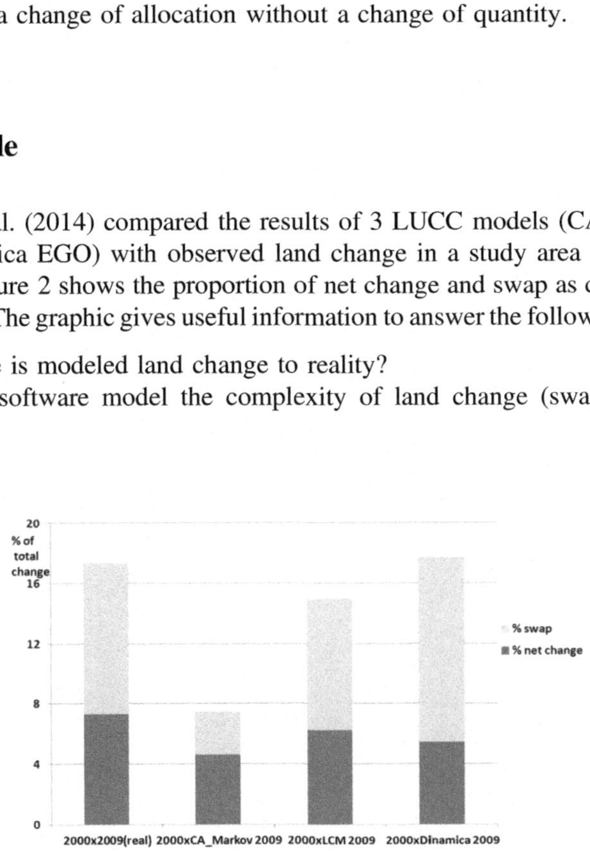

Fig. 2 LUCC budget components net change (*bottom*) and swap (*top*) resulting from comparison between observed LUCC (2000–2009) and comparisons between observed LUC in 2000 and simulated LUC in 2009 using three different modeling software programs: CA-Markov, LCM and Dinamica EGO

References

Paegelow M, Camacho Olmedo MT, Mas JF, Houet T (2014) Benchmarking of LUCC modelling tools by various validation techniques and error analysis. Cybergeo document 701, mis en ligne le 22 décembre 2014. http://cybergeo.revues.org

Pontius RG Jr (2000) Quantification error versus location error in comparison of categorical maps. Photogramm Eng Remote Sens 66(8):1011–1016

Pontius RG Jr, Huffaker D, Denman K (2004a) Useful techniques of validation for spatially explicit land-change models. Ecol Model 179(4):445–461

Pontius RG Jr, Shusas E, McEachern M (2004b) Detecting important categorical land changes while accounting for persistence. Agr Ecosyst Environ 101:251–268

Chapter 25
Markov Chain

M.T. Camacho Olmedo and J.F. Mas

Abstract The Markov chain estimates the quantity of land use change and persistence. Markov matrix is integrated into various LUC models and its use is generalized within the community of land change modelers. In this technical note we present the interest and technical details before illustrating it by an example of annualized Markov estimations.

Keywords Calibration · Simulation · Markov chain · Transition matrix · Estimated quantities · Land change

1 Short Description of Interest

Several approaches are used to estimate the quantity of land use change and persistence. The Markov chain computes the transition areas matrix and the transition probability matrix by cross tabulation between LUC categories from two maps (t0 to t1), which represent LUCC during the calibration stage, to project the estimated changes and persistence at the simulation stage (t1 to t2). The Markov probability matrix calculates the probability of each LUC category (row) changing to another category (different LUC in column) or persisting (same LUC in column). Host categories are located in rows and claimant categories in columns.

See Chap. 2 about calibration.

M.T. Camacho Olmedo (✉)
Departamento de Análisis Geográfico Regional y Geografía Física,
Universidad de Granada, Granada, Spain
e-mail: camacho@ugr.es

J.F. Mas
Centro de Investigaciones en Geografía Ambiental, Universidad Nacional
Autónoma de México (UNAM), Morelia, Michoacán, Mexico
e-mail: jfmas@ciga.unam.mx

M.T. Camacho Olmedo et al. (eds.), *Geomatic Approaches for Modeling Land Change Scenarios*, Lecture Notes in Geoinformation and Cartography,
https://doi.org/10.1007/978-3-319-60801-3_25

The fact that the Markov matrix is integrated into various LUC models has led to its generalized use within the community of land change modelers. Nevertheless, the Markov chain does not estimate non-linear behavior properly. It is a trend-based function, which is based on the hypothesis of stationary changes between the calibration and validation period (Mas et al. 2014).

2 Technical Details

In order to assess LUCC, the transition matrix for the calibration period t0-t1 is obtained by overlaying the two LUC maps dated t0 and t1 (Fig. 1). This matrix, which shows the area for each transition, can be transformed into a Markov chain probability matrix for the entire period (T), normalizing the values of each cell by the sum of the area of each row (total area of each category at t0, see Table 1). This Markov matrix indicates the proportion of each category that has been converted to another category or remained the same (diagonal of the matrix). This proportion is interpreted as the probability of transition from one category to another during a period of time T and allows us to project the estimated areas of each LUCC transition (Table 2). For instance, the matrix for the projected areas of each transition between t1 and t2 (t1 + T) is obtained by an element wise multiplication of the vector of areas at t1 (9, 10, 6) and the matrix of probability. Each row element of the matrix is multiplied by the corresponding element from the vector of areas (Table 3).

As it is often desirable to use a time path different from the original period T for projecting into the future, the transition probability matrix is transformed into an annual matrix as follows (Takada et al. 2010):

$$A = H \begin{pmatrix} (\lambda_1)^{1/t} & & 0 \\ & \ddots & \\ 0 & & (\lambda_n)^{1/t} \end{pmatrix} H^{-1}$$

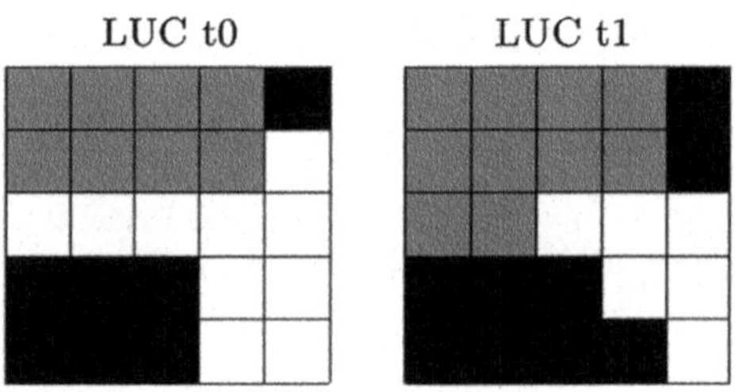

Fig. 1 LUC maps for t0 and t1. The matrix of transition is constricted by overlaying these two maps

Table 1 Matrix of transition for the period t0–t1

t0–t1	Black	Grey	White	Sum (area t0)
Black	7	0	0	7
Grey	0	8	0	8
White	2	2	6	10
Sum (area t1)	9	10	6	

Cells indicate the number of pixels

Table 2 Matrix of the probability of transition for the period t0–t1

t0–t1	Black	Grey	White	Sum
Black	1	0	0	1
Grey	0	1	0	1
White	0.2	0.2	0.6	1

Cells indicate the probability (proportion) of the transition from one category to another during a period T

Table 3 Matrix showing the transition areas estimated by the Markov projection

t1–t2	Black	Grey	White
Black	$1 \times 9 = 9$	$0 \times 9 = 0$	0×9
Grey	$0 \times 10 = 0$	$1 \times 10 = 10$	0×10
White	$0.2 \times 6 = 1.2$	$0.2 \times 6 = 1.2$	$0.6 \times 6 = 0.36$

where A is the annual matrix, t is the number of years, H is the eigenvector of the original transition matrix, and ʎi is the i-th eigenvalue of the original transition matrix.

The annual transition matrix can also be obtained by generating three transition matrices that cover the projection time period. This is done by powering the original matrix and fitting the probability value of the same transition from the three different dates by quadratic regression models (Mas et al. 2014). For instance, The Markov module available in the GIS TerrSet (Eastman 2015) uses interpolation for the entries in the Markov matrix in order to compute a Markov entry between 0 and 1 for the desired extrapolation year. If the same time steps are used, the transition probability obtained from the calibration period (t0 to t1) also applies for the simulation period (t1 to t2). To this end, the Markov extrapolation starts by considering the most recent date in the calibration period to itself (t1 to t1). This obviously produces total persistence (value 1.0000) and zero change (0.0000). It then computes the probabilities of changes for the simulation period by a constant annual rate to match the changes from the calibration period. If the time period in the calibration stage is greater than the time period in the simulation stage, an a priori constant annual rate of change is applied. Despite this, as all transitions must force equilibrium to sum to 1 for every host LUC, in the end some results are slightly different to the estimated amounts.

Some methods for matrix calculation can encounter problems when the calibration interval and the simulation interval have different durations (Takada et al. 2010; Flamenco-Sandoval et al. 2007).

3 Example

We did a case study (Camacho Olmedo et al. 2015) in which the calibration period (t0 to t1) was 1990–2000 and two Markov matrices were obtained for two simulation periods (t1 to t2): 2000–2006 and 2000–2010 (Table 4). In the calibration period, data persistence of LUC 1 was 0.9941% and data transition from LUC 1 to 2 was 0.0044%.

For the simulation period 2000–2010 (the calibration and simulation periods are identical), the estimated persistence of LUC 1 is 0.9941% and the estimated transition from LUC 1 to 2 is 0.0044%.

For the simulation period 2000–2006 (the calibration period is 10 years and the simulation period is 6 years), the estimated persistence of LUC 1 rises to 0.9965% and the estimated transition from LUC 1 to 2 falls to 0.0027%. Consequently, if the calibration period is longer than the simulation period, the estimated persistence is greater than the real persistence in the calibration period and the estimated number of changes is lower than the real changes in the calibration period.

Table 4 Annualized Markov estimations for LUC 1 persistence and for the transition from LUC 1 to LUC 2, projecting in every case from the calibration period 1990–2000 (t0 to t1) to successive years until 2010

Annualized Markov estimations from calibration period 1990–2000 (t0 to t1) Examples of simulation periods (t1 to t2): 2000–2006 and 2000–2010		
Date of estimations	Persistence LUC 1	Transition LUC 1 to LUC 2
t1 2000	**1.0000**	**0.0000**
2001	0.9994	0.0004
2002	0.9988	0.0009
2003	0.9982	0.0013
2004	0.9976	0.0018
2005	0.9970	0.0022
t2 2006	**0.9965**	**0.0027**
2007	0.9959	0.0031
2008	0.9953	0.0035
2009	0.9947	0.0040
t2 2010	**0.9941**	**0.0044**

Examples of simulation periods (t1 to t2): 2000–2006 and 2000–2010

References

Camacho Olmedo MT, Pontius RG Jr, Paegelow M, Mas JF (2015) Comparison of simulation mo-dels in terms of quantity and allocation of land change. Environ Model Softw 69:214–221

Eastman JR (1987–2015) TerrSet, Manual. Clark University 394 p. © 1987–2015. http://www.clarklabs.org/

Flamenco-Sandoval A, Martínez Ramos M, Masera OR (2007) Assessing implications of land-use and land-cover change dynamics for conservation of a highly diverse tropical rain forest. Bio-logical Conserv 138:131–145

Mas JF, Kolb M, Paegelow M, Camacho Olmedo MT, Houet T (2014) Inductive pattern-based land use/cover change models: a comparison of four software packages. Environ Model Softw 51:94–111

Takada T, Miyamoto A, Hasegawa S (2010) Derivation of a yearly transition probability matrix for land-use dynamics and its applications. Landsc Ecol 25(4):561–572

Chapter 26
Multi Criteria Evaluation (MCE)

M. Paegelow

Abstract This technical note presents multi criteria evaluation (EMC). EMC is, in the frame of modeling, a technique used to allocate simulated quantities to most probable or suitable space. First we present the interest and technical details of this technique before giving an example.

Keywords Calibration · Simulation · Multi criteria evaluation

1 Short Description of Interest

MCE aims to compute suitability maps on the basis of a multitude of criteria. MCE belongs to the family of multi-criteria analyses (MCA) (cf. also the French technique ELECTRE, Roy 1991) and in the context of LUCC modeling is applied to produce maps allocating simulated quantities to most probable or suitable space.

Implemented in TerrSet (former Idrisi) software since the 1990s, MCE (Eastman et al. 1993; Eastman 2015; Saaty 1987; Yager 1988) distinguishes between constraints and factors. The first have a Boolean character and indicate whether or not a land use is possible—their rule is to mask space. The latter express the continuous suitability of land to be used/covered by a particular LUC. MCE is used to produce a suitability map for each caption of the LUC maps involved.

MCE may be split into three steps (detailed below):

- Standardization: each factor, expressed in original units such as meters, percent, minutes, $US, has to be converted into an index using the same scale. Therefore, MCE provides a fuzzy membership-based standardization tool.

See Chap. 2 about calibration.

M. Paegelow (✉)
Département de Géographie, Aménagement, Environnement,
Université Toulouse Jean Jaurès, Toulouse, France
e-mail: paegelow@univ-tlse2.fr

M.T. Camacho Olmedo et al. (eds.), *Geomatic Approaches for Modeling Land Change Scenarios*, Lecture Notes in Geoinformation and Cartography,
https://doi.org/10.1007/978-3-319-60801-3_26

- Weighting: each standardized factor can be weighted by different techniques.

A parameterization of trade-off and risk-taking can be obtained by combining different factors used for the same suitability map.

2 Technical Details

Here are some details about the three steps of MCE referred to above.

- Standardization allows us to combine factors expressed in different units. The transformation may be done manually. Alternatively, MCE provides a tool called fuzzy set membership functions (sigmoidal, J-shaped, linear or user defined), which allows us to transform original values into index values (Fig. 1).

  MCE standardization showing three transformation options:

- Factor weighting can also be done manually on the basis of the available data. In cases with a lot of factors where the specification of each factor weight is not easy, MCE provides a matrix tool. The user estimates the relative weight of each factor compared to the others and the algorithm performs its eigenvector as factor weight.
- Most MCA techniques sum the weighted suitability scores and the user computes the average suitability score. This means total trade-off: a place with a critical suitability score for one important factor may be rescued by other factors with high scores at this place. Allowing full trade-off is risk-taking while limiting the impact of each factor on the final score is risk-averse. MCE therefore offers researchers the possibility of using order weights. The number of order weights is equal to the number of factors and their sum is equal to 1. Unlike factor weights, order weights are space specific: for each location (pixel) the weighted factors are ranked from the lowest (left) to the highest (right). Order weights are given for each position in the ranking. Giving the same weight to all positions means free trade-off (strategy 1 in Fig. 2). By contrast, putting all the weight on the left-most (the location specific lowest) factor is risk averse and

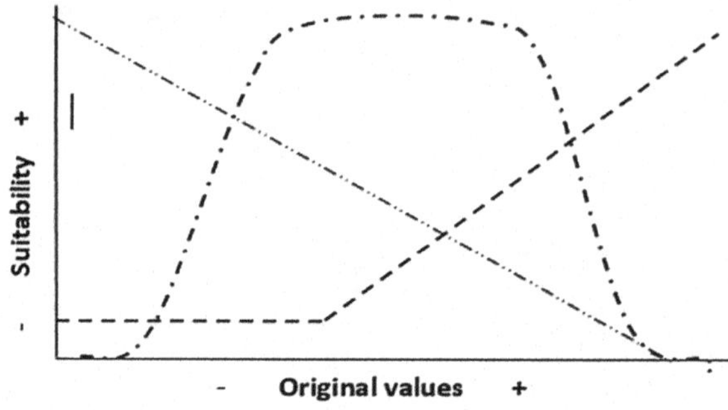

Fig. 1 Various fuzzy set membership functions used for standardization

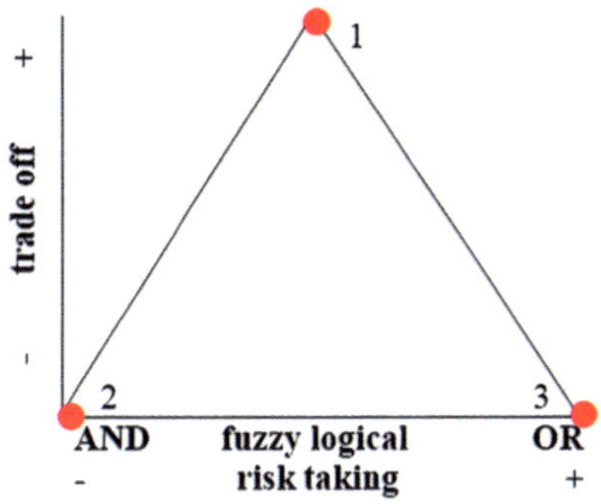

Fig. 2 Decision space offered by OWA technique

means no trade-off (strategy 2). The opposite (putting 100% of order weight on the locally highest factors) also excludes any trade-off but is the highest risk-taking option (strategy 3). The driving possibilities by combining these two criteria form a so-called decision space in the form of a triangle.

- By using order weighted averaging (OWA) we can make different designs for the suitability maps.

References

Eastman JR (2015) TerrSet, Manual. Clark University 394 p. © 1987–2015. http://www.clarklabs.org/

Eastman JR, Kyem P, Toledano J, Weigen J (1993) Explorations in geographic systems technology, vol 4. GIS and Decision Making. Unitar, Geneva

Roy B (1991) The outranking approach and the foundations of electre methods. Theory Decis 31 (1):49–73

Saaty RW (1987) The analytic hierarchy process–what it is and how it is used. Math Model 9 (3):161–176

Yager RP (1988) On Ordered weighted averaging aggregation operators in multicriteria decision making. IEEE Trans Syst Man Cybern 8(1):183–190

Chapter 27
Multilayer Perceptron (MLP)

H. Taud and J.F. Mas

Abstract Artificial Neural networks have been found to be outstanding tools able to generate generalizable models in many disciplines. In this technical note, we present the multi-layer perceptron (MLP) which is the most common neural network.

Keywords Calibration · Neural networks · Non-linear relationships · Back propagation

1 Short Description of Interest

Artificial Neural Networks (ANNs) are structures inspired by the function of the brain. These Networks can perform model function estimation and handle linear/nonlinear functions by learning from data relationships and generalizing to unseen situations. One of the popular Artificial Neural Networks (ANNs) is Multi-Layer Perceptron (MLP). This is a powerful modeling tool, which applies a supervised training procedure using examples of data with known outputs (Bishop 1995). This procedure generates a nonlinear function model that enables the prediction of output data from given input data.

See Chap. 2 about calibration.

H. Taud (✉)
Centro de Innovación y Desarrollo Tecnológico en Cómputo,
Instituto Politécnico Nacional, Mexico City, Mexico
e-mail: htaud@ipn.mx

J.F. Mas
Centro de Investigaciones en Geografía Ambiental, Universidad Nacional
Autónoma de México (UNAM), Morelia, Michoacán, Mexico
e-mail: jfmas@ciga.unam.mx

M.T. Camacho Olmedo et al. (eds.), *Geomatic Approaches for Modeling Land Change Scenarios*, Lecture Notes in Geoinformation and Cartography,
https://doi.org/10.1007/978-3-319-60801-3_27

2 Technical Details

In order to understand the MLP, a brief introduction to the one neuron perceptron and single layer perceptron is provided. The former represents the simplest neural network and has only one output to which all inputs are connected. Given $i = 0,1, \ldots, n$ where n is the number of inputs, the quantities $\{w_i\}$ are the weights of the neuron. The inputs $\{x_i\}$ correspond to features or variables and the output y to their predictive binary class. Figure 1 describes the three steps forming the perceptron model. Figure 2 shows its simplified representation. The weighting step involves the multiplication of each input feature value by its weight $\{x_i w_i\}$ and in the second step they are added together ($x_0 w_0 + x_1 w_1 + \cdots + x_n w_n$). The third is the transfer step where an activation function f (also called a transfer function) is applied to the sum producing an output y presented as:

$$y = f(z) \; and \; z = \sum_{i=0}^{n} w_i x_i \tag{1}$$

$x_0 = 1, w_0$ the threshold or bias, and y the output.

The activation function takes various forms. Their common functions are listed in Table 1.

A perceptron can only learn linearly separable functions from Eq. (1). Figure 3a shows an example of linear function $w_1 x_1 + w_2 x_2 + w_0 = 0$ that separates the data into two classes. In two dimensions with two features, the function is a line. In three dimensions with three features, it is a plane. In n dimensions, it is a hyperplane with equation:

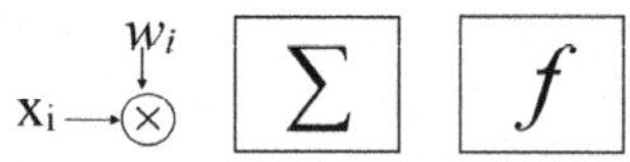

Fig. 1 Perceptron steps: from *left* to *right*, weighting, sum and transfer steps

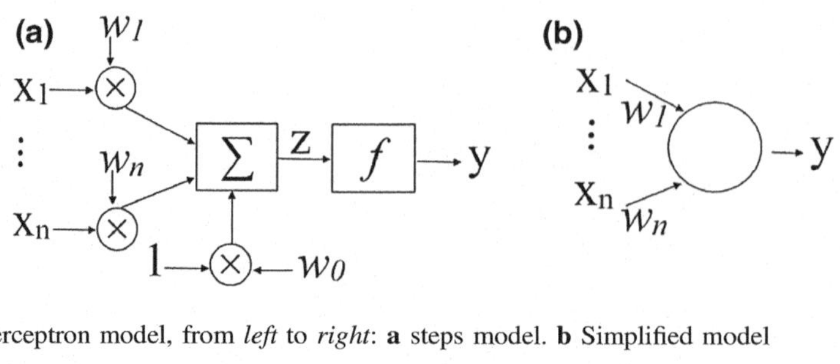

Fig. 2 Perceptron model, from *left* to *right*: **a** steps model. **b** Simplified model

Table 1 Some activation functions

Activation function	Equation	2D graph
Unit step (Heaviside)	$f(z) = \begin{cases} 1 z \geq 0 \\ 0 z < 0 \end{cases}$	
Linear	$f(z) = z$	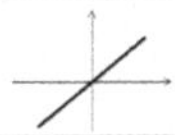
Logistic (sigmoid)	$f(x) = \frac{1}{1+e^{-x}}$	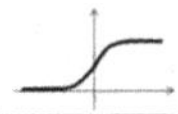

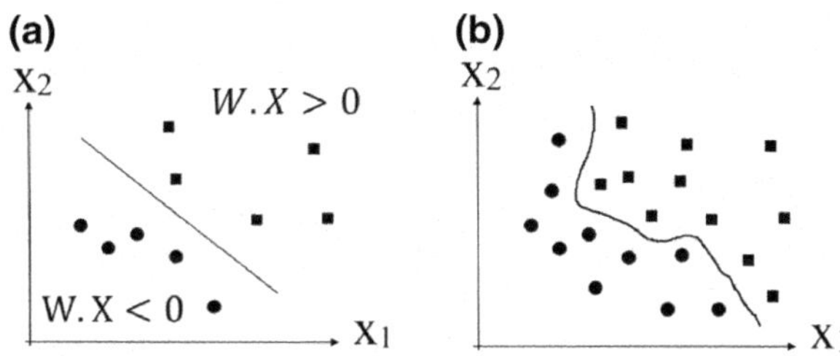

Fig. 3 Input patterns, from *left* to *right*: **a** linearly separable, **b** nonlinearly separable

$$\sum_{i=0}^{n} w_i x_i = 0 \tag{2}$$

The Equation (2) can be presented by the dot product between the weight vector W and the input vector X:

$$W \cdot X = 0 \tag{3}$$

With known responses of the input training data, the learning step (also known as the training step) is completed. The purpose of learning is to optimize the weights by minimizing a cost function, which is usually a square error between the known response and the estimated one. Analytical techniques such as gradient descent determine the optimum weight vector. The algorithm converges to a solution reaching an operational configuration network. The validation of the model is achieved using new data in order to show how the configuration can be generalized to new situations.

The parallel connection of many perceptrons generates a single layer perceptron (SLP) architecture, which is used in the case of various outputs. Figure 4a shows an example with an input and output layer serving in a linearly separable multiclass case.

The perceptron and the single layer perceptron do not resolve the nonlinearly separable problem (Fig. 3b). In this case, a solution can be found by adding any number of layers in successive arrangement and creating a MLP architecture (Fig. 4b). The output of one layer becomes the input of the next and so on. The first

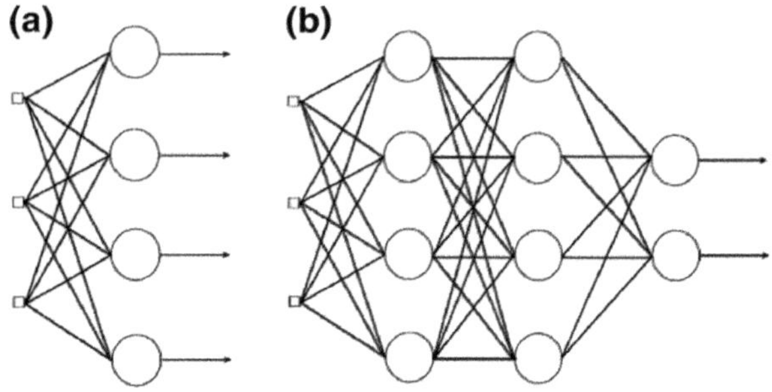

Fig. 4 Layer structure: **a** SLP with three inputs and four outputs. **b** MLP with three inputs, two hidden layers, and two outputs

and the last layers are called input and output layers respectively, while the others are the hidden layers of the neural network.

The MLP is a layered feedforward neural network in which the information flows unidirectionally from the input layer to the output layer, passing through the hidden layers (Bishop 1995). Each connection between neurons has its own weight. Perceptrons for the same layer have the same activation function. In general, it is a sigmoid for the hidden layers. Depending on the application, the output layer can also be a sigmoid or a linear function.

Among many other algorithms, the widely known MLP learning algorithm is a backpropagation, which is a generalization of the Least Mean Squared rule (Du and Swamy 2014). Weights can be corrected by propagating the errors from layer to layer starting with the output layer and working backwards, hence the name backpropagation.

The MLP model performance depends not only on the choice of the variables, the numbers of hidden layers, nodes, and training data but also on the training parameters such as learning rate, momentum controlling the weight change, and number of iterations. A MLP with one hidden layer identifies the nonlinear function with lower accuracies. Networks with more hidden layers are likely to overfit the training data. The learning rate and the momentum control the speed and effectiveness of the learning process.

In land change modeling, the analysis of the complex relationships between land transition and the large number of variables acting as drivers, needs advanced empirical techniques to find a nonlinear function that describes such a complex relationship (Mas et al. 2014). Variables such as distance, slope, type of soil, land tenure, etc. are presented at the input node of the network. Each output node represents a different land transition (e.g. forest to pasture, forest to cropland, and forest to urban, etc…) for which explanatory variable values are known, as well as the land transition observed in the past. After the training step, the MLP is able to predict the potential change of each transition when new input data is presented to the network (Pijanowski et al. 2002; Mas et al. 2004).

References

Bishop CM (1995) Neural networks for pattern recognition. Oxford University Press, Oxford 482

Du K-L, Swamy MNS (2014) Neuronal networks and statistical learning. Springer, Berlin

Mas JF, Puig H, Palacio JL, Sosa AA (2004) Modelling deforestation using GIS and artificial neural networks. Environ Model Softw 19(5):461–471

Mas JF, Kolb M, Paegelow M, Camacho Olmedo MT, Houet T (2014) Inductive pattern-based land use/cover change models: a comparison of four software packages. Environ Model Softw 51:94–111

Pijanowski BC, Brown DG, Shellito BA, Manik GA (2002) Using neural nets and gis to forecast land use changes: a land transformation model. Comput Environ Urban Syst 26(6):553–575

Chapter 28
Multi-objective Land Allocation (MOLA)

M.T. Camacho Olmedo

Abstract In this techincal note we present the Multi-objective Land Allocation (MOLA), an algorithm that solves concurrences between different uses or transitions to allocate the estimated changes in space in the simulation step. First we present the interest and technical details before giving an example using an a priori identical MOLA algorithm included in Land Change Modeler (LCM) and Cellular Automata Markov (CA_MARKOV), in TerrSet software.

Keywords Simulation · Multi-objective Land Allocation · Estimated quantities · Transition matrix · land change

1 Short Description of Interest

Once the estimated changes and persistence have been computed in the simulation stage, the next step is to allocate these changes in space. This process gives rise to hard simulation results, in which the simulated map has the same categories as those used in calibration. In order to perform the allocation, Land Change Modeler (LCM) and Cellular Automata Markov (CA_MARKOV) models, included in GIS TerrSet (Eastman 2015), use an a priori identical Multi-Objective Land Allocation (MOLA) algorithm that solves concurrences between different uses or transitions.

See Chap. 3 about simulation and the short presentations in Part V of this book about CA_MARKOV and LCM.

M.T. Camacho Olmedo (✉)
Departamento de Análisis Geográfico Regional y Geografía Física,
Universidad de Granada, Granada, Spain
e-mail: camacho@ugr.es

M.T. Camacho Olmedo et al. (eds.), *Geomatic Approaches for Modeling Land Change Scenarios*, Lecture Notes in Geoinformation and Cartography,
https://doi.org/10.1007/978-3-319-60801-3_28

2 Technical Details

In the calibration stage, a collection of ranked change potential maps are created, in which the pixels are ranked according to the greatest potential for the occurrence of change. The Markov extrapolation (see the technical note about Markov chain in Part IV of this book) specifies the estimated quantities of change, that is, it targets the number of pixels for transition during the validation interval. Through the Markov matrix, TerrSet's Multi-Objective Land Allocation (MOLA) creates a list of host classes (categories that will lose area, in rows) and claimant classes (categories that will gain area, in columns) for each host. The allocation is done for all claimant classes of each host class, but some pixels could transition to more than one claimant category. Therefore the MOLA algorithm solves the conflicts based on a minimum-distance-to-ideal-point rule using the weighted ranks, and the final result is the overlay of each host class reallocation (Eastman et al. 1995; Kangping 2014; Mas et al. 2014).

Nevertheless, MOLA works differently in LCM from in the CA_MARKOV models. In LCM, a simple easy procedure is followed: MOLA works only once and, consequently, LCM exactly simulates the quantities estimated by the Markov matrix per modeled transition.

In CA_MARKOV the user must incorporate the number of iterations, i.e., the number of time units in the simulation period, the last iteration being the prediction in the later date. We choose ten iterations if ten is the number of years in the simulation interval, and MOLA will run ten times. If we choose twenty iterations for a ten-year simulation interval, MOLA will run twenty times based on 6-month time increments (Eastman 2015). MOLA runs once for each iteration and allocates the divided quantities into equal intervals: e.g. the surface area for each claimant class, only within each host, is the same as estimated by Markov divided by the number of iterations. The final result is the overlay of each new simulation map after each MOLA reallocation. The procedure forces an adjustment of quantities in the last reallocation.

Besides, the MOLA for each iteration does not use the original ranked suitability maps to allocate the divided quantities, using the filtered suitability maps instead (see the technical note about Cellular Automata in CA_MARKOV in Part IV of this book) or a filter adjusted by the user. The filter is applied for each binary LUC map that is temporally extracted from the simulated LUC map. Both conditions in CA_MARKOV (iterations and the use of a Cellular Automata) affect the MOLA procedure. Consequently, the area of simulated transitions does not coincide with the area of transitions estimated by the Markov matrix.

3 Example

As an example, Table 1 presents a matrix with the quantities in hectares of each change/persistence per category estimated by a Markov matrix from the calibration period (t0–t1) to the simulation period (t1–t2) and simulated by LCM and CA_MARKOV for t2. We must make clear that only seven transitions—underlined values—have been modeled (Camacho Olmedo et al. 2015). Bold values in diagonal are persistence.

The Markov extrapolation for the seven modeled transitions matches the output from LCM, in other words MOLA runs once in LCM and respects the Markov matrix.

Alternatively, in CA_MARKOV when the number of iterations is equal to t1–t2, MOLA runs several times and the CA_MARKOV output does not respect the estimated transition quantities resulting from the Markov matrix. The use of CA can also partly explain the differences between simulated and estimated quantities.

Table 1 Quantities in hectares of each change/persistence per category estimated by a Markov matrix from the calibration period (t0–t1) to the simulation period (t1–t2) and simulated by LCM and CA_MARKOV for t2 using MOLA

	1	2	3	4
Markov matrix from t0–t1 (row) to t1–t2 (columns)				
1	**11,732**	32	11	0
2	<u>531</u>	**65,957**	<u>445</u>	*230*
3	<u>1006</u>	131	**70,002**	1772
4	<u>1111</u>	<u>150</u>	<u>12,727</u>	**61,052**
TOTAL estimated LUC at t2	14,380	66,270	83,185	63,054
LCM simulated transitions				
1	**11,775**	0	0	0
2	<u>531</u>	**65,957**	<u>445</u>	*230*
3	<u>1006</u>	0	**71,905**	0
4	<u>1111</u>	<u>150</u>	<u>12,727</u>	**61,052**
TOTAL simulated LUC at t2	14,423	66,107	85,077	61,282
CA_MARKOV simulated transitions (number iterations = t1–t2)				
1	**11,774**	0	0	1
2	<u>523</u>	**66,095**	<u>396</u>	<u>149</u>
3	<u>977</u>	18	**71,915**	1
4	<u>1127</u>	<u>39</u>	<u>11,174</u>	**62,700**
TOTAL simulated LUC at t2	14,401	66,152	83,485	62,851

References

Camacho Olmedo MT, Pontius RG Jr, Paegelow M, Mas JF (2015) Comparison of simulation models in terms of quantity and allocation of land change. Environ Model Softw 69:214–221

Eastman JR (1987–2015) TerrSet, Manual. Clark University 394 p. © 1987–2015. http://www.clarklabs.org/

Eastman JR, Jin W, Kyem PAK, Toledano J (1995) Raster procedures for multi-criteria/multi-objective decisions. Photogramm Eng Remote Sens 61(5):539–547

Kangping SI (2014) Multiple-objective spatial optimization with application to site search problems in a raster space. Dissertation Thesis. Clark University, Worcester, Massachusetts

Mas JF, Kolb M, Paegelow M, Camacho Olmedo MT, Houet T (2014) Inductive pattern-based land use/cover change models: a comparison of four software packages. Environ Model Softw 51:94–111

Chapter 29
The NASZ Model

F. Escobar

Abstract In this technical note we describe how under the cellular automata-based NASZ model the transition potential is computed after conditions related to Neighborhood (N), Accessibility (A), Suitability (S) and Zoning (Z).

Keywords Transition potential · Neighborhood · Accessibility · Suitability and Zoning

1 Short Description of Interest

In cellular automata-based land change models a transition potential (TP) determines the future state (land use) of a cell, within a raster space. Typically, TP for a particular cell is based on the neighborhood rules, in terms of attraction and repulsion, affecting surrounding cells. In some models, TP is also affected by three other parameters; Accessibility (A), Suitability (S) and Zoning (Z), which together with Neighborhood (N) form the NASZ model.

2 Technical Details

In the NASZ model, TP is computed as follows:

$$ {}^{t}\mathrm{Pf,c} = {}^{t}\,\mathrm{Nf,c} \; * \; {}^{t}\,\mathrm{Af,c} \; * \; {}^{t}\,\mathrm{Sf,c} \; * \; {}^{t}\,\mathrm{Zf,c} * \alpha \qquad (1) $$

See Chap. 2 about calibration and the short presentation in Part V of this book about Metronamica.

F. Escobar (✉)
Department of Geology, Geography and Environmental Sciences, University of Alcalá, Alcalá de Henares, Spain
e-mail: francisco.escobar@uah.es

M.T. Camacho Olmedo et al. (eds.), *Geomatic Approaches for Modeling Land Change Scenarios*, Lecture Notes in Geoinformation and Cartography,
https://doi.org/10.1007/978-3-319-60801-3_29

where:

^{t}Pf, c	is the transition potential at time t
^{t}Nf, c	is the neighbourhood potential
^{t}Af, c	is accessibility
^{t}Sf, c	is suitability
^{t}Zf, c	is zoning
α	is a stochastic factor applied to avoid over-determinism

The total amount of change for every land use class is controlled by the demand, which is calculated using factors exogenous to the model. Demand is calculated only for land-use classes that have experienced significant growth during the calibration period. In Metronamica they are called "function" classes (RIKS 2011; Hewitt et al. 2014; Escobar et al. 2015).

The NASZ model is presented graphically in Fig. 1. In the calibration step, land-use maps for two different dates, about 10 years apart, are required. The first map acts as a baseline map while the second is used to provide the demand for each function land-use class. For every single land-use class and at every time step, the neighborhood effect is calculated based on the established rules of attraction and repulsion. Accessibility to the main communication infrastructure is then computed by multiplying the result of the neighborhood effect. This is followed by the computation of the suitability for that particular land-use class and it is also multiplied by the previous product. We then calculate the zoning effect for the

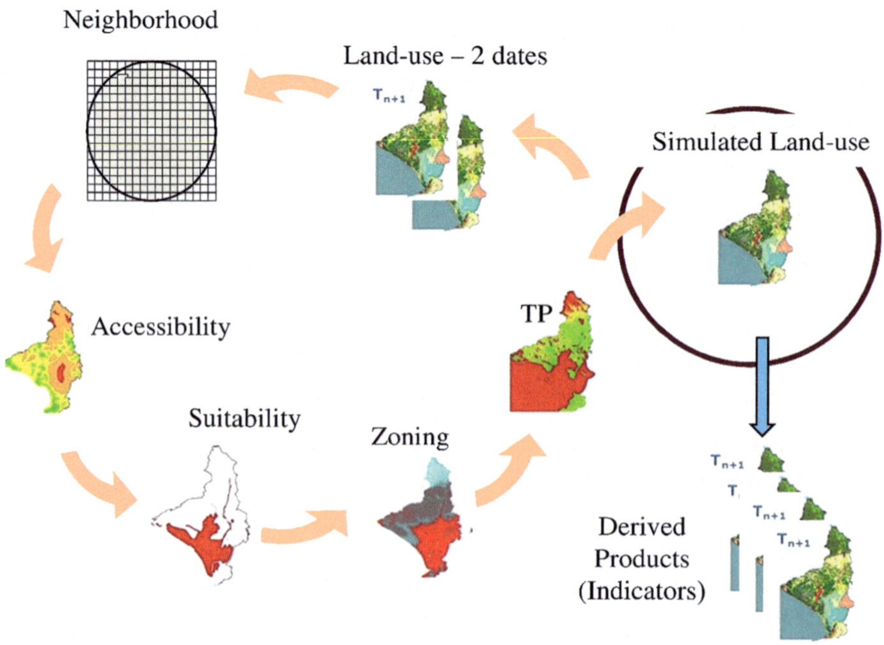

Fig. 1 Graphic description of the NASZ model

particular land-use class. Finally, a stochastic factor is applied in order to minimize over-determinism. The result is the total transition potential for land-use class 1. This process is repeated for each function land-use class. Considering the demand for each land-use class and the calculated transition potential, the model allocates cells to the land-use classes, so making up the simulated land-use map used for calibration. This map is then compared to the actual t2 land-use map and accepted as well-calibrated if the modeler observes enough similarity between them.

Once the calibrated map is produced, the same rules, with different land demands, can be applied for the production of future land use maps corresponding to different scenarios.

We will now describe each component of the NASZ model:

- Neighborhood potential

Each land use that occurs in a cell is influenced by the land uses that occur in a predefined neighborhood of cells. In Metronamica, this neighborhood is defined by a radius of 8 cells, i.e. every cell is affected by a neighborhood of 197 cells. The way this neighborhood influences the target cell depends on the specific rules of attraction or repulsion observed among land-use classes (RIKS 2011).

- Accessibility

Accessibility measures the effect of the proximity and importance of different types of transport networks (roads, railways, canals…) and transport points of interest (train stations, bus stops, entries to motorways…) on the possible future occurrence of a land use function on a particular cell.

- Suitability

Suitability refers to the influence that physical elements of the environment have on the possible future occurrence of land uses on a particular cell. Suitability adopts the form of a composite map made out of as many geo-physical variables as needed. The composite map for each of the function land-use classes includes values ranging from 0 (not suitable at all) to 10 (most suitable).

- Zoning

While suitability refers to the influence of physical elements, zoning refers to human-made elements. In other words, zoning measures the influence that legislation and planning exert over the occurrence or not of a particular land-use class at a certain location. Each function land-use class has a number of zoning maps, one for each different legislative and/or planning framework existing within the modeling period. Each of these maps includes four values; 0 (strictly forbidden), 1 (weakly forbidden), 2 (permitted) and 3 (actively encouraged).

References

Escobar F, Hewitt R, Hernandez V (2015) Usos del Suelo en los Parques Nacionales Españoles. Evolución y Modelado Participativo. In: Amengual & Asensio (eds) Proyectos de Investigación en Parques Nacionales: 2010–2013. Madrid: Organismo Autónomo Parques Nacionales, Ministerio de Agricultura y Medio Ambiente pp 175–211

Hewitt R, van Delde H, Escobar F (2014) Participatory land use modelling, pathways to an integrated approach. Environ Model Softw 52:149–165

Riks BV (2011) Metronamica documentation. Accessed March 2017. http://www.riks.nl/resources/documentation/Metronamica%20documentation.pdf

Chapter 30
Receiver Operating Characteristic (ROC) Analysis

J.F. Mas

Abstract The Receiver Operating Characteristic (ROC) is widely applied to assess the performance of spatial models that produce probability maps of the occurrence of certain events such as the land use / land cover changes, the presence of a species or the likelihood that landslides will occur. In this technical note, the construction of the ROC curve and the calculation of the Area Under the Curve (AUC) index are presented.

Keywords Validation · Accuracy · ROC curve · AUC · Uncertainty

1 Short Description of Interest

In LUCC modeling, the Receiver Operating Characteristic (ROC) analysis is applied to assess the performance of models that produce a probability map which indicates the sequence in which the model ranks the change potential of cells. For this, the probability map (Fig. 1, left) is compared with the map of the true binary transition (Fig. 1, right) in order to assess the spatial coincidence between the true transition and the probability values. A model with a high predictive power is able to produce a map of probability in which the highly ranked probabilities coincide with the true transition.

See Chap. 4 about validation.

J.F. Mas (✉)
Centro de Investigaciones en Geografía Ambiental,
Universidad Nacional Autónoma de México (UNAM), Morelia, Michoacán, Mexico
e-mail: jfmas@ciga.unam.mx

M.T. Camacho Olmedo et al. (eds.), *Geomatic Approaches for Modeling Land Change Scenarios*, Lecture Notes in Geoinformation and Cartography,
https://doi.org/10.1007/978-3-319-60801-3_30

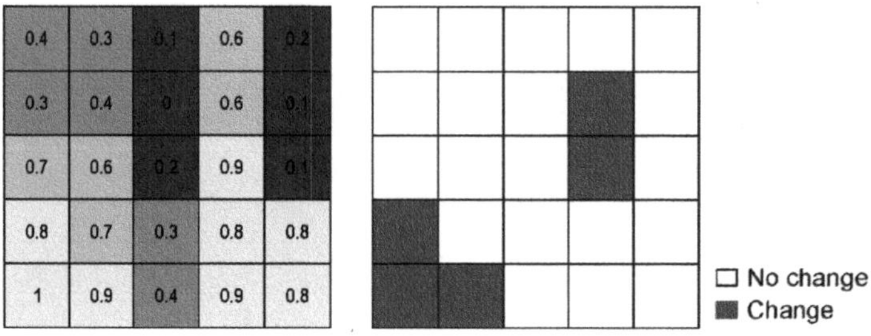

Fig. 1 Maps of change potential (the value in each cell indicates the probability of change) and true change (binary map)

Fig. 2 Maps of predicted change obtained by thresholding the probability map using different values (0.75, 0.5 and 0.25)

Table 1 False and true positive rates at different threshold values

Threshold	False positive rate	True positive rate	Point in the ROC curve (Fig. 3)
1	0/19	0/6	Bottom left corner
0.75	4/19	4/6	A
0.5	8/19	5/6	B
0.25	12/19	6/6	C
0	19/19	6/6	Upper right corner

2 Technical Details

Various thresholds are applied to the probability map to produce binary predicted change maps (Fig. 2). The coincidence between predicted and true transition is assessed by making a curve, called the ROC curve. In this curve, the horizontal axis represents the false positive rate, i.e. the proportion (Fawcett 2006; Mas et al. 2013) of no change cells predicted as change, and the vertical axis the true positive rate, which is the proportion of true change predicted as change. False and true positive

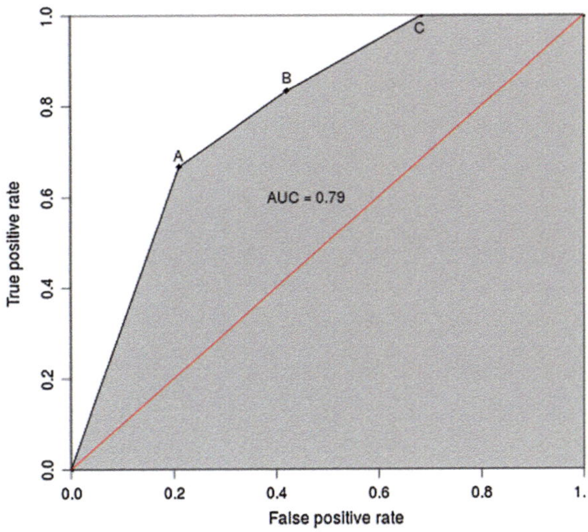

Fig. 3 ROC curve for the maps in Fig. 1. True and false positive rates are computed for each threshold applied to the probability map (see Table 1). The area under the curve (AUC) corresponds to the *grey shaded area*

rates are also referred to as one minus specificity and sensitivity respectively. Table 1 shows the values for false and true positive rates at different threshold values for the data in Figs. 1 and 2. Figure 3 is the ROC curve obtained with the same data.

The ROC curve is used to compute an index: the area under the curve (AUC) represented by the grey shaded area in Fig. 3. When the true change coincides perfectly with the higher ranked probabilities, then the AUC is equal to one because the curve begins at the point (0, 0), goes up the horizontal axis to the point (0, 1), and to the right to the point (1, 1). A random probability map produces a diagonal ROC curve in which the true positive rate equals the false positive rate at all threshold points. A ROC curve below the diagonal indicates a less predictive probability map than a random map (Pontius and Parmentier 2014).

References

Fawcett T (2006) An introduction to ROC analysis. Pattern Recogn Lett 27:861–874

Mas J-F, Soares Filho B, Pontius RG, Farfán Gutiérrez M, Rodrigues H (2013) A suite of tools for ROC analysis of spatial models. Int J Geo-Inf 2(3):869–887

Pontius G, Parmentier B (2014) Recommendations for using the relative operating characteristic (ROC). Landsc Ecol 29(3):367–382

Chapter 31
Weights of Evidence

J.F. Mas

Abstract The weights of evidence, a quantitative method for combining evidence in support of a hypothesis, is commonly used in pattern based models. It enables mapping the probability of the occurrence of a certain event such as, for example, a land cover change, a wildfire or a landslide using a map of the occurence of this event and ancillary data. In this technical note, the computing of the weights of evidence and the probability is presented.

Keywords Calibration · Conditional probability · Drivers

1 Short Description of Interest

The Weights of Evidence (WoE) method is based on conditional probabilities. The conditional probability of an event is the probability that this event will occur given the knowledge that another event has already occurred. In land change modeling, this method is used to produce maps of change probability taking into account spatial variables such as distance, slope or population density.

2 Technical Details

Figure 1 is a very simple map of a forest area. It shows where deforestation occurred (black cells) with regard to a protected area PA (shaded green) and the area close to a road R (shaded red). The figure shows for example that the prob-

---

See Chap. 2 about calibration.

J.F. Mas (✉)
Centro de Investigaciones en Geografía Ambiental, Universidad Nacional Autónoma de México (UNAM), Morelia, Michoacán, Mexico
e-mail: jfmas@ciga.unam.mx

M.T. Camacho Olmedo et al. (eds.), *Geomatic Approaches for Modeling Land Change Scenarios*, Lecture Notes in Geoinformation and Cartography,
https://doi.org/10.1007/978-3-319-60801-3_31

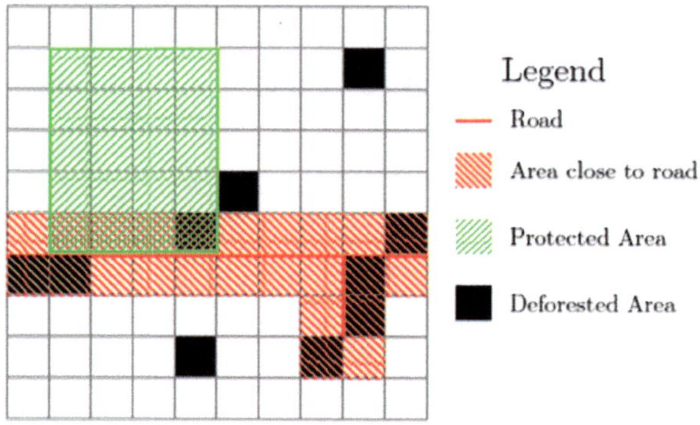

Fig. 1 Distribution of deforestation patches in an area with a road and a protected area. Deforestation is more frequent near the road and much less likely inside the protected area (only one deforestation cell), suggesting that the probability of deforestation varies depending on the distance from the road and location inside or outside the protected area

ability of deforestation P(D) for the entire area without any knowledge is 0.1 (10 deforestation cells out of a total of 100). However, the conditional probability of deforestation for the area close to the road P(D|R) is 0.29 (7 deforested cells/24 cells close to the road). The conditional probability of deforestation within the protected area P(D|PA) is 0.05 (1/20). The WoE method enables users to calculate the probability of an event (e.g. deforestation) taking into account different conditions simultaneously (in this example only two: PA and closeness to road).

The positive weight of evidence w^+ associated with the presence of a given condition is calculated according to Eq. 1.

$$w^+ = \left[\frac{P(C|E)}{P(C|\bar{E})}\right] \quad (1)$$

where P(C|E) is the probability of the condition C given the occurrence of the event E and $P(C|\bar{E})$ is the probability of the condition C assuming that the event E did not occur. When the condition is associated with a low occurrence of the event, the value of w^+ is negative. By contrast, when the occurrence of the condition tends to increase the likelihood of the event, the value of the weight w^+ is positive.

In practice, the weights are very easy to compute. For instance, if the event E we want to model is deforestation and the condition is closeness to the road, P(C|E) is the ratio between the number of cells for deforestation close to the road and the total number of cells for deforestation. $P(C|\bar{E})$ is computed in the same manner taking into account cells where deforestation did not occur. In the example in Fig. 2 weight values for the categories close to the road, far from the road, inside the PA and outside the PA are 1.31, −0.99, −0.75 and 0.13 respectively.

The conditional probability taking into account various conditions is also easy to calculate, by summing the weights of evidence together (Eq. 2). However, this calculation is based on the assumption of independence between the conditions.

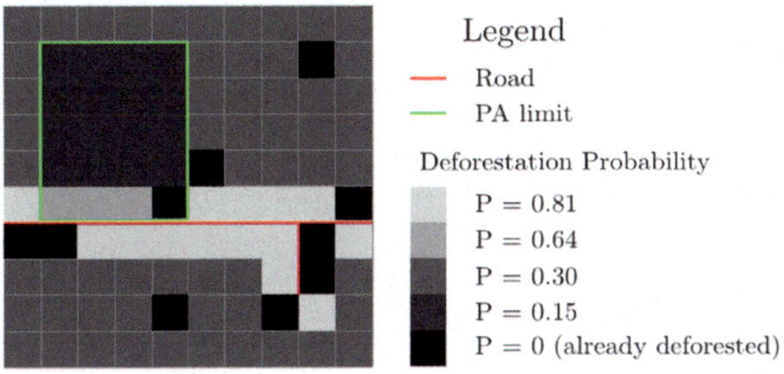

Fig. 2 Probability of deforestation obtained through the WoE method. Forest cells with two simultaneous "adverse" conditions (near the road and outside the PA) have the highest probability of deforestation, because the probability is computed with two positive weight values. By contrast, cells inside the PA and far from the road have a low probability of being deforested because probability is based on two negative weights

Therefore, the variables used as conditions should be tested for spatial dependence by using indices such as Cramer's coefficient.

$$P(E|C_1 \cap C_2 \cap \ldots \cap C_n) = \frac{e^{\sum w_i^+}}{1 + e^{\sum w_i^+}} \tag{2}$$

where $P(E|C_1 \cap C_2 \cap \ldots \cap C_n)$ is the probability of the event E taking place at a site presenting the conditions C_1, $C_2 \ldots C_{i \ldots} C_n$.

Figure 2 shows the probability of deforestation based on the WoE and taking into account the variables PA and closeness to road as represented in Fig. 1.

It is worth noting that weights of evidence can be calculated taking into account binary explanatory variables (e.g. inside/outside a protected area) or multiple categorical variables (e.g. types of soils or land tenure). Continuous variables, such as distance, elevation or slope, cannot be used directly to compute weights and must first be transformed into categorical variables by binning. For a more detailed description of the weights of evidence method, see Soares-Filho et al. (2010).

Reference

Soares-Filho BS, Moutinho P, Nepstad D, Anderson A, Rodrigues H, Garcia R, Dietzsch L, Merry F, Bowman M, Hissa L, Silvestrini R, Maretti C (2010) Role of Brazilian Amazon protected areas in climate change mitigation. Proc Natl Acad Sci USA (PNAS) 107(24):10821–10826 (Supporting Information)

Part V
Short Presentations About the Modeling Software Packages

Chapter 32
A Short Presentation of the Actor, Policy, and Land Use Simulator (APoLUS)

R.J. Hewitt

Abstract Land use change is a social-environmental process strongly influenced by the dynamic behaviour of key actors (e.g. land managers, regulators, policy makers). Existing frameworks for modelling land use change tend to under-represent the role of these actors, which makes it difficult to study strongly actor-driven land change processes, like renewable energy development or intensive agriculture. In this chapter we present the Actor, Policy and Land Use Simulator (APoLUS) model, a free-and-open-source (FOSS) geographical model for the R environment which allows the dynamic interaction of actors to be integrated with Neighbourhood (N), Accessibility (A), Suitability (S) and Zoning (Z) parameters found in a conventional cellular automata-based geographical model. The inclusion of actor dynamics in APoLUS makes it easier to model the effect of policy interventions on land use change and leads to more realistic simulation of land change processes than in non actor-driven models.

Keywords Actors · Land use models · Simulation · R Environment · Free-and-open-source software

1 Introduction

APoLUS (Actor, Policy and Land Use Simulator) is a free-and-open-source (FOSS) geographical computer model for the R environment, designed to simulate the effects of complex actor behaviour on land use (Fig. 1). APoLUS is a developed version of the experimental SIMLANDER model (Hewitt et al. 2013a, b) which is still available and which some users may prefer for simpler applications with a single active land use. The first working version of APoLUS was completed in

R.J. Hewitt (✉)
James Hutton Institute United Kingdom, Aberdeen, UK
e-mail: richardjhewitt@hushmail.com; richard.hewitt@observatorioculturayterritorio.org

R.J. Hewitt
Observatorio para una Cultura del Territorio, Madrid, Spain

M.T. Camacho Olmedo et al. (eds.), *Geomatic Approaches for Modeling Land Change Scenarios*, Lecture Notes in Geoinformation and Cartography,
https://doi.org/10.1007/978-3-319-60801-3_32

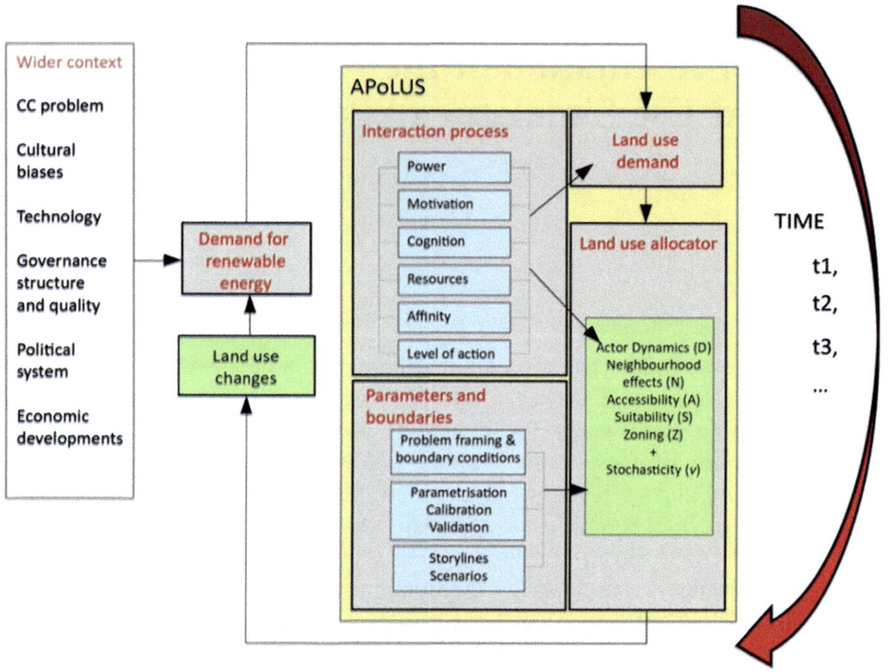

Fig. 1 APoLUS model conceptual diagram, as applied to the case of Navarre, Spain, under theCOMPLEX Project

September 2015. The model uses the Cellular Automata (CA) approach described by White and collaborators (e.g. White and Engelen 2000) to simulate land use change based on the interaction of 5 key parameters, Neighbourhood (N), Accessibility (A), Suitability (S), Zoning (Z) and Actor Dynamics (D). The model was developed under the EU FP7 Project COMPLEX to allow land use types that can be shown to follow an incremental cellular growth pattern (e.g. residential land, industrial land) and land use types that are strongly driven by the behaviour of actors like policy makers and planners (e.g. renewable energy, irrigated cropland) to be modelled together. The influence of actors in shaping land use change is generally under-represented in many existing land use simulation models, which makes it difficult to study the spatial consequences of transformative economic or policy actions like renewable energy implementation or variation in crop prices. Conventional CA land use models are popular, reliable and easy to calibrate against historical data, but they do not represent actor behaviour directly, rather by proxy through transition rules. Agent-Based Models (ABMs), on the other hand, can model actor behavior directly but tend to be data and processing intensive and hard to calibrate. By incorporating the effect of actor dynamics (D) into a conventional CA model structure, APoLUS simulates the influence of actor behaviour on land allocation without losing the advantages of conventional CA models. In this sense, APoLUS tries to achieve balance between two competing visions of land use modeling—on the one hand, as an emergent process arising from the interaction of human agents (as in an ABM, e.g. the FEARLUS model; Polhill et al. 2001), and on the other, as a deterministic process controlled by spatio-temporal geographical and socio-economic drivers, as in a CA model such as Metronamica.

The model comprises two basic components, described as follows:

1.1 Land Use Allocator

At the core of the model is the land use allocator. To simulate land use allocation, a digital map in Geographical Information System (GIS) format (ESRI ascii) is introduced which represents land use at a given moment in time (T_n) in the form of a grid of cells, in which each cell contains a single land use. Cells are able to transform from one land use to another over the course of a time sequence (T_1, T_2, $T_3 \ldots T_n$) according to their Transition Potential (TP), which is dependent on five key parameters; (1) *Actor Dynamics* (D), the actor dynamics influence score for given land uses in the map in the areas that correspond to the actors' area of influence, determined by the actor interaction process (see below); *Neighbourhood* interaction (N), the relationship between the cell's land use and the land use of adjacent or nearby cells; (2) *Accessibility* (A), the influence of lines of communication, e.g. transport, irrigation, electricity network; (3) *Suitability* (S), the biophysical characteristics (e.g. rainfall, slope) of a given cell that influence the land use that can be assigned to it; *Zoning* (Z), planning restrictions in place, e.g. protected areas. Finally, the stochastic uncertainty of the land allocation process is represented through the addition of a random factor (v).

1.2 Actor Interaction Process

The influence of actor dynamics on land allocation (D) is computed as an aggregate score for each model region on the basis of six key actor and process variables, defined according to the Contextual Interaction Theory (CIT) approach of De Boer and Bressers (2011), and the Participatory Action Research approach described in Hewitt et al. (2017).

(1) *Motivation*—the actor's degree of motivation to implement the modelled process for the relevant land use;

(2) *Cognition*—the actor's degree of awareness and knowledge that enable them to implement the modelled process for the relevant land use;

(3) *Resources*—the resources (monetary or otherwise) at the actor's disposal;

(4) *Power*—the power of the actor with respect to other actors in the model;

(5) *Affinity*—the degree to which the actor is sympathetic towards implementation of the modelled process for the relevant land use, and;

(6) *Level of action*—the typical administrative level of action (usually, but not necessarily their level of official competence) of this actor respect to the modelled process.

Actor characteristics can be defined either through a participatory process, in which stakeholders are tasked with analysing actor inter-relationships and behaviour with respect to the implementation goal (e.g. Hewitt et al. 2017), or through policy analysis carried out by researchers (e.g. Bressers and Dinica 2003).

2 Description of the Methods Implemented in the Model

2.1 Model Set up

The general procedure for simulation modeling in APoLUS is to proceed in order through the numbered steps in the main switchboard (Fig. 2), import regions and land use maps, define actors, calibrate neighbourhood, suitability, accessibility and zoning, and run the model. At the map import stage, each land use category in the input map (e.g. urban, agriculture, solar), must be assigned a status and an actor influence value. Status relates to the land use category's behaviour respect to other categories in the model, and can take one of three possible values: passive, active or static (see Hewitt 2015 for detailed explanation). The actor influence parameter takes value 1 or 0 which determines whether or not allocation and demand for a particular land use will be influenced by actor dynamics (D).

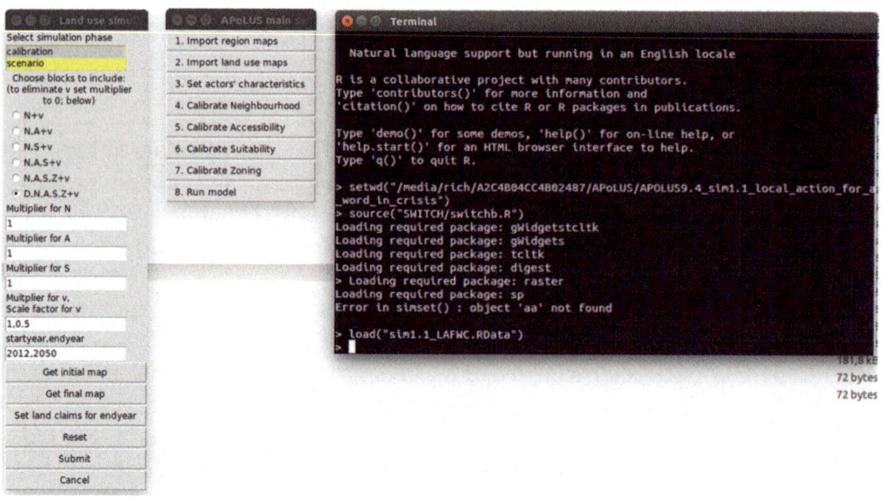

Fig. 2 APoLUS model screen capture, ubuntu linux environment. *Left* simulation dialog; *Centre* main switchboard; *Right* R command line interface

2.2 Calibration

Detailed calibration and validation procedure is the same as that used for the Metronamica model, and involves generating simulated land use maps for known historical dates, and then comparing the simulations against the real maps using standard statistical techniques such as Ksim, fractal dimension or clumpiness (see, for example, Van Vliet et al. 2013; Hewitt et al. 2014; Newland et al. 2015). The time period between t1 (the first map available) and t_2 (the second map available, posterior to t_1) is known as the calibration period. If data are available for a third date (t3), then it simulations can be evaluated for a second period (t2–t3), known as the validation period.

Actor dynamics are in general difficult to calibrate because the state of the actor interaction process for historic periods is usually unknown. For this reason, it is recommended to calibrate the model for N, A, S, Z and then subsequently add actor dynamics. The simulation panel (Fig. 2) is designed to facilitate the calibration process, allowing the user to calibrate one parameter at a time, and then run the model for this parameter only, selecting the appropriate radio button (Fig. 2). Once calibration and validation have been carried out successfully, the model is considered to be ready to generate simulations for future dates. What constitutes a "successful" calibration is hard to define, and highly case-dependent, but an overall Ksim score of >0.1, and an individual category Ksim score of >0.1 for the active land use categories is a useful benchmark. If this cannot be achieved, it may be acceptable to demonstrate that both the calibration and validation dates outperform a null or neutral model in which change areas are located at random (Hagen-Zanker and Lajoie 2008). One of the most important aspects of calibration is knowing when to stop; Ksim scores >0.2 may indicate overfitting.

3 Applications

APoLUS at present has been applied in two cases, to model renewable energy implementation in the region of Navarre (Hewitt 2015), and to model agricultural intensification in European member states (Pera 2016). APoLUS' sister model SIMLANDER also has a number of ongoing use cases with publications pending.

4 Final Considerations and Technical Summary

APoLUS is implemented in R, with the exception of the multiple land use allocation script which is written in C and called (automatically) from the R environment at runtime. It is recommended to use R version >3.0. APoLUS is run from the R command line, and user input is facilitated through a series of simple dialogues.

Installation of the raster and gWidgetstcltk packages is required. The model has been tested and functions correctly on both 32 and 64 bit Windows systems, 32 and 64 bit Linux systems and 64 bit Macintosh systems. The APoLUS model can be found in the COMPLEX project model repository at: http://owsgip.itc.utwente.nl/projects/complex/index.php/2-uncategorised/21-plus4-cmp.

APoLUS, and sister model SIMLANDER, can also be downloaded directly from https://simlander.wordpress.com.

References

Bressers JTA, Dinica V (2003) The implementation of renewable energy policies: theoretical considerations and experiences from Spain, Netherlands and United Kingdom. In: Krauter (ed) (2003) Renewables for a sustainable energy supply. pp 39–45

De Boer C, Bressers H (2011) Contextual interaction theory as a conceptual lens on complex and dynamic implementation process. In: Proceedings of conference: challenges of making public administration and complexity theory work

Hagen-Zanker A, Lajoie G (2008) Neutral models of landscape change as benchmarks in the assessment of model performance. Landsc Urban Plan 86(3–4):284–296

Hewitt R (2015) APoLUS model full system documentation. Project Report for EU FP7 Pro-gramme COMPLEX Project (deliverable 3.5). https://www.researchgate.net/publication/292047909_APoLUS_model_full_system_documentation

Hewitt R, Díaz Pacheco J, Moya Gómez B (2013a) A cellular automata land use model for the R software environment (weblog) https://simlander.wordpress.com

Hewitt R, Díaz Pacheco J, Moya Gómez B (2013b) A cellular automata land use model for the R software environment (presentation) https://av.tib.eu/media/15501?0

Hewitt R, Van Delden H, Escobar F (2014) Participatory land use modelling, pathways to an integrated approach. Environ Model Softw 52:149–16

Hewitt RJ, Winder NP, Jiménez VH, Alonso PM, Bermejo LR (2017) Innovation, pathways and barriers in Spain and beyond: an integrative research approach to the clean energy transition in Europe. Energ Res Soc Sci 34:260–271

Newland CP, Maier HR, Newman Van Delden H (2015) Relationships between cellular automata based land use models parameters and spatial metrics: enhancing understanding in a calibration context. http://web.mit.edu/cron/project/CUPUM2015/proceedings/Content/modeling/285_newland_h.pdf

Pera FA (2016) Tendencias en la relación de la agricultura intensiva en Europa con las regiones potencialmente vulnerables al cambio climático. Master's Thesis. Departamento de Geología, Geografía y Medio Ambiente, Universidad de Alcalá. Available at: http://observatorioculturayterritorio.org/wordpress/wp-content/uploads/2017/01/TFM.pdf

Polhill JG, Gotts NM, Law AN (2001) Imitative versus nonimitative strategies in a land-use simulation. Cybern Syst 32(1–2):285–307

van Vliet J, Naus N, van Lammeren RJ, Bregt AK, Hurkens J, van Delden H (2013) Measuring the neighbourhood effect to calibrate land use models. Comput Environ Urban Syst 41:55–64

White R, Engelen G (2000) High-resolution integrated modelling of the spatial dynamics of urban and regional systems. Comput Environ Urban Syst 24(5):383–400

Chapter 33
A Short Presentation of CA_MARKOV

J.R. Eastman and J. Toledano

Abstract CA_MARKOV is a combined Cellular Automata/Markov Chain/Multi-Criteria/Multi-Objective Land Allocation land cover prediction procedure. CA_MARKOV allocates land based on the suitability of the land for end covers along with a cellular automaton rule to promote spatial contiguity. CA_MARKOV works well when historical land cover data is not available or is not a good predictor of future land cover.

Keywords Land cover prediction · Land allocation

1 Introduction

CA_MARKOV is a combined Cellular Automata/Markov Chain/Multi-Criteria/Multi-Objective Land Allocation land cover prediction procedure that was developed as a precursor to the Land Change Modeler (LCM). The most fundamental difference between CA_MARKOV and LCM is that CA_MARKOV allocates land based on the suitability of the land for end covers along with a cellular automaton rule to promote spatial contiguity. In contrast, LCM models suitability for transition rather than suitability for the ending land cover. Although CA_MARKOV is superseded by LCM, it is still provided for experimental purposes in the TerrSet/IDRISI software system.

2 Descriptions of the Methods Implemented in the Model

Use of CA_MARKOV proceeds through three stages, relying on existing modules of the TerrSet/IDRISI software for each component.

J.R. Eastman (✉) · J. Toledano
Clark Labs, Clark University, Worcester, USA
e-mail: reastman@clarku.edu

M.T. Camacho Olmedo et al. (eds.), *Geomatic Approaches for Modeling Land Change Scenarios*, Lecture Notes in Geoinformation and Cartography,
https://doi.org/10.1007/978-3-319-60801-3_33

2.1 Stage 1: Development of the Suitability Images

In the first stage, the user is required to develop suitability images for each of the land cover classes. The suitability images are intended to represent the relative suitability of the land for each end land cover. It is envisioned that the user would develop these using Multi-Criteria Evaluation.

In TerrSet/IDRISI, Multi-Criteria Evaluation is achieved with the MCE module that permits both Weighted Linear Combination (WLC) and Ordered Weighted Averaging (OWA). The former combines evidence as a weighted average and thus provides full tradeoff between factors (i.e., poor qualities can be balanced by good qualities). OWA (Yager 1988; Jiang and Eastman 2000) is a refinement that allows the user to vary the degree of tradeoff (to as little as none) as well as the balance between opportunity and conservativeness in the aggregation of evidence. A companion module named WEIGHT provides the Analytical Hierarchy Process pairwise comparison procedure (Saaty 1987) for developing weights for factors that reflect the consensus of a participating group.

2.2 Stage 2: Calculation of the Transition Areas

The second stage is a determination of the amount of area that needs to go through each transition for the future prediction. For this CA_MARKOV relies on the MARKOV module in TerrSet/IDRISI. MARKOV takes two historical land cover images as input and performs a Markov Chain analysis to estimate both a transition probability matrix and a corresponding matrix of the expected quantities (in areal units) associated with each transition according to the prediction date. The estimation procedure is identical to that performed in LCM (see the short presentation in Part IV of this book).

2.3 Stage 3: Change Allocation

The third stage is allocation of the expected transitions. This is a form of cellular automaton process in the manner described by White and Engelen (1997). The user specifies the number of modeling steps (such as one step per year in the prediction). At each step, a contiguity filter (which can be user-modified) progressively down-weights the suitabilities of pixels distant from existing areas of each class (as of that iteration), thus giving preference to contiguous suitable areas. However, down-weighting never exceeds 90% allowing for the possibility of allocations at a distance if they were highly suitable to start with.

Within each time step, after filtering, each land cover is considered in turn as a host category. All other land cover classes act as claimant classes and compete for land (only within the host class) using the MOLA (multi-objective land allocation) procedure (Eastman et al. 1995) in the TerrSet/IDRISI system. The area requirements for each claimant class within each host are equal to the total established by the transition areas file divided by the number of iterations. The results of each MOLA operation are overlaid (using a COVER operation) to produce a new land cover map at the end of each step.

3 Applications

CA_MARKOV has been evaluated and compared to other models since its release in 1993 (Paegelow and Camacho Olmedo 2008; Memarian et al. 2012; Pontius and Malanson 2005). It has been applied across many disciplines in varied geographic areas, including environmental impacts on water pollution (Houet and Hubert-Moy 2006), managing Europe's heathlands (Mobaied et al. 2011); predicting land degradation in Zimbabwe (Kamusoko et al. 2009) and evaluating protected area policies (Adhikari and Southworth 2012; Mondal and Southworth 2010). CA_MARKOV has been employed for general land use prediction and scenario modeling (Subedi et al. 2013; Halmy et al. 2015; Ding et al. 2015) including urban (Sang et al. 2011) and coastal landscapes (Kityuttachai et al. 2013) as well as modeling the impacts of climate change (Tong et al. 2012) and sea-level rise (Shirley and Battaglia 2008). CA_MARKOV has even been used to model pollen-based land cover reconstruction over 4000 years in Estonia (Poska et al. 2008).

4 Final Considerations

It is the belief of Clark Labs that the logic of LCM generally presents a stronger approach since it models the suitability for transition rather than suitability for the ultimate cover class. In addition, in LCM the model is developed empirically using historical data as a guide. That said, historical data are not always available or it may be that the historical period is not a good predictor for what is expected to drive land cover change. In these cases, CA_MARKOV may prove to be very useful.

References

Adhikari A, Southworth J (2012) Simulating forest cover changes of Bannerghatta National Park based on a CA-Markov model: a remote sensing approach. Remote Sens 4(10):3215–3243

Ding H, Chiabai A, Tobar D (2015) Cost-benefit analysis of alternative land-use scenarios: a sustainability study for the Volcanic Central Talamanca Biological Corridor. In: Chibai A

(2015) Climate change impacts on tropical forests in Central America: an ecosystem perspective. Routledge, New York pp 152–178
Eastman JR, Jin W, Kyem PAK, Toledano J (1995) Raster procedures for multi-criteria/multi-objective decisions. Photogramm Eng Remote Sens 61(5):539–547
Halmy MWA, Gessler PE, Hicke JA, Salem BB (2015) Land use/land cover change detection and prediction in the north-western coastal desert of Egypt using Markov-CA. Appl Geogr 63:101–112
Houet T, Hubert-Moy L (2006) Modelling and projecting land-use and land-cover changes with a cellular automaton considering landscape trajectories: an improvement for simulation of plausible future states. EARSeL eProceedings 5:63–76
Jiang H, Eastman JR (2000) Application of fuzzy measures in multi-criteria evaluation in GIS. Int J Geogr Inf Sci 14(2):173–184
Kamusoko C, Aniya M, Adi B, Manjoro M (2009) Rural sustainability under threat in Zimbabwe-Simulation of future land use/cover changes in the Bindura district based on the Markov-cellular automata model. Appl Geogr 29:435–447
Kityuttachai K, Tripathi NK, Tipdecho T, Shrestha R (2013) CA-Markov analysis of constrained coastal Urban growth modeling: Hua Hin Seaside City, Thailand. Sustainability 5:1480–1500
Memarian H, Balasundram SK, Talib JB, Teh Boon Sung C, Mohd Sood A, Abbaspour K (2012) Validation of CA-Markov for simulation of land use and cover change in the Langat Basin, Malaysia. J Geogr Inf Syst 4:542–554
Mobaied S, Riera B, Lalanne A, Machon N (2011) The use of diachronic spatial approaches and predictive modelling to study the vegetation dynamics of a managed heathland. Biodivers Conserv 20:73–88
Mondal P, Southworth J (2010) Evaluation of conservation interventions using a cellular automata-Markov model. For Ecol Manage 260(10):1716–1725
Paegelow M, Camacho Olmedo MT, eds (2008) Modelling Environmental Dynamics. Springer-Verlag, Berlin
Pontius RG Jr, Malanson J (2005) Comparison of the structure and accuracy of two land change models. Int J Geogr Inf Sci 19(2):243–265
Poska A, Sepp E, Veski S, Koppel K (2008) Using quantitative pollen-based land-cover estimations and a spatial CA_Markov model to reconstruct the development of cultural land-scape at Rõuge, South Estonia. Veg Hist Archaeobotany 17:527–541
Saaty RW (1987) The analytic hierarchy process–what it is and how it is used. Math Model 9 (3):161–176
Sang L, Zhang C, Yang J, Zhu D, Yun W (2011) Simulation of land use spatial pattern of towns and villages based on CA-Markov model. Math Comput Model 54:938–943
Shirley LJ, Battaglia LL (2008) Projecting fine resolution land-cover dynamics for a rapidly changing terrestrial-aquatic transition in Terrebonne Basin, Louisiana, USA. J Coast Res 24 (6):1545–1554
Subedi P, Subedi K, Thapa B (2013) Application of a hybrid cellular automaton–Markov (CA-Markov) model in land-use change prediction: a case study of Saddle Creek Drainage Basin, Florida. Appl Ecol Environ Sci 1(6):126–132
Tong STY, Suna Y, Ranatungaa T, Hea J, Yangb YJ (2012) Predicting plausible impacts of sets of climate and land use change scenarios on water resources. Appl Geogr 32(2):477–489
White R, Engelen G (1997) Cellular automata as the basis of integrated dynamic regional modelling. Environ Plan 24:235–246
Yager RR (1988) On Ordered Weighted Averaging aggregation operators in multicriteria decision making. IEEE Trans Syst Man Cybern 8(1):183–190

Chapter 34
A Short Presentation of CLUMondo

J. van Vliet and P.H. Verburg

Abstract CLUMondo simulates changes in land systems in response to an exogenous demand, land system characteristics, and a series of biophysical and socioeconomic variables. Land systems are defined in terms of their land cover composition as well as land use intensity. As a consequence, land systems can multifunctional and thus provide multiple different goods or services. Moreover, an increase in demand for, say, crop produce, can lead to cropland expansion, cropland intensification, or both. Here we explain the model algorithm, and illustrate the advantage of the land system approach over traditional land use models at the national and the global scale. CLUMondo is available as a free and open source model.

Keywords Land use change · Land use intensity · Land systems · Land use model · Ecosystem services

1 Introduction

Changes in land use and cover are made in response to demands for various goods and services provided by the land, such as food produce of providing shelter. Changes in these demands can result in land cover conversion, for example an increase in food demand may lead to a conversion from forests to cropland, and a growing population may lead to an increase in built-up area. However, these demands can also be satisfied by increasing the land use intensity of a given area of land. For example, the conversion of subsistence agriculture to market-oriented production is characterized by an increase in agricultural yields, while the area under cultivation doesn't necessarily change. Hence both subsistence cultivation

J. van Vliet (✉) · P.H. Verburg
Environmental Geography Group, Faculty of Earth and Life Sciences,
VU University Amsterdam, Amsterdam, The Netherlands
e-mail: jasper.van.vliet@vu.nl

P.H. Verburg
e-mail: peter.verburg@vu.nl

M.T. Camacho Olmedo et al. (eds.), *Geomatic Approaches for Modeling Land Change Scenarios*, Lecture Notes in Geoinformation and Cartography,
https://doi.org/10.1007/978-3-319-60801-3_34

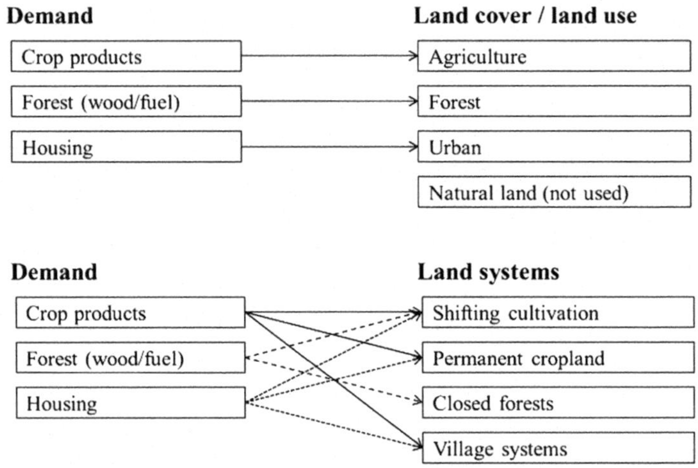

Fig. 1 Schematic representation of the relation between various demands, and changes in land uses or land systems. Conventional models (*top*) typically link one demand directly to one land cover or land use type, while CLUMondo (*bottom*) allows to link a demand to multiple different land systems and vice versa

and market based production are defined by the same land cover, namely cropland, while they differ in their land use intensity. These different intensities related the same land cover may have important, differential, impacts on climate (Luyssaert et al. 2014), biodiversity (Kleijn et al. 2009), water and soil quality (Keatley et al. 2011), and rural livelihoods (Cramb et al. 2009).

The CLUMondo model uses a land systems approach towards land change simulation. Land systems refer to typical combinations of land cover and their land use or management intensities (van Asselen and Verburg 2012), but may also contain information on the temporal and spatial configuration of the land system components. Each land system produces a specific combination of goods and services, such as tons of crop produce and head of livestock. Besides the provisioning of such commodities, also other services may be provided that are valued by society, such as water regulation and carbon sequestration (Wolff et al. 2015). Consequently, each good or service can be supplied by one or more different land systems, and one land system can supply one or more goods or services (see Fig. 1). This approach requires a different model representation than other models, where each land cover is typically driven by one (area) demand only.

2 Descriptions of the Methods Implemented

CLUMondo is a forward looking model that simulates land system changes in response to various types of exogenously defined demand and endogenously defined transition rules (van Asselen and Verburg 2013). Each simulation starts from an initial land systems map, which changes in yearly time steps. The user may

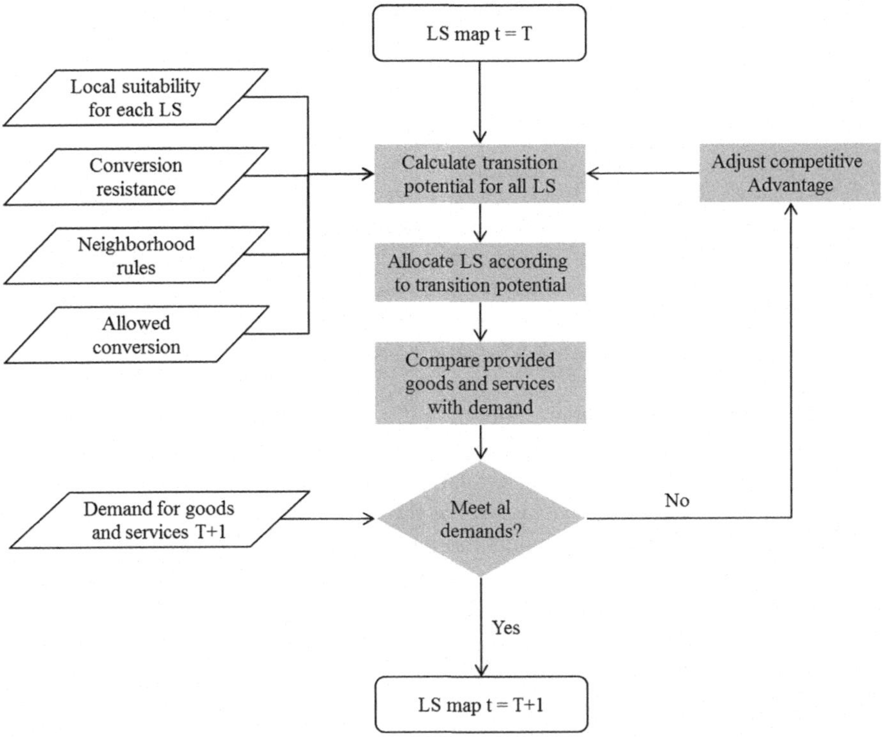

Fig. 2 Schematic overview of the land system (LS) allocation in CLUMondo. The *grey boxes* indicate the iterative loop for allocating LS changes within each time step

define whether demands for a particular year need to be met exactly (assuming an equilibrium) or whether they serve as minimum or maximum levels (such as indicating a minimum amount of carbon sequestration or a maximum amount of water extraction).

Within each yearly time step, land systems are allocated in an iterative procedure in which land systems are allocated according to the transition potential at time (*t*) and location (*i*) for each land system (*LS*), and the demands for goods and services for that specific year (see Fig. 2). The transition potential ($Ptrans_{t,i,LS}$) is calculated as the sum of the local suitability ($Ploc_{t,i,LS}$), the conversion resistance ($Pres_{LS}$), the neighborhood effect ($Pneigh_{t,i,LS}$), and the competitive advantage of a land system ($Pcomp_{t,LS}$) (van Asselen and Verburg 2013):

$$Ptrans_{t,i,LS} = Ploc_{t,i,LS} + Pres_{LS} + Pneigh_{t,i,LS} + Pcomp_{t,LS}$$

The local suitability of a location for a particular land system can be specified by the user or estimated based on current spatial patterns of different land systems. The latter employs one logistic regression model for each land system separately, where the occurrence of a land system is the dependent variable ($\beta_0, \beta_1, \cdots, \beta_n$) while the independent variables are a set of biophysical and socioeconomic conditions $(f_1, f_2, \cdots, f_n)$:

$$Ploc = \frac{1}{1 + e^{-(\beta_0 + \beta_1 f_1 + \beta_1 f_1 + \cdots + \beta_n f_n)}}$$

Conversion resistance is an indication of the costs of converting a particular land system into any other system. Conversion costs are typically high for land systems with high capital investments and systems that are difficult to remove physically, such as urban and peri-urban systems. Extensive agricultural systems and (semi-) natural systems, on the other hand, are relatively easy to convert and are therefore typically characterized by a low conversion resistance. The conversion resistance is calibrated manually, based on expert knowledge, with values between 0 and 1.

The neighborhood effect represents the influence that land systems in the direct surroundings exert on the allocation of land systems. While the neighborhood effect is commonly used to simulate the mutual attraction of urban land uses (van Vliet et al. 2013), it can also be used to express the influence of land availability in the trade-off between cropland expansion and intensification (van Asselen and Verburg 2013). In this case it is assumed that under conditions of high land availability cropland expansion is possible, while intensification is induced when this is not the case. The neighborhood effect in CLUMondo is calculated as a function of the number of cells in the user-defined neighborhood, with land systems that contribute (f_{LS}), a constant (a), and a weight (w). The weight may be determined by the fraction of a specific land cover in a land system, e.g. to differentiate between land systems with a low and a high share of urban land cover. Note that the weight and the constant can be positive as well as negative. Therefore, the neighborhood effect can represent attraction, for example in urban agglomerations, and repulsion, for example due to limited land availability for cropland expansion.

$$Pneigh = a + w * f_{LS}$$

The competition between land systems is simulated based on ability of land systems to supply the goods or services for which there is a demand. Initially, the competitive advantage is 0 for each time step. This value is subsequently adjusted in an iterative procedure, based on demands for goods and services that are not yet provided. When land systems have a competitive advantage in supplying multiple (undersupplied) demands, the competitive advantages are added. A solution is found when all demands are fulfilled by the allocated land systems. Hence, in contrast to some other land use change models, CLUMondo does not use a hierarchy or heuristic to handle trade-offs between competing demands, but simulates their competition dynamically.

Other constraints on land allocation can be implemented in CLUMondo, and overrule the calculation of the transition potential as described above. Two important examples are whether specific conversions are allowed and the restrictions posed by spatial layers. The first typically reflects practical constraints for conversion, for example to indicate that cropland cannot become a forest directly, as it takes several years and intermediate stages to grow trees (Verburg and Overmars 2009). The second represent specific constraints for the occurrence of land systems, such as natural parks that limit the expansion of urban land, or biophysical constraints that limit the expansion of cultivated areas (Eitelberg et al. 2015).

3 Applications

The CLUMondo model is flexible with regards to the scale, resolution, and the land systems to be considered. These model characteristics may be defined based on the needs of the study area and research questions at hand. However, the definition of land systems as typical combinations of land cover and land use intensities suggest a certain minimum resolution, as all components of a specific land system need to be included in the simulation unit. Consequently, CLUMondo is particularly well suited to simulate changes over relatively large areas. Current applications range from provincial to global scale. In this section we briefly present one national scale application and one global scale application.

Crop production in Laos takes place in a range of land systems with different intensities (Hett et al. 2012). Many villages, especially in less accessible places, are still dependent on subsistence agriculture, using swidden cultivation (Schmidt-Vogt et al. 2009). Other places, however, are characterized by permanent croplands, including paddy fields, and large scale plantations. As a consequence, an increased demand for food can be satisfied by changing relatively extensive swidden fields into more intensive permanent croplands, but also by cultivating new cropland areas. In our projections, intensification predominates in the near future (Ornetsmüller et al. 2016) (Fig. 3). This land change trajectory is a model result, as it was not specified a priori how the increased demand for food should be produced.

Globally, land systems differ in their land management intensity, but also in the goods and services they produce. Consequently, an increase in demand for crop products can lead to a cropland expansion but also to an intensification of existing cropland, depending on the local characteristics, the land system patterns to start with, as well as the availability of new land that can accommodate expansion (Eitelberg et al. 2015). However, land change is also increasingly driven by demands for other goods and services, for example carbon sequestration and biodiversity protection. As these demands are implemented through policy instruments, they are now drivers of land change, as well as consequences (Eitelberg et al. 2016). In the baseline scenario of this global application, land change is driven by a demand for crop production, head of ruminant livestock, and area of built-up land. Subsequently, we designed two alternative scenarios, that include an additional demand for carbon storage and biodiversity protection, respectively. As a result of the increased competition for land resources due to these additional demands, these scenarios yield more intensification and less expansion of cropland, and thus to more specialized land systems (see Fig. 4).

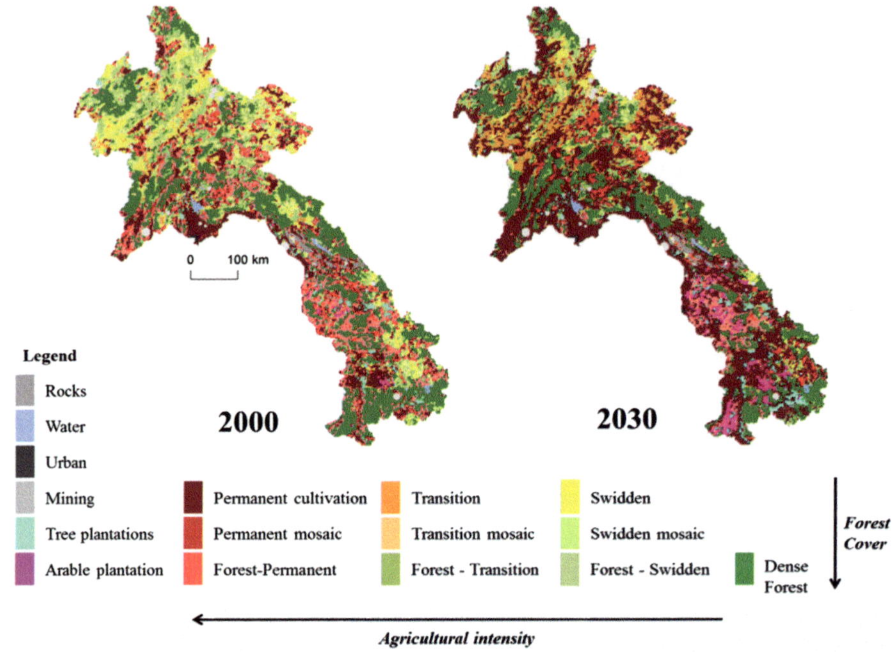

Fig. 3 Land systems in Laos in the years 2000 and 2030 (based on Ornetsmüller et al. 2016)

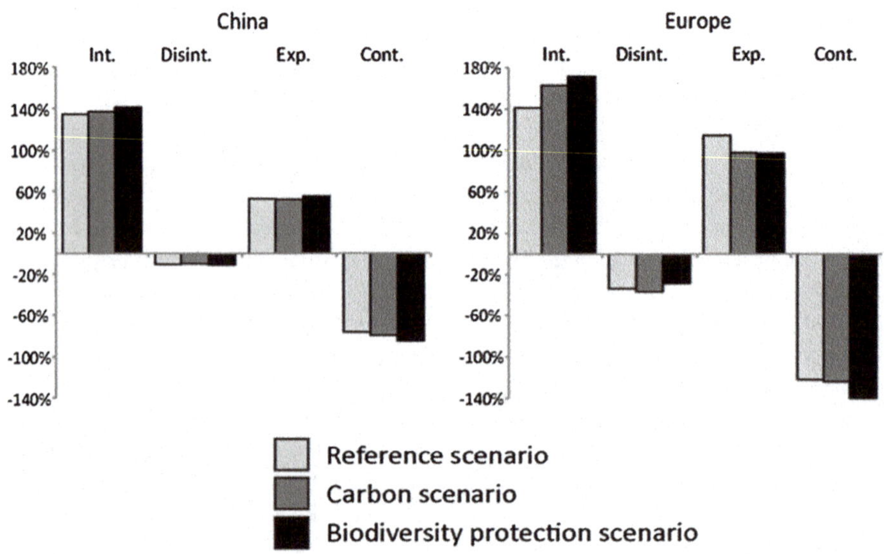

Fig. 4 Changes in crop production in various land change scenarios. The bars indicate the % change relative to the year 2000 for two selected model regions, illustrating the additional intensification caused by adding demand for carbon storage or biodiversity protection (based on Eitelberg et al. 2016)

4 Final Considerations and Technical Summary

Contrary to the representation in most models, land change is typically not driven by a demand for areas of land cover, but by a range of demands for goods and services provided by the land. These include food production, and housing, but increasingly also other demands, such as recreation, carbon storage, biodiversity, and disaster risk reduction (Wolff et al. 2015). The CLUMondo model is the first model that directly uses these demands as input for land change simulations. The examples provided in this chapter illustrate the two main advantages of this land systems approach. First, the representation of land systems allows for both expansion and intensification, in response to increased demand for food, as shown in Laos. Second, this approach allows to include multiple different demands, including those that are not linked to one land use strictly, such as carbon storage or biodiversity protection, as shown in our global application.

CLUMondo is available as a free and open source model from the dedicated webpage: http://www.environmentalgeography.nl/site/data-models/models/.

References

Cramb RA, Colfer CJP, Dressler W, Laungaramsri P, Le QT, Mulyoutami E, Peluso NL, Wadley RL (2009) Swidden transformations and rural livelihoods in Southeast Asia. Human Ecology 37:323–346. doi:10.1007/s10745-009-9241-6

Eitelberg DA, van Vliet J, Verburg PH (2015) A review of global potentially available cropland estimates and their consequences for model-based assessments. Glob Chang Biol 21:1236–1248. doi:10.1111/gcb.12733

Eitelberg DA, van Vliet J, Doelman JC, Stehfest E, Verburg PH (2016) Demand for biodiversity protection and carbon storage as drivers of global land change scenarios. Glob Environ Change 40:101–111. doi:10.1016/j.gloenvcha.2016.06.014

Hett C, Castella JC, Heinimann A, Messerli P, Pfund J-L (2012) A landscape mosaics approach for characterizing swidden systems from a REDD + perspective. Appl Geogr 32:608–618. doi:10.1016/j.apgeog.2011.07.011

Keatley BE, Bennett EM, MacDonald GK, Taranu ZE, Gregory-Eaves I (2011) Land-use legacies are important determinants of lake eutrophication in the anthropocene. PLoS ONE 6(1): e15913. doi:10.1371/journal.pone.0015913

Kleijn D et al (2009) On the relationship between farmland biodiversity and land-use intensity in Europe. Proc R Soc B Biol Sci 276:903–909. doi:10.1098/rspb.2008.1509

Luyssaert S et al (2014) Land management and land-cover change have impacts of similar magnitude on surface temperature. Nat Clim Chang 4:389–393. doi:10.1038/nclimate2196

Ornetsmüller C, Verburg PH, Heinimann A (2016) Scenarios of land system change in the Lao PDR: Transitions in response to alternative demands on goods and services provided by the land. Appl Geogr 75:1–11. doi:10.1016/j.apgeog.2016.07.010

Schmidt-Vogt D, Leisz SJ, Mertz O, Heinimann A, Thiha T, Messerli P, Epprecht M, Van Cu P, Chi VK, Hardiono M, Dao TM (2009) An assessment of trends in the extent of swidden in Southeast Asia. Human Ecology 37:269–280. doi:10.1007/s10745-009-9239-0

van Asselen S, Verburg PH (2012) A Land System representation for global assessments and land-use modeling. G Chang Biology 18:3125–3148. doi:10.1111/j.1365-2486.2012.02759.x

van Asselen S, Verburg PH (2013) Land cover change or land-use intensification: simulating land system change with a global-scale land change model. Glob Chang Biol 19:3648–3667. doi:10.1111/gcb.12331

van Vliet J, Naus N, van Lammeren RJ, Bregt AK, Hurkens J, van Delden H (2013) Measuring the neighbourhood effect to calibrate land use models. Comput Environ Urban Syst 41:55–64

Verburg PH, Overmars KP (2009) Combining top-down and bottom-up dynamics in land use modeling: exploring the future of abandoned farmlands in Europe with the Dyna-CLUE model. Landsc Ecol 24:1167–1181. doi:10.1007/s10980-009-9355-7

Wolff S, Schulp CJE, Verburg PH (2015) Mapping ecosystem services demand: a review of current research and future perspectives. Ecol Ind 55:159–171. doi:10.1016/j.ecolind.2015.03.016

Chapter 35
A Short Presentation of Dinamica EGO

H. Rodrigues and B. Soares-Filho

Abstract Dinamica EGO is a flexible software that allows the construction of many different types of environmental simulation models, including complex spatial dynamic ones. By using an intuitive, friendly and yet very powerful graphical interface, modelers can freely employ a combination of map algebra, cellular automata techniques, and table data manipulation to represent complex socio-economic and environmental systems, not being limited to the use of only predefined models.

Keywords Environmental simulation · Land use and change

1 Introduction

Dinamica EGO (EGO stands for Environment for Geoprocessing Objects) is a freeware for environmental modeling. Its modeling platform allows the design from very simple spatial models to very complex dynamic ones (Soares-Filho et al. 2002, 2006). Dinamica EGO favors usability, flexibility and performance, optimizing speed and computer resources. The software interface allows designing models using a graphical programming language in an intuitive and friendly way. Users build models by simply dragging geoprocessing operators and connecting them to represent the model visual diagram. While such a simplicity facilitates newcomers' learning, sophisticated and powerful features address the challenges posed by expert modelers. Advanced features include nested iterations, multi-transitions, dynamic feedbacks, multi-region approach, decision processes for bifurcating and joining execution pipelines, a complete series of spatial algorithms for the analysis

H. Rodrigues (✉) · B. Soares-Filho
Centro de Sensoriamento Remoto, Universidade Federal de Minas Gerais (UFMG), Belo Horizonte, Brazil
e-mail: hermann@csr.ufmg.br

B. Soares-Filho
e-mail: britaldo@csr.ufmg.br

M.T. Camacho Olmedo et al. (eds.), *Geomatic Approaches for Modeling Land Change Scenarios*, Lecture Notes in Geoinformation and Cartography,
https://doi.org/10.1007/978-3-319-60801-3_35

and simulation of space-time phenomena, model wizard, and high performance computing thanks to a 64-bit native and multiprocessor architecture that handles large raster datasets. Dinamica EGO also allows the user to break up the model into sub-models to simplify design and enhance communication, or to design new operators that can be stored in the software library or exchanged using an online repository. In addition, Dinamica EGO enables map operation combining raster maps in any geographic projection, spatial resolution, or extent, making it truly a multiple resolution and multi-scale software. The software environment also allows the online coupling with R studio taking fully advantage of Dinamica EGO high performance and R vast statistics capabilities in one integrated modeling environment.

The software environment, developed mainly in C++ and Java, contains a series of algorithms called operators or "functors".[1] Dinamica EGO operators include the most common spatial algorithms available in commercial GIS, and a series of algorithms especially designed for spatial simulations, including cellular automata transition functions, and calibration and validation methods. A special class of operator is the "container" that can envelop a series of operators and other containers, for example, to control the model dataflow, such as the "Repeat" container. Operators, including containers, are sequenced in a graph form to establish a visual data flow. With the help of a friendly graphical interface, users can create models by simply dragging and connecting operators via their ports; each port represents a connector to a data element, such as a map, table, matrix, mathematical expression, or constant. Thus, a model can be designed as a diagram, whose execution follows a data flow chain. Sub-models (an encapsulated part of a model) can be stored in the user's library as new operators to be reused in other models or shared through an online store, thus facilitating the exchange of models as well as new functionalities developed by Dinamica EGO's worldwide community of users. Models developed using the graphical interface are saved in EGOML (a form of Extensible Markup Language) or EGO programming script language; the latter format enables script writing using a text editor, which can be converted to EGO graphical diagram and vice versa. Dinamica EGO provides various tools for data visualization, including maps, tables, and graphs. Worthy of mention, modeler can build a wizard tutorial for communicating the model with end-users. In the forefront of environmental modeling, Dinamica EGO is a freeware and as such can be used at no cost for scientific, personal, and commercial purposes.

[1]http://www.csr.ufmg.br/dinamica/wiki.

2 Description of Some Methods Implemented in the Model

Most commonly, Dinamica EGO models employ some combination of map algebra, cellular automata technique, and tabular data manipulation to represent complex socioeconomic and environmental systems. The map algebra sub-library includes a vast set of predefined operators (assigning map categories, extracting map values, distance calculation, accumulated flow, etc.) and the calculate map operators whereby users can write any mathematical or logical expression using a combination of maps, tables, and constants. The map calculation operator "Calculate Map" includes local, zonal and neighborhood functions. Because operators can be sequenced forming parallel and bifurcated execution pipelines and loops, the user is free to connect any set of operators to form a visual data flow. Hence, any variable in Dinamica EGO can become dynamic receiving feedbacks from any model element.

Dinamica EGO comes with a set of pre-implemented cellular automata transition functions, but modelers can also implement their own cellular automata from scratch using the "Calculate Map" operator together with its neighborhood functions. Thanks to the set of cellular automata transition functions (named "Patcher" and "Expander"), which allow the definition of form and size of patches of changes, Dinamica EGO can simulate very intricate and complex landscape structures. Of relevance, these functions replicate the expanding and contracting landscape elements, thereby simulating edge processes. The software holds multiple transitions that are calibrated by employing the Weights of Evidence method to calculate the influence of spatial determinants on the location of changes, producing as result an integrated transition potential map, also known as the transition probability map. The transition probability determines the likelihood that a specific cell or spatial unit will change from one state to another over a time step. The transition probabilities are calculated in Dinamica EGO using an adapted version of the Bayesian method of conditional probability (Bonham-Carter 1994), known as the Weights of Evidence (WOFE). See Soares-Filho et al. (2004, 2006, 2009, 2010). In addition, a genetic algorithm tool available in Dinamica EGO is flexible enough to embrace a multitude of spatial models as well as their specific fitness functions, thus offering a practical way to optimize the performance of environmental models (Soares et al. 2013).

The cellular automata functions allocate the changes, whose rates are either passed by a coupled model or exogenously prefixed (e.g. Markovian chain). The spatial determinants represent proximate causes of land-use change (e.g. the opening or paving of a road) or are simply preferable (e.g. more fertile soil, low slope) or more restricted (land-use zoning, such as protected areas) sites (Soares-Filho et al. 2001, 2010).

Dinamica EGO can use any customized approach to validate a model. In addition, Dinamica EGO comes with a map comparison method named "Reciprocal Similarity Comparison" that compares the spatial matching of maps of changes (Almeida et al. 2008; Soares-Filho et al. 2009). Since this method was made available in Dinamica EGO, a series of studies have applied it to perform map comparison (e.g. Soares-Filho et al. 2010; Walker et al. 2010; Silvestrini et al. 2011; Lapola et al. 2011). A detailed mathematical description of this method is found in Dinamica EGO guidebook (Soares-Filho et al. 2009) and in Soares et al. (2013). Dinamica EGO features simultaneous multiple resolution simulation, implemented through its sub region approach, a functionality that allows the user to customize the model parameters or to perform a particular calculation for a map zone, i.e. a region in a map, such as a country or state. Regions themselves can also be dynamic, changing boundaries every time-step, or be nested allowing the models to aggregate different calculations for different region levels. For example, a model can perform certain calculations at a finer resolution, e.g. at municipality level, and others at a coarser resolution, e.g. at state level. A series of models to perform landscape metrics comes with the dataset. Other examples include a road constructor submodel, land-change simulations, as well as many image-processing algorithms.

3 Applications

Applications of Dinamica EGO are many.[2] They include, for example, simulation of urban growth and intradynamics (Almeida et al. 2005; Godoy and Soares-Filho 2008), land-use change (Stickler et al. 2009; Teixeira et al. 2009), agricultural expansion (Gouvello et al. 2010), fire (Silvestrini et al. 2011), deforestation (Soares-Filho et al. 2002, 2004, 2006; Maeda et al. 2010), rent models of logging (Merry et al. 2009) and cattle ranching (Bowman et al. 2012), and analyses of opportunity cost of reducing deforestation (Nepstad et al. 2009) and the effectiveness of protected areas (Soares-Filho et al. 2010). The software has made an important contribution to more than 150 peer reviewed papers by scholars worldwide and it is widely used by governmental organizations and planning bodies.

4 Final Considerations and Technical Summary

Dinamica is a very flexible modeling tool that can be run from the desktop to a high-performance computer. Thanks to its innovative techniques, the software provides a complete solution for calibrating, running, and validating space-time models, no matter the complexity.

Dinamica EGO is a freeware spatial modeling software, available for research and commercial use. Web page of the software package: http://dinamicaego.com.

[2]http://csr.ufmg.br/dinamica/publications/.

References

Almeida C, Monteiro M, Camara G, Soares-Filho BS, Cerqueira G, Pennachin CL (2005) GIS and remote sensing as tools for the simulation of Urban land use change. Int J Remote Sens 26:759–774

Almeida C, Gleriani JM, Castejon EF, Soares-Filho BS (2008) Using neural networks and cellular automata for modeling intra-urban land-use dynamics. Int J Geogr Inf Sci 22:943–963

Bonham-Carter G (1994) Geographic information systems for geoscientists: modelling with GIS. Pergamon, Oxford, UK

Bowman MS, Soares-Filho BS, Merry FD, Nepstad DC, Rodrigues HO, Almeida OT (2012) Persistence of cattle ranching in the Brazilian Amazon: a spatial analysis of the rationale for beef production. Land Use Policy 29:558–568

Godoy MMG, Soares-Filho BS (2008) Modelling intra-urban dynamics in the Savassineigh-bourhood, Belo Horizonte city, Brazil. In: Paegelow M, Camacho Olmedo MT (eds) Modelling environmental dynamics. Springer, Berlin, pp 319–338

Gouvello C, Soares-Filho BS, Nassar A, Schaeffer R, Jorge F, Nogueira W (2010) Brazil Low-carbon country case study. World Bank Report

Lapola DM, Schaldach R, Alcamo J, Bondeau A, Msangi S, Priess JA, Silvestrini R, Soares BS (2011) Impacts of climate change and the end of deforestation on land use in the Braziolian Legal Amazon. Earth Interact 15

Maeda EE, Clark BJF, Pellikka P, Siljander M (2010) Modelling agricultural expansion in Kenya's Eastern Arc Mountains biodiversity hotspot. Agric Syst 103:609–620

Merry F, Soares-Filho BS, Nepstad D, Aamacher G, Rodrigues H (2009) Balancing conservation and economic sustainability: The Future of the Amazon Timber Industry. Environ Manage 44:395–407

Nepstad D, Soares-Filho BS, Merry F, Lima A, Moutinho P, Carter J, Bowman M, Cattaneo A, Rodrigues H, Schwartzman S, Mcgrath D, Stickler C, Lubowski PP, Rivero S, Alencar A, Almeida O, Stella O (2009) The end of deforestation in the Brazilian Amazon. Science 326:1350–1351

Silvestrini RA, Soares-Filho BS, Nepstad D, Coe M, Rodrigues HO, Assunção R (2011) Simulating fire regimes in the Amazon in response to climate change and deforestation. Ecol Appl 21:1573–1590

Soares B, Rodrigues H, Follador M (2013) A hybrid analytical-heuristic method for calibrating land-use change models. Environ Model Softw 43:80–87

Soares-Filho BS, Assunção RM, Pantuzo A (2001) Modeling the spatial transition probabilities of landscape changes in an Amazonian Colonization frontier. Bioscience 51:1039–1046

Soares-Filho BS, Coutinho Cerqueira G, Lopes Pennachin C (2002) Dinamica–a stochastic cellular automata model designed to simulate the landscape dynamics in an Amazonian colonization frontier. Ecol Model 154:217–235

Soares-Filho BS, Alencar A, Nepstad D, Cerqueira G, Vera Diaz M, Rivero S, Solorzano L, Voll E (2004) Simulating the response of land-cover changes to road paving and governance along a major Amazon highway: the Santarém-Cuiabá corridor. Glob Chang Biol 10:745–764

Soares-Filho BS, Nepstad D, Curran L, Voll E, Cerqueira G, Garcia RA, Ramos CA, Mcdonald A, Lefebvre P, Schlesdinger P (2006) Modeling conservation in the Amazon basin. Nature 440:520–523

Soares-Filho BS, Rodrigues H, Costa WLS (2009) Modeling environmental dynamics with Dinamica EGO. www.csr.ufmg.br/dinamica

Soares-Filho BS, Moutinho P, Nepstad D, Anderson A, Rodrigues H, Garcia R, Dietschi L, Merry F, Bowman M, Hissa L, Silvestrini R, Maretti C (2010) Role of Brazilian Amazon protected areas in climate change mitigation. Proc Natl Acad Sci 107:10821–10826

Stickler C, Nepstad D, Coe M, Macgrrath D, Rodrigues H, Dabidson E, Soares-Filho B (2009) The potential ecological costs and co-benefits of REDD: a critical review and case study from the Amazon region. Glob Chang Biol 15:2803–2824

Teixeira AM, Soares-Filho BS, Freitas S, Metzger JPW (2009) Modeling landscape dynamics in the Atlantic rainforest domain: implications for conservation. For Ecol Manage 257:1219–1230

Walker WS, Stickler CM, Dellndorfer JM, Kirsch KM, Nepstad DC (2010) Large-area classification and mapping of forest and land cover in the Brazilian Amazon: a comparative analysis of ALOS/PALSAR and landsat data sources, vol. 3. Woods Hole Research Center, pp 594–604

Chapter 36
A Short Presentation of the Land Change Modeler (LCM)

J.R. Eastman and J. Toledano

Abstract The Land Change Modeler is a land change projection tool for land planning. It uses historical land cover change to empirically model the relationship between land cover transitions and explanatory variables to map future scenarios of change.

Keywords Land change · Land change prediction · Land planning

1 Introduction

The Land Change Modeler (LCM) was developed (Eastman 2006) as an empirically parameterized land change projection tool to support a wide range of planning activities. Based on an analysis of historical land cover change, the system develops an empirical model of the relationship between land cover transitions and a set of explanatory variables. Mappings of future change are then based on this empirical relationship and a projection of quantity derived from a Markov Chain. The result is a business-as-usual (BAU) projection of change without subjective intervention. It is designed to support applications with strict BAU baseline needs such as REDD (Reducing Emissions from Deforestation and forest Degradation) climate mitigation projects.

2 Description of the Methods Implemented in the Model

At present, three separate empirical model development tools are provided in LCM: a Multi-Layer Perceptron neural network (MLP), Logistic Regression (LR) and SimWeight (SW). The MLP procedure is the default, and is the most mature and

J.R. Eastman (✉) · J. Toledano
Clark Labs, Clark University, Worcester, USA
e-mail: reastman@clarku.edu

M.T. Camacho Olmedo et al. (eds.), *Geomatic Approaches for Modeling Land Change Scenarios*, Lecture Notes in Geoinformation and Cartography,
https://doi.org/10.1007/978-3-319-60801-3_36

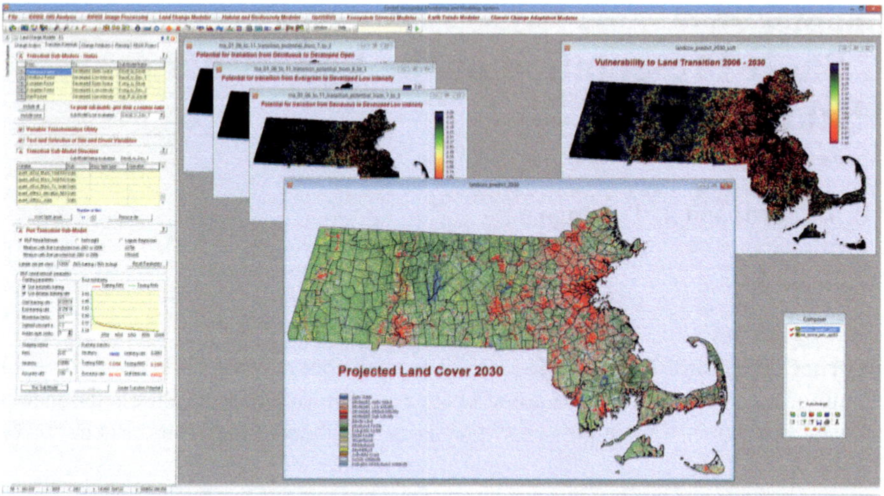

Fig. 1 In LCM, the process of land change modeling is organized into major stages embodied by tabs in the interface. The most important stage is that in which empirical models are developed relating historical changes to explanatory variables. In this example, the default Multi-Layer Perceptron Neural Network is used to develop transition potential maps (small maps, *upper left*)—empirically derived statements of the potential of land to undergo specific transitions. These are used in the subsequent change prediction tab to generate both future scenarios (large image, *center*) and maps of vulnerability to change (*upper left*)

primary focus of LCM (Fig. 1). MLP is also the only procedure that can model multiple transitions at the same time. Logistic Regression is provided primarily for pedagogic reasons while SimWeight is an experimental machine learning procedure based on a K-Nearest Neighbor variant (Sangermano et al. 2010). In each case, analysis of two land cover layers in the recent past is used to train and evaluate the model.

2.1 Training

For the Multi-Layer Perceptron, LCM examines each of the transitions over the historical period to determine the number of pixels that went through the transition being modeled (change pixels) and the number that were eligible to, but which did not (persistence pixels). The user is then required to specify the sample sizes to use for training the model. For MLP and SimWeight, equal-sized samples of change and persistence are required. However, the default sample sizes are very different—10,000 for MLP and 1000 for SimWeight. The difference relates to how they are used—for iterative learning in the case of MLP and characterization for SimWeight. For Logistic Regression, the sample chosen is proportional to the relative number of change and persistence pixels for the individual transition being modeled. The user is able to indicate the sampling proportion (the default is 10%) and the method of spatial sampling—stratified random sampling (the default) or systematic.

2.2 Simulation

In LCM, the simulation proceeds in three stages. The first is the development of transition potentials—mappings of the readiness of land to go through each of the transitions under consideration. The second is the estimation of the expected quantity of change and the third is the spatial allocation of the estimated change based on the transition potentials.

Transition Potentials
A transition potential is a continuous value from 0–1 that expresses the relative potential of a pixel to transition from one state to another. The metric varies from one empirical modeling procedure to another but in the final stage of the simulation, only the relative value of the metric matters.

If MLP is used as the modeling procedure, the transition potential is the activation level of the output neurons which represents the posterior probability of transition under an assumption of equal probability of change/persistence. With Logistic Regression, the transition potential is the probability of change assuming an identical quantity to that which transitioned during the historical period. With SimWeight, the value is unitless, but monotonic with the posterior probability of transition.

Estimation of the Quantity of Change
In the second stage, a Markov Chain analysis is used to determine the quantity of change for the forecast date selected. A Markov Chain assumes that the rate of change (but not the quantity of change) remains constant over time. The calculation proceeds (Eastman 2014) by first computing a cross tabulation of transitions between the land cover maps for the two historical dates. From this, the basic transition probability matrix (X) is calculated. If the date being projected forward is an even multiple of the training period, then the new transition probability matrix is calculated through a simple powering of the base matrix (Kemeny and Snell 1976). For example, if the training period is from 2002 to 2011 (9 years) then the transition probability matrix for 2020 from 2011 (9 years forward) is X^1, for 2019 (18 years forward) is X^2, for 2047 (36 years forward) is X^4, and so on. However, if the projected time period is in between even multiples of the training period, then the power rule is used to generate 3 transition matrices that envelop the projection time period (if the 3 time periods are times A, B and C, the period to be interpolated will be between A and B). The three values at each cell in the transition probability matrix are then fed into a quadratic regression (thus there will be a separate regression for each transition probability matrix cell). Given that a quadratic regression ($Y = a + b_1X + b_2X^2$) has 3 unknowns and we have three data points, it yields a perfect fit. This equation is then used to interpolate the unknown transition probability. From these transition probabilities, the projected quantity of change is determined for each transition being modeled.

Spatial Allocation

Given a set of transition potential maps and the projected quantity of change for each transition, LCM then allocates change based on a greedy selection algorithm. The greedy selection is based on the simple assumption that the areas with the highest transition potential will always transition first. Because a single pixel may be selected for multiple transitions, a competitive strategy is used whereby it will be assigned to the transition with the highest marginal transition potential. This will lead to some transitions being allocated less than the expected quantity of change. Thus the procedure iterates through a process of selecting pixels with lower transition potentials until all transitions achieve their required quantity.

LCM recognizes that some explanatory variables may be based on land cover, and thus change as transitions progress. For example, a model focused on deforestation may use a variable of proximity to existing agriculture. As agriculture expands, this variable will constantly be changing. Such variables are termed *dynamic* as opposed to *static*. Thus the user has the ability to predict in stages with dynamic variables being automatically recomputed at each stage. The procedure chosen for re-computation can be as simple as a single distance calculation to a complex macro. For example, a user may have empirically determined the potential for transition based on the age of the forest post-agriculture, and thus may use the macro option to add time and then re-compute the transition potential at each step.

2.3 Validation

Validation is handled differently for each of the empirical modeling procedures. For MLP, half of the training data are reserved for validation. Validation is a critical component of the training process. At each stage in the training, learning is refined with one half of the data and the quality of the model is assessed by comparison with the other half. Accuracy and model skill over all transitions and persistences combined are dynamically reported during the training process. Model skill is reported as a Heidke Skill Score (Heidke 1926), also known as Kappa (Cohen 1960), which ranges from −1 to +1 with 0 indicating a skill no better than random allocation. At the end of the empirical modeling procedure, LCM provides a detailed accounting of accuracy and skill for each transition and each persistence category. It also provides a wealth of information about the contribution of each variable to the model including a backwards stepwise assessment that allows for a very easy determination of the most parsimonious model.

For SimWeight, again, 50% of the training data are reserved for validation. From these, a Peirce Skill Score (Joliffe and Stephenson 2003) is evaluated—a value similar in nature to a Heidke Skill Score in that it ranges from −1 to +1 with 0 representing the point where the hit rate and false alarm rate are equal. SimWeight also reports the relevance of variables by examining the variability of a variable within historical samples of transition relative to all areas (Sangermano et al. 2010).

For Logistic Regression, validation is handled by a Goodness of Fit measure which expresses the degree to which the fitted values of the modelled regression match the training data. Relevance of variables is assessed by the slope coefficients and t-score values, although use of standardized variables is recommended for this purpose.

2.4 REDD

An important application of this kind of empirically-modeled projection tool is the climate change mitigation strategy known as REDD—Reducing Emissions from Deforestation and forest Degradation. To meet the special needs of these programs, LCM provides a special set of tools for the development of REDD projects. Tools provide for the definition of the project and leakage areas, specification of carbon pools to be considered, method of calculation and carbon density in the evaluation of CO_2 emissions. Non-CO_2 emissions are also considered. Leakage, success and effectiveness rates are then specified for each of the reporting stages. In the end, 19 tables are produced following the BioCarbon Fund methodology. However, these are easily re-formatted into any of the prevailing approved methodologies.

3 Applications

Land Change Modeler has been evaluated and applied across many disciplines in varied geographic areas since its release in 2007. It has been evaluated against other land change methods (Eastman et al. 2005; Fuller et al. 2011; Mas et al. 2014; Paegelow and Camacho Olmedo 2008). LCM has been applied to forest monitoring and deforestation (Khoi and Murayama 2011; Valle Jr et al. 2012) and its impacts on biomass (Eckert et al. 2011; Fuller et al. 2011; Saha et al. 2013), biodiversity (Dean and Salim 2012; Uddin et al. 2015) and species habitat impact assessment (Fuller et al. 2012). LCM has even been employed to model post-socialist land change in Eastern Europe (Václavík and Rogan 2010). LCM is an accepted tool by the Verified Carbon Standard (VCS 2014) and is extensively used in REDD (Areendran et al. 2013; Centro de Conservación 2012; Kim and Newell 2015) and REDD+ project planning (Moore et al. 2011; Sangermano et al. 2012; Scheyvens et al. 2014; VCS 2014).

4 Final Considerations

LCM has now been in public use for more than a decade. It was commissioned by Conservation International and the conservation community still constitutes the largest body of users. Companion software components have been developed to work closely with LCM including the Habitat and Biodiversity Modeler and the Ecosystem Services Modeler. Development of LCM continues with two new machine learning procedures (Weighted Normalized Likelihoods and a Support Vector Machine) currently in testing. Additionally, a cloud-based implementation is currently under development. Given the pace of anthropogenic land conversion, the ability to develop defensible and skillful land change models is of critical importance.

References

Areendran G, Raj K, Mazumdar S, Puri K, Shah B, Mukerjee R, Medhi K (2013) Modelling REDD+ baselines using mapping technologies: a pilot study from Balpakram-Baghmara Landscape (BBL) in Meghalaya, India. Int J Geoinf 9(1):61–71

Centro de Conservación, Investigación y Manejo de Areas Naturales (2012) Cordillera Azul National Park REDD Project, Peru

Cohen JA (1960) Coefficient of agreement for nominal scales. Educ Psychol Measur 20(1):37–46

Dean A, Salim A (2012) Technical background material, Kalimantan InVest and LCM modeling. Heart of Borneo: investing in nature for a green economy. WWF HoB Global Initiative

Eastman JR (2006) IDRISI 15.0: The Andes edn. Clark University, Worcester MA

Eastman JR (2014) TerrSet geospatial monitoring and modeling system. Clark University, Worcester, MA

Eastman JR, Solorzano L, Van Fossen M (2005) Transition potential modeling for land-cover change. In: Maguire DJ, Batty M, Goodchild MF (eds) GIS, spatial analysis and modeling. ESRI Press, Redlands, CA, pp 357–385

Eckert S, Ratsimba HR, Rakotondrasoa LO, Rajoelison LG, Ehrensperger A (2011) Deforestation and forest degradation monitoring and assessment of biomass and carbon stock of lowland rainforest in the Analanjirofo Region, Madagascar. For Ecol Manag 262(11):1996–2007

Fuller DO, Hardiono M, Meijaard E (2011) Deforestation projections for carbon-rich peat swamp forests of central Kalimantan, Indonesia. Environ Manag 48(3):436–447

Fuller DO, Parenti M, Hassan AN, Beier JC (2012) Linking land cover and species distribution models to project potential ranges of malaria vectors: an example using Anopheles arabiensis in Sudan and Upper Egypt. Malaria J 11(1):264

Guidolini JF, Batista Pedroso L, Neves Araújo MV (2013) Análise Temporal Do Uso e Ocupação Do Solo Na Microbacia Do Ribeirão Do Feijão, Município De São Carlos–SP, Entre Os Anos De 2005 e 2011. In: Anais XVI Simpósio Brasileiro De Sensoriamento Remoto–SBSR, Foz Do Iguaçu, PR, Brasil, 13 a 18 De Abril, 4503–4509. INPE 2013

Heidke P (1926) Berechnung des erfolges und der gute der windstarkvorhersagen im sturm-warnungsdienst. Geogr Ann 8:301–349

Houet T, Vacquié L, Sheeren D (2015) Evaluating the spatial uncertainty of future land abandonment in a mountainous valley (Vicdessos, Pyrenees–France): insights from model parameterization and experiments. J Mt Sci 12(5):1095–1112

Joliffe IT, Stephenson DB (2003) Forecast verification: a practitioners guide in atmospheric science. Wiley, Chichester, West Sussex, England

Kemeny JG, Snell JL (1976) Finite markov chains. Springer, New York

Khoi DD, Murayama Y (2011) Modeling deforestation using a neural network-Markov model. In: Murayama Y, Thapa RB (eds) Spatial analysis and modeling in geographical transformation process: GIS-based applications. Springer, New York, pp 169–192

Kim OS, Newell JP (2015) The 'Geographic Emission Benchmark' model: a baseline approach to measuring emissions associated with deforestation and forest degradation. J Land Use Sci 10 (4):466–489

Mas JF, Kolb M, Paegelow M, Camacho Olmedo MT, Houet T (2014) Modelling land use/cover changes: a comparison of conceptual approaches and softwares. Environ Model Softw 51:94–111

Moore C, Eickhoff G, Ferrand J, Khiewvongphachan X (2011) Technical Feasibility Assessment of the Nam Phui National Protected Area REDD+ Project in Lao PDR. Deutsche Gesellschaft für Internationale Zusammenarbeit (GIZ). http://theredddesk.org/sites/default/files/resources/pdf/2013/giz-clipad-technical-feasibiltiy-study-nam-phui-npa-lao-pdr-v1.4.pdf. Accessed 5 Dec 2016

Paegelow M, Camacho Olmedo MT (eds) (2008) Modelling environmental dynamics. Springer, Berlin

Saha GC, Paul SS, Li J, Hirshfield F, Sui J (2012) Investigation of land-use change and groundwater–surface water interaction in the Kiskatinaw River Watershed, Northeastern British Columbia, Parts of NTS. Geoscience BC Summary of Activities 2012. Geoscience BC, Report 2013-1:139–148

Sangermano F, Eastman JR, Zhu H (2010) Similarity weighted instance based learning for the generation of transition potentials in land change modeling. Trans GIS 14(5):569–580

Sangermano F, Toledano J, Eastman JR (2012) Land cover change in the Bolivian Amazon and its implications for REDD+ and endemic biodiversity. Landsc Ecol 27(4):571–584

Scheyvens H, Gilby S, Yamanoshita M, Fujisaki T, Kawasaki J (2014) Snapshots of REDD+ Project Designs. Institute for Global Environmental Strategies (IGES) https://pub.iges.or.jp/pub/redd-projects-snapshots-redd-project-designs. Accessed 5 Dec 2016

Uddin K, Chaudhary S, Chettri N, Kotru R, Murthy M, Chaudhary RP, Ning W, Shrestha SM, Gautam SK (2015) The changing land cover and fragmenting forest on the Roof of the World: A case study in Nepal's Kailash Sacred Landscape. Landsc Urban Plan 141:1–10

Valle RFD Jr, Evangelista Siqueira H, Ferreira Guidolini J, Abdala VL, Ferreira Machado M (2012) Diagnóstico de Mudanças e Peristência de Ocupação do Solo Entre 1978 e 2011 No IFTM-Campus Uberaba, Utilizando o 'Land Change Modeler (LCM)'. Enciclopédia Biosfera 8 (15):672–681

Václavík T, Rogan J (2010) Identifying trends in land use/land cover changes in the context of post-socialist transformation in central Europe: a case study of the greater Olomouc Region, Czech Republic. GISci Remote Sens 46(1):54–76

Verified Carbon Standard (2014) Carmen del Darien (CDD) REDD+ Project. 2014. 3rd ed. http://database.v-c-s.org/sites/v-c-s.org/files/CDD%20REDD+%20Project%20Description%20v5.9.pdf. Accessed 5 Dec 2016

Chapter 37
A Short Presentation of LucSim

J.P. Antoni

Abstract LucSim is a cellular automata (CA) dedicated to geographical analysis and spatial simulation for researchers and advanced planning institutes, providing user-friendly software in order to analyze and simulate land use changes and dynamics. Two complementary models are integrated in the CA: (1) a Markov Chain used to calculate transition matrices from a date to another, and (2) a Decision Tree able to automatically determine a set of transition rules to be applied on land use data. LucSim includes GIS compatibility functions allowing to display ESRI shapefiles and is based on raster georeferenced images saved in TIF format. It was mostly applied on French urban case studies.

Keywords Cellular automata · Markov chains · Transition rules · Decision tree · Urban development

1 Introduction

LucSim is a cellular automata model dedicated to geographical analysis and simulation for researchers and advanced planning institutes. The goal of the project is to provide user-friendly software in order to analyse and simulate land use changes and spatial dynamics (Fig. 1). It is currently being developed at laboratory ThéMA (University Burgundy Franche-Comté and CNRS) from the basis of the CWS/Camdeus project, in collusion with the MobiSim LUTI project, to provide a suit of simulation tools for decision making in urban and land planning.

J.P. Antoni (✉)
Laboratoire the MA, UMR 6049 CNRS, University of Burgundy Franche-Comté, Franche-Comté, France
e-mail: jean-philippe.antoni@u-bourgogne.fr

M.T. Camacho Olmedo et al. (eds.), *Geomatic Approaches for Modeling Land Change Scenarios*, Lecture Notes in Geoinformation and Cartography,
https://doi.org/10.1007/978-3-319-60801-3_37

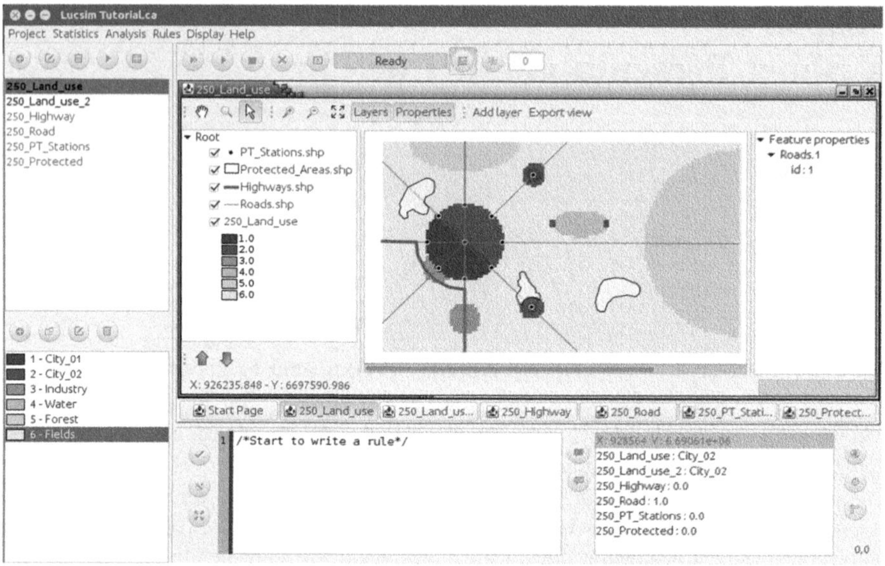

Fig. 1 Screenshot of LucSim tutorial

2 Description of the Methods Implemented in the Model

LucSim relies on the basics of cellular automata, involving grid mapped land-use and transition rules. Its major functionality is based on P. Torrens's definition (Torrens 2000). Technically, the application of a set of transition rules, where the state S of a cell i at step $t + 1$ depends on its state at step t and on its neighbourhood at the same step in a radius O, constitutes the main engine to simulate prospective scenarios of land use change. Land use transition rules can be defined manually, constrained by different techniques or determined automatically.

For rules constraints, two models are integrated in the CA. First, from the land use maps, LucSim is able to calculate transition matrices from a date to another, and to run a markovian process. This Markov chain is useful to calibrate the number of cells that can evolve in the future, and then to improve the temporal dimension of land use change simulation, which is often missing in classical CA tools. Second, LucSim integrates a potential model based on the main principles of spatial interaction. This potential can be used to integrate a specific value to cells and to weight their decreasing influence on the neighbourhood according to their increasing distance. Markov chains and potential modeling can be automatically combined into the CA engine to improve the relevance and the efficiency of the transition rules.

For transition rules automation, LucSim integrates a Decision Tree (DT) process to automatically determine a set of transition rules to be applied an land use data. According to users' parameters and calibration, this DT is based on learning machine and demands to split the initial data in two sub-datasets. The first one is used for training and the second one for testing the obtained results. Resulting transition rules can immediately be analysed and run through the regular CA process to test hypothesis or forecast future land use changes.

Spatial statistics (neighbourhood analysis) is also an advanced function of the model, allowing to extract specific neighbourhood, to compare sets of land use images, and to assess the relevance of the CA constraints and simulation results.

3 Applications

LucSim was mostly applied on French urban case studies (Belfort, Besançon, Montbéliard, Nantes, Rennes) and on the cross-border regions of Strasbourg-Kehl and Luxembourg.

4 Final Considerations and Technical Summary

As a geographical cellular automata, LucSim includes GIS compatibility functions allowing to display ESRI shapefiles (.shp) and is based on raster georeferenced images saved in TIF format. LucSim must then be connected and feed by GIS and Raster graphics editors. So far as LucSim is strictly defined as a geographical cellular automata (including diachronic land use analysis tools), it does not assume any image creation or modification.

LucSim is a .jar software developed in Java language and necessitates the installation of Java 8 at least to be executed on any system operator (Linux, Mac OS or MS Windows). LucSim can be downloaded here: https://sourcesup.renater.fr/lucsim/.

Reference

Torrens P (2000) How cellular models of urban systems work, Working paper series 28(1) Centre for Advanced Spatial Analysis 68 p

Chapter 38
A Short Presentation of Metronamica

H. van Delden and R. Vanhout

Abstract Metronamica is a generic and spatially explicit land use modelling framework integrating various drivers and processes relevant for understanding and assessing land use dynamics. As a decision support system, it lets users evaluate spatial planning and infrastructure development policy interventions and provides results in the form of spatial and non-spatial policy relevant indicators. With over a hundred applications worldwide it has demonstrated that the simulation of universal concepts can be tuned to local contexts across the world to cater for very different socio-economic, environmental and governance conditions. The full framework comprises a suite of components like land use, population dynamics, economics and transport, as well as powerful tailored data processing and analysis tools, which can be turned on or off based on the scale and purpose of the application. Metronamica components have been integrated into various tailor-made integrated models and have been enhanced to better represent the multifunctionality of our land, as well as the management and intensity of its use. Its wide user group has benefitted from its ongoing development, by highlighting scientific challenges and providing feedback on its usefulness and user-friendliness.

Keywords Land systems approach · Land use model · Cellular automata · Model calibration · Model integration

1 Introduction

Metronamica (RIKS 2017; Van Delden and Hurkens 2011, www.metronamica.nl) is a generic forecasting tool for planners and policy analysts to simulate and assess the integrated effects of policy measures on urban and regional development.

H. van Delden (✉) · R. Vanhout
Research Institute for Knowledge Systems, Maastricht, The Netherlands
e-mail: hvdelden@riks.nl

R. Vanhout
e-mail: rvanhout@riks.nl

M.T. Camacho Olmedo et al. (eds.), *Geomatic Approaches for Modeling Land Change Scenarios*, Lecture Notes in Geoinformation and Cartography,
https://doi.org/10.1007/978-3-319-60801-3_38

The system interactively simulates the impact of a variety of external influences (e.g. macro-economic changes, population growth, etc.) and policy measures (e.g. land use zoning, conservation policies, densification policies, etc.) on the regional development of a city, region, country or continent. With the integrated scenario support, what-if analyses can be performed that help evaluate alternative plans under various external conditions.

The core of Metronamica is a CA-based land use allocation component that simulates land use developments over time based on a 'competition for space' principle. Based on their economic and political power, actors will be able to occupy the locations which are most desirable for them. These behavioural dynamics can be facilitated or countered by planning and policy interventions in obtaining a more desirable future. Metronamica is equipped with a set of indicators to assess how autonomous developments shape long-term land use dynamics and how (combinations of) policy options impact on these future pathways.

Metronamica is developed using the Geonamica software environment (Hurkens et al. 2008) and includes a model library containing a range of models from various disciplines: land use, regional interaction, transport, economics and demographics. Applications can be set up with one, two or three spatial levels depending on their scope. Spatial resolution at local level varies for current applications between 25 and 1000 m. Temporal resolution is a year. Temporal horizon is 20–50 years into the future.

2 Description of the Methods Implemented in the Model

The models that are incorporated in Metronamica simulate activities that take place at three spatial scales: global, regional and local, where global refers to the entire simulated area. At global level, a macro-economic model is tied with an age-cohort model that simulates structural demographic changes and population levels. The age-cohort model incorporates immigration patterns and provides the labour force supply; the latter are used as an input for the economic model. Economic conditions, in return, have an impact on migration and mortality rates. Figures for population and jobs in main economic sectors are used as an input for the regional model.

At the regional level, socio-economic changes take place based on the relative attractiveness of regions and the costs required to travel from one region to another. These costs are provided by the transport model that uses information from the regional and local models to generate trips and calculate the speed, intensity and congestion on the network. This provides the basis for the distribution of national growth as well as migration of jobs and people over regions and is furthermore input for the allocation of activities within the regions.

On the local level, land use demands from the regional model are allocated to grid cells based on several elements including local accessibility, physical suitability, zoning regulations and the attraction, repulsion and competition between different land use functions. Finally, the local bio-physical and socio-economic characteristics feed back into the attractiveness at the regional level and the land use configuration is used as an input for the production of trips in the transport model.

For each application, the user can select one or more model components. Based on the selection of components, inputs for them come from other components or are defined as exogenous drivers.

With the focus of the book on land use modeling, the land use component is described below in a bit more detail. More information on all components including the equations used can be found in the Metronamica documentation (RIKS 2017).

2.1 Land Use Component

The land use model operates at local level and uses a grid of cells. A cellular automaton (CA) based land use model is used to determine the state of a cell within the overall growth for each of the regions calculated by the regional model (White and Engelen 1993). Changes in land use at the local level are driven by four important factors that determine the potential for each location for each actor (see also Fig. 1):

- Physical suitability, represented by one map per land use function modelled. The term suitability is used here to describe the aptness of a cell to support a particular land use function and its associated activity.
- Zoning or spatial planning, represented by one map per land use function modelled. For different planning periods the map specifies which cells can and cannot be taken in by the particular land use and how strict or flexible the various plans are.

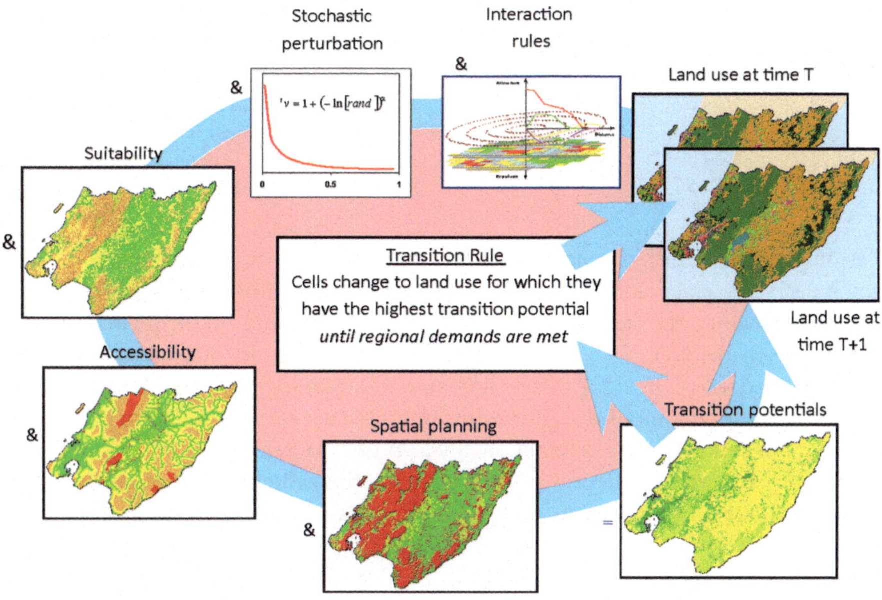

Fig. 1 Main drivers of the Metronamica land use model as shown for an application to Greater Wellington, New Zealand

- Accessibility, represented by one map per land use function modelled. Accessibility is an expression of the ease with which an activity can fulfil its needs for transportation, mobility and other facilities in a particular cell, based on the proximity to infrastructure networks.
- Human behaviour, represented by spatial interaction rules simulating the preferences of various actors for certain locations based on the land uses in the area surrounding the location, including their power to occupy the most desirable locations.

If the potential is high enough, the function will occupy the location, if not, it will look for more attractive places. New activities and land uses invading a neighbourhood over time will thus change the attractiveness for activities already present and others searching for space. This process constitutes the highly non-linear character of this model.

2.2 Land Systems and Multifunctional Land Use

To enhance the representation of the land dynamics, the local land use component has been complemented with intensity and management information to provide a full land systems approach. For the socio-economic functions this entails the incorporation of activity or density levels in addition to the land use and management of the location (Van Vliet et al. 2012), for agricultural and natural uses local suitability and management decisions are included for simulating intensity levels (van Delden et al. 2007). These developments enable the simulation of the multi-functionality of the land and also provide relevant intensity and density information for further impact assessment.

2.3 Indicators

Metronamica includes a range of socio-economic and environmental indicators which can be selected and configured based on a selection of algorithms. Examples of such indicators are the expansion of urban areas, habitat fragmentation and the distance from residential locations to the nearest recreation site. Other indicators can be added on demand by selecting one from a set of available algorithms, providing additional input data and adjusting model parameters. Examples of such indicators are urban development in areas prone to flooding—requiring a map indicating areas prone to flooding—and job potential, which is the ratio between the number of jobs and inhabitants in the vicinity of a residential location.

2.4 User Interface

An important task for the developer of an integrated spatial decision support system (ISDSS) is to bridge the gap from scientific tools to user-friendly systems, by creating a graphical user interface (GUI) that is easy to use and guides users in the steps that need to be taken to carry out a scenario or policy impact assessment study. In addition, as ISDSS often encompass complex models, the user interface should provide insight into the structure and functioning of the model and provide access to all relevant model inputs and outputs for updating the data, calibration and validation. Trying to incorporate both in one interface often leads to a malfunctioning system that is far from optimal for any user.

In the design of the Metronamica user interface we decided that the interface should be able to provide access to two different types of users: the policy analysts who use the system as part of their policy process and who carry out scenario and impact assessment studies with the model, and the scientists or modellers who can update the underlying data and parameters and possibly even the model equations. For the latter group of users, we created the modeller interface where elements are grouped per model; each individual model has its own access point through the system diagram. Access to settings for the policy user is structured according to their logical function in the policy interface. On a high level, access is organized by the steps that a user takes to carry out an impact assessment analysis: configure drivers, create integrated scenarios, run the simulation, review output through the indicators and do comparative analysis. Zooming in on those parts, we grouped settings and outputs by their type and their domain; for example, all economic policy measures together, all external factors together, all ecological indicators together etc. Example of the policy and modeller interface are provided on the website www.metronamica.nl.

2.5 Setting up and Calibrating the Land Use Component (Metronamica SL)

When setting up a new application, the following steps are generally applied for finding an appropriate parameter set and assessing its quality.

1. As part of the *data analysis* the current situation and historic developments are analysed. This includes analysing the temporal change in total area surface for various land uses as well as the change in landscape structure. Regarding the latter, metrics such as the clumpiness index (McGarigal 2014) and the rank size distribution (Gabaix 1999) are used in conjunction with a visual inspection of the developments. Furthermore, the enrichment factor is used to analyse the over- and underrepresentation of certain land uses in the neighbourhood of changed land uses (Van Vliet et al. 2013).

2. *Model set-up* includes a set of choices relevant for setting up the model to a specific region and context. In CA-based land use modeling, main choices are related to the decision on the area extent, the applied resolution and the selection of land use classes to be modelled, where finding a balance between providing additional information and creating a false sense of accuracy is often a crucial point of discussion (van Delden et al. 2011).
3. During the *calibration*, parameter values are set and fine-tuned and subsequently the model is assessed on its behaviour and results, frequently over a historic calibration period. Difficulties in calibrating CA-based land use models mainly relate to the large number of parameters that need to be set, the limited availability of time series of land use maps, and finding objective ways to assess the quality of the calibration. Regarding the latter, progress has been made over the past years, which has resulted in the use of neutral models to act as a benchmark for quality assessment (Hagen-Zanker and Lajoie 2008), together with the use of objective measures to complement the more subjective visual assessment. To assess the quality of the calibration we take into account the *predictive accuracy*, which is the ability of the model to accurately simulate actual land use patterns; and the *process accuracy*, the extent to which the modelled processes are consistent with real world processes (Brown et al. 2005). Main indicators used for assessing the quality of the calibration are indicators for location agreement, such as Fuzzy Kappa (Hagen-Zanker 2009) and Fuzzy Kappa Simulation (Van Vliet et al. 2013); indicators for landscape structure agreement, such as the clumpiness index (McGarigal 2014), the fractal dimension (Chen 2011), the rank size distribution (Gabaix 1999), and the enrichment factor (Van Vliet et al. 2013); and visual inspection.
4. During the *validation*, the model's behaviour and results, based on the parameters settings obtained during the calibration, are assessed over a data set independent from the one used as part of the calibration. This usually results in an evaluation of the model's behaviour over a different historic period; although other independent data sets are equally valid (see e.g. Van Vliet et al. 2010). Assessment criteria are the same as for the calibration.
5. Finally, the model is tested and evaluated on its *long-term behaviour*, which includes a long-term simulation with the calibration parameters, a number of tests with extreme scenarios to assess the robustness of the model, a number of tests to assess the sensitivity of model results on small changes to the parameter settings and some tests to assess the impact of the main perceived uncertainties.

2.6 Applications

Metronamica has over 100 applications worldwide. Some examples of applications using the land use component only include:

- Hewitt et al. (2014) focusing on a participatory calibration of the land use model
- Wickramasuriya et al. (2009) applying Metronamica to shifting cultivation in Sri Lanka
- Furtado et al. (2012) applying Metronamica with income-differentiated actors to Belo Horizonte, Brazil
- van Delden and Hagen-Zanker (2009) using the story line-and-simulation approach to simulate exploratory scenarios for Europe.

Metronamica has been incorporated in various integrated models, of which some examples are provided below:

- UNHARMED (Riddell et al. 2016)—a decision support system for disaster risk reduction, applied to Greater Melbourne, Greater Adelaide and Tasmania, Australia.
- RECARE IAM (van Delden et al. 2016)—a Europe-wide integrated assessment model for the impact of policy and land management options on soils and ecosystem services.
- LUMOCAP PSS (van Delden et al. 2010)—a policy support system for evaluating the impacts of agricultural policies on the European land use and landscape.
- MedAction PSS (van Delden et al. 2007)—an integrated policy support application to assess the main issues underlying the causes and effects of land degradation and develop integrated planning policy options and mitigation strategies to combat desertification in the Northern Mediterranean region.
- Environment Explorer (Engelen et al. 2003)—an integrated spatial decision support system in which social, economic and ecological processes are simulated to explore policy alternatives in relation to the quality of the environment in which Dutch citizens live, work and recreate.
- WISE (Rutledge et al. 2008), an integrated scenario explorer to support the development of a strategic vision for New Zealand regions by taking into account social, environmental and economic well-being.

2.7 Final Considerations and Technical Summary

www.metronamica.nl for the Metronamica modeling framework and www.riks.nl for integrated models that include Metronamica. The Metronamica modeling framework is generally not open source although in specific cases arrangements are made to share the source code. Metronamica is generally not free, but under certain conditions fees might be reduced or waived.

References

Brown DG, Page S, Riolo R, Zellner M, Rand W (2005) Path dependence and the validation of agent-based spatial models of land use. Int J Geogr Inf Syst 19(2):153–174

Chen Y (2011) Derivation of the functional relations between fractal dimension of and shape indices of urban form. Comput Environ Urban Syst 35:442–451

Engelen G, White R, de Nijs T (2003) Environment Explorer: spatial support system for the integrated assessment of socio-economic and environmental policies in the Netherlands. Integr Assess 4(2):97–105

Furtado B, Ettema D, Machado Ruiz R, Hurkens J, Van Delden H (2012) A celular automata intra-urban model with prices and income-differentiated actors. Environ Plan B 39:897–924. doi:10.1068/b37134

Gabaix X (1999) Zipf's law for cities: an explanation. Quart J Econ 114(3):739–767

Hagen-Zanker A (2009) An improved fuzzy kappa statistic that accounts for spatial autocorrelation. Int J Geogr Inf Sci 23(1):61–73

Hagen-Zanker A, Lajoie G (2008) Neutral models of landscape change as benchmarks in the assessment of model performance. Landsc Urban Plan 86(3–4):284–296

Hewitt R, Van Delden H, Escobar F (2014) Participatory land use modelling, pathways to an integrated approach. Environ Model Softw 52:149–165

Hurkens J, Hahn BM, Van Delden H (2008) Using the Geonamica® software environment forintegrated dynamic spatial modelling. In: Sànchez-Marrè M, Béjar J, Comas J, Rizzoli A, Guariso G (eds) Proceedings of the iEMSs Fourth Biennial Meeting: Integrating sciences and information technology for environmental assessment and decision making. Int Environ Model Softw Soc, Barcelona, Spain, pp 751–758. ISBN 978-84-7653-074-0

McGarigal K (2014) FRAGSTATS Help. University of Massachusetts, Amherst

Riddell G, Maier HR, Van Delden H, Newman JP, Zecchin AC, Vanhout R, Daniell J, Shäfer A, Dandy GC, Newland CP (2016) A spatial decision support system for natural hazard risk reduction, policy assessment and planning. AFAC16

RIKS (2017) Metronamica documentation. Maastricht, The Netherlands

Rutledge DT, Cameron M, Elliott S, Fenton T, Huser B, McBride G, McDonald G, O'Connor M, Phyn S, Poot J, Price R, Scrimgeour F, Small B, Tait A, Van Delden H, Wedderburn ME, Woods R (2008) Choosing regional futures: challenges and choices in building integrated models to support long-term regional planning in New Zealand, vol 1. Regional Science Policy & Practice, pp 85–108

Van Delden H, Hagen-Zanker A (2009) New ways of supporting decision-making: linking qualitative storylines with quantitative modelling. In: Geertman S, Stillwell JCH (eds) Best practice and new methods in planning support systems. Springer, Dordrecht, The Netherlands

Van Delden H, Hurkens J (2011) A generic integrated spatial decision support system for urban and regional planning. Keynote presented at MODSIM 11 international congress on modelling and simulation. Perth, Australia

Van Delden H, Luja P, Engelen G (2007) Integration of multi-scale Dynamic spatial models of socio-economic and physical processes for river basin management. Environ Model Softw 22 (2):223–238

Van Delden H, Stuczynski T, Ciaian P, Paracchini ML, Hurkens J, Lopatka A, Shi Y, Gomez Prieto O, Calvo S, Van Vliet J, Vanhout R (2010) Integrated assessment of agricultural policies with dynamic land use change modelling. Ecol Model 221(18):2153–2166

Van Delden H, Van Vliet J, Rutledge DT, Kirkby MJ (2011) Comparison of scale and scalingissues in integratedland-use modelsforpolicysupport. Agr Ecosyst Environ 142 (1–2):18–28

Van Delden H, Fleskens L, Irvine B, Kirkby M, Schwilch G, Teixeira da Silva R, Verzandvoort S (2016) Application of a soil-related ecosystem services framework to support European policies. In: european ecosystem services conference. Antwerp, Belgium

Van Vliet J, Vanhout R, Saretta A, Lavalle C, Van Delden H (2010) MOLAND-Light: Supporting planners with a generic land use model for European regions. In: Paper presented at the international conference on managing the urban rural interface 19–22 October. Copenhagen, Denmark

Van Vliet J, Hurkens J, White R, Van Delden H (2012) An activity based celular automaton model to simulate land use dynamics. Environ Plan B 39:198–212. doi:10.1068/b36015

Van Vliet J, Naus N, Van Lammeren RJ, Bregt AK, Hurkens J, Van Delden H (2013) Measuring the neighbourhood effect to calibrate land use models. Comput Environ Urban Syst 41:55–64

White R, Engelen G (1993) Cellularutomata and fractal urban form: a cellular modelling approach to the evolution of urban land use patterns. Environ Plan A 25:1175–1199

Wickramasuriya RC, Bregt AK, Van Delden H, Hagen-Zanker A (2009) The dynamics of shifting cultivation captured in an extended constrained cellular automata land use model. Ecol Model 220:2302–2309

Chapter 39
A Short Presentation of SLEUTH

K.C. Clarke

Abstract This chapter summarizes information about SLEUTH, a popular cellular automaton model that simulates urban growth and land use change. The model is supported in the public domain and all source code is open, including extensive documentation and discussion fora. The input data for SLEUTH are listed, the model's behavior and its control parameters explained, and methods described for model calibration, use in simulation, and for validation. Pointers to review papers are given as starting points for the reader to find SLEUTH applications, and the operating system and computer requirements are given. This volume includes a paper by the author that makes a substantial improvement to SLEUTH's calibration procedures.

Keywords SLEUTH · Cellular automaton · Land use change model · Urban growth · Simulation · Model calibration

1 Introduction

SLEUTH is a simulation model consisting of computer code written in the C programming language. Its purpose is to simulate urban growth over time, and to propagate change across a range of land use classes specified by the user. The model consists of two tightly coupled cellular automaton (CA) models: the Urban Growth Model and the Deltatron land use change model. Three main versions of the model exist, with three variants. Version 1 was experimental, version 2 used dynamic memory allocation, while version 3 adopted the Cray flat memory model and included support for the Message Passing Interface to allow parallel processing. SLEUTH-r used SLEUTH but changed some of the road handling routines to speed up the model, and simplified the calibration process (Jantz et al. 2010). SLEUTH*

K.C. Clarke (✉)
Department of Geography, University of California Santa Barbara,
Santa Barbara, USA
e-mail: kclarke@geog.ucsb.edu

M.T. Camacho Olmedo et al. (eds.), *Geomatic Approaches for Modeling Land Change Scenarios*, Lecture Notes in Geoinformation and Cartography,
https://doi.org/10.1007/978-3-319-60801-3_39

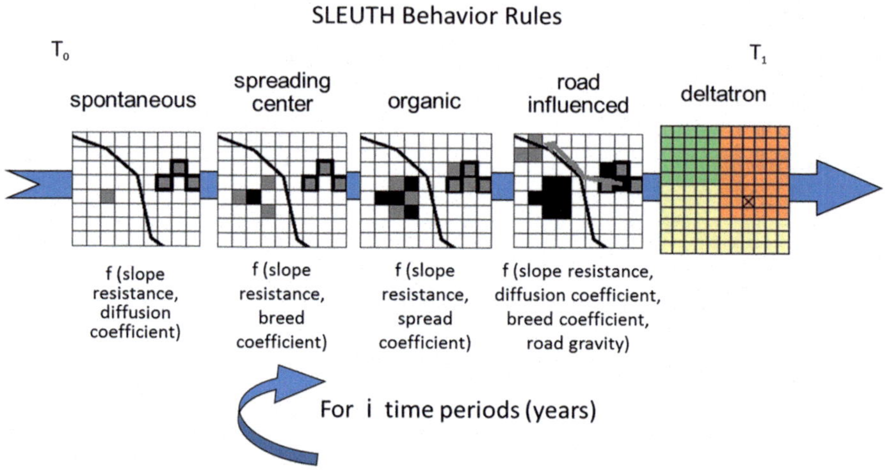

Fig. 1 SLEUTH behavior rules

included a user interface to support decision-making and scenario planning (Houet et al. 2016). The model ingests data in the form of raster images that give the model its name, topographic slope, land use, exclusions, urban extents, transportation and a hillshade layer for visualization. At a minimum, the model needs one GIS data layer at different periods of past time for slope, exclusion and hillshade, two or more transportation and land use layers, and four or more urban extent layers.

The model uses four types of CA behavior rules: diffusive growth, new spreading centers, organic growth and road influenced growth (Clarke and Gaydos 1998). These behaviors are determined by the values of five control coefficients that take integer values between 0 and 100 (Fig. 1). At zero, the behavior type is disabled, while at 100 it is uninhibited by probabilities determined by the input data and the other coefficients. To allow non-linear feedbacks in the model, the coefficients are also subjected to self-modification, in which the state of the entire system changes the values during a run. During the automated calibration process a single run starts at the earliest year for which data is available, and runs to the last year or "present." Thirteen performance metrics are then used to evaluate the coefficient values, averaged over a series of Monte Carlo iterations. Calibration consists of selecting the best five coefficients using brute force, i.e. trying combinations and permutations to select the one that best simulates the known data (Silva and Clarke 2002). These settings are then used for forecasting. More recently, the brute force calibration method has been replaced with a genetic algorithm (Clarke-Lauer and Clarke 2011, Clarke this volume). The land use change model uses four phases: initiate change, cluster change, propagate change, and age deltatrons respectively (Clarke 2008a, b). This ensures that spatial and temporal autocorrelation exists in the land use change patterns.

2 Description of the Methods Implemented in the Model

Calibration: The brute force calibration process is described in detail in Silva and Clarke (2002). The fact that so many coefficient combinations must be tried, with Monte Carlo iterations further increasing the computational time, means that the model calibration process can take days, weeks or even months of CPU time. While the speed of processors has taken up much of this burden, parallel computing has also decreased the calibration time (Guan and Clarke 2010; Clarke 2003). Most recently, SLEUTH-GA has simply replaced brute force with a genetic algorithm that "breeds" coefficient combinations that evolve toward a best solution (Clarke, this volume). Calibrations have for some time focused on the Optimal SLEUTH metric as the single best goodness of fit parameter to maximize during calibration (Dietzel and Clarke 2007).

Simulation: SLEUTH requires calibration to give reliable scientific results (Clarke 2004). Once the best coefficient values have been determined, the model is run over the calibration period with a large number of Monte Carlo iterations and the coefficients averaged at the end of the period, i.e. the start date for simulation. The model then takes the most recent data as inputs, and runs as far into the future as is desired. Model outputs include reports, accuracy statistics, maps, animations and uncertainty estimates. If no land use data are present, the model simply simulates urban growth. By varying the parameters and input data simulations of different scenarios can be created (Xiang and Clarke 2003). Others have used the exclusion layer, including incorporating other methods such as Multi-Criterion Evaluation into the scenarios (Mahiny and Clarke 2012).

Validation: SLEUTH is among the most validated of land use change models. Not only has the model been subjected to sensitivity analysis, its accuracy has been reported for about 100 different applications. In many cases its reported accuracy during calibration has been in the 80–90% range. At least one study has returned to areas forecast in the past to fully validate the model (Manca and Clarke 2012). Others have investigated temporal sensitivity (Akin et al. 2014; Chaudhuri and Clarke 2014; Peiman and Clarke 2014) and other factors. A full survey of these studies is contained in Chaudhuri and Clarke (2013).

3 Applications

Survey articles that cover a majority of the applications are Clarke et al. (2007, 2008a, b) and Chaudhuri and Clarke (2013). The Gigalopolis project website, cited in these publications, contains a more complete application survey and an inventory of data and results. To the author's knowledge, there have been over 100 applications on 6 continents, at a range of spatial scales and geographic extents and covering western cities, favelas, informal settlements and many other fields.

4 Final Considerations and Technical Summary

SLEUTH research remains active, with new applications and model refinements continuously appearing. The author thanks in remembrance Dr. Leonard Gaydos, who first funded the model's development at USGS and who remained a colleague and friend for two decades. His unfortunate death in a snorkeling accident in 2015 was a great loss to Geography.

SLEUTH is open source and available for free at: http://www.ncgia.ucsb.edu/projects/gig/. The model requires a UNIX-like operating system (such as Linux, Ubuntu or Cygwin). Test data with results are available on the website. A discussion forum exists that can answer the majority of questions users may have. Documentation is fully online at the Gigalopolis website.

References

Akin A, Clarke KC, Berberoglu S (2014) The impact of historical exclusion on the calibration of the SLEUTH urban growth model. Int J Appl Earth Obs Geoinf 27(B):156–168

Chaudhuri G, Clarke KC (2013) The SLEUTH land use change model: a review. Int J Environ Resour Res 1(1):88–104

Chaudhuri G, Clarke KC (2014) Temporal accuracy in urban growth forecasting: a study using the SLEUTH model. Trans GIS 18(2):302–320

Clarke KC (2003) Geocomputation's future at the extremes: high performance computing and nanoclients. Parallel Comput 29(10):1281–1295

Clarke KC (2004) The limits of simplicity: toward geocomputational honesty in urban modeling. In: Atkinson P, Foody G, Darby S, Wu F (eds) GeoDynamics. CRC Press, Florida

Clarke KC (2008a) Mapping and modelling land use change: an application of the SLEUTH model. In: Pettit C, Cartwright W, Bishop I, Lowell K, Pullar D, Duncan D (eds) Landscape analysis and visualisation: spatial models for natural resource management and planning. Springer, Berlin, pp 353–366

Clarke KC (2008b) A decade of cellular urban modeling with SLEUTH: unresolved issues and problems. In: Brail RK (ed) Planning support systems for cities and regions. Lincoln Institute of Land Policy, Cambridge, MA, pp 47–60

Clarke KC, Gaydos L (1998) Loose coupling a cellular automaton model and GIS: long-term growth prediction for San Francisco and Washington/Baltimore. Int J Geogr Inf Sci 12(7): 699–714

Clarke KC, Gazulis N, Dietzel CK, Goldstein NC (2007) A decade of SLEUTHing: lessons learned from applications of a cellular automaton land use change model. In: Fisher P (ed) Classics from IJGIS. Twenty years of the international journal of geo-graphical information systems and science. Taylor and Francis, CRC. Boca Raton FL, pp 413–425

Clarke-Lauer MD, Clarke KC (2011) Evolving simulation modeling: Calibrating SLEUTH using a genetic algorithm. In: Proceedings of 11th international conference on geo computation, University College London, London

Dietzel C, Clarke KC (2007) Toward optimal calibration of the SLEUTH land use change model. Trans GIS 11(1):29–45

Guan Q, Clarke KC (2010) A general-purpose parallel raster processing pro-gramming library test application using a geographic cellular automata model. Int J Geogr Inf Sci 24(5):695–722

Houet T, Aguejdad R, Doukari O, Battaia G, Clarke KC (2016) Description and validation of a 'non path-dependent' model for projecting contrasting urban growth futures. Cybergeo Eur J

Geogr, Systèmes, Modélisation, Géostatistiques, document 759. http://cybergeo.revues.org/27397, doi:10.4000/cybergeo.27397

Jantz CA, Goetz SJ, Donato D, Claggett P (2010) Designing and implementing a regional urban modeling system using the SLEUTH cellular urban model. Comput Environ Urban Syst 34: 1–16

Mahiny AS, Clarke KC (2012) Guiding SLEUTH land-use/land-cover change modeling using multicriteria evaluation: towards dynamic sustainable land-use planning. Environ Plan 39: 925–944

Manca G, Clarke KC (2012) Waiting to know the future: validating a SLEUTH model forecast of urban growth with real data. Cartographica 47:4250–4258. doi:10.3138/carto.47.4.1321

Peiman R, Clarke KC (2014) The impact of data time span on forecast accuracy through calibrating the SLEUTH urban growth model. Int J Appl Geospatial Res 5(3):21–35

Silva EA, Clarke KC (2002) Calibration of the SLEUTH urban growth model for Lisbon and Porto, Portugal. Comput Environ Urban Syst 26(6):525–552. doi:10.1016/S0198-9715(01)00014-X

Xiang W-N, Clarke KC (2003) The use of scenarios in land use planning. Environ Plan 30: 885–909

Printed by Printforce, the Netherlands